Heteroepitaxial Semiconductors for Electronic Devices

Heteroepitaxial Semiconductors for Electronic Devices

Edited by

G. W. Cullen
C. C. Wang

With contributions by

V. S. Ban   S. Berkman   J. Blanc
G. W. Cullen   M. T. Duffy
N. Goldsmith   W. E. Ham
C. C. Wang   P. J. Zanzucchi

Springer-Verlag
New York Heidelberg Berlin

6288976X

G. W. CULLEN
RCA Laboratories
Princeton, NJ 08540
USA

C. C. WANG
RCA Laboratories
Princeton, NJ 08540
USA

CHEMISTRY

Library of Congress Cataloging in Publication Data

Main entry under title:

Heteroepitaxial semiconductors for electronic devices.

   1.   Semiconductor films—Addresses, essays, lectures. 2. Crystallography—
Addresses, essays, lectures. 3. Semiconductors—Addresses, essays,
lectures. I. Cullen, Glenn Wherry, 1932–   II. Wang, Chih-Chun, 1932–
TK7871.15.F5H47   621.381'71   77-21749

© 1978 by Springer-Verlag New York Inc.

Printed in the United States of America

9 8 7 6 5 4 3 2 1

ISBN 0-387-90285-6   Springer-Verlag New York
ISBN 3-540-90285-6   Springer-Verlag Berlin Heidelberg

*To our management,
who feel that communication
with our scientific colleagues
is part of our job.*

Preface

Some years ago it was not uncommon for materials scientists, even within the electronics industry, to work relatively independently of device engineers. Neither group had a means to determine whether or not the materials had been optimized for application in specific device structures. This mode of operation is no longer desirable or possible. The introduction of a new material, or a new form of a well known material, now requires a close collaborative effort between individuals who represent the disciplines of materials preparation, materials characterization, device design and processing, and the analysis of the device operation to establish relationships between device performance and the materials properties. The development of devices in heteroepitaxial thin films has advanced to the present state specifically through the unusually close and active interchange among individuals with the appropriate backgrounds. We find no book available which brings together a description of these diverse disciplines needed for the development of such a materials–device technology. Therefore, the authors of this book, who have worked in close collaboration for a number of years, were motivated to collect their experiences in this volume. Over the years there has been a logical flow of activity beginning with heteroepitaxial silicon and progressing through the III-V and II-VI compounds. For each material the early emphasis on material preparation and characterization later shifted to an emphasis on the analysis of the device characteristics specific to the materials involved. It has been an exciting period of experimental activity for us, and we hope that the readers will share our enthusiasm. Some of the materials described are now being used in commercially available device structures. Some of the materials must be further developed in order to be useful for commercial application. It is also the hope of the authors that some readers will see potential in the less well developed heteroepitaxial materials systems, and identify applications which will justify the effort needed for further development.

Many of the authors' colleagues at RCA Laboratories, too numerous to mention, have contributed to this technology. Clearly there have been experimentalists active in other laboratories who have made major contributions, and we have made every effort to identify them in the text. We would like to recognize the efforts of Professor M. Schieber and Dr. S. Larach for their liaison with the publisher and reviewers. We are very appreciative of the many excellent suggestions offered by reviewers affiliated with other research centers. Prior to the delivery of the manuscript to the publisher, Barbara Kerler worked with the authors in the assembly and layout of the book. Her skill and dedication have been a tremendous help and are much appreciated.

G. W. Cullen

C. C. Wang

Princeton, N. J.

Contents

Heteroepitaxial Semiconductors for Electronic Devices

Chapter 1

Introduction

G. W. Cullen and C. C. Wang

In nature's infinite book of secrets
    a little I can read
SHAKESPEARE, *Antony and Cleopatra*

The authors' intent

Perhaps the most useful function that an introduction can serve is to let the reader know, more completely than can be indicated in the title, what the authors have set out to accomplish. Our intention is to describe in some detail the preparation and characterization, and to point out the applications, of heteroepitaxial thin films of silicon, III–V compounds and alloys, and II–VI compounds deposited on sapphire and spinel substrates. The methods of preparation and characterization have been developed specifically with applications in mind, and to a large extent the quality of the thin films has been determined by examination of the functional characteristics required for application in device structures. The development of this heteroepitaxial thin-film technology has followed a logical progression in the methods of synthesis employed, the nature of the materials prepared, and the device functions that can be served by the individual materials or combination of materials.

The real possibility for commercial application of the heteroepitaxial deposits started with silicon. As success was achieved in realizing usable semiconducting properties in the thin heteroepitaxial silicon films, it seemed logical to apply the heteroepitaxial growth technology to the III–V and II–VI compounds. The availability of the compound heteroepitaxial deposits offered the possibility of employing the photoconductive, electroluminescent, piezoelectric, and electrooptic properties of these materials. (These functions were not included in the book's title because we felt it would have complicated the title more than would have been justified, but they are discussed in Chapters 3 and 4.)

In conjunction with the preparative aspects of the heteroepitaxial film, methods have been employed for the chemical, physical, and electrical characterization of the deposits, some of which are specific to the nature of the thin films on the insulating substrates. The coalescence of these technologies is the topic of this book and is very briefly reviewed in the following section.

A brief overview of the technology* discussed in this book

It has been appreciated for some time that the availability of single crystal semiconductor thin films on an insulating substrate would permit some very significant advances in the characteristics of transistor structures fabricated in the thin films. More than ten years ago "single crystal" silicon deposits were realized on a number of refractory oxide substrate crystals. The crystalline nature of the deposits was determined by physical characterization techniques and gave very little information about the usefulness of the deposits in practical device structures. In fact, the semiconducting properties proved to be a strong function of the distance from the silicon/substrate interface, and usable electrical properties were observed only in deposits much thicker than could be processed (by available methods) into usable device structrues. The electrical properties also proved to be unstable during the thermal cycling required for device processing. These undesirable characteristics were associated, for the most part, with reaction between the deposition constituents and the substrate material. The products of the reaction with the substrate led to the degrada-

---

*Complete references are found in subsequent chapters.

tion of the films both by interference with nucleation on the substrate surface and by contamination of the silicon. The interface reactions were suppressed by substituting hydrides for halides as the silicon source and by decreasing the growth temperature and increasing the growth rate. The ability to improve the crystallinity of the deposits, while increasing the growth rate and decreasing the growth temperature, can be associated (at least in part) with the development of methods to remove the substrate surface crystallographic work damage introduced during the abrasive fabrication of the refractory oxide crystals. The finishing of the substrate surface proved to be more difficult than originally anticipated and the requirement of providing a single crystal refractory oxide surface free of mechanical work damage was new to practitioners in the field of ceramic fabrication. A good deal of reorientation of those skilled in this "art" was required before methods were developed for reproducible production of work damage–free heteroepitaxial substrates. As the contamination by substrate reaction products was minimized, the influence of impurities introduced from the source gases and reactor components was observed, and methods were developed to rigidly exclude these impurities. It proved to be necessary to reduce the level of the gaseous contaminants to a lower level in the heteroepitaxial growth than can be commonly tolerated in the homoepitaxial growth of silicon. At the relatively low heteroepitaxial growth temperatures undesirable species, volatile at higher growth temperatures, interfere with the semiconducting properties as the result of condensation on the substrate and deposit surfaces.

Having briefly outlined the course of development of the heteroepitaxial growth of silicon, an outline of the heteroepitaxial growth of the compound semiconductors can be very much briefer, because much of the wisdom developed in the growth of silicon was directly applicable to the synthesis of the other composite systems. Using halides as the source gases, the heteroepitaxial compound semiconductors were seriously degraded by reaction with the substrate. Under very specific conditions good films of several of these materials were obtained from halide sources, but it proved to be difficult to identify the critical variables and to maintain reproducibility. The reaction with the substrate was suppressed by introducing the cationic component as a volatile organometallic compound (such as trimethyl gallium) rather than as a halide. The hydride of the anion (such as arsine) was used in both systems. With the use of the organometallic source materials, it was possible to lower the deposition temperature while maintaining desirable growth rates. The crystallographic nature of the substrate surface also proved to be a critical factor in the achievement of desired electrical properties. In hindsight, a good deal of the irreproducibility observed in the early phases of the heteroepitaxial growth of both silicon and the III–V materials can be assigned to irreproducibility in the crystalline nature of the substrate surface.

In extending the heteroepitaxial growth on insulator technology from the III–V semiconductors to piezoelectric (AlN, GaN, ZnO) and dielectric (Si_3N_4, Al_2O_3) materials, the overall philosophy of deposition on work damage–free substrates, avoiding the introduction of halides into the system, and depositing at as low a temperature as possible has been followed. With the exception of the one II-VI compound of interest here (ZnO), all of these materials are prepared from the organometallic compound of the cation and the hydride of the anion. Although ZnO cannot be deposited using the organometallic process, the use of close-spaced transport, with hydrogen as the transporting agent, has made it possible to avoid the halides in this case as well.

Feasibility of constructing devices with various functional characteristics has been demonstrated in each of the heteroepitaxial materials individually deposited on the insulating substrates. High-speed low-power complementary pair MOS transistor (C-MOST) structures have successfully been constructed in 0.6-μm-thick heteroepitaxial silicon on sapphire. Light-emitting diodes and photocathode structures have been demonstrated in the III–V semiconductors on sapphire and spinel. Surface acoustic waves have been manipulated in AlN heteroepitaxial deposits on sapphire, and operational optical waveguide structures have been constructed in ZnO on sapphire. The challenge of the future is to integrate these materials on a single substrate wafer in order to provide the capability to manipulate electronic, light, and acoustic signals in devices bridged by single crystal interconnections where needed. Chemical compatibility problems must be solved in order to make such a technology a reality. The most likely heteroepitaxial material for each function to be considered for integration into a more complex system on a single wafer is summarized in Table 1.1.

Flexibility in integrating device functions is provided by layering the materials as well as depositing them side by side on the same substrate. The properties of the deposits can often be improved by providing an intermediate layer of a material that is functionally inactive in the particular device structure between the active layer and the refractory oxide substrate. For example, GaP and a thin layer of silicon are sufficiently transparent to visible light to serve as an intermediate buffer layer in a GaAs photocathode structure in which the light enters through the substrate. Materials that have been grown in side-by-side and layered structures are listed in Table 1.2.

Inherent in the development of any new technology involving electronically active materials is the ability

TABLE 1.1   Device Integration Through Heteroepitaxy

Material	Functional Characteristic	Device
Si	Semiconductor	Transistors
GaP	Electroluminescent	Light emitters
GaAs	Photoconductor	Photocathodes
ZnO	Electrooptic	Optical modulators
AlN	Piezoelectric	Surface acoustic wave devices
Al_2O_3 and $MgO \cdot Al_2O_3$	Inactive substrate Low-loss substrate	Support and orientation control Surface acoustic wave devices

to characterize the material. In the single crystal thin film technology, conventional electrical and physicochemical methods of analysis must be employed with caution. In some cases established techniques can be modified to take into consideration the peculiar properties of the thin films. In other areas of analysis, new techniques have had to be developed. Both the physicochemical and electrical characterization of the heteroepitaxial films are complicated by the small volume of material, the large changes in the material properties with distance from the deposit/substrate interface, and the proximity of the components sandwiching the film. In the physicochemical methods it has been necessary to employ techniques to sample small portions of the already thin film. It is highly desirable, in both the physicochemical and electrical methods, to characterize the materials nondestructively. The processing of complex integrated devices is a very expensive proposition, and therefore it is very important to develop the capability to qualify the heteroepitaxial deposits prior to processing the complex device structures. In the electrical characterization of the films it must be kept in mind that concepts derived from measurements on nearly perfect semiconductors may not be directly applicable to the relatively imperfect heteroepitaxial deposits. In addition to the structural imperfections, the electrical characterization is complicated by the "bulk" of the film being sandwiched between two closely spaced electrically active interfaces. These interfaces may have a strong influence on the film properties and must be taken into consideration both in the design and interpretation of electrical tests.

Preparative solid state chemists have had the tendency to focus on the chemical and crystallographic aspects of thin-film deposition and to treat the conditions of gas flow during deposition in an empirical manner. When one faces, however, the commercial utilization of the heteroepitaxial deposits, it is necessary to turn to some of the fundamentals of gas-flow dynamics in designing conditions to maintain the required uniformity of thickness, doping, and composition in scaled-up deposition apparatus. Therefore, a discussion of gas-flow dynamics is included in this book. The role of the "boundary layer" is discussed, and a model is developed in which the thermal gradients and gas constants can be treated in a general manner. Although the model has been designed to handle some of the conditions peculiar to heteroepitaxial growth (for instance, relatively steep thermal gradients), it is applicable to both hetero- and homoepitaxial growth. In order to handle the variables involved, it has been necessary to make some empirical simplifying assumptions concerning the averaging of the temperature gradients. Nonetheless, we have found that the model has brought to the foreground some relationships that may defy intuition and has been very useful in the design of scaled-up deposition apparatus.

In any treatment of heteroepitaxial growth one will almost always find a discussion included on lattice mismatch between the deposit and the substrate. We have included one particular view of this issue. The feeling is expressed that in such complex systems as discussed in this book, an understanding sufficiently accurate to lead to more than vague guidelines will not

TABLE 1.2   Composite Heteroepitaxial Deposits on Sapphire and Spinel

Side-by-Side Structures	Layered Structures
Si + GaAs/sapphire and spinel	Ge/Si/sapphire
Si + GaP/sapphire and spinel	GaAs/Si/sapphire and spinel
Si + GaN/sapphire	GaAs/Ge/Si/sapphire and spinel
Si + AlN/sapphire	GaP/Si/sapphire and spinel
Si + ZnO/sapphire	GaAs/GaP/Si/sapphire and spinel
GaAs + GaP/sapphire	Si/AlN/sapphire
GaAs + AlN/sapphire	

come in the near future. First, the lattice match in more simple systems, such as homoepitaxial silicon on doped silicon substrates, must be defined. This view is supported by the observation that even in the deposition of cubic III–V compounds on cubic spinel, the orientation of the overgrowth is not uniquely defined by the orientation of the substrate. A number of lattice-matching schemes have been presented to rationalize, for instance, the heteroepitaxial growth of cubic silicon on rhombohedral sapphire. None of these schemes have been sufficiently general to aid in the search for new and improved heteroepitaxial composite systems. The search for substrates on which better electrical properties in heteroepitaxial semiconductors might be achieved has been more Edisonian than the solid state chemist might have wished. It is not practical, however, to wait for the development of useful theory prior to lauching efforts directed toward further optimization of the heteroepitaxial systems. It is hoped that some of the experimental observations presented in this book, such as the lack of unique definition of orientation of GaP on spinel, will incite the interest of the more theoretically oriented crystallographers to develop useful models for well oriented heteroepitaxy.

Clearly we have made no effort to balance the length of the chapters. The intent has been to provide the information in the degree of detail that we feel would contribute to the overall theme of the book.

The content of the individual chapters

In Chapters 2, 3, and 4, the development of the preparative techniques for the various heteroepitaxial materials and the results and conclusions drawn from characterization are discussed. Current and potential applications in device structures, and some of the material aspects of device processing are outlined. Included in Chapter 2 is a discussion of the selection of substrate materials, the growth of single crystals, the shaping of single crystals into the substrate geometry, and the methods used to remove the surface work damage. This discussion was intended to serve all three of the preparative chapters, and therefore, the sections on substrate preparation in Chapters 3 and 4 are brief.

In Chapter 5, the characterization of the thin films is discussed, with emphasis on the problems encountered in the chemical and physical (crystallographic) methods. In Chapter 6, the electrical characterization specific to the nature of the thin heteroepitaxial silicon films is treated in some detail. A model is also presented of current transport through crystallographic-imperfect heteroepitaxial silicon.

The discussion of gas-flow dynamics and the development and application of a general model are presented in Chapter 7. Finally, some general comments are given in Chapter 8 on the current state of understanding of crystallographic misfits at the deposit/substrate interfaces.

The role of heteroepitaxial films in semiconductor technology

It is important to point out the degree to which the processing of the heteroepitaxial thin films is compatible with the processing of device structures in single crystal silicon. A tremendous amount of technology has been developed for the processing of device structures in bulk silicon. New materials must exhibit very large advantages to displace bulk silicon as the basic semiconductor material, particularly if the processing of the new material is incompatible with and cannot draw from bulk silicon processing technology. Essentially no new materials have made the grade in the past decade. We have long since passed the "if you can't fight them, join them" turning point. In order for a material or composite material system to be seriously considered as a complement to silicon technology it must be basically compatible with bulk silicon processing. In the case of silicon on sapphire, while it has been necessary to develop methods to minimize the thermal cycling, basically the same oxidation, masking, diffusion (and/or ion implantation), and metallization schemes can be applied using the bulk silicon processing equipment. In the case of the integration of the various functional materials on an insulating substrate, significant differences in chemical reactivity must be taken into consideration. On the other hand, it is anticipated that processing of the integrated materials can be developed to the point where it is compatible with modified silicon processing techniques.

The authors

The authors of the individual chapters in this volume are associates at the RCA Laboratories. For a number of years they have been working closely together on the various aspects of heteroepitaxial films. There has been unusually good cooperation and communication among the individuals involved in the preparation of the substrates and thin deposits, the chemical and physical analyses, and the electrical characterization of the materials. It has been the feedback between these individuals that has made it possible for us to have developed as complete a picture as is presented of the heteroepitaxial systems described in this volume. It is fair to say that the authors have drawn most heavily from their own work; clearly they know that best. It is certainly recognized, however, that there has been excellent work carried out in other organizations on the same topic. The authors have made every effort to include pertinent work done by other research groups. The intent has been to describe the course of development of the technology and the present capabilities rather than to construct a comprehensive review of all the work done. Particularly in the earlier work, examples are cited from the literature rather than presenting a review of the literature. A number of very complete review articles have been cited by the various authors.

Chapter 2

The Preparation and Properties of Heteroepitaxial Silicon

G. W. Cullen

2.1. Introduction

Motivation

The heteroepitaxial growth of silicon on refractory oxide substrates is an intriguing technology to the materials scientist. There is some question as to the degree we now understand (or indeed will in the near future understand) the factors promoting or impeding good crystallinity in heteroepitaxial growth.* There is no question, however, as to the value of realizing thin films of "device quality" silicon on insulating substrates. The question is not, "Is it useful?" Rather, the question is, "How similar to bulk silicon can we make heteroepitaxial silicon in the thickness of interest for device application (i.e., 0.5–1 μm)?"

In many areas of explorative materials synthesis, it is necessary to prepare the material or material composite system in order to examine its properties and evaluate its applicability. Under these circumstances, one may be uncertain as to whether or not it is worthwhile to pursue the effort prior to the achievement of the material. And if the effort is pursued to the point of successful preparation, one must face the question as to whether or not the properties achieved are the best fundamentally possible. In the area of heteroepitaxial silicon, we are freed of most of these apprehensions. The advantages can be anticipated by analogy to well known dielectrically isolated device structures, and one can assume that (at least at some thickness of the heteroepitaxial silicon) the fundamental limits are the same as those operative in processed bulk silicon technology.

---

*See Chapter 8.

Requirements for device application

It is difficult to concisely define what materials characteristics must be achieved for the application of heteroepitaxial deposits in device structures. This difficulty in part arises from the properties of the heteroepitaxial silicon being a strong function of the distance from the silicon/substrate interface and from the specific location in the film in which the properties must be realized depending on the type of device structure to be used and the specific method with which it is processed. It is fortunate that in MOS transistor circuitry only the carrier mobilities within approximately 1000 Å of the outer surface (the surface most remote from the silicon/substrate interface) play a role in the device performance. It is a necessary, but not sufficient, requirement that the mobilities in this thin layer be similar to the mobilities observed in processed bulk silicon. It is also fortunate that high minority-carrier lifetimes are not required for application of heteroepitaxial silicon in MOS transistor structures. The lifetime is more critically dependent on the crystalline structure than is the mobility. Aside from the need for "bulk-like" mobilities, the semiconducting properties must be maintained during the harsh treatment (from a chemical/crystallography point of view)that the composite structure endures during device processing. Temperatures and temperature gradients equal to, and in some cases exceeding, those experienced during deposition are imposed during device processing. Early in the development of this technology it was not unusual for the materials scientist to deliver to his device-processing colleagues deposits that appeared to satisfy all the known requirements but that did not retain their properties during processing. It is a particularly interesting point that the semiconducting properties of heteroepitaxial silicon measured after device processing are a stronger function of the deposition conditions than are the properties measured immediately after deposition. This necessitated a particularly close collaborative effort between the materials and device personnel in the development of the needed technology; it was essential that the materials personnel design the original preparative conditions to realize after-processing semiconducting properties, and it was also essential that the device personnel design the processing procedures to take into consideration the peculiar chemical/physical characteristics of the heteroepitaxial silicon/insulator composite system. The methods employed to qualify the deposits from a materials point of view are discussed in some detail in this chapter.

Advantages in the application of heteroepitaxial silicon in MOS transistor structures

The use of single crystal silicon films on insulating substrates, rather than on high-resistivity silicon substrates, offers the following advantages in integrated circuit technology:

- The total junction area is minimized, and the capacitances associated with isolation junctions in silicon on silicon devices are eliminated; this provides relatively high operational speed at low levels of power consumption.
- Dielectric isolation is provided between closely spaced units of integrated circuit arrays; this provides high device-packing density.
- Both active and passive devices may be fabricated on the same substrate.
- The minimization of the total junction area also contributes to increased radiation resistance.

In recent years the above advantages have been achieved in the heteroepitaxial silicon thin-film technology. An additional prospect, which will require a good deal of further development, is the integration of silicon, III–V, and II–VI thin heteroepitaxial films on a single insulating substrate wafer.

A brief chronological overview*

From the point of view of a materials scientist, the realization of 6-μm-thick cubic single crystal silicon on rhombohedral sapphire substrates would be considered to be a significant achievement. The single crystal nature was originally established by x-ray diffraction. From an applications point of view, however, these deposits are of very little use. In the efforts described in this chapter, the goals have been defined by application of the heteroepitaxial silicon in semi-

---

*For detailed references on these developments refer to subsequent sections of this chapter.

conducting device structures. There is no practical method to fabricate the desired devices in the 6-μm-thick film, and x-ray diffraction provides very little indication of the semiconducting properties. In fact, the semiconducting properties are a much more sensitive measure of the crystallinity and purity of the heteroepitaxial silicon than are the physicochemical characterization techniques commonly employed by materials scientists. Thus, the realization of x-ray diffraction patterns indicative of single crystal character in heteroepitaxial silicon was only the first step in a sequence of developments in which the semiconducting properties have been employed as a very critical measure of the quality and applicability of the thin deposits.

With the realization of single crystallinity (albeit not good crystallinity) in heteroepitaxial silicon, investigators associated with the electronics industries appreciated the potential in the silicon/insulator composite structure and rapidly launched materials efforts designed to optimize the semiconductor properties. In the early work, the silicon was deposited by the hydrogen reduction of silicon tetrachloride at relatively high temperatures. This resulted in the heavy contamination of the silicon as the result of the reaction of both silicon and halides with the substrate. Under these circumstances, measurable carrier mobilities were observed in only relatively thick deposits. The halides were excluded from the deposition atmosphere by the use of silane as the silicon source, and usable carrier mobilities were achieved in films of intermediate thickness (1.5–3 μm). It proved to be difficult to fabricate devices in such films, because the semiconducting properties changed during the thermal treatment involved in device processing and the dopant diffused rapidly in the horizontal direction in the crystallographically poor silicon adjacent to the silicon substrate interface. It was also difficult to deposit continuous metal contacts over the edge of the relatively thick films. With further development to improve the deposits for device application, the changes in semiconducting properties during thermal processing were minimized by using relatively low deposition temperatures and high deposition rates. The capability of employing such conditions was brought about, at least in part, by improvements in the methods used to free the substrate surface of work damage and the control of impurities from source gases. These preparative conditions also contributed to improving the crystallinity of the heteroepitaxial silicon adjacent to the interface and led to the overall improvement in the physical properties in the thinner deposits (0.5–1.0 μm).

In the course of this development, magnesium aluminate spinel was examined as a possible substitute for sapphire as a substrate material, mainly because the spinel appeared to be less reactive chemically and because the cubic structure offered a better lattice match to silicon. Unusually high hole mobilities were achieved in silicon on flame fusion spinel, but the best electrical properties were not achieved in the crystallographic orientation favored for device fabrication, and it proved to be difficult to obtain reproducible properties in the large-diameter spinel prepared by the Czochralski method. Therefore, specifically for commercial application in MOS transistor structures, sapphire proved to be the favored substrate material. There is little doubt, however, that the simultaneous examination of the two substrate materials with significantly different chemical and crystallographic properties contributed significantly to a practical understanding and an accelerated development of the heteroepitaxial silicon preparative technology. In the event that the integration of heteroepitaxial silicon, III–V, and II–VI materials becomes a reality, spinel substrates may play an important role. For these reasons this chapter discusses experimental work carried out on both substrate materials.

The development in the preparative technology has resulted in the availability of carrier mobilities similar to those of bulk silicon in MOS device structures processed in heteroepitaxial silicon 0.6 μm in thickness. The changes in semiconducting properties during processing are acceptable, as the result of the development of both the preparative and device-processing procedures. This is not to say that all problems have been solved or that the factors limiting the semiconducting properties are understood. The details of what determines the crystallographic structure in the heteroepitaxial deposit and the nature of the silicon/substrate interface are still topics of considerable effort. These studies continue not only to provide further control of such factors as stability and radiation hardness in MOS transistor structures, but they also have as a goal improving the lifetime sufficiently so that bipolar transistors can be fabricated in deposits 1 μm and less in thickness. Were this goal achieved, it is likely that the heteroepitaxial silicon would constitute a significant fraction of the total semiconductor technology.

Background literature

Although pertinent early literature has been cited in this chapter where the content bears on the discussion of current problems and developments, in general there has been no attempt to present an historic review of the silicon-on-insulator literature. The purpose is to describe developments that contribute to an appreciation of the current technology, to present an overview of the current problems and capabilities, and to suggest some direction where future efforts would be rewarding. Review articles specifically dealing with or including discussions of the silicon-on-insulating-

surface technology are Miller and Manasevit (1966), LaChapelle, Miller, and Morritz (1967), Filby and Nielsen (1967), Joyce (1968), Cullen (1971), Filby (1972), Manasevit (1974), and Joyce (1974). An attempt has been made to include sufficient background material to generate a reasonably complete story for the main body of readers. Those seeking more detailed information on the general subject of the growth of single crystal thin films are referred to literature which treats specific areas in detail. For instance, the classic article by Pashley (1965) provides perspective on the crystallography of composite systems. Chopra (1969) discusses the preparation and characterization of a variety of crystalline and noncrystalline thin films. The collected individually authored chapters in volumes edited by Schneider and Ruth (1971) and Matthews (1975) include discussions of many aspects of homoepitaxial growth applicable to heteroepitaxy. Because of the commercial significance of chemically vapor-deposited silicon, numerous journal articles are to be found on this topic. A broad view of some of the more practical aspects of chemical vapor deposition is found in a volume authored by Powell, Oxley, and Blocher (1966).

2.2. Selection of substrate materials

Overview

The criteria by which substrate materials are selected is qualitative, at best. In the final analysis, potential materials must be carefully qualified by experimentation. Even in the laboratory, the usefulness of a particular substrate crystal system may not be immediately obvious. It is interesting to note that usable semiconducting properties have been achieved on the most commonly used substrates only after a considerable effort to develop deposition conditions that minimized the problems associated with the substrate or after considerable effort to modify the nature of the substrate material.

In the analysis of the factors limiting the semiconducting properties of silicon grown on single crystal insulating substrates, investigators in the field have commonly focused on three factors: (1) the spatial match between the crystalline lattice of silicon and the lattice of the supporting substrate; (2) the contamination of the silicon as the result of reaction between the deposition constituents (including silicon) and the substrate material; and (3) the relative thermal contraction of the silicon and the substrate on cooling from the deposition temperature. In the early search for substrate materials on which single crystal (or, at least, oriented) silicon growth could be achieved, and in the continuing search for substrate materials that would lead to improvement of the semiconducting properties already realized, the three factors cited above have been seriously considered. This chapter is oriented toward the use of heteroepitaxial silicon for the fabrication of low-power field effect device structures. For other applications, considerations other than the three cited above become important. For instance, in high-power microwave circuit applications the thermal conductivity of the substrate must be considered for power dissipation. In geometries in which the silicon is used in conjunction with other single crystal materials in layered or side-by-side geometries, the functional properties of the associated materials may play a role in the choice of the substrate. In a layered structure in which the silicon/substrate composite serves as a substrate for the growth of optically active III–V compounds (Chapter 3), the optical properties of the substrate must be considered. In a side-by-side structure including silicon and AlN in surface acoustic wave device geometries (Chapter 4), the acoustic wave attenuation of the oxide substrate is a factor in the choice of the substrate.

In a practical sense, the properties of heteroepitaxial films are also determined by (1) the ability to grow good-quality crystals of the desired substrate material (see Section 2.3); (2) the ability to prepare the surface of the crystal for heteroepitaxial growth (see Section 2.4); and (3) the ability to develop deposition techniques that constitute a realistic compromise between the conditions most favoring single crystal growth and the conditions providing suppression of the reaction between the deposition constituents and the substrate material (see Section 2.8).

In the development of sapphire as a substrate material, availability probably ranked equal to reactivity and thermal expansion in the initial selection. It was fortunate that, as heteroepitaxial deposition techniques were refined, other commercial applications brought the development of methods for growth of high-quality sapphire crystals. In the case of spinel, after preliminary observation of positive results on crystals not adequate for substrate usage, it was necessary to modify the substrate crystal growth specifically for application as heteroepitaxial silicon substrates.

Beryllium oxide has been considered for substrate use because of its unusually high thermal conductivity, as well as for its low reactivity and cubic structure. There has been considerable effort to develop the growth of BeO into a practical manufacturing process (Austerman, 1964; Austerman et al., 1967, 1972). At this time the material is not readily available, however, nor is information on the properties of silicon deposited on the BeO. This substrate is therefore not discussed here.

The physical properties of sapphire and spinel which are useful for evaluation of sapphire and spinel as substrates for silicon growth are given in Table 2.1.

Orientation and lattice match: theory

Early in the development of heteroepitaxial silicon technology a good deal of emphasis was placed on lattice-matching considerations. This was probably based on the availability of detailed and accurate information on the various crystal structures of inter-

TABLE 2.1   Physical Properties of Silicon, Sapphire, and Spinel

Parameter	Silicon Si	Sapphire α-Al_2O_3	Spinel $MgAl_2O_4$
Crystal structure	Face-centered cubic	Rhombohedral	Face-centered cubic
Unit cell dimension (Å)	$a = 5.4301$	$a = 4.758, c = 12.991$	$a = 8.083$
Density (g/cm³)	2.33	3.98	3.58
Hardness (mohs)	7	9	8
Indentation No. (kG/mm²)	1000	1525–2000	1320
Melting point (C)	1412	2030	2105
Vaporization process, bp (C)	3145	2980	Unknown
Species	Si-Si$_7$	Al$_2$O and O$_2$	Unknown (Mg-rich)
Dielectric constant	11.7 (500 Hz–30 MHz)	9.4 (to c) (100 Hz–100 kHz)	8.4 (100 Hz–100 kHz)
Dissipation factor, tan δ	—	10^{-3}–10^{-4}	10^{-3}–10^{-4}
Refractive index	3.4975 (at 1.357 μm)	1.7707 (at 5461 Å)	1.7202 (At 5461 Å)
Optical transmission	Transparent in infrared	80 percent minimum (0.24–6.0 μm, 0.0175 in.)	80 percent minimum (0.31–5.1 μm, 0.0175 in.)
Thermal conductivity (cal/cm-s-°C) at 25°C	0.30	0.065 (60° to c)	0.035
Thermal expansion coefficient (1/°C) at 25–800°C	3.59×10^{-6}	8.40×10^{-6} (60° to c)	7.45×10^{-6}

Source: Cullen (1971)

est and on a confidence in recently developed theories of heteroepitaxy.

Lattice matching for the various orientations of interest of both sapphire and spinel has been dealt with in some detail by a number of investigators [see, for instance, the review articles by La Chapelle et al. (1967) and Filby and Nielsen (1967) and papers by Nolder and Cadoff (1965) and Manasevit et al. (1968)]. Using the rationale presented in the work cited, the lattice misfits for cubic silicon on rhombohedral sapphire and cubic spinel have been estimated. For instance, if one assumes coincidence of silicon and aluminum (La Chapelle et al., 1967), the mismatch between the (100) silicon on ($1\bar{1}02$) sapphire is ~12.5 percent in the [$11\bar{2}0$] direction and ~4.2 percent in the [$1\bar{1}01$] direction. This orientation has been the most extensively investigated. Assuming a coincidence of silicon with the substrate cations, and using a matching scheme wherein three unit cells of silicon are superimposed on two unit cells of spinel, the calculated mismatch between silicon and the $MgO \cdot 3.3Al_2O_3$ spinel is ~1.9 percent (Seiter and Zaminer, 1965; Manasevit and Forbes, 1966; Heywang, 1968). The lattice constant of spinel increases with decreasing alumina content, and thus the calculated mismatch between silicon and the $MgO \cdot Al_2O_3$ (stoichiometric) spinel is less than 1 percent (Wang, 1969).

Detailed lattice-matching schemes have been developed for other substrate types and orientations. These are not discussed here, however, because the calculated mismatch values have not proven to be useful in choosing substrate materials or in predicting the quality of the heteroepitaxial films. It is interesting to note that Pashley (1965) pointed out that a small lattice misfit is not an essential condition for the occurrence of an oriented overgrowth and that the orientation occurring in heteroepitaxial growth is not always that which corresponds to the best geometrical fit. Nor does a good lattice match assure oriented growth. Green and coworkers (1970) failed to observe oriented growth even for misfits as low as 0.4 percent in metals on lithium fluoride (the title of the paper is "The Insignificance of Lattice Misfit for Epitaxy").

The lack of success in formulating misfit guidelines to predict heteroepitaxial growth may at least in part be attributed to the fact that the real surface of the crystal may not include the same atomic spacings as the bulk of the crystal [see, for instance, Chang (1971)] and that, under the circumstances where the deposit reacts with the surface, more than one crystallographic plane may be exposed. Other factors involving misfit considerations are discussed in detail in Chapter 7. Although the effect of lattice misfit on the nature of the heteroepitaxial deposits is unpredictable, it is reasonable to assume that spatial mismatches at the silicon/substrate interface play a key role in determining the semiconducting properties of heteroepitax-

TABLE 2.2   Silicon on Sapphire Orientation Relationships

Relationship	Orientation Planes Silicon|Sapphire	Orientation Direction Silicon|Sapphire
I	(001)|(01$\bar{1}$2)	[100]|[$2\bar{1}\bar{1}$0]
II	(111)|(11$\bar{2}$4) (110)\|(11$\bar{2}$0)	[$1\bar{1}$0]|[$1\bar{1}$00]
III	(111)|(11$\bar{2}$0) (11$\bar{2}$)\|($\bar{1}$104)	[$1\bar{1}$0]|[$2\bar{2}$01]
IV	(111)|(10$\bar{1}$4)	[$1\bar{1}$0]|[$1\bar{2}$10] (11$\bar{2}$)\|[10$\bar{1}$1]
V	(001)|(10$\bar{1}$2)	[100]|[$2\bar{1}\bar{1}$0]
VI	(310)|(10$\bar{1}$1)	[1$\bar{3}$2]\|[$\bar{1}$2$\bar{1}$0], [1$\bar{2}$10] [$\bar{1}$35]\|[$\bar{1}$012]
VII	(111)|(0001)	[11$\bar{2}$]|[$\bar{1}$100]
VIII	Four multicrystalline relationships	

Source: Manasevit et al. (1968).

ial silicon. The significance of lattice matching is a complex and interesting issue, which bears further study. We share the viewpoint expressed in Chapter 7 (by J. Blanc) that initial understanding will come from the study of much simpler systems than silicon on sapphire.

Orientation relationships: experiment

The orientation relationships and the related crystallinity of the silicon deposited on the various sapphire and spinel orientations have been developed by experimentation. Manasevit and coworkers (1968) investigated the silicon-on-sapphire orientation relationships by depositing on rounded cylindrical surfaces of single crystal sapphire (Cullen et al., 1973A). It was found that ed in Table 2.2 and illustrated in the stereographic projection shown in Figure 2.1. Estimates of the crystalline quality of the various orientations have also been made by depositing on hemispheres of the single crystal sapphire (Cullen et al., 1973). It was found that the best crystalline quality, as determined by x-ray examination, could be correlated with the highly reflective areas of the deposit and with the thickness at which the highly reflective areas became rough. The surface of the (100) silicon on the ($1\bar{1}02$) sapphire remained reflective to a thickness as great as 60 μm. Good single crystal (100) silicon was observed only on or near the ($1\bar{1}02$) sapphire orientation. At 10 μm,

---

*Various investigators use different notations to describe the crystallographic planes of interest in sapphire. Note that in the rhombohedral system the ($1\bar{1}02$), ($\bar{1}012$), and ($01\bar{1}2$) planes are equivalent. The ($0\bar{1}12$), ($\bar{1}102$), and ($10\bar{1}2$) planes are also equivalent to one another but not to the above set. The symmetry associated with these planes may be visualized by reference to Figure 2.1.

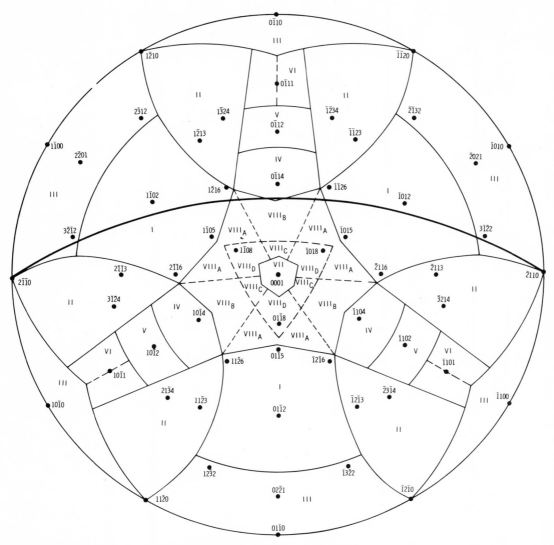

FIGURE 2.1   A complete stereographic projection map of the silicon orientation relationships on sapphire. From Manasevit et al. (1968)

reflective surfaces were observed in (111) silicon on the (11$\bar{2}$0) and (10$\bar{1}$3) sapphire. This correlates with the work of Manasevit and coworkers (1976), who have observed good semiconducting properties in (111) silicon on the (11$\bar{2}$0) sapphire, as well as on the (10$\bar{1}$4) surface which is very close to the (10$\bar{1}$3) orientation cited above. Prior to this work of Manasevit and coworkers, the highest mobilities in $\leq$ 1-μm-thick films had been observed in the (100) silicon deposited on the (1$\bar{1}$02) sapphire. In general, the (100) orientation has been favored for device application also because the surface state density at the silicon/silicon dioxide interface decreases in the order (111) > (110) > (100) (Grove, 1967).

In silicon on spinel, the orientation of the overgrowth matches the orientation of the substrate. This

has not been the case, however, for all cubic materials deposited on the cubic spinel (see Table 3.6). The mobilities observed in silicon on spinel have consistently been higher in the (111) silicon deposited on the (111) spinel than in the (100) silicon deposited on the (100) spinel (Cullen, 1971). There has been no clear correlation developed between this observation and the crystallinity of the heteroepitaxial silicon. The effect may be related to the ease of preparing the substrate surface, which has proven to be highly orientation dependent in both spinel and sapphire.

Chemical reactivity of the substrate

Probably more materials have been rejected for use as substrates because of chemical reactivity than for the other reason. Silicon is very reactive at the deposition temperatures and leads to the chemical reduction even of the highly stable aluminum oxide (see Section 2.6). It may be assumed that any material that adheres well to silicon also reacts with silicon. It must be kept in mind that 1 ppm is equivalent to a concentration of 6 $\times$ 10^{16}/cm^3 in silicon, and that for device application the concentration of active carriers must be well controlled to levels as low as 1×10^{15}/cm^3. It is not practical to think in terms of holding the total impurity level to well under 1 ppm. Only specific species, however, are electrically active in the heteroepitaxial silicon. Even when the substrate contains compounds of elements which are well known dopants in silicon (such as Al), it is difficult to anticipate whether or not the reaction products will be electrically active. It is clear that not all of the aluminum-containing reaction products introduced into the silicon from either sapphire or spinel are electrically active. The aluminum appears to be present in an oxygen-containing complex which has not to date been identified (see Section 2.12). The point to be made here is that, a priori, it is not possible to choose substrate materials on the basis of whether or not impurities introduced from the substrate will be electrically active. The concentration of the nonactive impurities must be at a level that does not degrade the film crystallinity, particularly in the region of the silicon directly adjacent to the silicon/substrate interface [see, for instance, the nature of early growth of silicon on Czochralski spinel (Section 2.6)].

Thermal expansion of the substrate

In regard to the thermal expansion of the substrate, the first consideration is that the substrate contract the same as or more than the silicon on cooling from the deposition temperature. Under these conditions the film does not crack since it is either not stressed or is under compression. Compression can influence the transport properties in the silicon (see Section 6.12). Tensile stress in the substrate may lead to cracking of the substrate. The critical ratio of film thickness to substrate thickness that results in the substrate cracking is a function of the elastic constants and critical stress constants of the various substrate materials. The ratio of film thickness to substrate thickness in practical MOS device structures is at least 1/350 (1 μm of silicon on 14-mil-thick 2 inch-diameter substrates). With this ratio, there is no cracking problem with either sapphire or spinel substrates as the result of differences in the thermal expansion of the silicon and the substrate. Beyond the requirement that the film be under compression, it is not possible to state quantitatively an acceptable difference between the contraction of the substrate and the deposit, particularly under the circumstances where the stress-relieving

mechanisms and the influence of stress on electrical properties have not been completely identified (see Section 2.11).

2.3. Growth of substrate crystals

The development of silicon heteroepitaxy on flame fusion and Czochralski-grown substrate crystals

The early work on the heteroepitaxial growth of silicon on sapphire (Manasevit and Simpson, 1963, 1964) was carried out using substrate wafers fabricated from commercially available crystals prepared by the flame fusion method. At that time flame fusion–grown sapphire, which is prepared in large quantities for gemstones and watch bearings, was more readily available and considerably less expensive than crystals grown by the Czochralski method. During the late 1960s most commercial suppliers in the United States shifted from an emphasis on flame fusion to Czochralski growth as the demand for highly perfect single crystals arose to meet the requirements imposed by laser, electrooptic, and heteroepitaxial substrate technologies. In part, because of the development of automatic growth control, the cost difference between Czochralski and flame fusion sapphires was considerably decreased.[*] Greater control can be exercised over the thermal gradients and nature of the growth interface in Czochralski growth than in flame fusion growth, and the defect density of the Czochralski material is typically less than that of flame fusion crystals. The possibility arose, therefore, that the crystalline quality and associated semiconducting properties of silicon on Czochralski sapphire would be superior to the properties of silicon on the flame fusion material. A number of investigators, however, reported that the properties of heteroepitaxial silicon on the two types of substrate were similar (Hart et al., 1967; Dixon and Willis, 1967). It was proposed at that time that the crystalline nature of the silicon was more dominated by lattice mismatch and chemical reaction between the silicon and sapphire than by the crystalline quality of the substrate within the range of crystalline quality investigated. These conclusions were reached, however, using thicker films (>2 μm) than are of current interest and test structures (such as Hall patterns) that spanned larger areas of the substrate than the individual transistors in integrated arrays. The conclusion may not be applicable, therefore, to current technology. Trilhe and coworkers (1975) have more recently reported that the MOS transistors do operate when fabricated in heteroepitaxial silicon over grain boundaries in the substrate. The distribution in the electrical

---

[*]Union Carbide Corporation, Crystal Products Dept., San Diego, Ca.

characteristics, however, of MOS transistor arrays fabricated in silicon on flame fusion substrates proved to be greater than in devices fabricated in silicon on the Czochralski grown material. The information on the effect of substrate grain boundaries on device characteristics is incomplete, and this must be considered to be an open question. Attempts are currently being made to grow grain boundary–free flame fusion sapphire as large as three inches in diameter.

In the development of new methods for the growth of sapphire, emphasis has been placed on cost reduction. Sapphire grown in a "ribbon" form by the edge-defined film-fed growth (EFG) technique (LaBelle and Mlavsky, 1971; LaBelle, 1971A,B) more closely approximates the shape required for silicon substrates and therefore requires less fabrication. The heat-exchanger technique (Schmid and Viechnicki, 1970) requires less sophisticated control than either the Czochralski or EFG methods for the growth of large boules. The use of costly iridium crucibles has been successfully avoided in both the EFG and gradient furnace* methods. The Hall mobilities of 1-μm-thick (100) silicon grown on the Czochralski, EFG, and gradient furnace (1$\bar{1}$02) sapphire are similar (Corboy, 1974). The three types of sapphire are currently under evaluation for commercial application.

The first heteroepitaxial growth of silicon on the magnesium aluminate spinel was also carried out on crystals commercially grown by the flame fusion process (Seiter and Zaminer, 1965; Manasevit and Forbes, 1966). The ease of growth of spinel crystals is directly proportional to the alumina content within the compositional range $MgO \cdot Al_2O_3$ to $MgO \cdot 3.5Al_2O_3$. It was observed, however, that the thermal stability of the spinel is inversely proportional to the alumina content (Wang, 1969). The commercially available crystals, of nominal composition $MgO \cdot 3.3Al_2O_3$ to $MgO \cdot 3.5Al_2O_3$, proved to degrade at the temperatures employed for epitaxial growth and device processing (900–1200°C). In order to avoid the problems related to the thermal instability, methods were developed to grow flame fusion spinel within the compositional range $MgO \cdot 1.5Al_2O_3$ to $MgO \cdot 2.5Al_2O_3$ (Grabmaier and Watson, 1968A,B,C; Wang, 1969). Below an alumina content of $MgO \cdot 1.5Al_2O_3$, it proved to be difficult to grow crystals of the diameter desired (at that time 1–1.5 inch). Very promising semiconducting properties were achieved on the "low-alumina-rich" flame fusion spinel [Seiter and Zaminer (1965), Cullen et al. (1969); see also Section 2.12]. As with sapphire, focus was placed on achieving between 2- and 3-inch-diameter boules grown by the Czochralski process as

the interest in silicon on insulator integrated devices increased. The similarity in the development of silicon on sapphire and silicon on spinel ends here, for whereas the characteristics of silicon on sapphire appeared to be relatively independent of the method used for the substrate growth, the properties of silicon on spinel proved to be a strong function of the method of substrate growth. When the Czochralski sapphire became commercially available, it was substituted for the flame fusion material with no modification in the method of substrate surface preparation or silicon deposition conditions. The Czochralski spinel, however, proved to be more reactive than the flame fusion spinel in contact with both hydrogen and silicon. It is interesting to note that, from the standpoint of utilization as a silicon substrate material, there is more similarity between sapphire (either flame fusion or Czochralski) and flame fusion spinel than between flame fusion spinel and Czochralski spinel. As has been previously mentioned, the high-alumina-rich spinel is most readily prepared by the flame fusion method (Wang, 1969). The near-stoichiometric flame fusion spinel boules commonly craze on cooling from the growth temperature. By contrast, the near-stoichiometric spinel is most readily prepared by the Czochralski process; it has been reported that second-phase precipitates appear in the alumina-rich Czochralski spinel (Cockayne and Chesswas, 1967; Grabmaier and Watson, 1968A,B,C; Cockayne et al., 1969; Hammond, 1969). The origin of the relationship between the ease of growth by the two methods and the MgO/Al_2O_3 ratio is not understood in detail, but it may well be related to the large differences in the thermal gradients characteristic of the various growth methods.

Single crystal spinel has also been grown by the EFG (LaBelle et al., 1973) and gradient furnace (Schmid and Viechnicki, 1973) techniques. As applied to spinel, these crystal growth techniques are in the early stage of development, and the reactivity and stability of the materials for use as silicon substrates have not been characterized.

The current emphasis is on the optimization of the synthesis of sapphire by the various crystal growth techniques. Because of the problems that have been faced in the synthesis of spinel by methods other than flame fusion growth and the improvement in semiconducting properties only in (111) silicon on spinel, only sapphire has been commercially employed as a substrate material for MOS transistor circuitry fabricated in the heteroepitaxial silicon. Spinel may be favored as the substrate material in applications wherein heteroepitaxial materials other than silicon are required on the same substrate (see Chapters 3 and 4).

The techniques employed to grow substrate crystals, the advantages and disadvantages of the various growth methods, and the characteristics of the crys-

---

*The terms heat-exchanger and gradient furnace techniques have been used interchangeably in the literature, and more recently the method has been referred to as the Schmid-Viechnicki technique. See the section on heat-exchanger growth, which follows.

tals specific to the growth method are discussed in the following sections.

Flame fusion growth

In the flame fusion process (also referred to as the Verneuil method), the melt is held on the cap of the growing crystal by surface tension. Heat energy is introduced with a hydrogen–oxygen flame, and new melt is supplied by dropping powered material through the flame onto the molten cap (Djevahirdjian, 1961; Adamski, 1965; Grabmaier and Watson, 1968A,B,C; Wang, 1969; Ricard and Cisccolaini, 1972; Falckenberg, 1972, 1975, 1976). The key feature of the flame fusion process is that no crucible is required. As in the case of Czochralski growth, the crystalline quality of the growing crystal is a strong function of the thermal gradients in the system and the nature of the melt–solid interface. There is less latitude in controlling these factors in flame fusion growth as compared with Czochralski growth. The thermal gradients can be controlled to some degree by the location of the flame in relation to the molten cap, the interior shape of the furnace, and the length of the furnace muffle below the growth interface. The system has a relatively high thermal inertia; not only the crystal but the inner portion of the massive refractory around the crystal are heated by the flame. In relation to the furnace geometry, the position of the liquid–solid interface is held constant as new material is supplied from a hopper and the crystal is lowered during growth.

Therefore, after growth to the desired diameter, the heat input is held constant during the growth of a crystal. Because the thermal inertia is high, the flame relatively stable, and the position of the melt–solid interface fixed, sophisticated closed-loop control mechanisms are not necessary. This is fortunate, since the hydrogen–oxygen flame is cumbersome to control with closed-loop temperature-control systems. Usually the flame is stabilized with simple gas input-pressure regulators. It is interesting to watch a single operator monitoring as many as 50 flame fusion growth stations operated on common gas feed lines.* After initial alignment of the seed crystal in the flame and growth to the desired final diameter, the feed rate and pull rate are fixed, and the growth of the crystal progresses essentially unattended.

Flame fusion sapphire and spinel crystals prepared in large quantities for gemstone and bearing application are commonly about one inch in diameter. In commercial production for these applications, quantity is achieved by operating many simple growth stations rather than by enlarging the individual crystals.

For application as heteroepitaxial silicon substrates

---

*Hrand Djevahirdjian SA, Industrie De Pierres Scientifiques, Monthey, Switzerland.

it has been necessary to develop the technology for the growth of larger crystals. The problems faced in attempts to grow the larger crystals are cracking during cool down and melt runover from the cap. Falckenberg (1972, 1975, 1976) has introduced a number of flame fusion furnace and burner design features in an effort to grow relatively large crack-free spinel and sapphire crystals specifically for application as heteroepitaxial silicon substrates. In the most refined version, retractable concentric tubes are positioned around the growing crystal in such a manner that the tubes can be retracted as the diameter of the crystal increases. A schematic representation of this configuration is shown in Figure 2.2. Falckenberg (1975) also found it advantageous to employ a three-tube burner for growth of the larger crystals. (A two-tube burner is shown in Figure 2.2.) In the three-tube design, an

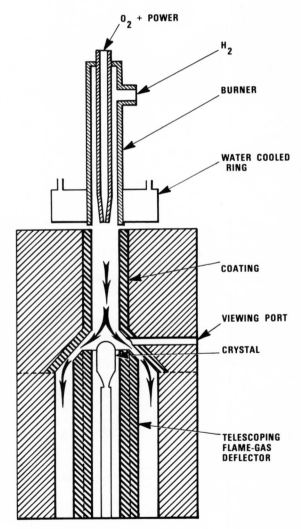

FIGURE 2.2   A schematic presentation of a flame fusion growth apparatus as described by Falckenberg (1976)

FIGURE 2.3
A flame fusion sapphire crystal ~80 mm in diameter.
Courtesy Le Rubis Synthetique Des Alpes (Groupe
Pechiney-Ugine-Kuhlmann)

outer concentric tube carries an additional oxygen flow. The gas flow in the furnace was designed to maintain the radial thermal gradient at the level of the melt–solid interface below a critical value. Using this scheme $\langle 100 \rangle$ spinel crystals can be enlarged to ~50 mm before runoff of the melt. The specific mechanism by which the crystal is enlarged in flame fusion growth has been described in some detail by Falckenberg (1975). In order to grow crack-free $\langle 1\bar{1}02 \rangle$ sapphire as large as ~45 mm it was necessary to provide additional heating to the side of the muffle.

Falckenberg (1976) has reported that the large flame fusion spinel crystals contain subgrains with deviations (from the $\langle 100 \rangle$ growth axis) up to 2°. The large flame fusion $\langle 1\bar{1}02 \rangle$ sapphire crystals contain strain but no discrete grain boundaries and axis wander not exceeding 1°.

Ricard and Cisccolaini (1972) as well as Falckenberg (1976) found that it was possible to grow larger $\langle \bar{1}010 \rangle$ than $\langle 1\bar{1}02 \rangle$ sapphire crystals by the flame fusion method. It is clearly necessary, however, to grow the crystals in the direction normal to the desired $(1\bar{1}02)$ substrate plane in order to economically extract substrate wafers equal to the diameter of the boule. Ricard (1975) recently reported the growth of $\langle \bar{1}010 \rangle$ sapphire ~80 mm in diameter (see Figure 2.3). It has

been possible to grow some crystals this size grain-free, but they typically contain ~2.4 grains/cm². Efforts are still in progress to realize the growth of 80-mm $\langle 1\bar{1}02 \rangle$ sapphire.

Although the technology for producing 2-inch-diameter $\langle 1\bar{1}02 \rangle$ flame fusion sapphires with adequate crystalline quality for substrate usage is at hand, it appears that a reasonably large development in the technology will be required to satisfy the orientation and crystalline perfection requirements in 3-inch-diameter crystals.

Czochralski growth

The fundamentals involved in Czochralski growth may be found in a number of publications [see, for instance, Paladino and Roiter (1964); Rubin and Van Uitert (1966); Charvat et al. (1966); Cockayne et al. (1967)]. The details involved in making this method a viable commercial process are largely proprietary, however, and not to be found in the open literature. The basic arrangement commonly employed in Czochralski growth is shown schematically in Figure 2.4. The single crystal is pulled directly from the melt contained in a refractory metal crucible, usually iridium. Heat energy is lost by radiation and conduction through the growing crystal, causing solidification at the melt–solid interface as the crystal is pulled away from the melt. The energy for melting is most commonly introduced by inducing currents in the crucible in an RF (450-kHz) field. The quality of the Czochralski-grown crystals is strongly dependent on the shape of the melt–solid interface, which in turn is a function of the thermal gradient at the interface and the nature of fluid flow across the solid–liquid interface. The thermal gradient is precisely adjusted with heat shields; the fluid flow across the surface is controlled to some extent by rotation of the crystal and by the geometry of the crucible. For the growth of low-defect-density crystals, it is essential that the conditions at the growth interface be held as constant as possible for the duration of the growth. A fundamental problem in Czochralski growth is that the thermal conditions at the solid–liquid interface change as the melt recedes and the grown crystal becomes exposed to irradiation from the hot crucible walls. This necessitates a change in the power supplied to the system as the crystal grows. Short-term power fluctuations are generally suppressed by a closed-loop control in which the control signal is proportional to the RF power supplied to the coil, the temperature of the crucible, or a combination of the two. In simple systems, the power level is adjusted as the melt recedes by an operator who watches the shape of the crystal, or it is programmed on the basis of experience. In the more sophisticated systems employed in the commercial growth of sapphire, automatic diameter control is

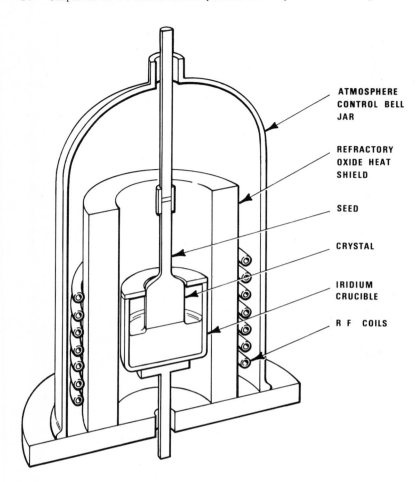

ATMOSPHERE
CONTROL BELL
JAR

REFRACTORY
OXIDE HEAT
SHIELD

SEED

CRYSTAL

IRIDIUM
CRUCIBLE

R F   COILS

FIGURE 2.4
A schematic presentation of a typical
Czochralski growth apparatus

commonly employed. The input to the control system is a signal proportional to the increase in weight of the growing crystal (Bardsley et al., 1972), the decrease in weight of the crucible (Zinnes et al., 1973; Kyle and Zydzik, 1973), or a change in position of the outer edge of the meniscus of the growing crystal as detected by a sighting device (Gross and Kersten, 1972). Since only the crucible and the melt are held above the melting point of the material, the thermal inertia is relatively low, and fast-acting temperature controllers must be employed. Diameter control provides uniformity of boule shape and thus minimizes loss of material during the grinding of the boule into a straight-wall cylinder. Even more important, the use of diameter control minimizes crystalline defects that may be introduced into the crystal under the conditions where the boule diameter is changing. The use of diameter control also minimizes the labor involved in growing the crystal, so that an experienced operator may simultaneously maintain a number of Czochralski growth stations. An unfabricated 2.5-inch-diameter Czochralski crystal is shown in Figure 2.5.

The pull rate of Czochralski sapphire and spinel is approximately linearly proportional to the inverse of boule diameter for crystals up to 2 inch in diameter. For the growth of low-defect crystals, rates of 1.0 inch/hr are typically used for crystals 0.5 inch in diameter, and 0.25 inch/hr for crystals 2 inches in diameter. These rates may differ significantly between various organizations, depending on the specific growth conditions employed, the purity of the starting material, and the crystallographic orientation. Voids resulting from constitutional supercooling appear in crystals pulled at excessive rates. For substrate use, voids must be completely avoided since they will intersect the surface of substrates cut from the crystal.

The latitude in control of the thermal gradients at the growth interface in Czochralski growth constitutes one of the main advantages of this growth method. The crystal can be observed during growth, estimates can be made of the progress and quality, and immediate corrections can be made. With such control, it is possible to produce crystals relatively free of defects. The density defect observed in the rhombohedral plane of "substrate quality" Czochralski crystals is typically about $5 \times 10^4/cm^2$ (Mustachi, 1974).

The main disadvantage in the Czochralski method for the preparation of sapphire for silicon substrates is that the crystal is not in the shape useful for substrates. Prior to slicing the boules are ground into a cylindrical form. The cross section of crystals grown in the direction normal to the $(1\bar{1}02)$ plane is elliptical, and thus there is some material loss involved in forming the cylinder. For cylinders more than 2 inches in diameter, typically 50 percent of the crystal is lost during slicing (if OD saws are used, see Section 2.4), and an additional 50 percent of the remaining material is lost during lapping and polishing. An additional disadvantage is that iridium, which is commonly used as the crucible material, is expensive and difficult to form. Structural deterioration of the crucible results from the formation of hot spots and local melting as the result of flow of RF-induced currents through physical heterogeneities. Note that there is a relatively small difference between the melting point of the crucible and the sapphire (Ir: 2440°C; Al_2O_3: 2030°C). Resistance and gas fire heating of the iridium crucible have been employed in an effort to minimize nonuniform heating and increase crucible life. There are problems involved, however, in maintaining reproducibility in the gradients of such heaters continually operated at temperatures in excess of 2000°C, and these modes of heating have not been widely accepted. Iridium is lost from the crucible during the crystal growth by chemical transport. To minimize this loss, small amounts of oxygen (~0.5 percent by volume) are introduced into the inert gas growth ambient. The presence of the oxygen suppresses the formation of volatile intermediate-oxidation-state oxides of iridium.

FIGURE 2.5
An unfabricated boule of Czochralski-grown sapphire, 2.5 inches in diameter. Note the very uniform geometry obtainable with automatic diameter control. Courtesy Crystal Products Department, Union Carbide Corporation, San Diego, Ca.

The iridium platelets, which are apparently formed as the result of disproportionation of the volatile oxides, are found both on the surface and in the bulk of the boule if the crystal is grown in a reducing or neutral ambient.

Appreciable effort is currently being directed toward reducing the cost of the Czochralski sapphire substrates by increasing the pull rates and minimizing the fabrication loss. Both Czochralski growth and the fabrication of ceramic bodies are, however, relatively mature technologies. Innovative techniques will be required to realize significant cost reductions. The more immediate cost reductions will probably result from more efficient use of capital facilities with volume production.

Heat-exchanger growth

The heat-exchanger growth method (Schmid and Viechnicki, 1970; Viechnicki et al., 1972) is the single crystal variation of the directional solidification-casting technique. A single crystal seed is located at the bottom of a molybdenum crucible charged with the source material (see Figure 2.6). The charge is melted by resistance heating, using a carbon heating element enclosed in a vacuum chamber. The seed is partially melted back to assure contact between the melt and the seed in order to nucleate the growth of a single crystal; complete melting of the seed is avoided by careful temperature control and by cooling the bottom of the crucible with a jet of helium gas directed through the high-temperature heat exchanger. The thermal gradient is maintained and controlled by adjusting the flow of helium. As the crucible is slowly cooled, the melt solidifies as a single crystal from the seed. The thermal inertia of the system is high, and therefore the temperature control and the lowering of the temperature are relatively simple processes which can be carried out over a long period of time with little supervision. In order to partially melt back the unobserved seed crystal, it is necessary to establish a highly reproducible vertical thermal gradient across the crucible. Since the carbon heating element deteriorates during use, care must be exercised to maintain the reproducibility in thermal conditions.

The heat energy is introduced through the sides of the crucible and extracted through the bottom (sapphire near its melting point is an excellent thermal conductor). The melt–solid interface is roughly hemispherical as it moves from the bottom to the top of the crucible, and therefore some impurities (depending on the segregation coefficient) are zoned to the outer edge of the boule. For substrate use cylinders are extracted by core drilling from the less contaminated portion of the crystal. The outer 0.25 inch of the periphery of the crystal is not used.

A 10-inch-diameter single crystal grown by the heat-

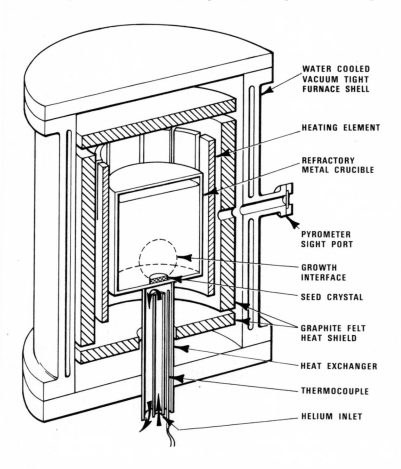

WATER COOLED
VACUUM TIGHT
FURNACE SHELL

HEATING ELEMENT

REFRACTORY
METAL CRUCIBLE

PYROMETER
SIGHT PORT

GROWTH
INTERFACE

SEED CRYSTAL

GRAPHITE FELT
HEAT SHIELD

HEAT EXCHANGER

THERMOCOUPLE

HELIUM INLET

FIGURE 2.6
A schematic presentation of the heat-exchanger growth apparatus

exchanger method is shown in Figure 2.7. The point of attachment of the seed can be seen through the clear cap of the crystal. Nineteen 2-inch-diameter cylinders can be core drilled from the 10-inch-diameter boule. With the exception of the outer skin, the remaining alumina can be recycled.

The early crystals prepared by the heat-exchanger method typically contained small amounts of the crucible material and grain boundaries. The incorporation of crucible materials is associated with the presence of uncontrolled impurities in the growth system, and the grain boundaries are associated with the seeding technique. The development of the method has been rapid, however, and crystals have been grown that are free of both grain boundaries and refractory metal contamination (within the detection limits of emission spectroscopy). The acceptance of the heat-exchanger sapphire as silicon substrate material will depend, to a large extent, on the development of methods to routinely exclude grain boundaries in the material. The crystals grown by the heat-exchanger method may be relatively free of dislocations because the crystal is grown slowly (with a typical diameter increase of 1.2 cm/hr) in a comparatively low thermal gradient and in a system with high thermal inertia. The dislocation

density observed in the rhombohedral plane of a grain boundary–free gradient freeze crystal is typically $\sim 2.2 \times 10^4/cm^2$ (Mustachi, 1974).

In order to avoid confusion in the terminology employed here, it should be pointed out that until recently this method was referred to in the literature as the "gradient furnace" growth method. Schmid (1974) points out that the technique is more appropriately called the heat-exchanger method. His rationale for initiating a change in the terminology is that in the gradient furnace method it is commonly assumed that the temperature gradient in both the solid and liquid is determined by the furnace configuration. In the heat-exchanger method, to a large extent the thermal gradient in the solid is determined by the gradient in the furnace, whereas a certain amount of independent control of the gradient in the liquid is possible by controlling the flow of helium through the heat exchanger.

Edge-defined film-fed growth

In the edge-defined film-fed (EFG) method of crystal growth (LaBelle, 1971A), the cross-sectional shape of the growing crystal is determined by the shape of a refractory metal die (LaBelle and Mlavsky, 1971; LaBelle, 1971B; Chalmers et al., 1972). The general configuration of the EFG apparatus is similar to that of a Czochralski growth station, with the exception of the shaping die (Figure 2.8). For the growth of sapphire substrates, the shaping die is constructed of two parallel refractory metal plates. The spacing between the plates provides a passage for the source material. The outer edges of the plates define the cross-sectional shape of the crystal. The bottom of the die extends into the molten source material. The source liquid rises to the top of the die by capillary action. For this to take place, the molten source must wet the surface of the die. To initiate crystal growth, a single crystal seed of the desired orientation, commonly cylindrical, is inserted into the molten source held in the slot between the die plates. Material solidifies at the end of the seed as heat is piped up the seed rod. As the crystal solidifies on the seed, it is pulled away from the die, and the liquid held above the die (by solid–liquid cohesive forces) spreads in width to the side of the die and in thickness to the outer edge of the horizontal portion of the top of the die. In order for the liquid to flow over the edge of the horizontal top surface of the die the contact angle would be forced away from the equilibrium value; this is resisted by natural forces and thus the spreading of the melt stops at the edge of the horizontal portion of the die. A thin liquid film is maintained between the growing crystal and the die as the cross section of the crystal assumes the configuration of the upper edges of the die. Under carefully controlled thermal gradients and pull rates, a steady state condition is established in which the solidifica-

tion of the melt proceeds downward from the crystal at a rate matching the pull rate, and the source material is drawn through the die capillary to replenish the melt between the die and the crystal. The solid–liquid interface can be observed at the edge of the die. LaBelle and Mlavsky (1971) have proposed that small fluctuations in the pull rate and temperature are accommodated by changes in the thickness of the liquid layer between the die and the crystal. The pull mechanism and temperature control can therefore be relatively simple. Since the position of the melt–solid interface is fixed, there is no need to program the temperature as the liquid level in the crucible recedes. Solid source material may be added to the melt in the crucible without perturbing the growth interface.

The crystallinity of the as-grown ribbon surface is adequate for the heteroepitaxial growth of silicon, but the topology is not appropriate for the precision photomasking required for the fabrication of device structures in the thin silicon film. At the current state of the art, the ribbon surface must be lapped and polished. The interior of the EFG ribbons contain an array of voids which appear in sheets parallel to the surface of the crystal. The position of the void planes is related to the die geometry, and therefore the location of the voids can be controlled to some extent. The current procedure is to position the voids near the surface and remove them during the surface finishing operation. Voids remaining in the interior that do not intersect the finished surface do not influence the properties of the heteroepitaxial silicon. The defect density in the immediate vicinity of the voids is higher than in the matrix of the ribbon. The defect density in a fabricated

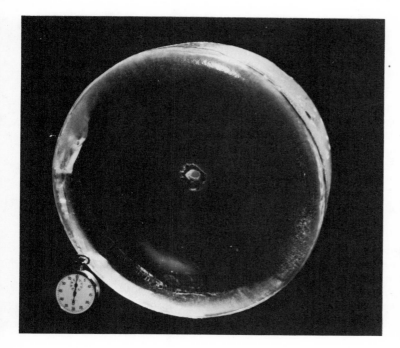

FIGURE 2.7
An unfabricated single crystal grown by the heat-exchanger technique (also referred to as the Schmid–Viechnicki technique). The boule is 10 inches in diameter and 4 inches thick. The point of attachment of the seed can be seen through the clear cap of the crystal. Courtesy Crystal Systems, Inc., Salem, MA.

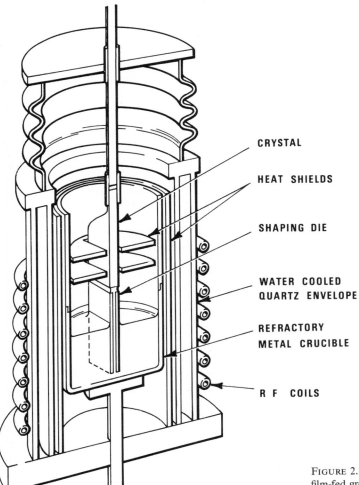

CRYSTAL

HEAT SHIELDS

SHAPING DIE

WATER COOLED
QUARTZ ENVELOPE

REFRACTORY
METAL CRUCIBLE

R F COILS

FIGURE 2.8   A schematic presentation of an edge-defined
film-fed growth (EFG) apparatus

($1\bar{1}02$) surface free of intersected voids is typically about $1.4 \times 10^5/cm^2$ (Mustachi, 1974).

The growth direction of the EFG crystals is in the plane of the substrate surface, whereas that of Czochralski growth is normal to the substrate surface. Therefore, to prepare the same number of substrate wafers per unit of the growth apparatus on-time, the rate of pull of the EFG ribbon must exceed that of Czochralski by a factor equal to the product of the wafer diameter/wafer thickness ratio and the linear fractional yield of the Czochralski crystal after fabrication (see Section 2.4 for a discussion of the fabrication yield). At present, approximately the same number of 2-inch-diameter substrates can be produced by the two methods of growth during the same time period. It is important to note that the rate of pull of the EFG ribbon is inversely proportional to the ribbon thickness but is relatively independent of the ribbon width.

Procedures have recently been developed to grow more than one EFG ribbon simultaneously from the same crucible (LaBelle et al., 1973/74). A group of three ribbons attached to a single seed is shown in Figure 2.9. The properties of the three ribbons are essentially identical. It is clear that this development has a major impact on the economics of the production of substrate by the EFG process and offers the possibility of significant reductions in the cost of substrate materials prepared by the EFG method.

A disadvantage of the EFG process is that it is a relatively new and complex technology, and a high degree of skill and experience are required to establish a growth system.

Substrate crystal growth: summary

The salient features of the crystal growth techniques discussed in this section are summarized in Table 2.3.

2.4. Fabrication of the heteroepitaxial substrates

General considerations

The driving force for orienting the deposited film in heteroepitaxial growth is derived (in an as yet not well understood manner) from the periodicity of the substrate surface. Clearly, the substrate single crystal lattice must be exposed to the deposition constituents in order to nucleate and grow a single crystal overlayer. Defect structures in the substrate surface are propagated into the deposited film and lead to degradation of the semiconducting properties and anomalous diffusion of dopants. Therefore, the characteristics of devices fabricated in heteroepitaxial films are critically dependent on the nature of the substrate surface finish. The objective in surface finishing is to present to the growth ambient an abruptly terminated single crystal lattice free of defect structure. A good deal of research effort has been expended to develop practical methods to achieve this objective. The most effective of the methods currently employed are time consuming and wasteful of material. Reproducibility is difficult to hold in a full scale commercial operation. Indeed a primary motivation in the development of the

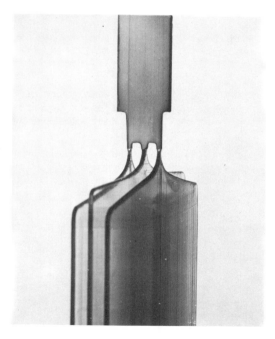

FIGURE 2.9   The upper portion of three 3-inch-wide single crystal ribbons grown simultaneously from the same seed by the edge-defined film-fed (EFG) growth method. The ribbons are ~3 feet long. The growth was carried out at RCA Laboratories under license with the Saphikon Division of Tyco.

edge-defined film-fed method for the growth of sapphire (LaBelle and Mlavsky, 1971) has been to reduce the effort involved in fabrication and finishing.

The influence of surface finishing on the film properties is discussed later in this chapter. A very graphic example of the influence of surface finishing on the crystalline nature of a heteroepitaxial deposit is shown in Figure 2.10. A 2-μm-thick silicon film was deposited on a sapphire sample in which the quality of the substrate surface finish decreased from the center to the edges (Cullen and Scott, 1972). A diffused horizontal junction was then formed in the silicon, the sample was lapped at a 3° angle to give mechanical magnification to the cross section of the silicon, and the cross section was stained to make the junction visible. On examination of the as-deposited silicon surface, it can be seen that the surface is relatively smooth over the well polished substrate but rough on the poorly finished substrate. In the intermediate areas, lines of substrate work damage have been propagated to the surface of the silicon, and are seen in the unetched silicon as a line of indentations in the surface (D in Figure 2.10). In the angle-lapped area a dopant diffusion spike (C in Figure 2.10) can be seen in the junction directly in line with the row of surface indentations. It is clear that the substrate surface work damage has locally influenced the crystallinity of the heteroepitaxial silicon, resulting in anomalously high diffusion of the dopant in the silicon. In the silicon deposited on the substrate relatively free of work damage, the junction is a straight line (A in Figure 2.10). As the damage increases in the substrate surface, the junction in the silicon becomes ragged (B in Figure 2.10), and finally in the silicon deposited over the most heavily work-damaged substrate the dopant has diffused completely to the silicon/substrate interface. It should be mentioned that the differences in substrate surface quality shown here are extreme and seldom observed in substrates fabricated using current technology. Such extremes have been observed, however, during the development of the technology.

In addition to the quality of the substrate surface, a major consideration in substrate fabrication is the loss of material involved in slicing and polishing operations. It is not unusual to lose about 50 percent of a 2-inch-diameter crystal during the slicing operation (kerf loss) and then to lose an additional 50 percent of the remaining wafer thickness during the removal of work damage. As larger-diameter crystals are cut, the yield decreases, both because the saw blades must be thicker and because the finished wafer thickness must be increased to avoid breakage during fabrication of the substrate and subsequent device processing.

The technology of shaping and finishing the brittle refractory oxides is currently in the transition from an art to a science [see, for instance, Schneider and Rice

TABLE 2.3   A Comparison of the Methods Employed to Grow Substrate Crystals

	Flame Fusion	Czochralski	Heat Exchanger	Edge Defined
Growth method	Melted source powder dropped onto boule cap	Withdrawal of crystal from the melt	Slow cooling of a melt in a thermal gradient	Melt supplied by capillary feed to a shaped die; crystal pulled from melt on the die
Container	None	Iridium	Molybdenum (or other refractory metals)	Molybdenum, tungsten, rhenium
Heat input	H_2-O_2 flame	RF-induced current	Resistance heating of graphite elements, gradient established partially with gas cooling	RF-induced current
Atmosphere	Air and H_2O (product of combustion) + excess H_2 or excess O_2	Neutral or slightly oxidizing	Vacuum	Inert gas
Control	Pressure control of source gases; a relatively stable system, minimum control required, heat input held constant	Fast-acting closed-loop system in conjunction with automatic diameter control; power input must be lowered as material is removed from the crucible	Conventional resistance furnace control with programmed cooling	Regulated power input, conditions held constant after growth stabilized
Thermal inertia	Relatively high	Low	High	Very low
Observation during growth	Poor visibility	Good visibility	Crystal cannot be observed during growth	Excellent visibility; the interface shape can be observed
Size limitation	Crystals larger than 2 inches difficult to grow with the desired orientation	Larger than 4-inch boules have been grown	Cylinders core drilled from 10-inch boules	No fundamental limitation on width
Crystalline quality (rhombohedral plane)	$>10^6$/cm²	$\sim 5.2 \times 10^4$/cm²	$\sim 2.2 \times 10^4$/cm² in crystals free of grain boundaries	$\sim 1.4 \times 10^5$/cm² on a fabricated substrate surface, internally higher in area of voids
Disadvantages	Little control in adjustment of the thermal conditions of the growth interface, and therefore the crystallographic quality is relatively poor; difficult to grow large crystals	Employs expensive and difficult to fabricate iridium; boule not in useful shape	Care must be exercised to exclude grain boundaries, crystal cannot be observed during growth	A relatively new and unestablished technology, higher density of defects than Czochralski material
Salient advantages	The expense and contamination associated with crucibles avoided	Good flexibility in adjusting thermal gradients at growth interface; a well established technology	Very large boules can be grown with little supervision, crucible material inexpensive, fast effective growth rates possible, crystals do not require annealing prior to fabrication	Minimum fabrication necessary, output high with multiple ribbon growth

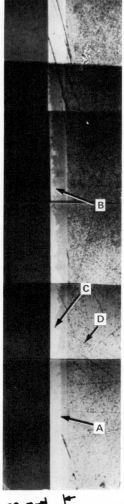

FIGURE 2.10 An optical micrograph of an angle-lapped and stained deposit of silicon on sapphire in which a diffused horizontal junction had been formed. The quality of the surface of the substrate increases, top to bottom. The nature of the junction is closely related to the quality of the substrate finish. Letters are defined in the text. From Cullen and Scott (1972).

(1972)], and a good deal is to be gained in the near future from a detailed examination of the fundamental processes involved and a break from tradition. Since the properties and economics of heteroepitaxial semiconductors are so closely related to the preparation of the substrate, this topic is discussed in some detail.

Substrate shaping and finishing procedures

Prior to fabrication, the crystals grown by the flame fusion, Czochralski, and EFG methods must be annealed to avoid shattering caused by stresses introduced by solidification in steep thermal gradients. In effect the heat-exchanger crystals are annealed during slow cooling and require no further thermal treatment prior to core drilling. Cylinders are formed from the as-grown crystals by grinding (Czochralski boules) and core drilling (heat exchanger boules).

Substrate wafers are fabricated from the cylinders by a sequence of sawing, grinding and/or lapping, polishing, and chemical/mechanical polishing or chemical etching. The steps are schematically shown in Figure 2.11. The procedures employed, specifically as related to the finishing of sapphire and spinel surfaces, are outlined in the following paragraphs.

Sawing. Substrate "blanks" are sawed from the bulk crystals with outside-diameter (OD) or inside-diameter (ID) blades. The OD blades are made up of diamond, silicon carbide, or alumina particles bonded with resins or ceramics. Boules between 1.5 and 2.0 inches in diameter are typically cut with 0.032-inch-thick OD blades. The width of the cut (the kerf) is typically ~0.035 inch, due to the saw chatter and the abrasive action of particles (removed from the blade surface and the boule) entrapped between the blade and the crystal. A 2.0-inch substrate blank is typically 0.025 inch thick; thus, the kerf loss represents more than 50 percent of the total volume of the crystal. The amount of blade chatter is a function of the unsupported length of the blade. If the saw is translated across a stationary crystal (or vice versa) in a direction normal to the growth axis, the unsupported saw length must be greater than the diameter of the crystal. If the crystallographic orientations are desired such that the cut must be made at an angle to the growth axis, the unsupported portion of the blade, and thus the kerf loss, is greater than if the cut is made normal to the boule axis. The kerf loss may be minimized by rotating the crystal on the growth axis and cutting halfway through the crystal with a blade with an unsupported length slightly greater than the radius of the boule. In order to exercise this option, the crystal must be grown in the direction normal to the desired crystallographic substrate surface plane. Since the ease of growth is a function of the growth direction, the economics of fabrication and growth are interrelated.

Inside-diameter blades are thin metal sheets electrolytically plated with metal-coated abrasive particles. The blades are stretched from a supporting rim. The crystals are cut at the edge of a hole formed in the center of the blade. Boules 1.5 to 2.0 inches in diameter may be cut with 0.008-inch-thick ID blades, with a resulting kerf loss as little as 0.010 inch. The consider-

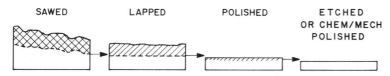

FIGURE 2.11
A schematic presentation of the fabrication of substrate wafers.

ations of blade chatter, in relation to the distance between the edge of the hole and the blade support, are similar in ID and OD cutting. The relatively soft magnesium aluminate spinels (1500 on the knoop scale) can be routinely cut with ID saws, but ID cutting of the harder sapphire (2000 on the knoop scale) is more difficult because of blade wander and because the edges of the diamond become dulled. The diamond bonded to the edge of ID saws extends about 1/8 inch from the outer circumference toward the axis of the saw, and therefore dull diamonds are stripped and new diamonds exposed during cutting or by dressing the blade. The plated diamond on the ID saws is only several diamond layers deep, and therefore the cutting surface cannot be continually renewed by exposure of new diamonds. The diamonds cannot be applied to ID blades by the same techniques used for OD blades because the resin- or ceramic-bonded abrasive will break away from the blade as it is stretched under tension. The use of ID saws for the fabrication of sapphire constitutes a trade-off between the cost of the blade replacement and the value of the single crystal material saved by using the thinner ID blade. At present it is uneconomical to cut sapphire crystals larger than 2 inches in diameter with ID saws. Development which would lead to the use of the ID saws to fabricate 3-inch-diameter boules could result in considerable reduction in the cost of the large diameter substrates.

Blade chatter in both OD and ID sawing leads to grooving of the substrate surfaces. The contours of the subsurface crystallographic work damage in the interior of the crystal follow the contours of the surface grooves (see Figure 2.11), and unless the damage is carefully removed in subsequent finishing steps, defects introduced during sawing will appear on the final etched surface. Residues of saw damage are readily identified by the arc shape with the radius of the saw.

Blanks cut by OD saws are typically wedge shaped, and blanks cut by ID saws are "dish" shaped due to blade wander and vibration of the unsupported portion of the crystal against the saw face. The plane of the sawed crystal surface supported by the bulk crystal is usually more normal to the axis of the crystal than the surface of the unsupported portion of the crystal as the result of the vibration of the unsupported piece. The amount of wedging or dishing is clearly a function of the diameter of the crystal. This must be taken into consideration when a precise crystallographic orientation is required. In the deposition of the semiconduct-

ing materials, small deviations from a specific crystallographic orientation may be tolerated, or in some cases, even desirable (Chapter 2). But for application of piezoelectric materials (Chapter 3) in surface acoustic wave device geometrics, the surface orientation must be held to less than ±0.25°.

In an attempt to minimize the effects of chatter of the circular blades, the refractory oxide crystals have been sliced with metal wires* and linear metal blades† held in tension at each end. Fine abrasive powders and low removal rates are employed in order to leave a surface more characteristic of fine lapping than sawing. Since the blades in this geometry can be readily arranged in a parallel array, compensation may be made for the low cutting rate by making many cuts simultaneously. Good results have been obtained with the relatively soft materials, but blade wander and wear have proven to be difficult to control when sapphire is cut by these methods.

When the relatively soft materials are cut, the abrasive may be supplied by flow of an abrasive-containing slurry across the surface being cut which is discarded after one pass. When the harder materials, such as sapphire, are cut by these methods, it becomes uneconomical and wasteful to supply the more expensive abrasive (such as diamond) in a flow system. Therefore, the promise in using these methods appears to lie in the development of abrasively loaded blades. Since the metal support is under tension, similar problems are faced as in the fabrication of the circular ID blades.

Grinding/lapping. After initial shaping by sawing, the substrate blanks are ground and/or lapped in order to produce plane parallel surfaces and to remove material to the bottom of the deepest saw groove. Clearly the amount of material that must be removed is a function of the quality of the sawing operation. In grinding, the abrasive is resin, metal, or ceramic bonded in a wheel geometry. In lapping, the abrasive is loose, and the operation is carried out against hard glass, hardened steel, or cast iron. Grinding results in the rapid removal of material, but high local forces are imposed on the crystal surface. Work damage as deep as 20 μm has been observed (using transmission electron microscopy) in sapphire surface ground with 325-mesh (44-μm) resin-bonded diamond wheel (Hockey, 1972A,B). It is not meaningful to derive a relationship between the abrasive particle size of a grinding wheel

---

*Geoscience Institute Corporation, Mt. Vernon, NY.
†Varian Vacuum Division, Newton, MA.

and the depth of damage since the abrasive is partially buried in the resin matrix. It has been proposed that grinding introduces more work damage into the crystal than does sawing. Although the manner in which the abrasive is contained is similar, the forces are more normal to the surface in grinding than in sawing, and in sawing the most heavily damaged material is removed from the crystal. Using 325-mesh diamond on an ID saw, a damage depth of 1.8 μm has been observed (Ehman, 1973). Although other variables have not been specified, it is nonetheless interesting that slicing with the same abrasive size can lead to an order of magnitude less damage than lapping. Further evidence that extremely high local forces (and possibly high temperatures) are involved in grinding is the observation of smearing on a sapphire surface (Koepke, 1972). Because of the depth of the work damage characteristic of the grinding operation, this procedure must be employed with caution in the finishing of expitaxial substrate surfaces.

In lapping, the abrasive particles are loose, and although relatively large particle sizes ($\sim$5–25 μm) are employed, the forces are more evenly distributed over the substrate surfaces. The emphasis in lapping is on material removal to shape the substrate rather than on the removal of work damage introduced during slicing.

A variation of the art of lapping is to use a resilient abrasive backing material such as aluminum, tin (Lindstrom, 1967), or acrylic plastic. Part of the abrasive becomes embedded in the backing material. Using this procedure the depth of the residual work damage is decreased with relatively little sacrifice in removal rate, for reasons that are not completely understood at this time.

Mechanical polishing. Grinding and lapping should be classified with sawing as shaping operations whereas polishing and chemical etching are carried out to remove the work damage remaining after (i.e., introduced during) the shaping of the substrate blank. The polishing of sapphire and spinel is carried out with loose abrasive particles of alumina, silicon carbide, or diamond, typically less than 3 μm in diameter. Relatively soft backing materials, such as Corfam* or Pad-K,† are used. In order to conserve time, material is removed with successively smaller abrasive particles. During each successive step, the work damage introduced by the previously used larger-diameter abrasive must be completely removed (see the later discussion of depth of damage in this section). Contamination of the abrasive with foreign particles larger than the abrasive must be avoided. It is particularly difficult to exclude chips removed from the edge of the crystal.

---

*George Newman and Co., Boston, MA.
†Pellon Corporation, New York, NY.

It is noted that in the period during which work damage remaining from the previous step is being removed, the removal rate is relatively high and smooth surfaces are realized. As the polishing approaches the undamaged bulk material, the removal rate decreases and the surface texture is likely to change as particles are removed along cleavage planes rather than from work damage fissures. In the final steps of polishing, as the work-damaged region becomes thinner, the ease of polishing and the topology of the surface become a function of the crystallographic orientation. At this point the uninitiated operator of a polishing apparatus may come to the conclusion that the finish is degrading rather than improving as polishing continues.

It has been commonly observed that the removal rate during polishing decreases rapidly with increasing diameter of the substrate, even when the pressure per unit area of surface is maintained constant. This effect may in part be related to the problem of distributing the abrasive uniformly over the surface of the substrate. A similar effect has also been observed during slicing.

The removal of residual mechanical work damage by thermal and chemical treatment

Subsurface crystallographic work damage cannot be completely removed by mechanical processing. The damage remaining after the final polishing must be eliminated by rearrangement of the disordered lattice at elevated temperatures by chemical action of corrosive liquid or gaseous etchants or by chemical-mechanical polishing. Usually some combination of these methods is employed.

Annealing. A detailed low-energy electron diffraction (LEED) study of the effect of vacuum annealing of mechanically polished sapphire surfaces has been carried out by C. C. Chang (1971). A summary of his findings is presented in Table 2.4. Most of the mechanically polished sapphire surfaces which had been recrystallized by vacuum annealing for 30s at temperatures between 1000 and 1200°C showed equilibrium LEED patterns. After annealing at 1400°C for 30 s all the samples exhibited sharp equilibrium patterns. This definitive study has put to rest the argument commonly used by proponents of chemical etches that crystallographic rearrangement cannot take place in single crystal lattices at temperatures below two thirds of the melting point (in Kelvin). Considering the brief heating durations used by Chang, removal of material (other than impurities) by sublimation cannot be considered to be a significant contributing factor. Chang futher concluded that, from the standpoint of surface crystalline perfection, no advantage is gained by chemically etching the surface prior to annealing. He

TABLE 2.4   The Effect of Vacuum Annealing Mechanically Polished Sapphire Surfaces

Temperature (°C)	Observations	Conclusions
25	Weak LEED patterns from chemically etched surfaces, diffuse or no pattern from unetched; Auger data poor, large Al, O, and C peaks.	Most of the surface highly disturbed, except for a few relatively perfect patches; major contaminant C.
500	LEED patterns deteriorate, pressure bursts to 10^{-8} torr; Auger data poor, partly due to large number of peaks.	Contaminants migrate to surface and distribute over entire surface; crystal degasses.
600	LEED patterns improve, degassing almost complete; Auger data better, large C peak.	Surface starts to anneal, most adsorbates desorb; major contaminant C.
700–900	Series of different LEED patterns appear; some etched samples show weak equilibrium patterns (EP), patterns visible from all samples, large C Auger peak.	Gross surface anneal nearly complete, contaminants produce superstructures, clean areas form equilibrium structures (ES).
1000–1200	LEED patterns sharpen, most surfaces show equilibrium patterns; C Auger peak decreasing rapidly.	Surface contaminants desorbing rapidly, equilibrium structure on most surfaces, small amounts of C left.
1400	Sharp equilibrium patterns form all surfaces; only Al and O Auger peaks detectable, C concentration below $5 \times 10^{13}/cm^2$.	Minimum temperature at which all surfaces form equilibrium structure in 30 s. Surface clean but may not be stoichiometric.
1700	Highest temperature investigated.	No new effects found above 1400°C.

Source: C. C. Chang (1971).

pointed out that etching prior to annealing is to be avoided in the preparation of heteroepitaxial substrates since it brings out a "scratch"-like surface structure that is not altered topologically by annealing. Electron mobilities similar to bulk values were observed in silicon vacuum evaporated on the annealed substrates (see Section 2.8).

Hockey (1972,A B) also observed alteration of the lattice structure in sapphire at relatively low annealing temperatures. He examined the material adjacent to diamond-abraded surfaces by transmission electron microscopy. The samples were annealed in air. Hockey reported that "dislocation activity" and stress relief took place after annealing (up to 4 hr) at 900°C, large dislocation networks developed at 1200°C, and the subsurface damage was completely eliminated at 1500°C. The removal of the damage at 1500°C was assigned to thermal etching. In discussions of whether the removal of work damage is associated with removal of material or crystallographic rearrangement, one must be careful to take into consideration the orientation involved. Moss (1973) observed thermal etching

of the basal plane but not the prismatic plane after annealing sapphire for 2 hr at 1500°C in both air and argon. Patterson (1974) detected no weight loss of rhombohedral plane substrates after annealing in air for 4 hr at 1600°C.

The sapphire and flame fusion appeal substrates have commonly been annealed in hydrogen (see the following). The Czochralski spinel, however, proved to be relatively reactive in the presence of hydrogen (Cullen and Wang, 1971; Green, 1972; Cullen and Dougherty, 1972), but has been successfully annealed in air.

Annealing in conjunction with gas etching. Hydrogen annealing/etching of mechanically polished surfaces has been commonly used by investigators working in the field (Manasevit and Simpson, 1964; Robinson and Mueller, 1966; Dumin, 1967B; Hart et al., 1967; Seiter and Zaminer, 1969; Cullen et al., 1969; and Mercier, 1970A,B). Several of these workers have reported that semiconducting properties of silicon deposited on the hydrogen-annealed surface are equivalent to or exceed the properties of silicon deposited on surface etched in liquid reagents (Hart et al., 1967; Mercier, 1970A). From a practical standpoint, hydrogen annealing is attractive because the reaction products are gaseous, and hydrogen is relatively easy to purify. There is little possibility, therefore, of contaminating the substrate surface with species which are electrically active in silicon.

There has been a good deal of discussion as to whether hydrogen firing leads to improvement of the mechanically polished substrate surface mainly by recrystallization or by removal of the work-damaged layer. This is an important point to consider in the development of effective and reproducible surface-finishing techniques and warrants examination of the relationships among the etch rate as a function of temperature, the duration of the hydrogen firing, and the depth of work damage. The etch rates at the extremes of the commonly employed substrate firing temperatures can be derived from the data of Zeveke and coworkers (1968); 0.01 μm/min at 1300°C and 0.1 μm/min at 1500°C. Their straight-line plot of log of the rate vs. reciprocal temperature, and the observation that the rate is insensitive to the hydrogen flow, suggest that the removal rate is kinetically limited. Allison and coworkers (1969) and Schmidt and Davey (1972) have also reported an etch rate of 0.1 μm/min at 1500°C. The agreement between these investigators may in part be attributed to the rate insensitivity of a kinetically controlled reaction to variations in the gas flow dynamics of the various reaction vessels. The agreement is nonetheless remarkable considering the inevitable differences in the extent of surface damage in substrates prepared at various locations. Gottlieb (1971) observed etch rates between 0.06 and 0.09 μm/min for substrates etched for 15 min at 1500°C, but the rate decreased to less than 0.03 μm/min when integrated over a 1 hr period. As is the case with liquid etches, it is anticipated that the more heavily work-damaged area near the surface etches more rapidly than the bulk of the crystal.

In the early work on silicon on sapphire, the substrates were annealed in hydrogen at 1500°C for 30 min (Mueller and Robinson, 1964). It is convenient, however, to heat treat the substrate and deposit the silicon in the same RF-heated carbon susceptor system. At 1500°C the carbon susceptors deteriorate and outgas; therefore, lower annealing temperatures were sought. As the methods for mechanical finishing were improved, it was found that the hydrogen firing temperature could be reduced. Using 0.30-μm alumina as the abrasive for the final finish, no advantage is seen in firing carefully polished substrates at temperatures in excess of 1300°C for 30 min (Gottlieb, 1971). Using Hockey's (1972A,B) estimate of 2-μm-deep work damage after 0.30-μm alumina finishing, and the commonly reported etch rate of 0.10 μm/min at 1500°C, it

appears that at 1500°C the residual damage is completely removed in a period less than 30 min. On the other hand, assuming a rate of 0.01 μm/min at 1300°C, it is clear that the improvement in surface observed after a 30-min firing must be attributed mainly to recrystallization. It has already been mentioned that the movement of dislocation networks at temperatures below 1300°C has been well established (Chang, 1971).

Schmidt and Davey (1972) reported that after hydrogen firing commercially polished "substrate grade" sapphire at 1500°C for 45 min, the surfaces were ordered crystallographically (as indicated by reflection electron diffraction) but were topologically poor. The surface roughness of the fired substrates is a function of the quality of the final polishing. Clearly, hydrogen firing is not a panacea for all substrate finishing processes and is acceptable only when all steps of the shaping and finishing processes are well controlled.

Further evidence for the effectiveness of hydrogen firing of the substrate surface is provided in the micrograph shown in Figure 2.12 (Mustachi, 1974). A polished wafer was broken into two parts, one half was exposed to hydrogen for 30 min at 1200°C, and then both halves were etched in KOH. It can be seen that the abrasive surface damage (A in Figure 2.12) has been completely eliminated by the hydrogen firing. This particular crystal contained grain boundaries (B in Figure 2.12); a boundary was employed to index the two halves of the sample. As was anticipated, the hydrogen firing had no noticeable influence on the

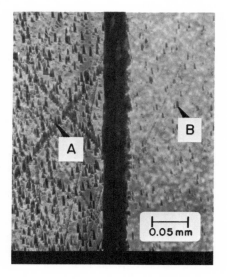

FIGURE 2.12   An optical micrograph of KOH-etched sapphire surfaces. Half of the polished wafer was exposed to hydrogen for 30 min at 1200°C. The grain boundary used to index the halves was unaffected at this temperature. Letters are defined in text. From Mustachi (1974)

gran boundary or the density of the grown-in defects (C in Figure 2.12). Dobrovinskaya and coworkers (1970) report that grown-in defects are not relieved in sapphire below 1800°C.

The optimum hydrogen annealing treatment for alumina-rich flame fusion magnesium aluminate spinel was found to be 1150°C for 30 min (Wang et al., 1969; Cullen et al., 1969). Although the flame fusion spinel is less reactive than the Czochralski spinel (Cullen and Wang, 1971), it is more reactive than sapphire, and higher firing temperatures were avoided. It must be assumed that the removal rate at temperatures of about 1250°C is low and that improvement of the spinel surface also results mainly from annealing. The polished surface of the relatively reactive stoichiometric Czochralski spinel cannot be improved by hydrogen firing (Cullen and Dougherty, 1972). Green (1972) has proposed that the surface of the stoichiometric material degrades as the result of the preferential loss of magnesium oxide.

Gas etching.   The etching of sapphire with fluorinated hydrocarbons, originally proposed by Manasevit (1968B), has been more recently reinvestigated by Schmidt and Davey (1972). The latter investigators adjusted the content of fluorotrichloromethane (Freon 11) in helium to obtain an etch rate of 7.5 μm/min. Smooth, scratch- and work damage–free surfaces are obtained on sapphire surfaces (60 and 90° to the c-axis) etched for 5 min. Other fluorine-containing gases have been proposed as etchants for sapphire (Manasevit and Morritz, 1967) but do not appear to have been applied to any great degree.

At temperatures above the critical condensation temperature of silicon on sapphire, the substrate is etched by silane (Filby, 1966) and silicon vapor (Reynolds and Elliott, 1966; Naber and O'Neal, 1968). Since the etchants are themselves the sources of silicon for CVD and evaporation processes and are available in high purity, they would appear to be convenient reagents to use. Little information is available in the literature, however, on the quality of the substrate surface or the properties of silicon deposited on silane- or silicon-etched surfaces.

Liquid etching.   The liquid reagents—for the most part molten salts or high-temperature acids—which have been used to etch sapphire include V_2O_5, $KHSO_4$, $Na_2B_4O_7$, and PbF_4. The use of these etchants has been reviewed by Filby (1972). More recently, Reisman and coworkers (1971) have used mixtures of H_2SO_4 and H_3PO_4, and Robinson and Wance (1973) have employed $Na_2B_4O_7$ for the chemical finishing of both sapphire and spinel. There have been no reports, however, that the semiconducting properties of silicon on chemically polished sapphire are superior to the properties of silicon deposited on the hydrogen-annealed substrates. The liquid etches are in general

more difficult to employ than the gaseous etches. The etch rates change because of accumulation of reaction products in the liquid or decomposition of the etchant (or both). The liquid flow over the substrate surface is difficult to control reproducibly, particularly in viscous liquids. In some cases difficult to remove reaction products form on the substrate surface (Robinson and Mueller, 1966). A general problem with the removal of subsurface work damage by etching is that the surface may be roughened as the result of preferential etching of the high-energy crystallographic defects. With materials such as silicon, the difference between the etch rate of the more perfect crystal and the defected areas is minimized by using high etch rates. With the less-reactive refractory oxides, high etch rates are difficult to achieve.

The chemical polishing of the stoichiometric Czochralski spinel has received attention in recent years because of the surface reactivity observed during hydrogen annealing (Green, 1972). In this case there has been some motivation for overcoming the difficulties commonly encountered in liquid etching. Aeschlimann and coworkers (1970) have achieved flat, scratch-free surfaces of spinel after removal of ~20 μm by exposure to orthophosphoric acid (H_3PO_4) for 20 min at 390°C. Temperature control is important. Above 400°C phosphates are precipitated on the surface, and below 370°C the surface is pitted. Good semiconducting properties have been achieved in silicon deposited by vacuum evaporation on these etched surfaces (Gassmann et al., 1972). Reisman and coworkers (1971) have been successful in removing residual work damage form (100) and (110) spinel by etching the surface in 3:1 H_2SO_4:H_3PO_4 for 15 min at 250°C. If mechanically polished substrates are etched in the H_2SO_4:H_3PO_4 mixture, deep scratch-like patterns are observed on the surface. If the polished sapphire is air annealed at 1500°C for 1 hr prior to etching, the substrate surface is featureless. Reisman concluded that the work damage is completely removed by the high-temperature treatment and that the chemical etching serves only to remove residual debris and impurities.

Robinson and Wance (1973) have investigated the chemical finishing of the spinel with $Na_2B_4O_7$. Work damage–free and smooth surfaces have been realized in (100) stoichiometric and alumina-rich spinel after exposure to borax at 800–900°C. As in the case with the H_2SO_4:H_3PO_4 etching, the (111) spinel surfaces are pitted in borax. Carrier mobilities in 1-μm-thick silicon deposited on the borax-etched (100) Czochralski spinel surfaces are similar to mobilities observed in silicon deposited on hydrogen-annealed alumina-rich spinel.

Chemical-mechanical surface removal.   It has been demonstrated recently that the subsurface work damage remaining after the mechanical polishing can be removed from sapphire surfaces by polishing with colloidal silica in alkaline aqueous solutions. This chemical mechanical method was originally developed for the finishing of silicon (Walsh and Herzog, 1965) and was successfully applied to the preparation of the surfaces of gadolinium gallium garnet prior to the liquid-phase epitaxial growth of magnetically active garnet films for application in "bubble" memory devices. The colloidal silica preparations are made up of 10 to 40 percent by weight, 10 to 75 nm silica (Walsh and Herzog, 1965) and used at pHs greater than 7. Various formulations are available under the trade names of Syton* and Quso.†

Corboy (1974) has found that the semiconducting properties of silicon deposited on colloidal silica–finished ($1\bar{1}02$) sapphire are similar to the properties of silicon deposited on sapphire annealed in hydrogen at 1300°C for 30 min. The advantage of avoiding a high-temperature annealing treatment must be balanced against the time consumed in finishing with silica sols and the difficulty in achieving reproducibility in the polished surfaces. Removal rates, depending on the specific conditions employed, are ~4 μm/hr (Messineo, 1975). The long-range smoothness of silica-polished sapphire is superior to that of sapphire etched in liquid reagents.

The mechanism of removal of sapphire in the alkaline silica preparations has not been systematically investigated, and it is difficult to explain some of the relationships that have been observed between removal rate and polishing conditions. A qualitative mechanism of the action of alkaline silica on sapphire is that the reaction products of the dissolution of alumina in the basic solution are removed by the abrasive action of the silica. The abrasive action takes place primarily at high points on the surface and not only results in removal of products but also leads to high local stirring and possibly a local increase in temperature. Because the chemical action is accelerated at high points, the surface is maintained flat over the entire surface of the substrate.

A major advantage in the use of chemical-mechanical, as compared to chemical, removal of work damage is that the surface is not roughened by preferential removal of high-energy crystallographic defects. However, it is difficult to achieve removal rates comparable to those commonly used in chemical etching. Investigating the processes involved in an effort to increase the material removal rates should prove advantageous.

The modification of the surface hardness of alumina in the presence of various chemical species may prove to have a significant impact on the shaping and finish-

*Monsanto Chemical Corporation, St. Louis, MO.
†Philadelphia Quartz Co., Phila., PA.

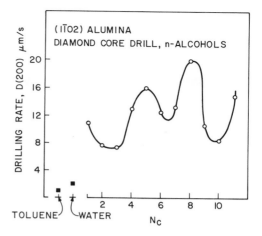

FIGURE 2.13   Rate of drilling of Al_2O_3 monocrystals in n-alcohols, toluene, and water. N_C = number of carbon atoms in the alcohol chain. From Westwood et al. (1973)

ing of the heteroepitaxial silicon substrates. Mackenzie (1971) has pointed out that consideration of the basic physicochemical processes involved in material removal has already found application in various refractory oxide processing procedures. To date, the effects and applications of the modification of the surface hardness has been more clear cut in shaping operations (such as core drilling) than in surface finishing operations (such as polishing). Westwood and coworkers (1973) have observed that, in the presence of octyl alcohol, the rate of core drilling of sapphire (in the rhombohedral plane) is an order of magnitude greater than the rate of drilling in the presence of water (Figure 2.13). In the presence of the n-alcohols, the rate of drilling was related to the alcohol chain length. It was found that the surface hardness could be correlated with changes in the zeta potential with the maximum hardness associated with the point of zero charge in the zeta potential. In turn, the highest core-drilling rates could be associated with the maximum surface hardness. Swain and coworkers (1975) later confirmed these results but found that abrasive wear rate of alumina on a diamond-impregnated wheel was least at the point of zero charge. These authors explain this apparent anomaly by pointing out that in core drilling material is removed by a brittle fracture process (i.e., impacting and chipping), whereas in grinding material is removed by plastic flow and flow-dependent fracture process (i.e., plowing and shearing). In respect to the mechanism of abrasive removal, the scanning electron microscopic observations of Swain and coworkers are in agreement with the earlier work of Koepke (1972). Thus, material removal by brittle fracture in enhanced by local surface hardness; if the surface is soft, the impact energy will be dissipated by plastic flow. The opposite is true of material removed by a shearing action (Latanision, 1973). Swain and coworkers also have observed that the relationship between the surface removal and the zeta potential do not hold under conditions where hydrodynamic lubrication effects are operative (relatively high viscosity of the fluid present, light loading, and high wheel speeds). In the grinding of alumina against a diamond-impregnated wheel in the presence of the *n*-alcohols, the removal rate is dependent only on the viscosity of the alcohol.

The depth of abrasive work damage

The surface of mechanically polished substrates, which appear to be featureless after the final surface finishing at magnifications as high as 10,000× (scanning electron microscopy), may exhibit evidence of heavy work damage after exposure to chemical attack. It is commonly believed that the gouging has been obscured by smearing of the surface or by filling the voids with polishing debris. The surface of brittle materials, however, is smoothed by removal of material rather than smearing (Rabinowicz, 1968) except under the extreme conditons of grinding (Koepke, 1972), and it is unlikely that polishing debris trapped in surface grooves would be undetected by scanning electron microscopy. A more feasible explanation of the grooved appearance of etched surfaces is the preferential removal by rapid etching of the high-energy locally work-damaged crystal lattice. This is schematically presented in Figure 2.14. It is common "art" to lap and polish surfaces until the last traces of grooving introduced by the previous operation are removed, as observed by low-power optical microscopy. If it is assumed that the depth of work damage is at least equal to the diameter of the abrasive particle, it is clear that removal of material just to the bottom of the scratch is not sufficient to clear the surface of damage introduced by the previous (larger) abrasive used. It is quite possible that all subsequent polishing is carried out within the region of the heavy work damage introduced by the first shaping operations. Indeed this is often the situation, and unless the finishing procedure is carefully designed and carried out, etching of highly polished substrates will bring out features with the characteristic geometry of saw damage.

The depth of damage introduced into sapphire and spinel is a function of the hardness of the material, the particular crystallographic plane being polished, the size, shape, and hardness of the abrasive particle, the resiliency of the carrier or backing of the abrasive, the pressure applied, and the lubricant employed. Because of the number of variables involved, it is difficult to develop relationships that lead to the ability to predict the depth of damage under a specified set of experimental conditions. It is useful, however, to examine the range of depth of damage previously

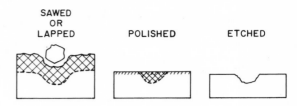

FIGURE 2.14   A schematic presentation of the source of etched-out work damage on the surface of a substrate which was topologically smooth prior to etching

reported. Hockey (1972A,B) has observed dislocation networks 1 μm below (0001) sapphire surfaces polished with 0.25-μm diamond or 0.3-μm alumina. Ehman (1973), however, has measured damage depths of only 0.09 μm on the same orientation of sapphire after polishing with 0.25-μm diamond. The order of magnitude difference in damage depth reported by these two investigators may in part be attributed to the differences in the conditions of polishing and of measuring the depth of damage. Ehman employed vibratory polishing, which is commonly used in conjunction with relatively low loading and results in low removal rates and relatively shallow damage, while Hockey employed conventional polishing techniques. Ehman inferred the depth of damage from etch rate vs. depth data; Hockey measured the depth of damage directly with transmission electron microscopy. It is reasonable to assume that localized crystalline defects are observable by electron microscopy at a density level that would not lead to detectable differences in the macroscopic etch rate. Short of determining the crystallographic nature of the surface specific to a particular finishing method employed, it would be dangerous to assume a depth of damage less than the diameter of the abrasive particle.

Characterization of surface finishing

The evaluation of the effectiveness of surface finishing for the heteroepitaxial growth of thin semiconductor films is a difficult task. To date, no procedure has been developed that is amenable to quality control in a production facility. The methods of characterization currently available are destructive or impractical for routine use (e.g., electron diffraction requires vacuum conditions). At present, it is necessary to establish finishing procedures using destructive methods of evaluation and then to hold the procedures as constant as possible. Unfortunately, because of the number of variables involved, surface finishing operations are inherently difficult to reproduce. Generous safety factors must be built into each step in the process. The characterization of substrates and of the heteroepitaxial film is discussed in detail in Chapter 5. Methods of characterization specifically as related to the evaluation of surface finishing are briefly summarized here.

Surface Microscopy: Substrate. It has previously been pointed out that surface microscopy of unetched substrates, even at magnifications available with scanning electron microscopy, is ineffective in revealing subsurface damage. The subsurface damage must be converted into topological effects by chemical erosion of local high-energy damaged regions. Examination of etched surfaces at magnifications as high as 10,000× (SEM) is an effective and commonly employed procedure [see, for instance, the papers in Schneider and Rice (1972)]. Of the various etches that have been employed to bring out the subsurface work damage in the rhombohedral plane of sapphire, concentrated KOH (at 300–310°C) appears to have been the most successfully employed (Mustachi, 1974; Maurits, 1974). A rigid test of the care with which the damage introduced during sawing and lapping has been removed by the final polishing operation is to examine the etched surface prior to the final thermal anneal. It is unusual to find a featureless surface after this treatment. A KOH etched substrate, before and after the final anneal, is shown in Figure 2.12. Note that grown-in defects, as well as the damage introduced by the shaping and finishing, are brought out by the KOH etch.

Surface Microscopy: Silicon. The physical characteristics of the surface of the deposited silicon may be used as a qualitative measure of the substrate surface quality. Unless the substrate finish is unusually poor, however, the effect of substrate work damage on the physical appearance of the heteroepitaxial silicon cannot be observed in as-deposited films 1 μm and less in thickness. The influence of the substrate surface damage may be brought out by depositing thick films ($\geq$2 μm) and by etching the thick film. The appearance of topological features in thick silicon deposited on damaged substrates can be attributed to the dependence of the growth rate of silicon on the crystallographic orientation. The silicon deposit on the damaged substrate will be misoriented (or polycrystalline) and thus grow at a different rate than the surrounds. Since the differences in the rate are small, the effect must be enhanced by the growth of relatively thick films.

It is difficult to derive information on the substrate surface finish by exposing thin (1 μm and less) silicon deposited on annealed substrates to the conventional dislocation etches. The overall high density of defect structures masks localized defects propagated from work-damaged areas, and etched surfaces are generally rough and unfaceted. As the thickness of the silicon increases, and if the overall substrate finish is good and the damage is localized, the crystallinity of the matrix improves more rapidly than the material in which substrate defects have been propagated, and dislocation etches can be employed. On unannealed substrates, however, the effect of the substrate finish is readily observable even in 1-μm-thick Sirtl-etched silicon deposits (Sirtl and Adler, 1961).

Transmission Microscopy. The examination of the crystalline nature of the surfaces by transmission electron microscopy provides information not only on the depth of damage but on the specific nature of the damage (Hockey, 1972A,B). This method has not been commonly employed due to the difficulties involved in thinning the sample.

X-ray Topography. Berg–Barrett x-ray topography has been employed to examine residual surface damage in sapphire and spinel substrates (Maurits and Hawley, 1971). An advantage to x-ray topography is that the entire surface of substrate wafers can be examined, and thus the homogeneity of the finishing is evident. The disadvantages are that, using the normal methods, the imperfections deep in the substrates are superimposed on surface defects, and magnification results only from photographic enlargement.

Low-Angle Electron Diffraction. Electron diffraction is a particularly useful surface evaluation technique because, since the penetration of the electrons (at the angles commonly employed) is only ~50 Å, the structure near the surface is emphasized. (In x-ray diffraction, sensitivity to surface effects is lost as more of the material is sampled.) Care must be taken to sample a sufficiently large portion of the surface to assure the detection of finishing heterogeneities. This method was commonly employed during the early development of methods to finish the substrate surface. The results are qualitative, however, and it is not clear that the method is sensitive enough to detect the small differences now being experienced in the better substrate surfaces.

Low-Energy Electron Diffraction (LEED) Studies. LEED studies provide detailed information on the nature of the atomic arrangement at the surface and of the presence of foreign materials (Estrup and McRae, 1971; Chang, 1971). Information on the subsurface effects can be derived, however, only to within several atomic spacings of the surface. Meaningful interpretation of LEED patterns requires considerable experience in the field.

IR Reflection. Lattice-related stress, which can result from mechanical work damage, deforms the normal modes for the sapphire vibrational states and gives rise to forbidden modes in the specular reflection spectrum (Barker, 1963). This effect is being used in the development of a method to monitor the substrate surface quality by IR reflection (Duffy et al., 1975; Duffy and Zanzucchi, 1976). The effect is not large, and care must be exercised to establish correlations between the observed shifts in the reflection spectra and the specific types of surface work damage. The method is attractive because it is nondestructive and can be rapid. With further development, it offers the possiblity of commercial in-line control of the quality of the surface finish.

Constancy of Etch Rate. Since the rate of dissolution of single crystal materials is a function of the crystalline perfection, the depth of damage may be determined by measuring the etch rate as a function of thickness. The etch rate is commonly determined by weight loss. This technique has commonly been employed in the examination of semiconductor single crystals (Faust, 1969) and has recently been applied to the refractory oxides (Ehman, 1973). Care must be taken to maintain the chemical conditions constant and to avoid pitting. In order to maintain control over the removal of very thin layers of material, low etch rates are employed, and therefore, the weight losses must be measured very accurately.

Surface Microhardness. It is interesting to note that the work of Ehman (1973) suggests a correlation between microhardness and depth of damage for refractory oxides. It is difficult to explain, however, why the relationships observed are internally consistent only for one type of preparation (sawing, lapping, or polishing); such correlations were not observed when measuring the depth of damage introduced by the different surface treatments. Although microhardness would be a convenient method of measurement of depth of damage, it would appear that a further understanding of the mechanism involved must be developed in order for this to become a method of general use.

Electrical Properties of the Deposited Silicon. The Hall mobility of $1.0–1.5$-μm-thick silicon on sapphire and spinel has been employed for the evaluation of the surface finishing (Gottlieb, 1971). The Hall patterns generally cover a relatively large area of the substrate (>0.5 cm^2), and it must be assumed that the surface finish under evaluation is homogeneous. Druminski and coworkers (1974) have examined the correlation between the MOS transistor mobility, the substrate quality (as measured by electron microscopy and diffraction), and the silicon quality (as measured by ion backscattering, optical absorption, and light microscopy). While qualitative relationships were observed between the results of all of the measurements cited, quantitative correlations with the mobility were observed only with optical absorption and ion backscattering of the heteroepitaxial silicon.

In the final analysis the question that must be asked is what influence local substrate defects, both grown-in and introduced during polishing, have on the electrical characteristics of individual components of integrated device structures. It is extremely difficult to isolate material effects from device processing effects. This has been the subject of a great deal of investigation in devices fabricated by diffusion and/or homoepitaxial growth on bulk silicon [see for instance papers in Huff and Burgess (1973)]. Efforts are now in progress at a number of organizations to establish relationships between local defect structures and device performance in silicon on sapphire.

Future trends in surface finishing

It has been common practice to minimize shaping costs by using rapid material removal procedures (high pressure, large abrasive) with little regard to the introduction of work damage. The assumption has been that work damage introduced during shaping can be removed at a later stage. Recently, more detailed examination of the microstructures of lapped and polished surfaces has brought an appreciation of the depth to which damage can be introduced and of the difficulties involved in reproducibly removing the damage. It seems logical that a future trend will be to avoid the use of such damaging shaping procedures in the first place and to compensate for the time consumed in using less damaging (slower) techniques with automatic equipment and multiple wafer processing techniques. As the demand for substrates increases, it becomes economically feasible to employ such equipment.

It is also anticipated that the chemical-mechanical methods of surface finishing will be further developed. With improvements in the cutting methods, and increases in the rate of chemical-mechanical polishing, the intermediate grinding and lapping could be eventually eliminated. The motivation for further development is clear. The time expended in substrate shaping and finishing, and the material loss, are significant factors in the economics of semiconductors on insulating substrates; and the crystallographic properties of the heteroepitaxial silicon are closely associated with the nature of the substrate surface.

2.5. The gaseous silicon source in heteroepitaxial growth

Silicon tetrachloride, which is commonly employed as the gaseous source of silicon in the homoepitaxial growth of silicon, was also used in the early attempts to realize deposition of single crystal silicon on sapphire and spinel (Manasevit and Simpson, 1963, 1964; Seiter and Zaminer, 1965). Trichlorosilane was also considered as the silicon source (Bicknell et al., 1966; Heywang, 1968). It was soon realized, however, that the halogen-bearing starting materials react with the sapphire and spinel (Bicknell et al., 1966). As a result, the impurity content of the silicon near the silicon substrate was high, the crystallinity was poor, and it proved possible to achieve usable semiconducting properties only in relatively thick films. More recently, Druminski and Wieczorek (1975) investigated di-chlorosilane as the source gas for the deposition of silicon on both sapphire and spinel but observed no advantage over silane. Therefore, investigators in the field turned to the use of the less corrosive (in relation to alumina) silane as the silicon source (Mueller and Robinson, 1964; Manasevit et al., 1965; Bicknell et al., 1966). The erosion of the substrate by the chlorinated silanes can be associated not only with the presence of the halogen, but also with the relatively high temperatures required for the chemical reduction of the chlorinated silanes ($\geq 1050°C$). The pyrolysis of silane at acceptable rates can be achieved at 950°C. At the higher temperatures, silicon, as well as the reaction products from the reduction of the chlorinated silanes, react with the substrate. The first silicon growth islands are more dense, and surface coverage is achieved more rapidly in the deposition of silicon from silane than from silicon tetrachloride (Manasevit et al., 1965) or trichlorosilane (Bicknell et al., 1966). This is because (at least in part) the reverse reaction is negligible in the pyrolysis of silane whereas in the reduction of the chloride the reverse reaction results in the removal of silicon from the surface. In the deposition of silicon from the chlorinated silanes, the silicon islands must form on a surface which is simultaneously being eroded and from which gaseous reaction products must escape.

In order to achieve acceptable semiconducting properties in silicon deposited at temperatures less than $\sim 1100°C$, atmospheric oxygen and moisture must be rigorously excluded from the deposition system. Above 1100°C silicon oxide is reduced by silicon to form gaseous intermediate-oxidation-state oxides, and thus oxygen at low levels can be tolerated. Below 1100°C, however, silicon oxide is stable in the presence of silicon. At oxygen levels too low to monitor with conventional analytical instrumentation (<0.1 ppm), the presence of oxygen results in the degradation of the semiconducting properties of heteroepitaxial silicon. It is for this reason that the relatively low-temperature silane pyrolysis reaction is not commonly employed in production facilities for the homoepitaxial growth of silicon, except for the preparation of device structures in which dopant diffusion must be minimized by the use of the low deposition temperatures. The current expense of silane is also a consideration for use in production facilities. It should be noted that silane spontaneously ignites on contact with air, and therefore, silane/hydrogen mixtures must be handled with care to avoid explosion.

Whereas the nonreversibility of the silane pyrolysis reaction is an advantage in the realization of rapid substrate surface coverage, in the homoepitaxial growth of silicon from silicon tetrachloride the operation of a reverse (etching) reaction results in the

enhancement of the crystalline perfection of the grown layer as a result of the removal of high-energy (defect) silicon and redeposition of a more ordered silicon structure. In order to take advantage of the positive features of both source gases, schemes have been developed in which the surface of the insulating substrate is covered using silane as the source (heteroepitaxy) and the silicon is then grown to the desired thickness (homoepitaxy) by substituting a chlorinated silane or by adding hydrogen chloride to the silane stream (Druminski and Schlötterer, 1972).

Silane is less stable than any of the chlorinated silanes. The standard heat of formation (ΔH_f) of silane is between 7.3 and 7.8 kcal/mole (Brimm and Humphreys, 1957; Gunn, 1961). Thus, silane is thermodynamically unstable at room temperature. The decomposition rate, however, is sufficiently low that silane can be stored at room temperature without time restraint. The standard heat of decomposition, standard free energy of decomposition, and equilibrium constant of silane as a function of temperature have been plotted by Mayer and Shea (1964). Also included in the plots are phosphine and diborane, which are commonly used as dopant sources.

Silane is less stable than any of the chlorinated silanes. The standard heat of formation (ΔH_f) of silane is between 7.3 and 7.8 kcal/mole (Brimm and Humphreys, 1957; Gunn, 1961). Thus, silane is thermodynamically unstable at room temperature. The decomposition rate, however, is sufficiently low that silane can be stored at room temperature without time restraint. The standard heat of decomposition, standard free energy of decomposition, and equilibrium constant of silane as a function of temperature have been plotted by Mayer and Shea (1964). Also included in the plots are phosphate and diborane, which are commonly used as dopant sources.

Although it has commonly been assumed that the silicon deposition reaction is surface catalyzed, this appears to occur more in the decomposition of silicon by the hydrogen reduction of the chlorosilanes than in the pyrolysis of silane, and measures must be taken in developing the deposition conditions to avoid homogeneous gas-phase decomposition and the subsequent deposition of solid silicon particles on the substrate surfaces. The amount of homogeneous gas-phase decomposition is a function of the thermal gradient normal to the substrate surface and is thus strongly dependent on the reactor geometry (see Section 2.10).

In an effort to completely avoid the presence of hydrogen (even as a reaction product) in the deposition ambient, attempts have been made to deposit silicon on insulating substrates using helium as the carrier gas and organosilanes (such as tetramethyl silane) as the silicon source. To date, the use of the organosilanes as a silicon source has led to the deposition of mixtures of silicon and silicon carbide (Avigal and Schieber, 1971, Avigal et al., 1974).

2.6. Chemical reactions between the substrate and the deposition constituents

The incorporation of substrate reaction products

One of the factors limiting the semiconducting properties of thin heteroepitaxial silicon on insulating substrates is the presence in the films of the products of reaction between the substrate and the deposition constituents. The incorporation of the reaction products during the growth of the silicon influences the semiconducting properties in a number of ways. Both substitutionally and interstitially incorporated reaction products may lead to degradation of the carrier mobility by distortion of the single crystal silicon lattice. If the contaminants are charged, degradation of the mobility may also result from scattering of the carriers by the charged centers. The impurities may also be ionized and lead to a different carrier concentration from that anticipated on the basis of the amount of intentionally added dopant material. The latter effect is commonly called "autodoping," by analogy with effects seen in the homoepitaxial growth of silicon wherein electrically active species are transferrred from the substrate into the epitaxially grown deposit as the result in diffusion, volatilization, or chemical transport during the growth of the epitaxial silicon. In the case of the heteroepitaxial growth of silicon, reaction of the deposition constituents with the substrates is the main source of autodoping. The erosion of the substrate surface may also result in roughening of the surface on a microscopic scale. This can lead to the exposure of undesired crystallographic orientations in substrate and the propagation of spurious orientations and microfaceting in the heteroepitaxial silicon.

The effect of the incorporated substrate reaction products on the film semiconducting properties is altered during the thermal processing (in neutral and oxidizing ambients) used in device fabrication. This must be taken into account in the design of processing procedures for the fabrication of device structures in the heteroepitaxial silicon. Because of this effect, a quality factor in the evaluation of heteroepitaxial growth conditions is the percentage change in the carrier concentration and mobility during a standard thermal oxidation. It is interesting to note that the "as-deposited" semiconducting properties of heteroepitaxial silicon are relatively insensitive to the deposition parameters, whereas the "after-oxidation" properties are a strong function of the deposition parameters (Cullen et al., 1969). This effect is discussed in further detail in a later portion of this chapter (see Section 2.12). Because of the undesirable effects of the presence of substrate reaction products, considerable effort has been expended in the development of deposition conditions that suppress the reaction between the deposition constituents and the substrate. At the same time an effort has also been directed toward minimizing the high-temperature processing required during the fabrication of devices in silicon on the insulating substrates.

A number of investigators have cited the reaction

$$2Si + Al_2O_3 \rightleftarrows Al_2O \uparrow + 2SiO \uparrow$$

as the primary source of unintentionally added impurities in both silicon on sapphire (von Grube et al., 1949; Manasevit et al., 1965; Bicknell et al., 1966; Hart et al., 1967) and silicon on spinel (Seiter and Zaminer, 1965; Cullen et al., 1969). Although the chemical reduction of alumina by silicon is cited as a source of autodoping of silicon on spinel, it is assumed that similar reactions involving the reduction of magnesia are involved.

The reaction

$$3Si + Al_2O_3 \rightleftarrows 2Al + 3SiO \uparrow$$

has also been cited as a source of autodoping after the observation of elemental aluminum as a reaction product (Chang, 1971). The presence of free aluminum is somewhat surprising, considering the relative reducing strength of silicon and aluminum ($E°_{Si} = 0.84$ V; $E°_{Al} = 1.67$ V). The escape of the intermediate-oxidation-state oxide of silicon apparently leads to a shift in the reaction toward the production of aluminum. In considering to what extent reactions must be shifted from equilibrium to account for the production of significant amounts of dopants, it should be kept in mind that 1 ppm of an electrically active species is equivalent to a carrier concentration (in silicon) of about $6 \times 10^{16}/cm^3$. Thus, the generation of reaction products in quantities normally considered to be insignificant in preparative inorganic chemistry can lead to serious alteration of the semiconducting properties of the heteroepitaxial silicon.

The identification of elemental aluminum as a product of the reduction of sapphire with silicon suggests the possibility that autodoping is the result of the direct substitution of aluminum in the silicon lattice. In previous discussions of the autodoping mechanism, the oxidation state of the aluminum had not been specified. In order to explain the donor activity of oxidized aluminum, interaction with crystalline defects or the formation of aluminum–oxygen complexes have been invoked.

Since gaseous reaction products must escape from the growth surface in order for the chemical reduction of the substrate by silicon to proceed, it is reasonable to assume that the reaction is terminated after the surface of the substrate is completely covered. On the other hand, the reaction of silicon on the back of the substrates as a source of autodoping must not be overlooked (Hart et al., 1967). Silicon may be in contact with the back surface of the substrate as the result of placing the substrate on a silicon-coated susceptor, as the result of transport of the silicon from the susceptor to the substrate, or as the result of the pyrolysis of silane that diffuses to the back of the substrate. Whereas evolution of gaseous reaction products from the front surface of the substrate is rapidly blocked by complete coverage of the surface with silicon, the back surface does not become completely covered with silicon and therefore may be a source of gaseous contamination during the entire deposition run. Also, the back surface is at a higher temperature than the front surface. Therefore, doping may be decreased by coating the back surface on the substrate, particularly for the preparation of very low carrier concentration films and when relatively low deposition rates are used (Mercier, 1968, 1970B). For carrier concentrations greater than $1 \times 10^{16}/cm^3$ and for deposition rates of $\geqslant$ 2 μm/min it appears that little advantage is gained by sealing the back of the substrate.

Above the critical condensation temperature of silicon ($>1200°C$), elemental silicon (Reynolds and Elliott, 1966; Naber and O'Neal, 1968) and silane (Filby, (1966) erode the sapphire at a sufficiently high rate to be useful for the removal of work-damaged sapphire. The latter author invokes the formation of a silicon–aluminum alloy as an intermediate step in the removal of sapphire as gaseous reaction products.

The reduction of both sapphire (Dumin, 1967B; Mercier, 1968, 1970B) and spinel (Robinson and Dumin, 1968; Green, 1972) by hydrogen has also been proposed as a source of reaction products which lead to the autodoping of heteroepitaxial silicon:

$$2H_2 + Al_2O_3 \rightleftarrows Al_2O + 2H_2O$$
$$H_2 + MgO \rightleftarrows Mg + H_2O$$

The reduction of sapphire, however, can be assigned primarily to silicon as the reducing agent, since no advantage is realized by depositing silicon on sapphire in a helium atmosphere (Chiang and Looney, 1973; see also following section). In the case of spinel, the chemical reactivity of the substrate surface is strongly dependent on the method used to prepare the substrate crystal (Cullen and Wang, 1971). The chemical reactivity of the $MgO \cdot 1.7Al_2O_3$ spinel substrates prepared by the flame fusion process is similar to that of sapphire, and no advantage is observed in the use of helium in place of hydrogen as a carrier gas for the deposition of silicon on the flame fusion spinel (Cullen and Dougherty, 1972). It has been observed (by SEM) that if the deposition of silicon on the flame fusion alumina-rich spinel ($MgO \cdot 1.7Al_2O_3$) is stopped prior

to complete coverage of the surface, the substrate is eroded at the edge of the silicon deposit but not in the open areas exposed to hydrogen. This demonstrates that with flame fusion spinel as well as sapphire, chemical reduction by silicon (rather than hydrogen) is the main source of reaction products leading to autodoping of the silicon deposited (see Section 2.9).

The more reactive stoichiometric spinel ($MgO \cdot Al_2O_3$) prepared by the Czochralski process is reduced by both silicon and hydrogen. In this case erosion of the substrate surface by hydrogen at the silicon deposition temperature can be clearly observed by microscopic examination of hydrogen-annealed surfaces (Cullen and Dougherty, 1972; Green, 1972). It has been proposed that the crazing of the Czochralski spinel on exposure to hydrogen is the result of the preferential reduction of the magnesia component of the spinel (Green, 1972). Because of the reactivity of the Czochralski spinel to hydrogen at temperatures in excess of 1000°C, the subsurface damage resulting from mechanical surface finishing cannot be removed by annealing under the conditions that are commonly used prior to deposition on sapphire (30 min at 1300°C in hydrogen). The Czochralski spinel can be effectively annealed, however, in air [Reisman et al. (1971); see also Section 2.4]. The Hall mobilities measured in silicon deposited in a hydrogen ambient on Czochralski spinel annealed in air for 5 hr at 1500–1600°C are similar to the mobilities observed in silicon deposited on the hydrogen annealed flame fusion spinel (Cullen and Corboy, 1974).

Helium as a carrier gas

The feasibility of using helium as a carrier gas has been explored. This approach was originally directed toward lowering the temperature for the homoepitaxial growth of silicon (Richman and Arlett, 1969; Mercier, 1971). Degradation of the silicon deposited on the Czochralski spinel can be minimized by employing helium as the carrier gas. The mobilities measured in silicon deposited in helium on the Czochralski spinel are similar to the mobilities observed in silicon deposited on flame fusion spinel in a hydrogen ambient (Cullen and Dougherty, 1972). It is interesting to note that in the helium ambient, the optimum growth temperature appears to be 40–60°C below that used in a hydrogen ambient (Chiang and Looney, 1973). Under these conditions, improved mobilities have been realized (see Figure 2.41).

The reactivity of spinel substrates

It is tempting to explain the different reactivities of the sapphire (Al_2O_3), the Czochralski stoichiometric spinel ($MgO \cdot Al_2O_3$), and the flame fusion alumina-rich spinel ($MgO \cdot 1.7Al_2O_3$) on the basis of the alumina content. A reasonable proposition would be that alumina is more resistant to chemical reduction (by silicon and hydrogen) than magnesia and that therefore the degree of autodoping is proportional to the alumina content. It has been observed, however, that the surface reactivity of the flux-grown stoichiometric spinel is more similar to that of the flame fusion alumina-rich spinel than of the Czochralski-grown stoichiometric spinel (Cullen and Wang, 1971). In an attempt to stimulate the growth conditions of the flame fusion process (wherein water is present as the product of combustion of the H_2-O_2 flame), moisture was added to the ambient during Czochralski growth. This procedure did indeed lead to the chemical stabilization of the Czochralski-grown stoichiometric spinel (Cullen et al., 1975B; Cullen and Corboy, 1974). Thus, the chemical reactivity within the $Al_2O_3/MgO \cdot 1.7Al_2O_3/MgO \cdot Al_2O_3$ system must be related to more subtle effects such as the presence of OH^- ions and crystallographic site occupancy than to the more obvious MgO/Al_2O_3 ratio.

Interface compounds

A number of investigators have proposed that the reaction between silicon and the substrate results in the formation of a compound at the interface with a composition, and possibly a crystalline structure, distinct from that of either the substrate or silicon. In early work, Heiman (1966) referred to the presence of a "glassy" material at the interface. His hypothesis was based primarily on electrical results. More recently, Ipri (1972A,B) reported the existence of a layer adjacent to the substrate surface which is essentially insulating. This conclusion was based on a rather long extrapolation of mobility vs. thickness data within about 0.2 μm of the silicon most remote from the substrate in a 1-μm-thick deposit. Stein (1972), on the basis of optical absorption measurements, concluded that spinel substrates are chemically modified by the deposition of silicon. Cullis (1972) examined the early growth of silicon on sapphire by electron microscopy. At low deposition rates he observed evidence of an interface compound, which he assumed (on the basis of the species present) to be an alumina silicate. More recently, Kühl and coworkers (1974, 1976A) have supplied the most direct evidence for the existence of an interface compound. Their work included both silicon on sapphire and silicon on spinel, but most of the data presented were gathered from silicon on spinel. They have proposed the presence of an alumina silicate layer ~400 Å thick, based on a difference in optically and mechanically measured thicknesses, differences in etch characteristics, optical absorption, and the

presence of an $AlSiO^+$ species in SIMS profiles of the deposit. Spots were observed on the electron diffraction pattern of the 400-Å-thick layer which could not be assigned to either the silicon or the substrate (in this case, spinel). It should be noted, however, that diffraction spots characteristic of single crystal silicon were observed in the 400 Å layer. The work of the Siemens investigators on the characterization of the interface layer has been reviewed by Schlötterer (1976). There is not agreement among investigators in the field on the nature or the existence of an interface compound. Cullen and coworkers (1975A) assigned the extra spots observed in the electron diffraction patterns of 0.2-μm-thick silicon on sapphire to misoriented silicon rather than to an additional phase. Abrahams and Buiocchi (1975, 1976) have observed no evidence for a discrete interface compound in the transmission electron microscopic examination of cross sections of the interface region of the silicon-on-sapphire composite. Although it is difficult to assign quantitatively a thickness below which one cannot observe an interface layer by TEM, it is anticipated that a layer between 200 and 400 Å would be seen if present. We have taken the position that the material directly adjacent to the substrate is heavily contaminated (see Section 2.7.) silicon with some degree of crystallographic misorientation in relation to the "bulk" of the silicon deposit. The electrical and physical properties of this contaminated/misoriented silicon differ markedly from the "bulk" of the thin film, and it is understandable that it is identified as a spurious phase, particularly when the investigation is based on electrical properties. Some of the discrepancy between the various investigators may be accounted for by differences in deposition conditions or substrate preparation.

Impurity profile in heteroepitaxial silicon

It has been assumed for many years that the dependence of the semiconducting properties on the heteroepitaxial silicon thickness can be associated with changes in the impurity content (and the associated change in crystalline perfection) with distance from the silicon/substrate interface. The direct measurement of the impurity profile is a difficult task because of the small volume of material available for analysis and the relatively low (in the ppm level) impurity content. Early attempts to derive such profiles by concentration of etchants used for the controlled removal of the deposit and by spark-source mass spectrometry did not lead to conclusive results. As secondary ion mass spectrometry (SIMS) became available, it became possible to develop impurity profiles as a function of deposition parameters. The SIMS is commonly used in conjunction with sputter removal of the deposit. This method is not, however, free of difficulties in execution and interpretation of results. Standards must be carefully established, the sensitivity to the aluminum content is strongly dependent on the presence of oxygen and the sample becomes charged as the silicon is removed and the insulating substrate exposed.

The earliest SIMS analysis of silicon on insulating substrates was carried out by Mercier (1971). He examined the aluminum and oxygen content of silicon deposited on sapphire substrates in helium (at 900°C) and hydrogen (at 1100°C) atmospheres. It was observed that the aluminum content of the silicon deposited in the helium atmosphere at the lower temperature was less than that of the silicon deposited in hydrogen at the higher temperature by a factor of 10^4. The relationship of these results to current technology is difficult to assess, however, because of a lack of calibration and because of the unusually low growth rate (0.1 μm/min) and the relatively high temperature employed for the deposition of the silicon on sapphire.

Garcia and coworkers (1973), on the basis of SIMS analysis of silicon on sapphire, observed that the aluminum content is a function of the distance from the substrate surface and the exposure of the composite to thermal oxidation. Aluminum contents as high as 10^{19}/cm^3 were observed near the silicon/substrate interface. These investigators proposed that Al_xO_y complexes are present. Rai-Choudhury and coworkers (1974B) have examined the aluminum distribution (using SIMS) of silicon on sapphire that contained a heavily phosphorus-doped (10^{20}/cm^3) layer at the interface to enhance the carrier lifetime in the "bulk" of the deposit. A peak in the aluminum profile at a position approximately one third of the thickness of the deposit (overall thickness unspecified) was associated with the presence of the phosphorus layer. Kühl and coworkers (1976A) and Schlötterer (1976) have observed an AlSio$^+$ signal in a SIMS profile which peaks near the interface of silicon on sapphire. A long tail of this signal on the substrate side of the interface cannot be explained by these investigators. Kühl and coworkers (1976B) also developed SIMS profiles of silicon on spinel and of the surface of the spinel after removal of the silicon by acid etching. The aluminum signals were observed to decrease by about a factor of 100 over a distance of approximately 100 Å from the substrate surface. The presence of silicon was observed in the material remaining on the substrate surface after the acid removal of the silicon. The presence of silicon/oxygen species was assigned to oxidation occurring after removal of the silicon layer. These investigators point out that the SIMS data are consistent with the presence of an Al/Si/O interface compound.

Cullen and coworkers (1975A) and Harrington and

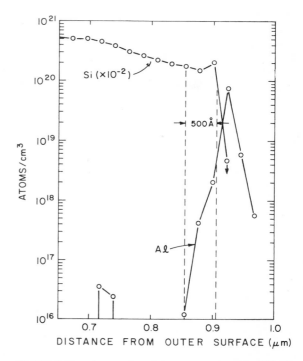

FIGURE 2.15   A secondary ion mass spectrometer (SIMS) profile of the aluminum concentration in the interface region of silicon deposited on sapphire at 2 μm/min (1000°C)

coworkers (1976), also using SIMS, have compared the aluminum profile in silicon deposited on sapphire at 2 μm/min and deposited using an initial growth rate (for the first 0.15 μm) of >5 μm/min (see Section 2.8.). A SIMS profile of the interface region of silicon deposited at 2 μm/min is shown in Figure 2.15. The x axis is the distance from the outer (most remote from the substrate) silicon surface. The y axis is the aluminum content, calibrated with ion implantation (aluminum) of bulk silicon. The silicon/substrate interface is taken, for the purpose of the analysis, as the point at which the silicon content decreases to 50 percent of the "bulk" value. The assignment of the interface in this manner is commonly done in sputter-removal analytical technique. It can be seen that the aluminum content decreases by approximately three orders of magnitude to a "bulk" level over a distance of 500 Å. A similar decrease with distance from the substrate was observed in a number of samples. In all of the samples examined, the aluminum content in the deposit more than 500 Å from the substrate is ≤5 × 10^{16}/cm^3. The shape of the SIMS profile within 500 Å of the interface was similar in the silicon deposited at a constant growth rate and the silicon deposited using a rapid early growth rate. The use of the rapid early growth technique did, however, lead to a reduction of the aluminum content in the "bulk" of the deposit (i.e., the deposit more than ~500 Å from the interface). It is clear that most of the aluminum is electrically inactive, since the level of autodoping of the silicon deposited under the conditions described is typically ≤ 10^{15} holes/cm³. In attempting to explain the insensitivity of the aluminum distribution in the interface region to growth rates in excess of 2 μm/min, it should be kept in mind that the thickness at which the substrate is covered has proven to be independent of growth rate for rates between 0.3 and 2 μm/min (in silicon on spinel) (Cullen and Dougherty, 1972). It is reasonable to associate the distance over which the aluminum abruptly decreases with the thickness at which the substrate is completely covered (300–500 Å in silicon on sapphire; see Section 2.7).

The effects of reaction between silicon and the substrate surface

Focus has been placed in the past on the reaction between deposition constitutents and the substrate in relation to autodoping effects. Using deposition rates ≥ 2 μm/min, and deposition temperatures ≤1000°C, the autodoping has been reduced to acceptable levels with regard to current device processing procedures. Of perhaps greater concern (at least under current deposition conditions) is the physical roughening of the substrate surface as the result of reaction between silicon and the substrate and the exposure of undesired substrate orientations. The roughening of the substrate surface leads to the nucleation of orientations that differ from the "bulk" of the heteroepitaxial deposit. Chang (1971), in connection with the evaporation of silicon on sapphire (see Section 2.9), has pointed out that faceting of the surface determines the upper limit of temperatures at which good crystallinity can be achieved in the evaporated heteroepitaxial silicon. Cullen and coworkers (1975A) have proposed that the improvement of the semiconducting properties in thin (<0.9 μm) heteroepitaxial silicon associated with rapid early growth is more related to maintaining the smoothness of the substrate surface than with suppression of autodoping per se. It is evident from electron diffraction of 0.2-μm-thick deposits that the surface of the silicon deposited using the rapid early growth method is smoother than the surface of the silicon grown at 2 μm/min. This is consistent with suppression of substrate surface roughening. We speculate that the rapid early growth may force the continuation of growth of the first nucleated silicon islands, rather than the nucleation and growth of some islands at the expense of other islands already on the surface of the incompletely covered substrate (see Figure 2.20). When the latter process is operative, some areas of the substrate are exposed more than one time to the expanding edge of a silicon growth island. It is at the edge of the growth island that the reaction between silicon and the substrate takes place.

2.7. The nature of silicon growth prior to complete coverage of the substrate surface

General considerations

It has been informative to examine the nature of the early growth of silicon on the insulating substrates as a function of the chemical and crystallographic character of the substrate, the method used for the preparation of the substrate surface, and the deposition temperature and rate. In this section the phrase *early growth* is taken to mean the growth of the silicon prior to complete coverage of the substrate surface. A qualitative measure of the extent of reaction between the substrate and silicon and the carrier gases can be derived from the examination of the early growth islands.

Manasevit and coworkers (1965) compared the nature of early growth of silicon deposited from silicon tetrachloride and silane on sapphire using replication electron microscopy. They noted that the substrate surface was covered more rapidly when silane was employed as the silicon source than when the silicon tetrachloride was the source. Some faceting of the early growth islands was observed in silicon deposited from the halide. Bicknell and coworkers (1966), also using replication techniques, compared the early growth of silicon deposited on sapphire from trichlorosilane and silane. At their relatively high deposition temperature (1200°C), reaction with the substrate was observed using the halogenated silane as the source, and the early growth islands proved not to be single crystalline. Reaction with the surface was detached when silicon was deposited from silane only at deposition rates less than 0.5 μm/min. Mercier (1971), using deposition rates less than 0.5 μm/min and silane as the source, also observed erosion of the sapphire substrate between the silicon islands. In a TEM study of silicon evaporated onto (0001) sapphire platelets, Cullis (1972) reported evidence of reaction between silicon islands and the substrate at an evaporation temperature of 900°C. Less reaction was observed at a deposition rate of 20 Å/s than at 0.5 Å/s.

Using flame fusion spinel as the substrate and silicon tetrachloride and trichlorosilane as the source for deposition at 1200°C, Heywang (1968) observed evidence for reaction between silicon and the incompletely covered substrate by TEM. In the presence of the corrosive halides, it is not clear how reaction with silicon was distinguished from reaction with the halides (or halide reduction product; e.g, HCl). Because of the reactivity of silicon tetrachloride and the halogenated silanes, in recent work only silane has been used as the source in the deposition of silicon on the refractory oxide substrates.

Early growth as a function of the chemical nature of the substrate and the depositon ambient

The influence of the chemical composition of the substrate on the nature of the growth of the silicon prior to complete surface coverage has been examined by Cullen and Wang (1971). Figure 2.16 shows the early growth of silicon simultaneously deposited—in the same growth run—on (111) alumina-rich flame fusion spinel, (111) stoichiometric Czochralski spinel, and (1102) flame fusion sapphire. The substrates were hydrogen annealed, and the surfaces were exposed for 5 s to growth conditions that would yield (if the films were grown to ~1.5 μm in thickness) an overall growth rate of 0.8 μm/min at 1100°C. It can be seen that the silicon on the flame fusion spinel is at a later stage of coalescence than the silicon on the flame fusion sapphire, and there is no evidence of the presence of reaction products. On the Czochralski spinel, the edges of the growth islands are not distinct, and what appears to be reaction products are observed on the bare substrate and on the deposited silicon. The appearance of the early growth corresponds with the experience that silicon may be deposited on the flame fusion spinel and sapphire using similar surface conditioning and deposition conditions, but that in order to utilize the Czochralski spinel it has been necessary to devleop surfacing (Reisman et al., 1971) and deposition (Cullen and Dougherty, 1972) conditions specific to the chemical nature of the substrate material. This may be attributed to the reactivity of the stoichiometric spinel in the presence of hydrogen (Green, 1972).

After observation of reactivity between hydrogen and the stoichiometric Czochralski spinel, attempts were made to use helium as the carrier gas in the deposition of silicon on the Czochralski spinel (Cullen and Dougherty, 1972; Chiang and Looney, 1973). Figure 2.17 shows scanning electron micrographs of silicon (just prior to complete coverage of the surface) deposited simultaneously on (a) flame fusion and (b) Czochralski spinel in a helium ambient (Cullen and Dougherty, 1972). Both substrates were exposed to the growth atmosphere for ~15 s at 1000°C. It can be seen that the surface of the silicon on the Czochralski spinel is slightly more mottled than on the flame fusion material, indicating the possibility of some interference reaction. On the other hand, the surface of the silicon is considerably clearer on the Czochralski spinel if helium rather than hydrogen is used as the carrier gas and there is no evidence for the condensation of a reaction product (compare with Figure 2.16b). There is little difference seen in the appearance

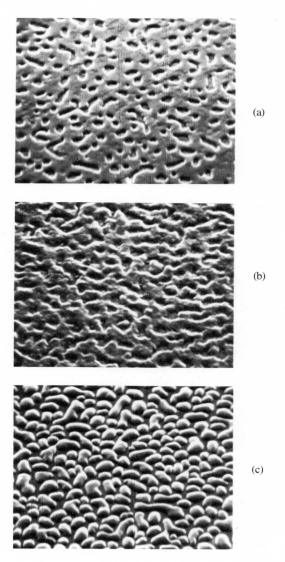

(a)

(b)

(c)

FIGURE 2.16   Scanning electron micrographs of silicon after deposition for 5 s at 0.8 μ/min on (a) flame fusion spinel, (b) Czochralski spinel, and (c) flame fusion sapphire. Magnification, 6,500×; beam-sample angle, 45°. From Cullen and Dougherty (1972)

of the surfaces of the silicon on the flame fusion spinel deposited in a hydrogen or helium carrier gas.

The observation of the coalescence of growth islands in silicon on spinel

The development and coalescence of the silicon growth islands have been followed by microscopic examination of the surfaces after abruptly terminating the silicon deposition following various predetermined deposition periods (Cullen and Dougherty, 1972).

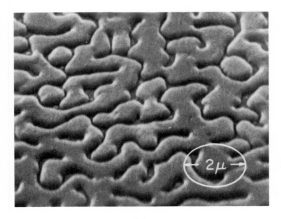

(a)

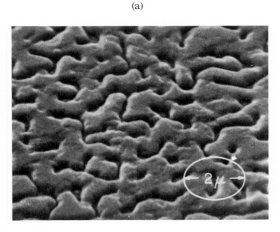

(b)

FIGURE 2.17   Scanning electron micrographs of the early growth of silicon deposited for 45 s at 0.2 μm/min in helium on (a) flame fusion and (b) Czochralski spinel. Original magnification 10,000×; beam-sample angle, 45°. From Cullen and Dougherty (1972)

Typical scanning electron micrographs of the various stages of surface coverage of (111) silicon deposited on (111) flame fusion spinel are shown in Figure 2.18. (The ellipse superimposed on the micrograph indicates the foreshortening resulting from the 70° observation.) The surface shown in Figure 2.18a was exposed for 1 s to deposition conditions which would (if continued to ~1 μm) lead to an overall growth rate of 2.0 μm/min. The deposition temperature was 1100°C. The surface shown in Figure 2.18b has been exposed to the same growth ambient for a period of 3 s. After 5 s of growth at a rate of 2.0 μm/min the surface is covered. In order to examine other intermediate stages of incomplete surface coverage, it was necessary to shift to a lower deposition rate. The surface shown in Figure 2.18c was grown for 20 s at an overall growth rate of 0.3 μm/min. At this growth rate the surface of the substrate is completely covered in

30 s, as shown in Figure 2.18d. The early growth was examined of silicon deposited at rates of 0.3, 0.8, and 2.0 μm/min. The thickness of the silicon normal to the substrate surface, as measured from the 70° scanning electron micrographs, is shown as a function of the growth duration in Figure 2.19. The broken line vectors drawn from the last thickness points represent the overall growth rates.

It can be seen in Figure 2.18 that after 1 s of growth at 2.0 μm/min the islands are already beginning to coalesce. The smallest islands are approximately hemispherical in shape; as the growth areas join, the silicon flattens into irregular plateau shapes. At this point about 60 percent of the surface is covered; no new islands are formed after the first 1 s of growth. The early growth islands are widely scattered, with further surface coverage proceeding by growth at the edges of the islands rather than by the formation of new islands. In other chemical systems where the deposit does not react with the substrate, it is not unusual to see surface finishing damage (scratches) act as preferential nucleation centers. In this case, however, the high-energy microcracked ("scratched") areas are etched out by chemical reduction of the silicon. Linnington 1974 observed denuded areas around surface work damage during the early growth of silicon (evaporated) on sapphire.

After 3 s of growth at 2.0 μm/min the growth plateaus have become completely connected. The area covered has increased during the 2 s of growth from about 60 to about 85 percent coverage, but the thickness of the plateaus normal to the surface has increased only from ~0.13 to ~0.17 μm. At this point in time the rate normal to the substrate surface (1.2 μm/min) is similar (within the error of the thickness measurements) to the overall growth rate of 2.0 μm/min. Considering the amount of surface covered from widely scattered growth islands, it is clear that the rate of growth in the plane of the substrate has been higher than the rate normal to the substrate. Growth normal to the substrate has been fed by the arrival of nutrient to the tops of the islands, whereas growth at the edge of the islands is presumably fed by nutrient impinging on the bare substrate and moving to the edge of the island. During the period in which the area of the exposed surface exceeds the surface area of the islands, the horizontal growth is more rapid than the vertical growth, assuming that little silicon is lost by reaction with the surface.

After 20 s of growth at a rate of 0.3 μm/min, the islands have coalesced to the point where only isolated holes exposing the substrate remain. The surface coverage is about 90 percent, and the thickness of the film is about 0.16 μm. It is interesting to note that there is no topological effect where the islands have joined. After 30 s of growth at 0.3 μm/min, the substrate is completely covered, the surface is very smooth, and no evidence of the island formation can be seen.

The shape, distribution, and manner in which the islands grow are similar for the three growth rates studied. It is only the duration of exposure of the substrate to the gaseous ambient that is altered by changing the growth rate. A useful conclusion of the examination of the mode of surface coverage of spinel is that the surface is completely covered at a thickness of ~0.2 μm and that this thickness is independent of growth rate within the range of rates studied.

If silicon is removed from the incompletely covered flame fusion spinel substrates by dissolution in an aqueous HF-HNO$_3$ etch, one finds that the shape of the silicon islands is imprinted on the substrate in the form of recessions which correspond to the edge of the removed silicon islands (Cullen and Dougherty, 1972). Such a pattern is not observed unless the growth is

interrupted, and thus the erosion of the substrate at the edge of the silicon island must have occurred at deposition temperature immediately after termination of the silane and during cooldown in hydrogen. This type of erosion is consistent with other evidence that reaction with the substrate is more the result of the chemical reduction of silicon than by hydrogen. Had chemical reduction by hydrogen been responsible for the substrate erosion, the recessions in the substrate would have been deepest in the center of the openings in the deposit rather than at the edge of the openings.

A series of (111) flame fusion spinel substrates was prepared simultaneously with the sapphire samples described in the following section. When examined by replication electron microscopy, the morphology of the silicon growth on spinel at 1000 and 1100°C was essentially identical to the morphology observed in silicon on sapphire shown in Figure 2.20. (The differ-

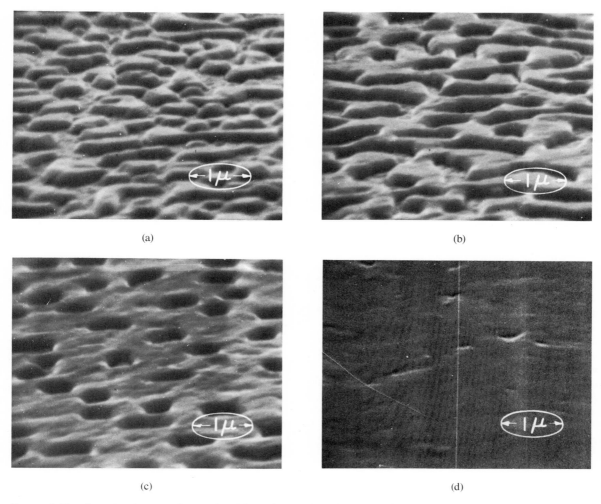

(a)

(b)

(c)

(d)

FIGURE 2.18   Scanning electron micrographs of the early growth of silicon deposited in hydrogen on flame fusion spinel. (a) After 1 s at 2 μm/min; (b) after 3 s at 2 μm/min; (c) after 20 s at 0.3 μm/min; (d) after 30 s at 0.3 μm/min. Original magnification 20,000×; beam-sample angle 70°. The ellipse shows the effect of foreshortening. From Cullen and Dougherty (1972)

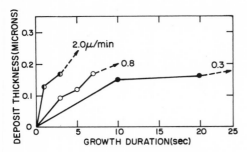

FIGURE 2.19   The thickness of silicon on spinel (prior to complete coverage of the substrate surface) as a function of the growth duration and rate. After Cullen and Dougherty (1972)

ence between this observation and the results described earlier can be attributed to refinements in the finishing and annealing procedures, see Figure 2.21.)

The early growth of silicon on sapphire

Replication electron micrographs of the early growth of silicon on Czochralski sapphire substrates at 1000 and 1100°C are shown in Figure 2.20. The deposition was abruptly stopped after growth durations of 0.5, 2, and 4 s using conditions that would yield an overall growth rate of 0.3 μm/min if carried beyond the point of complete surface coverage. The two temperatures were employed to examine the nature of silicon growth at the temperature used as the optimum growth temperature on sapphire and at a temperature 100°C above this, both to examine the influence of temperature on early growth and to make direct comparison with the growth of silicon on spinel at 1100°C.

It is clear that for a particular growth duration, the density of the islands is higher and the island size smaller at the lower growth temperature. After 0.5 s, at a growth temperature of 1000°C the silicon islands are too small to be readily resolved, whereas at a growth temperature of 1100°C the islands are widely scattered and easily resolved. A similar relationship between growth island size and temperature has been observed in systems (such as gold on sodium chloride) where reactivity between the deposit and the substrate is not invoked as a key factor influencing the nature of the growth [see, for instance, Chopra (1969)].

It is also clear that after 4 s of growth at 1000°C the surface is almost completely covered (there are small voids which are difficult to resolve), whereas at 1100°C the surface coverage is about 50 percent. At 1000°C a surface coverage at a thickness of ~300 Å is estimated from the measurement of round islands just prior to complete coverage (from observation of higher magnification replicas), assuming that the round islands are hemispherical in shape. On the basis of the

growth duration and overall rate, the thickness of the surface coverage is ~200 Å. There are clearly large errors involved in both estimates, but it is safe to say that the coverage on sapphire at 1000°C is between 200 and 400 Å.

It can be seen that between 0.5 s and 2.0 s at 1100°C the surface coverage increases by growth of the original widely distributed islands rather than by the formation of new islands. The island density at 2 s is lower than at 0.5 s as the result of some coalescence, as evidenced by the irregular shapes of some of the islands. It is interesting to note, on comparison of the silicon after 2 and 4 s of growth at 1100°C, that as the islands coalesce, the space between the islands becomes larger. Thus, it is seen that there are portions of the substrate that are covered by early island formation, that are subsequently exposed as the islands coalesce, and then are covered again as the coalesced islands grow. Under the conditions where the silicon reacts with the substrate mainly at the edge of the island, this mode of film development may have an effect on the semiconducting properties of the silicon adjacent to the substrate surface.

The thickness at which the sapphire surface is covered at 1100°C cannot be estimated from the micrographs shown in Figure 2.20. Since the replicas of the silicon on spinel at 1100°C were so strikingly similar to the silicon on sapphire at the same temperature, the value derived from the SEM study of silicon on spinel is taken as the thickness of coverage for silicon on sapphire at 1100°C (i.e., ~2000 Å).

The silicon is strongly bonded to the substrate surface, and it is unlikely that coalescence takes place by movement of even the smallest islands. The silicon must be transferred by a chemical transport mechanism which is not as yet well characterized. The sparse distribution of the first islands and the growth to relatively large islands at the higher temperature can be associated with the mobility of the atoms on the substrate surface and the kinetics of the transport reaction.

Linnington (1974) followed the formation and coalescence of evaporated silicon on sapphire by scanning electron microscopy. He achieved a resolution of ~200 Å by operating with the collector in the backscattered secondary-electron mode and by scanning with as short a working distance as possible. Using the short working distance the spherical aberration of the final lens was reduced by running with a high excitation current. In this study the silicon was evaporated into the sapphire at a rate of ~0.25 μm/min and at temperatures between 960 and 1100°C. The pressure during evaporation was 5×10^{-8} torr. Erosion of the substrate was observed only in surface work damage areas (polishing scratches) at the lower temperature. At the higher deposition temperature, some erosion of the substrate was observed at the edge of the growth islands. It is interesting to note that the general liquid-like three-dimensional nucleation and coalescence observed by Linnington is very similar to the nature of growth observed in silicon deposited from silane at similar temperatures and rates. The low pressure and absence of silane do not appear to have played a significant role.

Linnington points out that in the early stages of growth the nature of development of the silicon film can be characterized by the size and distribution of the islands. As the islands coalesce into nonequilibrium shapes, the size-measuring parameters are no longer useful, and the growth can be expressed only in terms of the film thickness and percentages of surface coverage. In order to circumvent this restriction, Linnington characterized the distribution of silicon on the surface in terms of various spatial frequencies present by digitizing the island pattern and numerically extracting its Fourier transform. The transform was found to be circularly symmetric, and thus the silicon distribution was reducible to a radial mean frequency distribution. In order to carry out this analysis, the micrographs of the early growth were digitized using a high-resolution television camera, and the digitized data processed with a computer. The results were analyzed in terms of existing theories of nucleation and coalescence.

Early growth as a function of surface preparation and the chemical nature of the substrate

A number of investigators have adopted the procedure of "annealing" the substrate surface in a hydrogen ambient for 20 min at 1150°C (see Section 2.3). This procedure is compatible with the silicon deposition systems and can be carried out immediately prior to the silicon growth. This avoids the possibility of contaminating the surface between annealing and deposition. The influence of annealing the substrates in air for 2 hr at 1600°C has also been investigated. This latter procedure received some attention in attempts to use the Czochralski-grown spinel substrates.

No difference in the nature of the early growth is seen between hydrogen-annealed or air-annealed Czochralski sapphire and flame fusion spinel. Micrographs of the silicon growth on these four substrates are essentially identical to Figure 2.20 for both deposition temperatures (Corboy, 1976).

Differences have been observed, however, between silicon deposited on hydrogen- and air-annealed Czochralski spinel (Cullen et al.). Scanning electron micrographs of the early growth of silicon after 1 s of growth at an overall rate of 2 μm/min and at 1100°C is shown in Figure 2.21. The deposition was carried out simultaneously on a polished surface (Figure 2.21a), a surface which had been annealed in hydrogen for 20

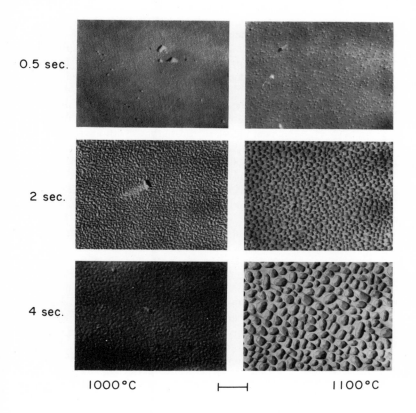

0.5 sec.

2 sec.

4 sec.

1000°C   ⊢——⊣   1100°C

FIGURE 2.20
Replication electron micrographs of the early growth of silicon on sapphire after 0.5, 2, and 4 s of growth at 1000 and 1100°C. From Corboy (1976)

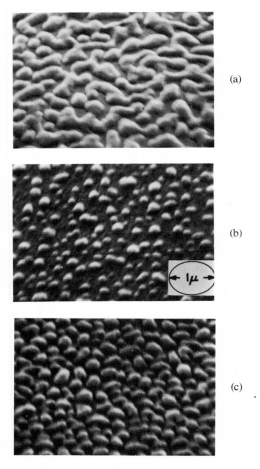

FIGURE 2.21   SEM micrographs of the early growth of (100) silicon on (100) spinel. Growth rate: 2 μm/min for 1 s; Growth temperature: 1100°C; substrate type: Czochralski (RCA); original magnification: 20,000×; beam-sample angle: 45°. Substrate conditioning: (a) no pregrowth heat treatment; (b) H_2, 1150°C, 20 min; (c) air, 1600°C, 2 hr. From Cullen et al., (1973)

and the most difficult to use. It is not surprising, therefore, that the largest differences are observed in the nature of early growth as a function of surface treatment on this material.

General trends in the nature of early growth

On comparison of the temperature and rate of deposition with the nature of early growth and the film semiconducting properties, it becomes clear that the rate of surface coverage and the thickness at which the surface is covered are important parameters in the heteroepitaxial silicon technology. In general, it is observed that the highest carrier mobilities in the thin (<1 μm) silicon deposits can be associated with conditions that lead to rapid coverage of the surface and to surface coverage at a minimum thickness. This is consistent with the observation previously discussed that the optimum electrical properties are realized at relatively high growth rates and low growth temperatures. From the observation of early growth, it is clear that the low growth temperature is advantageous not only because of the kinetics of the interface reactions but also because the thickness at which the substrate is covered decreases with decreasing deposition temperature.

The nature of the early growth is a function of the chemical reactivity of the substrate material, the manner in which the polishing work damage was removed, and the temperature of deposition. It is interesting to note that the manner in which the growth islands form is strikingly similar on the relatively unreactive Czochralski sapphire and flame fusion spinel at a given growth temperature. Clearly the crystallographic makeup of the (1$\bar{1}$02) rhombohedral sapphire and the (111) cubic spinel are significantly different, as are the number, distribution, and nature of defect sites. The chemical nature of the surfaces with appreciable different occupancy by elements is also significantly different. It must be concluded, however, that on the surfaces which have been effectively freed of crystallographic work damage, the early growth islands are not nucleated at crystallographic features. On both the Czochralski sapphire and flame fusion spinel, neither the electrical properties nor mode of early growth are a function of the two surface-annealing methods employed. On the more reactive Czochralski spinel, an influence of the method of annealing is observed. It is concluded that, unless the surface is sufficiently reactive in the presence of silicon to create topological effects, the nature of early growth is not influenced by the chemistry of the surface. The work damage–free smooth substrate surface must appear to be essentially homogeneous to the arriving silicon atoms.

A schematic overview of the nature of development of the silicon on the insulating substrates, derived

min at 1150°C (Figure 2.21b), and a surface which had been annealed in air for 2 hr at 1600°C (Figure 2.21c). It can be seen that the density of nucleation islands on the spinel is highest on the sample which had been air annealed (Figure 2.21c). The thickness of complete surface coverage on the air-annealed spinel is approximately half that observed on the hydrogen-annealed spinel. This effect has been observed with both the (111) and the (100) orientations, and for both orientations, the highest Hall mobilities are observed in the air-annealed samples. Of the substrates recently investigated for use as silicon substrates (i.e., sapphire and the spinel prepared by the various growth methods), the Czochralski spinel has proven to be the most reactive in the presence of the deposition constituents

from microscopic observation, is shown in Figure 2.22. The first agglomerates that can be observed are relatively widely spaced (Figure 2.22A); the spacing is a function of the growth temperature. The smallest islands are hemispherical liquid-like formations, and no faceting is observed at any stage prior to complete coverage (for growth from silane). Growth proceeds by enlargement of the first islands formed, rather than by formation of new islands (Figure 2.22B). This is clearer at the higher growth temperature. As the islands coalesce, flat-topped mesas are formed. As the mesas are formed, the edges of the islands that have coalesced into the mesas appear to draw in, leaving more space between the mesas than originally existed between the hemispherical islands (Figure 2.22C). Finally, the surface is completely covered as the edges of the cylindrically shaped holes in the deposit grow together. Just after complete coverage of the substrate, the silicon film is remarkably smooth and there is no topological evidence of the location where the islands have met and filled in (Figure 2.22D). It is reasonable to propose that the substrate reaction products are incorporated into the silicon primarily prior to complete surface coverage. After coverage, reaction between the silicon and substrate is terminated by blockage of the gaseous reaction products, and the residual reaction products are swept out of the reaction zone. After complete coverage, the level of contamination decreases and the crystallographic properties improve rapidly with distance from the silicon/substrate interface.

The semiconducting properties near the interface can be analyzed in terms of the thickness of and distance from the contaminated interface layer. For instance, it has been observed that the Hall mobilities in thin (<0.5 μm) silicon on spinel are lower than in silicon on sapphire, even though in the thicker film (>1.0 μm) the reverse is true (see Table 2.10). This can be associated with complete coverage of the hydrogen-annealed flame fusion spinel and sapphire being realized at silicon thicknesses at ~2000 Å and 300 Å, respectively (at the optimum deposition temperatures for the two materials). It is also interesting to note that the aluminum contamination in silicon on sapphire decreases by three orders of magnitude over a distance of ~500 Å from the silicon/sapphire interface (see Figure 2.15). It also seems likely that the 400-Å-thick "optically effective intermediate layer" described by Schlötterer (1976) and Kühl and coworkers (1976B) is related to the silicon formed prior to complete coverage of the substrate surface.

In discussing the effect of thermal processing on the film properties, it should be kept in mind that the nature of bonding and the oxidation state of the aluminum incorporated during growth of the silicon must be different from aluminum introduced by diffusion.

2.8. Epitaxial growth conditions: temperature and rate

Crystalline perfection and deposition conditions

In the deposition of homogeneous single crystals from the vapor phase, it is commonly assumed that crystalline perfection is achieved under near-equilibrium growth conditions. Relatively high growth temperatures provide the mobility, and low growth rates provide the time, for ad-atoms to reach equilibrium positions in the crystal lattice. However, in the growth of heteroepitaxial silicon on the refractory oxide substrates, reaction of the substrates with the deposition constituents interferes with the formation of good crystallinity in the overlayer. The local concentration of reaction products in the silicon near the silicon/substrate interface may be sufficiently high to distort the lattice, and erosion of the substrate may lead to faceting of the surface and formation of unwanted crystallographic orientations in the deposit. Reaction of the substrate with deposition constituents is clearly suppressed by epitaxial growth at low temperatures and high rates. In establishing the optimum epitaxial growth conditions, a compromise must be reached between the use of low-temperature/high-rate conditions to suppress the incorporation of reaction products and formation of spurious orientations in the growing film, and the use of high-temperature/low-rate conditions to provide the energy and time for the formation of a good single crystal lattice. The development of the optimum conditions for the heteroepitaxial growth of silicon has been the subject of a good deal of research effort (Filby and Nielsen, 1967; Cullen et al., 1970B; Dumin et al., 1970; Cullen, 1971; Gottlieb and

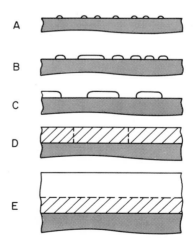

FIGURE 2.22   A schematic presentation of the nature of early growth and the incorporation of substrate reaction products in the deposition of silicon on insulating substrates

Corboy, 1972; Druminski and Schlötterer, 1972; Manasevit, 1974; Manasevit et al., 1976; Druminski, 1976).

Growth temperature

In open-flow (atmospheric pressure) systems, good crystallinity has been observed in heteroepitaxial silicon deposited from silane at growth temperatures as low as 920°C (Druminski, 1976). Below 920°C, the crystallinity of the deposit degrades. Considerably lower temperatures can be used in the vacuum deposition of heteroepitaxial silicon; this is discussed in some detail in Section 2.9. In general, temperatures between 950 and 1000°C (at rates between 2 and 3 μm/min) have been adopted for the routine growth of silicon on sapphire (Cullen et al., 1970A,B; Dumin et al., 1970; Gottlieb and Corboy, 1972). These investigators found that the crystalline quality (as inferred from electrical properties) decreased below deposition temperatures of 950°C. It is interesting to note that Druminski (1976) observed that the sensitivity of the crystalline quality (as determined by optical absorption measurements) to gaseous impurities increases with decreasing temperature. This suggests the possibility that deposition in the 950–1000°C range constitutes a trade-off between suppression of reaction with the substrate and sensitivity to source gas impurities. Within this temperature range autodoping has proven not to be a serious problem, and the carrier concentration can be controlled to levels above $10^{15}/cm^3$ by the addition of dopants in the gas stream. As the deposition temperature is increased above $\sim 1050°C$, the as-deposited carrier concentration becomes independent of the intentionally added dopant as the result of incorporation of electrically active substrate reaction products (aluminum in various oxidation states), and the effect of the added dopant is seen only after the unintentionally added species are deactivated or removed (at least partially) by a thermal oxidation of 1 hr at 1100°C (Dumin and Robinson, 1968A; Cullen et al., 1970B). Electrical properties usable for device application have been realized in silicon deposited on the temperature range 1050–1100°C (growth rate 2 μm/min) if the deposits are "homogenized" at 1100°C in oxygen for 30 min and nitrogen for 2 hr (Manasevit et al., 1976). As the deposition temperature is increased above $\sim 1150°C$, the crystallinity is degraded, unwanted orientations appear, and the films are heavily autodoped by substrate reaction products. At $\sim 1300°C$ only gaseous products of reaction between silicon and the substrate are formed, resulting in no deposition of silicon and only erosion of the substrate surface (Filby, 1966). At equal deposition temperatures the silicon on flame fusion spinel is less autodoped than silicon on sapphire (Robinson and Dumin, 1968), and therefore the optimum temperature for the growth of silicon on the flame fusion spinel (1100°C)

has proven to be $\sim 100°C$ higher than that commonly employed for the deposition of silicon on sapphire (Cullen et al., 1969).

Growth rate

At low deposition rates (<0.5 μm/min) autodoping is a problem for deposition temperatures above $\sim 950°C$. The autodoping decreases with increasing growth rates. As previously discussed (Section 2.7), it is reasonable to assume that incorporation of substrate reaction products takes place, for the most part, prior to complete coverage of the substrate surface. Apparently the largest effect of a high growth rate in the suppression of autodoping is to rapidly seal off the substrate surface and minimize the time during which the evolution of gaseous reaction products takes place. Under these circumstances, the optimum growth rate is different prior to complete coverage of the substrate from that after the substrate has been completely covered. The growth rate (unlike the growth temperature) can easily be abruptly changed, and therefore procedures have been adopted wherein the first portion of the silicon is deposited at a relatively high growth rate, and the rate is then quickly decreased and the film grown to the desired thickness. Gottlieb and Corboy (1972) deposited 1 μm of silicon at a rate of 2 μm/min (at 1000°C), and then continued the growth at a rate of 0.3 μm/min to a total film thickness of 1.5 μm. Improvement in mobility over those realized when the entire film was deposited at 2 μm/min was observed. Although the influence of the final growth rate on the film properties was not investigated between deposition rates of 0.3 and 2.0 μm/min, there is the implication that subsequent to the substrate surface coverage the optimum silicon-on-silicon growth rate by the pyrolysis of silane at 1000°C is less than 2 μm/min. It seems possible that the optimum growth rate (after coverage of the surface) is relatively low because the defect content of the film near the substrate is high, and a low growth rate provides the condition necessary for the improvement of the film with thickness (as inferred from electrical properties). Similar growth rates are employed, however, for the homoepitaxial growth of silicon from the chlorinated silanes. In the deposition of silicon on spinel at 1100°C, after-oxidation mobilities in 1.5-μm-thick films have been improved by employing overall growth rates in excess of 5 μm/min (Cullen, 1971). Since the optimum growth temperature for the deposition of silicon on spinel is higher than for the deposition of silicon on sapphire, no meaningful conclusion can be drawn on the effect of the crystallographic orientation of the substrate on the optimum deposition rate.

Although it appeared from Gottlieb and Corboy's work (1972) that the mobilities of the 1.5-μm-thick

silicon films on sapphire are degraded at initial growth
rates in excess of 2 μm/min, higher rates have since
been used to advantage. In the application of a slow–
fast technique to the deposition of thinner films, Cullen et al (1975A) used a growth rate of 5 μm/min for
deposition of the first 0.15 μm of 0.5- and 1-μm-thick
films. After deposition of the 0.15-μm-thick layer, the
films were grown to the desired thicknesses at a rate of
0.3 μm/min. The low final growth rate was used as
much to extend the duration of the deposition as it was
to improve the crystalline quality, and therefore the
possibility of using higher final growth rates was not
investigated. At an overall rate of 2 μm/min in the
deposition of a 0.5-μm-thick film, an error of several
seconds can make a significant error in the film thickness. The 5-μm/min first-growth rate resulted in the
improvement of the electron mobilities in the 0.5-μm-
thick silicon on sapphire film. It was also noted that
the crystallinity of the first 0.2 μm of silicon (as determined by electron diffraction) was improved by rapid
early growth. These results suggest that the growth
rate of silicon on the insulators, at least during the
period prior to complete coverage of the substrate
surface, is more limited by practical considerations
(duration of deposition) than by degradation of the
crystalline structure. The initial growth rate at which
the crystallinity of the deposit is degraded has not
been established. [Note that at deposition pressures
less than 1 atmosphere, single crystal films have been
achieved on silicon substrates at growth rates as high
as 24 μm/min (Joyce and Bradley, 1963).]

Using a procedure in which hydrogen chloride gas is
added to the silane after coverage of the sapphire
surface, Druminski and Schlötterer (1972) also
decreased the growth rate (from 2 to 0.3 μm/min)
during the halide introduction. The combination of
these two growth processes led to more homogeneous
films with better crystalline perfection than when a
single-source/single-rate process was employed.

It is interesting to note that Druminski (1976) has
found that even in the temperature range 920–980°C,
the rate at which good crystalline properties is
achieved increases with increasing growth
temperature.

Since growth rate has proven to be a critical parameter in the deposition of silicon on the insulator, the
rates have been continuously monitored by an inter-
ferometric method during experimental deposition
runs [Dumin (1967A); see also Chapter 5]. The development of the interference peaks is shown for silicon
deposited at a constant rate of 2 μm/min (Figure
2.23a); at a dual rate of 2 μm/min followed by 0.3 μm/
min (Figure 2.23b); and a dual rate of >5 μm/min
followed by 0.15 μm/min (Figure 2.23c). The initial
growth rate in the latter method is integrated over the
first 0.15 μm of silicon growth; a steady state rate is
not established during the brief (1–2 s) early portion of

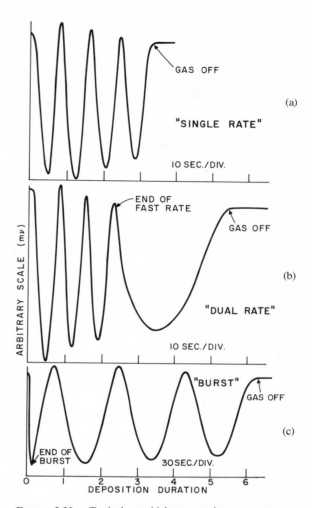

FIGURE 2.23   Typical rate-thickness monitor output traces
for films grown at a single rate, using two rates, and using the
burst growth technique

the growth run. We have labeled this the "burst"
growth method. Using traces such as shown in Figure
2.23, the growth rate at any period of the run (with the
exception of the burst period) can be established, as
well as the thickness at which the rate is changed and
at which the run is terminated. The continuous measurement of the film thickness also makes it possible
to grow to a desired thickness regardless of the rates
involved, which is particularly convenient since thickness can be held as a constant during the investigation
of the influence of the various growth rates on the film
semiconducting properties. The continuous monitoring of the growth rates and film thicknesses also gives
an indication of the lag times characteristic of the
various reactor geometries and provides information
on the reproducibility of growth rates from run to run,
as well as the constancy during any one growth run.
Using electronic closed-loop gas-flow monitors and

timers to control electrically actuated values, the growth rates and sequence of growth rates are very reproducible.

The trend toward high-rate low-temperature deposition conditions

Historically one finds that as the technology for the heteroepitaxial growth of silicon developed, there was a clear trend toward the use of lower deposition temperatures and higher growth rates. It is interesting to ask why the low-temperature/high-rate conditions were not established during early development of the conditions for the deposition of heteroepitaxial silicon. The following comments in this regard may contribute to an appreciation of the chronology of the development.

1. Early investigations were carried out on relatively thick (>3 μm) deposits; as device technology advanced, the demand arose for the thinner films, to the point where MOS transistor structures are now being fabricated in films as thin as 0.5 μm. The semiconducting properties of 0.5-μm-thick films are considerably more sensitive to deposition parameters than are the properties of the >3 μm deposits.

2. The after-oxidation (i.e., after device processing) semiconducting properties are considerably more sensitive to the deposition parameters than are the as-deposited properties. (This is strikingly shown in Table 2.10.) The chemists suffered some very frustrating times in delivery of what appeared to be films with excellent semiconducting properties to be processed into devices, only to find that the properties were seriously degraded during processing. Until the autodoping was suppressed to the point where little change was observed during device processing, a standard technique was adopted of evaluating the electrical properties both before and after exposure of the films to dry oxygen at 1100°C for 1 hr. (In early work, 1200°C was employed, but as the maximum device-processing temperature was decreased, so was the temperature of the standard test.)

3. The technology for the construction of gas-tight systems improved, and the purity of the commercially available source gases also improved (in particular, the silane–hydrogen and dopant gas–hydrogen mixtures). As has been previously mentioned, below about 1100°C the surface of the silicon cannot readily be freed of the oxide by reaction between silicon and SiO_2 to form volatile SiO, and therefore oxidizing impurities must be rigorously excluded from the deposition system.

4. The deposition temperature at which the optimum electrical properties are realized appears to be related to the amount of work damage present on the substrate surface. As the methods improved to remove the work damage from the substrate surface, it was possible to employ lower growth temperatures.

5. With the introduction of the deep-depletion MOS transistor fabricated in silicon on sapphire, the need arose for the preparation of thin films (<1.0 μm) doped to carrier levels between 1 and 5×10^{15}/cm³. This provided the need to further suppress the autodoping below that which had been achieved to date and thus led to the effort to further decrease the growth temperatures and increase the growth rates.

We have recently noted that acceptable heteroepitaxial deposits may be achieved at lower deposition rates than had previously been thought possible. On work-damage-free substrates, and in a clean gas ambient, at 1000°C films with acceptable semiconducting properties may be deposited at rates as low as 0.5 μm/min (Cullen, 1977). The range of acceptable temperatures has also increased, but to a lesser extent ($\sim \pm 15$°C). The conclusion reached is that the deposition rate, temperature, and surface finish are highly interactive. The presence of work damage in the substrate surface may be compensated for, to some degree, by deposition rate and temperature. At the higher growth temperatures it is possible that the substrate subsurface work damage is removed by reaction between the silicon and the substrate (Cullen and Dougherty, 1972). But at higher temperatures, the higher growth rates must be employed. If work-damage-free substrates are employed, the silicon deposit properties become less sensitive to rate within the low range of deposition temperature. Conversely, at the higher growth rates, some level of substrate work damage appears to be tolerable. These conclusions are based primarily on device mobility measurements. The effects of the preparative parameters on the more subtle device properties, such as channel leakage, have not yet been unambiguously established. Some useful clues on the rather complex interaction between the substrate surface condition, and the deposition temperature and rate come from examination of the properties of heteroepitaxial growth of silicon by vacuum evaporation. Under vacuum conditions, heteroepitaxy is realized over wider ranges of temperature and rate than has been the case in the chemical vapor deposition of silicon. The vacuum heteroepitaxial technology is therefore discussed in some detail in Section 2.9. Since the promise of the use of chemical vapor deposition temperatures significantly less than 1000°C in a helium ambient has not been realized (at least with sapphire substrates), lower growth temperatures appear to be achievable only at deposition pressures less than 1 atmosphere. Less than atmospheric pressures have been used to advantage in the chemical vapor deposition of homoepitaxial silicon. The low pressure has been employed both to modify the thickness of the boundary layer and to deposit at lower growth temperatures. This approach has been used to achieve better uniformity of the deposit (Tanikawa et al., 1973) and to suppress auto-doping (Deines and Spiro, 1974; Lucarini et al., 1974).

While the general trend has been toward the use of low temperatures, some investigators have adopted the approach that the autodoping can be sufficiently removed or deactivated by long-duration exposure to oxygen and inert gases at temperatures of approximately 1100°C. This procedure has been used in conjunction with relatively high growth temperatures for the optimization of electrical properties in deposit/substrate orientations other than (100)silicon/(1̄102)sapphire (Ruth et al., 1973; Manasevit et al., 1976)

Factors that influence growth rate

The deposition rate of silicon from silane is a function of the concentration of silane in the carrier gas (hydrogen or helium), the substrate temperature and temperature gradient normal to the substrate surface, the linear gas flow geometry, and the geometry of the gas passage over the substrate. The latter three factors are intimately related. Unfortunately, the deposition rate is sufficiently sensitive to the reactor geometry that it is difficult to compare the conditions employed by the various investigators in the field, since prior to the use of commercially available equipment the reactor configuration was seldom specified. This is further confused because of the custom in previous literature of specifying the total gas-flow rate rather than the linear gas-flow velocity across the substrate surface. The reactor geometries commonly employed for the deposition of heteroepitaxial silicon are discussed in Section 2.10.

The deposition rate of silicon from silane (for one silane concentration) as a function of temperature is shown in Figure 2.24. A horizontal water-cooled reactor was employed. The values for the growth rate are specific to the apparatus and the flow conditions employed and have little absolute significance unless these factors are described in detail. The issue of importance here, however, is the dependence of rate on temperature in the various temperature regimes. Other investigators have observed similar growth rate vs. temperature behavior for temperatures below 1100°C. It is generally accepted that for temperatures below about 900°C the deposition is surface reaction rate controlled with an activation energy of ~1.6 eV, and for temperatures between ~950 and 1100°C the deposition is mass-transfer controlled (Joyce and Bradley, 1963; Mayer and Shea, 1964; Kamins and Cass, 1973; Horiuche, 1975). Joyce and Bradley (1963), who deposited the silicon at reduced pressures, proposed that the decrease in the deposition

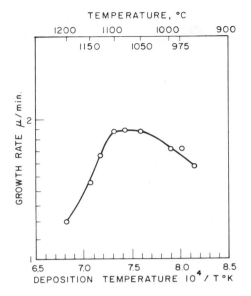

FIGURE 2.24   Deposition rate as a function of deposition temperature for silicon on sapphire. After Gottlieb (1971)

rate above ~1100°C is associated with the counter flow of hydrogen released by the pyrolysis of silane. The other possibility is that the deposition rate decreases rapidly above 1100°C due to depletion of the silane before it reaches the substrate surface as a result of homogeneous gas phase decomposition. The latter hypothesis is supported by the observation of the formation of solid material suspended in the gas stream (see Figure 2.28) and the sensitivity of the deposition rate above 1100°C to a particular reactor geometry (i.e., the thermal gradient normal to the substrate surface). When the system is operated in the mass-transfer control regime the thickness of the deposit is relatively insensitive to small differences in the temperature of the susceptor surface. This eases the control of the uniformity of the thickness over one substrate wafer and from wafer to wafer over a large susceptor. From Figure 2.24 it can be seen that the deposition rate changes only about 10 percent between 1000 and 1100°C. On the other hand, small differences in the supply of the reactant to various positions in the reactor system lead to variations in the deposit thickness, and consideration must be given to the gas-flow dynamics of the reactor system in an effort to achieve the desired thickness uniformity (see Chapter 7).

At the deposition temperatures of interest (i.e., in the mass-transport control region) the deposition rate is readily adjusted by control of the silane concentration introduced into the reactor. A typical rate vs. silane concentration for the deposition of silicon (on spinel) at 1100°C in a water-cooled horizontal reactor

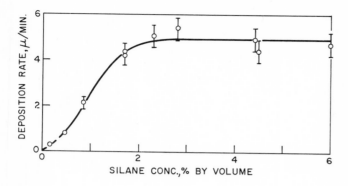

FIGURE 2.25
Silicon deposition rate at 1100°C as a function of the silane concentration of the gas stream. After Cullen et al. (1970A)

is given in Figure 2.25. The leveling off of the rate at about 5 μm/min is assigned to the onset of homogeneous gas-phase decomposition of the silane in the turbulent region of the gas passage. This conclusion is supported by the observation of some deposit of reaction products (not silicon) on the cold reactor walls at the higher silane concentrations and by the fact that a slight increase in the temperature results in a decrease in the maximum deposition rate. It is also noted that the deposition rate levels off at a value of ~6 μm/min if the total gas throughput is increased from 5000 to 7000 cm^3/min.

In the small-passage flow-through–type reactor systems there is no difficulty in achieving growth rates desired for the deposition of the thin heteroepitaxial deposits. In the "bell jar" reactor system the input silane is strongly diluted by the large volume of recirculating gases, and some effort in the adjustment of the input system may be required to achieve the desired growth rate (see Section 2.10).

2.9. Vacuum growth of silicon on sapphire and spinel

The choice between open-flow CVD and vacuum deposition

The commercial growth of both homo- and heteroepitaxial silicon is at present carried out exclusively in open-flow atmospheric-pressure "chemical vapor deposition" systems. Probably the main reason why vacuum deposition has not received wider acceptance is that very clean vacuum conditions are required to deposit silicon as free of contaminants as is achievable in open-flow systems. This may be attributed both to the purity of available source (SiH$_4$) and carrier (H$_2$) gases used in the CVD process and to the fact that at the higher deposition temperatures commonly employed in CVD many of the undesirable impurities are volatile and therefore present at the growth interface in low concentrations. Recently, more attention

has been directed toward the exclusion of specific impurities in "moderate" (10^{-8}–10^{-7} torr) vacuums [for instance, Weisberg and Miller (1968)], but in general "clean" conditions are related to "ultrahigh" ($<10^{-8}$ torr) vacuum conditions. The costs, complexity, and time required for vacuum evaporation increase rapidly with decreasing pressure below a level of ~10^{-7} torr. A number of investigators have argued that, under certain deposition conditions, low-pressure vacuum systems are not required for the heteroepitaxial growth of silicon. For instance Stadnik and coworkers (1972) deposited thick (>5 μm) silicon with useful semiconducting properties at pressures as high as 5×10^{-4} torr. The thinnest films (~2 μm) that exhibit (as-deposited) carrier mobilities similar to bulk silicon have, however, been deposited at a pressure of ~10^{-10} torr (Chang, 1971).

One finds a good deal of disagreement in the vacuum heteroepitaxial growth literature on the effect of the various deposition variables on the semiconducting properties of the silicon. This can be attributed, at least in part, to an unfortunate emphasis on the preparation of films ≥2 μm in thickness and on the measurement of the electrical properties only in the as-deposited condition. In the development of CVD heteroepitaxial silicon, it has been observed that the electrical properties of the thicker (>2 μm) as-deposited films are relatively insensitive to the deposition conditions. By contrast, the properties of deposits ≤1 μm in thickness that have been oxidized during device processing are a strong function of the deposition conditions (see Section 2.4). There is every reason to believe that the same effect is operative in vacuum-deposited silicon.

In one of the earlier investigations of the vacuum deposition of heteroepitaxial silicon, Fraimbault and coworkers (1965) employed a 1000-Å-thick layer of evaporated silicon to minimize the reaction with the sapphire during the chemical vapor deposition of an overlayer using a halide silicon source. Unfortunately, the thin-vacuum-deposited silicon was not individually characterized. Although this approach was not pursued using evaporated silicon as an intermediate layer, Druminski and Schlötterer (1972) covered the substrate with a thin layer of silicon deposited from silane prior to deposition of silicon from a chloride source.

Because of the emphasis on the preparation and characterization of the thick unprocessed silicon, one cannot glean from the available literature vacuum deposition conditions appropriate for the preparation of silicon films useful in MOS transistor applications. The research, nonetheless, provides useful and interesting information that contributes significantly to an understanding of the silicon/insulator composite system. In view of the results that have been achieved, one cannot exclude the possibility that further development of the vacuum deposition technology will lead to the preparation of heteroepitaxial silicon useful in device applications within the thickness range of current interest (0.5–1.0 μm).

The use of vacuum conditions for the heteroepitaxial growth of silicon offers several inherent advantages, particularly in relation to developing an understanding of the basic processes involved in the heteroepitaxial growth of silicon. The crystal structure of the substrate surface can be studied by methods such as low-energy electron diffraction, and the presence of trace constituents can be examined with Auger spectroscopy. Deposition can be carried out in the same system without exposure of surface to contaminants after characterization of the surface. These techniques have been employed by Chang (1971) to develop some of the most definitive information available on the preparation of the substrate surface (removal of work damage), the specific crystallographic nature of the substrate prior to deposition, and the effect of substrate orientation on the nature of the deposited silicon. It should also be noted that in clean vacuum deposition conditions the only constituents present are the substrate and the silicon, with no involvement of carrier gases, by-products of the pyrolysis of silane, or impurities present in the source and carrier gases.

The influence of vacuum deposition condition on the properties of the heteroepitaxial silicon

The variables involved in the vacuum deposition of silicon are the deposition rate, temperature, and pressure; the type of pumping system as related to the presence or absence of hydrocarbons; the purity of the source silicon; and the manner in which the substrate surface is prepared (if done under vacuum conditions). Common to vacuum and chemical vapor deposition are the substrate-related variables; the manner in which the substrate surface is prepared (if done outside the vacuum chamber); and the chemical nature and crystallographic orientation of the substrate. The

conditions employed for vacuum deposition, and the electrical properties realized by a number of investigators, are briefly summarized in Table 2.5.

Pressure, rate, and temperature. The pressure, rate, and temperature of vacuum deposition are sufficiently interdependent that it is impractical to discuss the influence of these variables separately. In the following discussion, an emphasis is put on semiconducting properties as a measure of the film quality both because of the sensitivity of the semiconducting properties to deposition conditions and because the orientation here is toward application in electronic device structures.

The rate of deposition is more limited in a vacuum than at atmospheric pressure by the mechanics of exposing the substrate to the source silicon. Sublimation rates are inherently limited by holding the silicon at temperatures below the melting point. In the evaporation of silicon, there are problems involved in the containment of silicon at temperatures above the melting point. Crucibles are universally avoided because, in the presence of the highly reactive molten silicon (chemically reducing), they are a major source of contamination. At low pressures, the rate of evaporation is limited by the rate of removal of heat (Salama and coworkers, 1967), and thus in general the evaporated rate available is proportional to the deposition pressure. Yasuda and Ohmura (1969) state that they were able to maintain an evaporation rate of 1500 Å/min (at a relatively high pressure of 2×10^{-6} torr) with an electron bombardment heater specifically designed to achieve high evaporation rates. It is the opinion of this author that the very much higher rates, particularly at the lower pressures, reported by several authors (see Table 2.5) are open to some question.

Low deposition rates are associated with low pressures and low temperatures. At the lower growth temperatures, the low rates are permissible because of the suppression of the reaction between silicon and the substrate and desirable because of the relatively low mobility of silicon on the substrate surface. Using low growth rates, the low pressures are needed to exclude from the growth surface contaminants which are volatile at the higher growth temperatures. The most significant aspect of low-pressure heteroepitaxial growth is the possibility of minimizing the silicon/substrate reactions by the use of low deposition temperatures.

Chang (1971), using ultrahigh (10^{-10} torr) vacuum conditions and low deposition rates (~60 Å/min) realized single crystal growth at unusually low temperatures on carefully characterized vacuum-annealed substrates. The temperature ranges within which single crystal growth was observed, for the various substrate orientations, are shown in Figure 2.26. It can be

TABLE 2.5　A Summary of the Conditions Employed and the Electrical Properties Realized in Vacuum-Deposited Heteroepitaxial Silicon by a Number of Investigators

Literature reference	Type of Vacuum System; Background Pressure (torr)	Pressure During Deposition (torr)	Rate of Deposition (Å/min)	Substrate Temperature (C)	Substrate Surface Conditioning
F. Gassmann et al. (1972)	"Ultrahigh"	5×10^{-10}	3600–18,000	900	Pyrophosphoric acid at 400°C
A. V. Stadnik et al. (1972)	Oil diffusion 10^{-7} Ion getter 5×10^{-8}	5×10^{-4}–5×10^{-8}	1000	850–1350 (deposits as temperature decreasing)	Si etched 1350°C
Chuan C. Chang (1971)	"Ultrahigh" 2–5×10^{-10}	$\sim 10^{-10}$	Capability: 80 Used: ~60	500–1000	Vacuum anneal 1400°C
F. Gassmann et al. (1971)	"Ultrahigh"	5×10^{-10}	3600–18,000	800–1000 (900 optimum)	Pyrophosphoric ($H_4P_2O_7$) at 400°C
Yukio Yasuda (1971)	Oil diffusion	2×10^{-6}	~25	900–1000	Sapphire: H_2 anneal 1300°C Spinel: H_2SO_4: HNO_3 at 180°C
R. W. Lawson and D. M. Jefkins (1970)	"Ultrahigh" and oil diffusion	5×10^{-8}–5×10^{-7}	1000–30,000	930–1150	Diamond polished and H_2 annealed at 1500°C
T. Itoh et al. (1969)	3×10^{-8}	5×10^{-8}–5×10^{-7}	120	850–1000 (950 optimum)	Vacuum annealed 1000°C for 1 hr
Y. Yasuda and Y. Ohmura (1969)	Oil diffusion	2×10^{-6}	1500	1050	3:1 H^2SO_4: HNO_3 at 100°C or H_2 anneal 1300°C for 30 min
C. T. Naber and J. E. O'Neal (1968)	"Ultrahigh"	10^{-8}–10^{-9}	50–700	500–1000	Vacuum anneal 1200°C H_2 anneal 1350°C or Si etch 1200–1300°C (Si etch 1300°C best)
L. R. Weisberg and E. A. Miller (1968)	"Ultrahigh" 10^{-9}	5×10^{-8}	~1000	780–1075 (1000 optimum)	H_2 anneal 1200°C
F. H. Reynolds and A. B. M. Elliott (1967)	"Ultrahigh"	10^{-7}–10^{-6}	140–250	880–975	Mechanically polished
C. A. T. Salama et al. (1967)	Oil diffusion 2–3×10^{-8}	7×10^{-7}	Up to 400	850–1100 (poly at 850; reevaporates at 1100)	Mechanically polished
S. Namba et al. (1967A,B)	"Ultrahigh"	5×10^{-8}	600	800	
S. J. Strepkoff (1966)	"Ultrahigh" 4×10^{-9}	2×10^{-8}	10,000	850–1300	Phosphoric acid etched
J. L. Fraimbault et al. (1965)	"Ultrahigh" 6×10^{-10}	5×10^{-8}	100	1060	

Substrate Orientation and Type	Film Thickness (μm)	Carrier Mobility (μ in cm²/V-s) (n in I/cm³)
Spinel (111)	2–4	μ_a = 350–380 at n_a = 1–2 × 10¹⁶ (t = 2–4 μm) no change in thermal oxidation: 1 hr at 1100°C oxidation improves low temperature μ
Sapphire (1010) (1$\bar{1}$02)	2.5–8	μ_a = 250 (t = 5.4 μm) μ_d = 525 (t = 6.0 μm) (both n ~5 × 10¹⁷ from Si source)
Sapphire (1$\bar{1}$02) (0001) (11$\bar{2}$3) (1120)	2–20	Above 2 μm in thickness films deposited under optimum condition μ similar to bulk to n as low as 5 × 10¹⁵
Spinel 3:1 flame fusion 1:1 Czochralski (111)	2.1–5	μ = 60–100 percent of bulk on flame fusion spinel μ = 100 percent of bulk on Czochralski spinel Si source B doped; deposit n = 1–2 × 10¹⁶
Sapphire (1$\bar{1}$02) (11$\bar{2}$0) Spinel (111)(100) (113)	Examined discontinuous and complete films	No electrical properties reported
Sapphire, Czochralski, and flame fusion (1$\bar{1}$02) and (0001)	5–12	μ_a = 233 at n_a = 4 × 10¹⁶ (t = 7 μm) on (0001) Al_2O_3 μ_a = 140 at n_a = 6 × 10¹⁶ on (1$\bar{1}$02) Al_2O_3
Spinel (111)	8	μ_d > 800 (75 percent bulk) for n_d = 1.3 × 10¹⁶ to 2.1 × 10¹⁸ μ_d = 100 percent bulk for n_d > 3 × 10¹⁷ (evaporated Sb simultaneously) best μ_a = 353 cm² at n_a = 2.5 × 10¹⁶
Sapphire (0001)	0.7–6.1	μ_a ~ 5 cm² at n_a = 3 × 10⁷ (t = 0.7 μm) μ_a = 250 cm²/V-s at n_a = 4 × 10¹⁶ (t = 6.1 μm) (not intentionally doped)
Sapphire (0001) (1$\bar{1}$02)	0.8–1.6	No electrical properties reported
Sapphire (0001) (1$\bar{1}$02)	4–20	μ_a = 290 at n_a = 5 × 10¹⁶ (t = 8 μm)
Sapphire (0001)	7.4–9	μ_a = 154 at n_a = 5.2 × 10¹⁵ (t = 7.4 μm)
Sapphire (0001) (1$\bar{1}$02)	0.35–1.7	μ_a ~ 10–30 for films as thin as 0.35 μm
Sapphire (0001)		μ_d = 600
Sapphire (0001)	0.6–40	μ_a = 200 μ_d = 425
Sapphire (0001)	0.1 by vacuum deposition, then the thickness increased by CVD	μ_a = 150

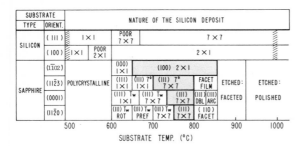

FIGURE 2.26   The crystalline nature of vacuum-deposited silicon on sapphire and silicon as a function of the substrate temperature and orientation. The deposition rate was 1 Å/s on a substrate with an equilibrium structure as determined by LEED. The abbreviations used are as follows:

$7°$: $(111)\|(11\bar{2}4)$
T_W: Primary (111) twin
Dbl: Double domains rotated $\pm 9°$
Arc: Diffraction spots appear as arcs
Rot: All rotations about the surface normal
Pref: Several preferred orientations

After Chang (1971)

seen that the temperature range for single crystal growth is a function of orientation. The largest temperature range was observed on the $(1\bar{1}02)$ sapphire. Below a substrate temperature of 600°C all films were polycrystalline, and between 800 and 925°C the substrates were etched, leaving a polished surface. Between 650 and 850°C, the deposits were single crystal, twinned, or faceted, depending on the substrate orientation. At a temperature of 775°C, single crystal films were observed on the $(1\bar{1}02)$, $(11\bar{2}3)$, (0001) and $(11\bar{2}0)$ substrates. It is interesting to note that (under the conditions employed by Chang) the lowest temperature for the homoepitaxial growth of silicon is only 50°C lower than the minimum temperature for the growth of heteroepitaxial silicon [both (100)] on the $(1\bar{1}02)$ sapphire.* In deposits 2 to 20 μm in thickness, the hole and electron mobilities in the vacuum deposited silicon were similar to bulk (to carrier concentrations as low as $\sim 5 \times 10^{15}$/cm³).

On the opposite extreme of the temperature–rate–pressure scale is the work of Stadnik and coworkers (1972). These investigators deposited silicon at a rate of ~ 1000 Å/min on the substrate as it cooled from 1350°C at pressures in excess of 1×10^{-5} torr. In the presence of silicon vapor, the conditions were shifted

---

*It is recognized that Jona (1966) observed homoepitaxial growth of (111) silicon at 400°C, and the growth of an "ideal" (no superlattice spots) structure in (100) silicon deposited presumably at considerably lower temperatures (the substrate was heated only by the evaporated silicon in a LEED system). The films examined were, however, 4 to 5 monolayers thick deposited at a rate of 0.3 monolayer/min. The impact of these interesting results on practical silicon epitaxy is still to be developed.

from etching the substrate surface with silicon to deposition of silicon on the surface as the temperature decreased. An interesting feature of this work is the insensitivity of the properties of the deposited silicon to the presence of contaminants. Stadnik and his colleagues claimed that if this "high-temperature nucleation" procedure were employed, the films could be deposited without degradation of the crystallographic properties even in the presence of 1×10^{-5} torr of oxygen and in the presence of hydrocarbons characteristic of oil diffusion pump systems. The crystallographic nature of the films was derived from x-ray rocking curve data. Stadnik et al. point out that the background defect density of the heteroepitaxial silicon was $\sim 1 \times 10^9$/cm² and that the presence of oxygen at a level resulting in a defect density of $\sim 1 \times 10^7$/cm² in homoepitaxial silicon does not have an observable effect against the background of the highly defected heteroepitaxial silicon. Although silicon as thin as 2.5 μm was deposited, the semiconducting properties were reported only for films thicker than ~ 5 μm. Hole mobilities of 250 cm²/V-s and electron mobilities of 525 cm²/V-s were realized at high carrier concentrations ($>1 \times 10^{17}$/cm³).

In the work of Stadnik and his coworkers it can only be stated that the silicon was deposited at some temperature between 850 and 1350°C. A high constant deposition temperature of 1050°C and a relatively high deposition rate of 1500 Å/min were employed by Yasuda and Ohmura (1969). As with the work of Stadnik et al. the carrier concentrations in the deposit were high, and similar hole mobilities at similar thicknesses were observed (250 cm²/V-s at 6 μm). An oil diffusion pump system was employed, with background pressures of $\sim 2 \times 10^{-6}$ torr.

It is clear that high deposition rates must be employed at the high substrate temperatures. Salama and coworkers (1967) reported that they were not able to achieve deposition of silicon at 1100°C with a deposition rate of 400 Å/min (with an oil diffusion system background pressure of 7×10^{-7} torr). Under their vacuum conditions, at 850°C the properties of the deposits were dominated by defects, at 950°C the carrier concentration could be controlled by the doping level of the source, and at 1050°C all the films were p-type as the result of autodoping. It is interesting to note that Salama and his coworkers found it difficult to explain the observation that the mobility increased with decreasing deposition temperatures (within the single crystal range) but that the crystalline quality (derived from optical absorption) increased with increasing temperature. The implication here is that autodoping dominates the carrier mobility more than the crystalline quality above some minimum level of crystalline quality. This is analogous to compensation effects observed in bulk silicon. This effect, however, has not been reported by other investigators.

Reynolds and Elliot (1967) found that at every temperature investigated (within the range 880 to 975°C) there was a rate at which the conditions shifted from deposition of silicon to etching of the substrate with silicon vapor. This rate was, for instance, 120 Å/min at 950°C (with a background pressure from a sorption/vacuum-ion system of 10^{-7}–10^{-6} torr). As did Chang (1971), these investigators found that the temperature range for single crystal growth on the (0001) sapphire was less than 100°C, and that even within this range, the films were faceted at the higher temperatures. The hole mobilities in films deposited at temperatures near the etching condition (975°C) were very low and increased to ~150 cm²/V-s (in 7.4-μm-thick films, n_a ~5 × 10^{15}/cm³) at a deposition temperature of 880°C. The achievement of the optimum mobilities at the low-temperature end of the single crystal growth temperature range is in agreement with the work of Salama and his coworkers (1967) but appears to be in disagreement with the work of Stadnik and coworkers (1972). A possible explanation is that under "clean" vacuum conditions the optimum semiconducting properties are realized at the lower end of the single crystal–growth temperature range, while in the presence of contaminants (such as hydrocarbons and oxygen) the upper end of the temperature range is favored.

Lawson and Jefkins (1970), in a manner similar to the work of Reynolds and Elliot (1967), also investigated the etching/deposition boundaries. The boundaries were found to be a function of the substrate orientation. The optimum deposition regimes were displaced from the etching/deposition boundaries to higher rates and lower temperatures. These investigators proposed that the crystalline quality of the silicon decreased as the etching boundary was approached because random nucleation resulted from deposition on a silicon-etched surface. In silicon deposits >5 μm in thickness, hole mobilities as high as 75 percent of bulk were observed in (111) silicon on (0001) sapphire, while hole mobilities of 45 percent of bulk were observed in (100) silicon on (1$\bar{1}$02) sapphire (n_a = 4–6 × 10^{16}/cm³). The electrical properties in the (100) silicon proved to be considerably more reproducible than in the (111) deposits. These films were typically deposited at a rate of 1 μm/min and at pressures between 5 × 10^{-7} and 5 × 10^{-8} torr.

It is interesting to note that several experimentalists who used a range of deposition pressures have noted no improvement in the film properties if the lower pressure of their range was employed. The most dramatic claim is that of Stadnik and coworkers (1972), who stated that no difference was observed (using the "high-temperature nucleation conditions") at pressures between 5 × 10^{-8} and 5 × 10^{-4} torr. Weisberg and Miller (1968) found no difference in film properties with deposition pressures of 10^{-9} and 10^{-8} torr if oxygen-free silicon was used as the source. Yasuda (1971), who employed crystallinity as a measure of the quality of the deposited silicon, reported no difference in the film properties between pressures of 1 × 10^{-8} and 2 × 10^{-6} torr, and Lawson and Jefkins report that films deposited at a pressure of 2 × 10^{-5} torr were indistinguishable from those produced at 5 × 10^{-7} torr.

Cleanliness of the vacuum atmosphere. In the vacuum heteroepitaxial growth work, oil-free vacuum systems have, for the most part, been employed. A number of investigators, however, used oil diffusion pump systems. Stadnik and coworkers, who employed both vac-ion and oil diffusion pump systems, state that the formation of SiC always occurred (at temperatures between 800 and 1200°C) in the oil diffusion–type system unless the "high-temperature nucleation" procedure (already discussed) was employed. Even if this technique was used, at low deposition rates (50 Å/min), the silicon deposits contained inclusions of polycrystalline silicon carbide. They stated that the influence of the hydrocarbon background cannot be avoided by baking and trapping.

Yasuda and Ohmura (1969), Yasuda (1971), and Salama et al. (1967) also employed oil diffusion–type vacuum systems (at pressures between 7 × 10^{-7} and 2 × 10^{-6} torr). As has already been discussed, the deposition rates and temperatures used by these two groups are among the highest reported. It appears that the high-temperature high–growth rate conditions make the properties of the deposit as insensitive as is possible to the presence of hydrocarbon contaminants. It is not clear, however, that oil diffusion pump systems can be tolerated at all in the heteroepitaxial growth of silicon. Both of the groups of investigators who employed the oil pump systems reported carrier mobilities in ~1-μm-thick films considerably lower than those observed in CVD films of similar thickness. Yasuda and Ohmura observed that the silicon adjacent to the substrate surface (within 800 Å) contained more than one orientation. Although it is tempting to relate this to the presence of silicon carbide, Yasuda later reported that only one orientation was observed in the growth of silicon on spinel in the same vacuum system. In silicon deposits thicker than 5 μm, Lawson and Jefkins (1970) did not observe a difference in the properties of silicon deposited in ion-pumped (5 × 10^{-7} to 5 × 10^{-8} torr) and oil-pumped (2 × 10^{-5} torr) systems.

Weisberg and Miller (1968) (who employed an oil-free vacuum system) emphasized the importance of the total exclusion of oxygen. They reported that the quality of heteroepitaxial silicon increased dramatically if oxygen-free silicon was used as the source. These investigators observed hole mobilities as high as 290 cm²/V-s in 8-μm-thick films deposited at 1000°C at a rate ~6 μm/hr. In light of the independence of the presence of oxygen reported by Stadnik and coworkers (1972), it should be noted that Weisberg and Miller deposited at temperatures below which silicon reduces silicon dioxide to a volatile intermediate-oxidation-state silicon oxide, while Stadnik et al. deposited at the higher temperatures at which this reaction can take place.

Lawson and Jefkins (1970) reported that no influence of the presence of oxygen was observed to oxygen partial pressures as high as 10^{-8} torr using a standard (i.e., not oxygen-free) silicon target. The temperature at which this study was carried out was not provided.

The influence of substrate surface conditioning on the properties of vacuum-deposited heteroepitaxial silicon

Chang (1971) employed low-energy electron diffraction to monitor the crystallinity of the substrate surface and Auger spectroscopy to monitor the constituents present on the surface. He observed equilibrium diffraction patterns from the surface after vacuum annealing at 1400°C for 30 s. After the 1400°C vacuum anneal, only Al and O Auger peaks were detectable, and he therefore concluded that the carbon concentration was less than 5×10^{13}/cm². No advantage was observed in annealing the substrate above 1400°C. Chang observed the onset of annealing at temperatures as low as 600°C. At this temperature, most contaminants other than carbon were removed. As the annealing temperature was increased from 600 to 1400°C, the equilibrium diffraction patterns became more dominant. At 900°C, however, a large carbon Auger peak was still observed, and even at 1200°C, small amounts of carbon remained on the surface. It is clear from this work that care must be taken to exclude carbon contamination of the surface even in an oil-free vacuum system. This study also brought out the point that the smoothness of a mechanically polished surface can be maintained while the subsurface work damage is removed by annealing. If the work damage is removed from the as-polished substrate by chemical etching (such as with phosphoric acid), the surface may be roughened by preferential etching of the more heavily damaged areas.

Naber and O'Neal (1968) investigated the effect of vacuum annealing at 1200°C, H₂ annealing at 1350°C, and etching of the substrate surface in silicon vapor at temperatures between 1200 and 1300°C. Based on the microscopic examination of the topology of the substrate surface and reflection electron diffraction examination of the deposited silicon, these investigators concluded that the optimum substrate surface preparation was etching with silicon at 1300°C with a silicon impingement rate less than 100 Å/min. Single crystal silicon was achieved on silicon-etched (1$\bar{1}$02) and

(0001) sapphire at deposition temperatures between 700 and 900°C and growth rates between 150 and 450 Å/min. In the "high-temperature nucleation" process used by Stadnik and coworkers (1972) it appears that the substrate surface is conditioned by etching with silicon.

Weisberg and Miller (1968) prepared the substrate surface by annealing in H₂ at 1200°C, and Yasuda and Ohmura (1969) and Yasuda (1971) annealed in H₂ at 1300°C. This procedure is similar to that which has been most commonly used to condition the substrate surface prior to CVD growth of silicon.

Gassmann and coworkers (1971, 1972) removed the work damage from flame fusion and Czochralski-grown spinel surfaces by etching in pyrophosphoric acid ($H_4P_2O_7$) at 400°C. Acid etching is not peculiar to the vacuum deposition technique and has been discussed in Section 2.4 of this chapter. The main point to be made here is that Gassmann and coworkers cited the crystallographic quality of the etched substrate surface as a key factor in achieving good semiconducting properties (in 2-μm-thick films) at a deposition temperature 200°C lower than the optimum CVD deposition temperature (900°C as compared to 1100°C). It must also be pointed out, however, that a contributing factor could have been the low deposition pressure (5×10^{-10} torr).

It is interesting to note that Lawson and Jefkins (1970), in their rather detailed study of the vacuum deposition of silicon on sapphire, concluded that the crystalline perfection of the substrate surface had little influence on the deposit quality. Most of the films studied, however, were between 5 and 12 μm thick. They did observe that the nature of the surface finishing influenced both the autodoping level and the reproducibility in the electrical properties, at least in the (100) silicon deposited on the (1$\bar{1}$02) sapphire. Unfortunately, it is difficult to draw conclusions on the influence of the deposition temperature from this study since the electrical results are expressed as a function of the ratio of the substrate temperature to the temperature of the etching–deposition boundary (for a particular deposition rate).

A number of the experimentalists involved in the vacuum heteroepitaxial growth of silicon made no effort to remove the subsurface work damage from the mechanically polished substrates. Yasuda and Ohmura (1969) were able to grow single crystal films on both H_2SO_4-HNO₃–cleaned surfaces (180°C) and H₂-annealed surfaces (1300°C), even though clear differences were seen in the crystallographic quality of these surfaces on examination with reflection electron diffraction. They concluded that the substrate surface quality did not play an important role in the growth of single crystals, but the semiconducting properties of silicon deposited on the substrate surfaces conditioned by the two methods was not provided.

Although a number of investigators active in the vacuum deposition of heteroepitaxial silicon have reported that the film properties are insensitive to substrate surface condition, investigators in the field of the chemical vapor deposition of heteroepitaxial silicon are in general agreement that the deposit properties are a strong function of the crystallographic condition of the substrate surface (see Sections 2.4 and 2.12). This apparent discrepancy can be explained by the observation that essentially all of the deposits studied in the vacuum deposition literature have been "as-deposited" films more than 2 μm in thickness, even in work carried out since 1970. During the same period, the emphasis in the CVD heteroepitaxial growth literature has been on the preparation and characterization of thermally oxidized deposits ≤ 1.5 μm in thickness which are used in device application. It is in the thin, oxidized deposits that the semiconducting properties are a strong function of the deposition conditions and the substrate crystallinity.

The influence of the substrate type and orientation on the properties of vacuum-deposited heteroepitaxial silicon

Chang (1971) found that the widest temperature range for single crystal growth was observed on the (1$\bar{1}$02) sapphire, and that therefore the conditions for single crystal growth are most easily satisfied on this orientation (see Figure 2.26). This has been the orientation most commonly employed in the CVD of silicon. At a temperature of 775°C, Chang realized single crystal growth on all of the four orientations investigated. Under the optimum deposition conditions, the semiconducting properties of the silicon (for thickness between 2 and 20 μm) on the four orientations investigated were similar. Weisberg and Miller (1968) found that the mobilities in the (111) Si on (0001) sapphire were higher than the mobilities observed in the (100) Si on (1$\bar{1}$02) sapphire (for film thicknesses between 4 and 20 μm).

Itoh and coworkers (1969) realized hole mobilities of 353 cm²/V-s and electron mobilities of 800 cm²/V-s in relatively thick (8 μm) (111) silicon on (111) spinel (at carrier concentrations ~2 × 10¹⁶/cm³). The deposition was initiated at a temperature of 950°C, and the temperature was allowed to drop to 900°C during the deposition. A rate of 120 Å/min was employed at a pressure between 5 × 10⁻⁸ and 5 × 10⁻⁷ torr.

Gassmann and coworkers (1971) found that the mobilities of silicon deposited on the Czochralski-grown (111) spinel were typically 100 percent of bulk silicon mobility (in films between 2 and 5 μm in thickness) but that on the flame fusion grown spinel the mobilities were less reproducible and fell between 60 and 100 percent of bulk. Although it has commonly been assumed that the grain boundaries characteristic

of the flame fusion substrate materials would lead to the degradation of the electrical properties of the heteroepitaxial silicon, this is the only report of the realization of higher mobilities (measured by the Hall method) on the Czochralski-grown spinel. Cullen and Wang (1971) had found that Czochralzki-grown spinel was less stable in the presence of the hydrogen carrier gas used in the CVD of silicon on spinel and that the electrical properties of the silicon were degraded by the incorporation of substrate reaction products. Under these circumstances, it is not surprising that better films have been achieved by avoiding the presence of hydrogen through the use of either vacuum conditions or helium as the carrier gas in atmospheric pressure CVD (Cullen and Dougherty, 1972). In the analysis of the differences between silicon deposited on the spinels grown by the two methods it must be kept in mind that the chemical compositions of the two spinels are different (see Section 2.3).

The semiconducting properties of vacuum-deposited heteroepitaxial silicon as a function of thickness and thermal oxidation

There is little information to be found in the vacuum silicon heteroepitaxial growth literature on the semiconducting properties of thin silicon films (≤ 2 μm in thickness) as a function of the film thickness or thermal oxidation. Yasuda and Ohmura (1969) measured the hole mobility and hole concentration (of unintentionally doped films) as a function of thickness. The films were deposited with a substrate temperature of 1050°C at a rate of 1500 Å/min in an oil diffusion vacuum system at a pressure of 2 × 10⁻⁶ torr. These investigators employed both "cleaning" and H₂ annealing for surface preparation but did not specify which method was used to prepare the substrates used for examination of the semiconducting properties. At a thickness of 6.1 μm, Hall hole mobilities of ~250 cm²/V-s were observed in (111) silicon on (0001) sapphire. The mobility decreased linearly with decreasing thickness to a value of ~150 cm²/V-s at ~1.5 μm. Below ~1.5 μm the mobility decreased rapidly to a value of ~10 cm²/V-s at 0.7 μm. The hole concentration decreased from a value of 3 × 10¹⁷/cm³ at 0.7 μm to ~4 × 10¹⁶/cm³ at 6.1 μm. From these results it appears that even at the rate of 1500 Å/min, which is relatively high for vacuum evaporation, at a substrate temperature of 1050°C films less than 1 μm in thickness are heavily autodoped.

Salama and coworkers (1967) reported hole mobilities between 10 and 30 cm²/V-s (for hole concentrations <1 × 10¹⁶/cm³) in films between 0.35 and 1.7 μm in thickness. The silicon was deposited at a rate of 400 Å/min with a substrate temperature of 850°C in an oil diffusion vacuum system at 7 × 10⁻⁷ torr. The mobilities are surprisingly independent of thickness within the range studied. The substrates used in this study were mechanically polished. Therefore, the low hole mobilities, as well as the insensitivity of the mobility to film thickness, may have been the result of poor film crystallinity associated with a work-damaged substrate surface. The relatively low autodoping observed can be associated with the relatively low substrate temperature.

Gassmann and coworkers (1972) have observed hole mobilities of 350–380 cm²/V-s in 2–4-μm-thick (111) boron-doped (n_a = 1–2 $\times$ 10¹⁶/cm³) silicon deposited (in a vacuum of 5 $\times$ 10⁻¹⁰ torr) on phosphoric acid–etched spinel. It would appear that the suppression of the autodoping and stability of the electrical properties during the thermal processing observed in this study were the result of the use of a substrate temperature (900°C) lower than the optimum CVD deposition temperatures.

An overview of the vacuum heteroepitaxial growth of silicon

The most significant feature of the vacuum growth of heteroepitaxial silicon is the realization of single crystal films at lower substrate temperatures than has been possible to date in the heteroepitaxial growth of silicon by CVD. At the higher growth temperatures (greater than ~950°C), it is not clear that there is any advantage to be gained by vacuum evaporation, and there is the disadvantage that the rate at which silicon can be supplied by evaporation is less than rates of supply ordinarily obtainable by the pyrolysis of silane. At vacuum deposition temperatures below 950°C, there is the clear advantage that the reaction between silicon and the substrate is suppressed. At intermediate temperatures (850–950°C), the films are less autodoped, and in the lowest temperature range [650–850°C for (1ĪO2) sapphire (Chang, 1971)], there is an added advantage that the erosion of the substrate is sufficiently low so that the crystallographic surface of annealed substrates is maintained during the early growth of the silicon. The lower substrate temperatures are associated with low deposition rates; as the surface mobility of the silicon decreases with temperature, the growth rate must also be lowered to maintain the growth of single crystal silicon. At the lower deposition temperatures, low pressures are required to avoid the condensation of contaminants which would interfere with the positioning of the silicon ad-atoms on the proper lattice sites.

There are a number of analogies that can be made between the progress made in heteroepitaxial silicon growth by CVD and by vacuum evaporation. Early in the development of the CVD technology, deposition temperatures in excess of 1100°C were considered to be optimum (see Section 2.8). As the technology developed, contaminants were excluded by building

vacuum-tight metering and flow systems (He leak tested at 10⁻⁵ torr), the purity of the commercially available silane improved, and methods were developed to reproducibly remove the substrate work damage. With these improvements, it was possible to lower the deposition temperature with no loss in crystalline quality but with improved properties as the result of the suppression of the reaction between silicon and the substrate.

A similar trend is observed in the development of conditions for vacuum heteroepitaxial silicon growth. In most of the early work little attention was given to the preparation of the substrate surface, the deposition temperatures were similar to those used in CVD, and the rates employed were as high as could be achieved at the relatively high pressures. Under these conditions, reaction between silicon and the substrate was a major cause of degradation of the electrical properties of the films. At the relatively high evaporation pressures and low deposition rates, the electrical properties of the thinner films were inferior to those achievable with CVD. As the vacuum deposition technology progressed, attention was given to the preparation of the substrate surface, "cleaner" (lower-pressure) vacuum conditions were employed, and it was found that lower temperatures could be used to advantage. The key difference between the course of the progress in the vacuum and CVD technology was that with "ultrahigh" vacuum equipment it was possible to deposit silicon with usable electrical properties at deposition temperatures significantly lower than obtainable by CVD. The limited growth rate available in the low-pressure systems, which is a disadvantage at the higher substrate temperatures, is a necessity at the low substrate temperatures. In the examination of the crystallographic and chemical nature of the surface by LEED and Auger methods, the vacuum conditions became an advantage rather than a liability, and important contributions were made to an understanding of the heteroepitaxial interface. An unanswered question at the current state of the technology is: At what pressure/rate combination can the low temperatures be maintained for the deposition of thin, thermally stable, silicon layers? For application in commercial heteroepitaxial growth systems this is a key question, since the expense and time involved in vacuum deposition are strongly related to the deposition pressure required.

It has been argued that the low-temperature limit of the CVD process is determined by the kinetics of the silicon source reaction. It has been observed, however, that at temperatures at which the rate of the pyrolysis of silane exceeds the rate of evaporation of silicon in clean vacuum systems [i.e., 900–950°C; see, for example, Gottlieb and Corboy (1972)], the properties of vacuum evaporated silicon are superior to the properties of CVD silicon. Extrapolation of the available data on the kinetics of the pyrolysis of silane (Joyce and Bradley, 1963) suggests that the crossover point of the available deposition rate of silicon by pyrolysis of silane and by evaporation is at pyrolysis temperatures well below 900°C. It can be concluded, therefore, that the low-temperature limit of the CVD of heteroepitaxial silicon is more defined by the presence of contaminants on the substrate surface and by the maintenance of the proper substrate surface crystal structure than by the kinetics of the pyrolysis of silane. Although use of the lowest vacuum heteroepitaxial growth temperature reported [650°C on the $(1\bar{1}02)$ silicon; Chang (1971)] would be precluded in CVD by the rate of the pyrolysis of silane, the achievement of silicon heteroepitaxy by CVD at temperatures well below 900°C is not outside the realm of possibility.

NOTE. Vacuum-deposited silicon has been discussed in some detail in the thesis work of P. F. Linnington (under the direction of A. Howie), Cambridge University (1971) and A. G. Cullis (under the direction of R. Booker), Oxford University (1972). The former investigation was carried out in collaboration with Lawson and Jefkins (1970). This investigation has been discussed in this section. The emphasis in both of the thesis investigations was on the microstructure of the heteroepitaxial silicon. These results are discussed in Section 2.11.

2.10. Deposition techniques

The development of deposition techniques

The techniques developed for the homoepitaxial growth of silicon from silicon tetrachloride are not directly applicable to the heteroepitaxial growth of silicon from silane. Due to the differences in the chemical nature of the two source materials, and the comparatively low deposition temperature of the heteroepitaxial growth, it has been necessary to develop methods specific to heteroepitaxial deposition on the insulating substrates. The different procedures are required mainly because of the tendency for silane to pyrolyze in the gas stream as well as on a hot surface and because oxygen must be rigorously excluded when deposition temperatures are used at which silicon dioxide is stable in contact with the silicon.

Also, because of the differences in the kinetics of the silane pyrolysis as compared to the silicon tetrachloride reduction, different reactor geometries are required to realize acceptable uniformity of the deposit thickness and carrier concentration. The effort to achieve uniformity in the deposits less than 1 μm in thickness doped to well under 10^{16}/cm^3 has resulted in a new appreciation of the role of gas-flow dynamics. The fundamentals of gas-flow dynamics as applied to

these deposition systems are discussed in some detail in Chapter 7. Aspects of the deposition techniques specific to the growth of the heteroepitaxial silicon are discussed in the following sections.

Types of deposition reactors

The four types of deposition chambers which have commonly been employed for the heteroepitaxial growth of silicon on the insulating substrates are schematically shown in Figure 2.27. In all of the systems shown the substrate is heated by contact with (and radiation from) a susceptor in which the heating currents are induced by an RF field. Reaction chambers heated externally by conventional resistance tube furnaces have also been employed, but particularly when silane is used as the source gas, it is desirable to suppress homogeneous gas phase reactions by holding the substrate (and susceptor) as the hottest part of the system. The simple pedestal-type reactor (Figure 2.27a) has been, and still is, used for experimental runs when it is not necessary to prepare more than one sample at a time and when the maintenance of thickness uniformity across the sample is not critical.

In small ($< \sim$1.5 inch-diameter) pedestal reactors, and in the horizontal-type (Figure 2.27b) and barrel-type (Figure 2.27c) reactors, the gas makes essentially one pass over the growth surface, and recirculation over the growth surface due to thermal effects is held to a minimum. The nature of flow in the horizontal and barrel reactors is amenable to analysis on the basis of known fluid flow relationships (see Chapter 7). The geometry of the gas passage between each face of the barrel susceptor and the reactor wall is essentially the same as the geometry of the gas passage of the horizontal reactor, and thus from the standpoint of fluid flow dynamics these two reactors are very nearly (with the exception of edge effects) equivalent. We have experimentally verified that gas-flow relationships developed in a horizontal reactor [for instance, by Eversteijn et al. (1970)] are applicable to the barrel reactor. Figure 2.28 is a photograph of a barrel reactor during a run in which the temperature was held above the normal deposition temperature to purposely induce gas phase decomposition in order to observe the gas-flow pattern. The silicon "smoke" can be seen

---

*Both the small pedestal (generally less than 2 inch-diameter) and the larger pancake (generally larger than 8 inch-diameter) have been referred to as "vertical" reactors (i.e., the gas input is normal to the growth surface). In some foreign literature (for instance, Japanese) the terms "horizontal" and "vertical" is assigned to the systems to designate the direction of gas flow with respect to a horizontal surface; using this terminology, both the barrel and pancake systems are vertical reactors. To avoid this confusion, the terms pedestal, horizontal (the one reactor with the horizontal gas flow), barrel, and pancake are used here.

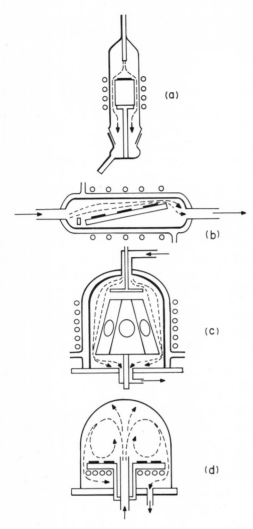

FIGURE 2.27   A schematic of the (a) pedestal; (b) horizontal; (c) barrel; and (d) pancake-type RF-heated silicon deposition systems

growth of silicon from halide source gases without the use of water-cooled walls. Silane, however, may pyrolyze in the gas phase at temperatures below the optimum temperature for the growth surface, and therefore, a steeper temperature gradient is required over the growth surface when silane is used as the source than when a halide is used. Particulate matter falling on the surface from gas phase pyrolysis may result in a roughening of the surface and interfere with the propagation of the single crystal lattice of the underlying silicon. The thermal gradient normal to the growth surface is a function of the temperature difference between the susceptor and the reactor wall, the proximity of the susceptor to the wall, the local gas velocity, and the arrival temperature of the gas. The temperature of the wall, as well as being a function of the above variables, is dependent on the nature of cooling and the degree to which the wall transmits, absorbs, or reflects radiation from the susceptor.

In the pedestal and pancake reactors, the growth surface is relatively remote from the reactor wall. The susceptor is close to the wall only at the edges, and the products of gas phase pyrolysis generated in this region are swept downstream away from the growth surface. The pedestal reactor has been used successfully both with (Dumin and Robinson, 1968A) and without (Manasevit, 1975) water-cooled walls. The larger pancake reactor has, to date, been used without water cooling. As a practical matter, the relatively large quartz bell jar is fabricated from opaque quartz.

In order to achieve the desired linear gas-flow velocities with reasonable gas flows, the distance between the susceptor and the reactor wall in the horizontal and barrel reactors is as small as 0.5 cm at the closest point, and gas phase pyrolysis products formed upstream may deposit on the downstream growth surfaces. Therefore, water cooling of these types of reactors has proven to be beneficial (Cullen, 1971). Kroko and Shaw (1969) observed a cloud of reaction products in the space between the susceptor and the wall of an uncooled horizontal reactor. These investigators also reported that the upper growth rate was defined by the occurrence of dendritic growth on the susceptor and substrates. Mayer and Shea (1964) claim to have observed no difference between the nature of growth of silicon from silane in water-cooled or non–water-cooled horizontal reactors. On the other hand, the maximum rates observed were relatively low (<1.5 μm/min), and dendritic silicon formation was observed at the higher growth rates. These results can be related to depletion of the source gases by gas phase pyrolysis and to the deposition of particulate matter on the surface. In our water-cooled horizontal and barrel reactors we have observed no evidence of the formation of solids in the gas stream or the onset of dendritic growth (at the optimum growth temperature) to deposition rates as high as 7 μm/min.

streaming from the downstream end (bottom) of the susceptor with very little turbulence. In the large pedestal and in the pancake* (Figure 2.27d) types of reactors the gases may recirculate over the growth surfaces prior to being expelled from the system. The gas flow in these systems is therefore less amenable to analysis on the basis of fluid flow dynamics.

Air- and water-cooled reactors

There has been some disagreement among investigators in the field concerning the necessity of water cooling the reactor walls when silane is used as a source gas. All of the reactors illustrated in Figure 2.27 have been employed for the homoepitaxial

Cooling of the walls also suppresses the deposition of pyrolysis products on the walls (see Figure 2.28). Wall deposits are to be avoided since they may be swept onto the growth surface and because they change the thermal gradient by reflection of radiation from the susceptor.

The influence of reactor types on deposition variables

The optimum conditions for heteroepitaxial growth are a function of the reactor geometry. In the horizontal and barrel reactors the uniformity of the deposit in the direction of the gas flow is strongly dependent on the angle between the susceptor surface and the reactor wall and on the linear gas-flow velocity. In the pancake reactor, the uniformity of the deposit is strongly dependent on the nature (number and size of orifices) of the gas inlet and on the input gas velocity. In the barrel reactor the susceptor is rotated to average out differences in the spacing between the susceptor and the wall which occur due to the difficulties involved in fabricating precise quartz parts. The uniformity of the deposit thickness is more commonly achieved through manipulation of the geometry of the gas passage and of the gas-flow velocity rather than by imposing thermal gradients in the direction of the gas flow, mainly because the former procedures are inherently more reproducible.

In the horizontal and barrel reactors, because of the narrow gas passages, efficiency is relatively high and deposition rates in excess of 7 μm/min are readily achieved (see Section 2.8). In the pancake reactor the input reactant gases are diluted by mixing with the relatively large volume of recirculating gas within the chamber. Thus, while high linear gas-flow velocities are used to achieve thickness uniformity in the horizontal and barrel reactors, the high flows are necessary in the pancake reactor to achieve the desirable growth rates ($\geqslant 2$ μm/min).

It had been originally predicted that the recirculation of gases in the pancake type of vertical reactor would lead to higher, but more uniform, autodoping in the silicon on sapphire. There are compensating effects, however, such as turbulence and a relatively high overall gas throughput. Differences in the level of autodoping, if present, have not to date been unambiguously established. The differences in semiconductor properties of the heteroepitaxial silicon deposited (under the optimum growth conditions) in the various reactors are subtle, and in a practical sense they can be characterized only by collecting rather extensive data on devices fabricated in the silicon deposited in the various systems. The problem here is that not many organizations have more than one type of reactor on stream to produce a sufficient number of devices to develop meaningful trends, and data on reproducibility and yield are usually proprietary.

One finds that details of the reactor geometry (of noncommercial reactors) are seldom defined in the literature, and thus it is not surprising that the various investigators report different properties in films deposited under what might appear to be similar growth conditions. It would be highly desirable if, short of providing the details of the reactor geometry, the investigators in the field would specify at least the linear gas-flow velocity across the growth surface, in addition to the deposition temperature and rate. In the pancake reactor the linear gas-flow velocity across the growth surface is not easily extracted from a knowledge of the input velocity and the reactor geometry, and under these circumstances a more complete description of the system geometry is necessary to uniquely define the growth conditions.

The general relationships bearing on gas-flow dynamics and thermal gradients just discussed may serve as a background for the quantitative analyses of these factors presented in Chapter 7.

RF Susceptors

Since the susceptor is the hottest part of the reactor system, efforts must be taken to assure that the sus-

FIGURE 2.28   A water-cooled RF-heated barrel reactor, photographed during a run in which the temperature was held above the normal growth temperature in order to produce a visible silicon "smoke." Note the nonturbulent streaming from the downstream (bottom) end of the susceptor and the absence of deposit on the water-cooled reactor walls

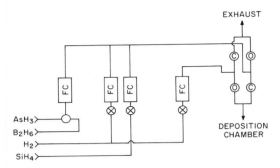

FIGURE 2.29   A schematic of a simple gas metering system for the growth of silicon from silane. FC indicates a flow controller, which may be electronic or manual. With the distribution system valves in the open or closed positions indicated, the reactants are exhausted as the growth chamber is flushed with hydrogen

ceptor is not a major source of impurities. In an effort to exclude all foreign materials from the hot part of the system, silicon susceptors with doped-silicon cores have been used in the more simple (pedestal and horizontal) reactor geometries (Mueller and Robinson, 1964; Hart et al., 1967). The core of the susceptors is more commonly fabricated of carbon, but even when dense high-purity carbon is used, out-gassing of the susceptor during deposition interferes with the properties of the deposit. One solution, in small susceptor systems, has been to cover the carbon with quartz (Hart et al., 1967; Mercier, 1968). In the barrel or pancake reactors, evolution of gases from the carbon susceptors has been successfully suppressed with silicon carbide (commonly used in homoepitaxial reactors) or pyrolytic carbon coatings (Cullen et al., 1969). Susceptors made entirely of bulk pyrolytic carbon have also been used (Benzing, 1974). When bulk pyrolytic carbon is employed, the anisotropies in the electrical and thermal conduction of the material must be taken into consideration in the susceptor design.

Gas metering systems

The flow scheme of the most simple gas metering and manipulation system is schematically shown in Figure 2.29. While the flow control can be effectively carried out with conventional manual valves and rotameters, the electronic flow control devices are particularly useful to introduce the very low volumes (1–10 cm³/min of a 10 ppm mixture) of dopant gas mixtures. The electronic flow controllers consist of an electrically actuated metering valve and a mass flow meter connected in a closed-loop system (Applied Materials Inc., Santa Clara, CA and Tylan Corporation, Torrance, CA). Use of the flow controllers also provides reproducibility and the possibility of automated sequencing.

The outlet network shown in Figure 2.29 is designed

to allow the growth chamber to be flushed with the carrier gas while the gas metering system equilibrates and is flushed through the exhaust. To initiate growth, the four valves are actuated simultaneously to switch the source gases into the reaction chamber and the chamber flushing gas to the exhaust. To terminate the run the procedure is reversed. The flow conditions are stabilized by using equal flow rates in the reactor and exhaust lines. This procedure provides reproducibility in the very brief growth runs (i.e., as brief as 15 s for the deposition of a 1-μm-thick film at 4 μm/min).

A single flow controller can be used for both the n- and p-type dopant gas mixtures (commonly 10 ppm arsine or diborane in hydrogen) to prepare films with carrier concentrations greater than ~1 × 10¹⁶/cm³ (as is done in the double-epi process, see Section 2.13). To assure reproducibility in the more lightly doped films (as is needed for the deep depletion type devices) it is advisable to use separate dopant gas metering systems. It is convenient to qualify the source gases with homoepitaxial silicon growth. This requires an HCl source to free the silicon substrates of the natural oxide. A schematic of a metering system with separate dopant lines and an HCl source is shown in Figure 2.30. In the metal (stainless steel) gas metering systems all of the valves are packless bellow valves, and all the connections are made either by welding or with a type of fitting that deforms a ferrule or O-ring onto the tube to form a vacuum-tight seal. At the relatively low growth temperature (typically 1000°C) it is essential that the system be vacuum leak-tight (tested to 10⁻⁵ torr).

The use of the electronic closed-loop gas-flow control devices has resulted in a very different type of manipulation system. Figure 2.31 is a photograph of the typical system used at the RCA Laboratories until about 1973. Figure 2.32 is a photograph of a system designed at RCA Laboratories and put into service in early 1974. It is clear that the gas manipulation system can be very much simplified if all doping is carried out by ion implantation.

Gas sources and purification

It has not been possible to maintain adequate purity of hydrogen carrier gas stored in metal cylinders. It is common practice to purify the hydrogen by passage through a hot palladium–silver membrane. Using the palladium purifiers moisture can be tolerated in the unpurified hydrogen, but a combination of moisture and trace amounts of halides results in damage to the purifier. The presence of hydrocarbon impurities in the unpurified hydrogen leads to a decrease in the efficiency of the metal membrane, but this can generally be recovered by burning the hydrocarbon reaction products off the metal at higher than normal operating temperatures.

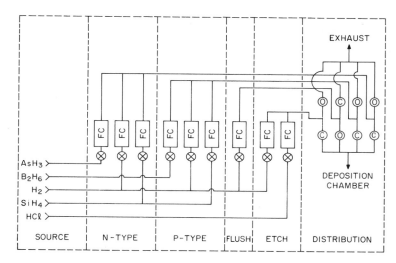

FIGURE 2.30   A schematic of a gas metering system for the deposition of silicon from silane in which separate piping is used for the n- and p-type dopants to avoid cross contamination during the growth of lightly doped deposits. An HCl source is provided so that the source gases may be qualified with homoepitaxial growth. With the distribution valves in the open or closed positions indicated, the reactant and dopant are exhausted as the reaction chamber is flushed with hydrogen

Hydrogen from liquid hydrogen sources has been successfully used directly as the carrier gas in the heteroepitaxial growth of silicon. Precaution must be exercised to avoid condensing atmospheric constituents in the storage tank, since condensates will volatilize and become incorporated in the hydrogen as the liquid level decreases.

It is necessary to depend on the purification and gas-handling procedures of the commercial suppliers for pure dopant gas–hydrogen mixtures and silane. While it is recognized that the purity of the hydrogen of the dopant gas mixture (stored in steel tanks) does not equal the purity of the carrier gas, this can be tolerated because the dopant–hydrogen mixtures make up a very small volume percentage of the total source gas flow (typically 1 to 10/25,000). While very pure silane is commercially available, for optimum reproducibility it is advisable to qualify the silane in a new tank with a standard silicon deposition run.

Thickness control

Using the flow control system described above, in routine deposition runs the overall thickness of the deposit can be adequately controlled with the deposition duration. In experimental growth runs, however, it is desirable to exercise independent control of the deposit thickness when variables such as the growth rate, temperature, or gas-flow conditions are being studied. The maintenance of a constant thickness is particularly important for investigating the deposition parameters of the films less than 1 μm in thickness, in

FIGURE 2.31   A gas manipulation system built around manual control and glass floating ball gas-flow meters. System designed at the RCA Laboratories

FIGURE 2.32   An automatic gas manipulation system built around electronic closed-loop gas-flow devices. System designed at the RCA Laboratories

which the semiconducting properties are a strong function of the thickness. In the horizontal reactor the thickness and growth rate can be continually monitored by sighting a conventional IR pyrometer, with an appropriate filter, through the deposit and substrate and onto the hot susceptor (Dumin, 1967A). As the film increases in thickness, the spacing between the two surfaces passes through values which lead to the interference of the radiation from the susceptor. The constant level (at temperature) output of the pyrometer is nulled out, and the small changes in the pyrometer signal are fed to a time-based recorder. This results in time-dependent ''interference'' patterns as shown in Figure 2.23. This method is described in more detail in Chapter 5.

The thickness uniformity across the substrates included within a single deposition run is established through manipulation of the gas-flow patterns. Appreciation of gas-flow dynamics discussed in Chapter 7 is useful in maintaining the thickness variation within the desired limits. It is clear that the difficulty in achieving the desired thickness uniformity increases with the length of the susceptor in the horizontal and barrel reactors and with the diameter of the susceptor in the pancake-type reactors.

2.11. The crystalline quality of heteroepitaxial silicon

An overview of the characterization of the crystal character of heteroepitaxial silicon

From a materials point of view the development of a detailed understanding of the silicon-on-insulator composite system depends on establishing correlations among the deposition parameters, the physical properties, and the semiconducting properties of the silicon. This has been particularly difficult to do in the case of heteroepitaxial silicon technology. A large body of quantitative information has been generated on the relationship between the deposition (and substrate parameters) and the semiconducting properties; some semiquantitative relationships have been established between the deposition (and substrate parameters) and the crystal quality; but little definitive information is to be found in the literature on the relationship between the crystalline structure and the semiconducting properties of the thin silicon film. As with bulk and homoepitaxial silicon, the relationship between crystalline structure and semiconducting properties in device structures is complex and difficult to establish. It is highly desirable to develop such relationships, both to isolate and understand the key variables involved and to have available nondestructive in-line quality control of the materials prior to the fabrication

of devices in which the electrical properties can be used for quality control.

In the relatively poor heteroepitaxial deposits, correlation can be observed between the semiconducting properties and the crystalline quality as measured by methods such as x-ray and electron diffraction and optical absorption. In this regime of crystalline quality, however, the carrier mobilities (as compared to bulk silicon) are low and not useful for application in device structures. In an intermediate range of crystal quality, it is reasonable to relate improvements in the semiconducting properties to improvements in the crystalline quality, but the small changes in the crystallinity are not easily measured by established techniques. It is clear that in this regime the semiconducting properties are more sensitive to the structure than are most of the analytical techniques. Above some level of crystalline quality (difficult to define quantitatively), the semiconducting properties of the film appear to be more dominated by impurity effects than by the crystalline quality. In this regime the crystalline perfection may be essentially constant while the semiconducting properties vary considerably. It is in this regime that the observation has been made that the after-oxidation (after device processing) semiconducting properties are a strong function of the deposition and substrate parameters, whereas the semiconducting properties of the as-deposited films are independent of the deposition and substrate parameters over a wide range. This has been an important consideration in the development of the technology, and has been discussed earlier in some detail (see, for instance, Table 2.9). It has even been suggested that above some minimum level of crystalline perfection the carrier mobility may be increased under conditions that may lead to some sacrifice in the crystal quality (Salama et al., 1967). If this is true, it must be peculiar to the situation in which a compromise must be made between deposition conditions leading to the suppression of contamination by substrate reaction products and conditions conventionally leading to good crystalline perfection (see Sections 2.6 and 2.8).

The crystalline nature of the heteroepitaxial silicon on sapphire and spinel and the nature of the substrate surface have been studied by a variety of analytical techniques: X-ray diffraction, reflection (high-energy) electron diffraction (RHEED), low-energy electron diffraction (LEED), optical absorption, replication electron microscopy (REM), transmission electron microscopy (TEM), and ion channeling. Most of these methods are discussed in some detail in Chapter 5. The crystalline quality has also been inferred from the film semiconducting properties, particularly in the case where the semiconducting properties have been measured as a function of distance from the silicon/substrate interface (see Section 2.12).

The ease of making the measurement and the information content of the physical characterization techniques are important considerations, particularly where reproducibility has been a problem and large numbers of samples must be characterized in order to accumulate sufficient data to develop real trends. For the most part the information content is inversely proportional to the ease of making the measurement. X-ray diffraction and optical absorption are particularly convenient in that they are used at atmospheric pressure and can be nondestructive. For the film thicknesses of interest here, the crystalline quality is integrated over the entire thickness of the film in both x-ray diffraction and optical absorption. Examination of the crystal structure by electron diffraction has the disadvantage that a vacuum environment is required, and the method is therefore time consuming and destructive. An advantage in examining very thin films and the substrate surfaces is that only about 50 Å of the material near the surface is sampled. This is particularly useful in the examination of the nature of the substrate surface, where bulk defects play a role only if they intersect the surface. Optical reflectivity measurements are useful in that they can be made at atmospheric pressure and are nondestructive. In establishing relationships between the microstructure of the composite structure and the electrical properties, all of these methods must be classified as semiquantitative in their present states of development. These semiquantitative optical, x-ray, and electron diffraction techniques have contributed significantly to the heteroepitaxial silicon technology. The optical methods offer the possibility of in-line quality control (Duffy et al., 1976).

Low-energy electron diffraction has been useful in establishing substrate surfacing procedures and in characterizing the overall structures of the substrate and the heteroepitaxial orientation relationships. Ion-channeling spectroscopy stands in a classification of its own both from the standpoint of method and nature of information derived. Correlations have been observed between results obtained from ion channeling and the more established methods.

Of the well established methods, transmission electron microscopy appears to offer the greatest possibility of accumulating information on the physical structure of the thin heteroepitaxial films in sufficient detail to develop correlations between preparative variables, structure and semiconducting properties. TEM is, however, the most difficult method to employ because of the need to thin the samples for penetration of the electron beam ($\sim$5000 Å for silicon, <3000 Å for sapphire using a 100-kV beam). The skill required for gathering and interpreting the data is matched only by low-energy electron diffraction. Some of the more pertinent information gathered using the various analytical techniques is discussed in the following sections.

The characterization of heteroepitaxial silicon by x-ray and electron diffraction

In the early work on the heteroepitaxial growth of silicon it was not unusual to verify the nucleation of single crystal films on the various substrate materials and substrate orientations and to provide rough indication of the crystalline perfection with Laue x-ray back-reflection patterns and x-ray diffractometer traces [for instance, Manasevit et al. (1965), Manasevit and Forbes (1966), Wang et al. (1969)]. In a more detailed examination of the microstructure in which x-ray diffraction was correlated with surface structure (as examined by replication electron microscopy) twins were observed to be the dominant defect structure in thick films deposited on sapphire from silicon tetrachloride (Nolder et al., 1965).

Reflection (high-energy) electron diffraction patterns were also commonly displayed in the literature to establish the existence of heteroepitaxy [for instance, Joyce et al. (1965), Salama et al. (1967), Naber and O'Neal (1968), Wang et al. (1969)]. Whereas diffraction techniques have been useful in establishing orientation relationships, in silicon more than $\sim$0.5 μm in thickness deposited by current technology they are not sensitive to defect structure at a level significantly influencing the semiconducting properties. Electron diffraction has proven particularly useful, however, in the examination of the microstructure of the silicon adjacent to the silicon/substrate interface. Yasuda and Ohmura (1969) examined the crystallinity of silicon evaporated onto (0001) sapphire. Prior to complete surface coverage, a number of orientations were observed [all with (111) parallel to the substrate surface] in $\sim$300 Å growth islands. At 800 Å, but still prior to complete coverage of the substrate, the islands recrystallized into a single crystal.

Kühl and coworkers (1974, 1975A) have concluded, on the basis of electron diffraction (in conjunction with optical absorption), that an intermediate layer of aluminum silicate is present at the substrate/silicon interface. Cullen and coworkers (1975A) have observed that the crystalline perfection of the silicon within $\sim$0.2 μm of the interface is strongly dependent on the rate of deposition of the silicon during the period that the substrate surface is not completely covered. These latter two studies are discussed in more detail in Section 2.6.

Some correlation appears to have been seen between x-ray diffractometer linewidths and Hall mobilities measured in silicon deposited on spinel. Gassmann and coworkers (1971), using a conventional (i.e., single crystal) diffractometer, measured "inte-

TABLE 2.6 θ:2θ and ω-mode X-ray Half-widths of the Heteroepitaxial Silicon and the Sapphire Substrate

| | | Silicon Deposit | | Sapphire Substrate | |
Growth Technique	Film Thickness (μm)	Half-width[a] θ:2θ (degrees)	Half-width[a] ω (degrees)	Half-width θ:2θ (degrees)	Half-width ω (degrees)
CR	0.2	0.3188	0.8000	0.0416	0.1150
REG	0.2	0.2500	0.5000	0.0438	0.1134
CR	0.3	0.1983	0.4000	0.0449	0.1169
REG	0.3	0.1306	0.2488	0.0434	0.1225
CR	0.5	0.2266	0.9000	0.0441	0.1163
REG	0.5	0.1500	0.2875	0.0417	0.1200
CR	0.75	0.1231	0.2875	0.0438	0.1175
REG	0.75	0.1106	0.2025	0.0413	0.1256
CR	1.0	0.1147	0.2188	0.0441	0.1200
REG	1.0	0.1063	0.2250	0.0395	0.1156
CR	2.0	0.0969	0.1831	0.0436	0.1275
REG	2.0	0.0906	0.1706	0.0422	0.1331

Source: Cullen et al. (1975A).
CR: constant rate growth.
REG: rapid early growth.
Orientation: (100)Si/(1102) sapphire.
[a]Silicon half-widths given as functions of film thickness and growth method

gral linewidths'' [of the (111) diffraction trace] of ~3.1 min for bulk silicon, between 7.5 and 10 min for good deposits of silicon on spinel, and as high as 15 min on poorer deposits. The integral linewidth was defined as the width of a rectangle determined by setting the height at the maximum trace intensity and setting the area of the rectangle equal to the area under the trace for the (111) reflection. Within the range characteristic of the better deposits, a linewidth of 7.8 min was associated with the hole mobility of 373 cm²/V-s (in a 3.60-μm-thick film) while a linewidth of 13.2 min was associated with a mobility of 252 cm²/V-s (in a 2.15-μm-thick film). In the θ:2θ scan mode employed here (i.e., the sample rotated at the angle θ and the detector at 2θ in relation to the beam) the broadening of the diffraction profile can be assigned essentially to a variation in lattice spacing.

Attempts have also been made to correlate deposition parameters with the crystalline nature of silicon on sapphire by the x-ray rocking curve technique. In this method the axis of the sample is held in a fixed diffraction angle θ in relation to the beam, and the sample is rotated (rocked). The method is also referred to as the ω-scan mode. The broadening of the diffraction profile in the ω mode is related to the presence of a ''mosaic'' structure in crystal (i.e., localized areas that are misoriented as compared to the surroundings). Using a double-crystal spectrometer, Stadnik and coworkers (1972) have observed linewidths of ~10 min for bulk silicon and ~800 min for vacuum-deposited silicon on sapphire. These workers

calculated background dislocations ~10⁹/cm² from the rocking curve data. The details of their analysis, however, were not provided. Although it is reasonable to assume a relationship between dislocation density and the ω-mode linewidths (Kane and Larrabee, 1970), experimental correlations in the heteroepitaxial silicon are yet to be experimentally established.

Both the θ:2θ and ω-scan modes were employed by Cullen and coworkers (1975A) to examine the effect of the early growth rate (i.e., growth prior to complete surface coverage; see Section 2.8) on the change in crystal perfection with distance from the silicon/sapphire interface. The half-widths of the silicon are shown as a function of the early growth rate and thickness in Table 2.6. Also shown in the table are the half-widths of the substrates employed. It can be seen that the half-widths decrease significantly with increase in the film thickness to a thickness of ~1 μm. The relatively small differences observed between the 1- and 2-μm-thick films are attributed to the entire deposit being sampled by the technique. In the deposits less than 0.5 μm in thickness, the half-widths are also a function of the rate of growth. No correlation is observed between the half-widths of the deposit and of the substrate. This is attributed to the bulk (rather than the surface crystallinity) of the substrate being sampled by this method. In heteroepitaxial silicon, variations in the lattice spacing (θ:2θ mode) and local misorientation (ω mode) appear to be closely related, and essentially the same conclusions can be drawn on examination of the x-ray data collected by the two scan modes. Because of the difficulties involved in making mobility measurements in silicon less than 0.5 μm in thickness (see Chapter 8), attempts to establish a relationship between the mobilities and the x-ray results were not fruitful.

Weitzel and Smith (1976) have been able to show a correlation between the θ:2θ half-width and transistor mobilities fabricated in 0.6-μm-thick silicon. The silicon was deposited at a rate of 1.8 μm/min at 1000°C on hydrogen-annealed (1200°C) sapphire substrates. A particularly interesting aspect of this investigation was the observation that the θ:2θ half-widths measured in the silicon deposit are a function of the magnitude and direction of misorientation of the substrate surface. Between misorientations of +2° and −2° parallel to the mirror plane of the sapphire, the θ:2θ half-width values change monotonically. Thus, the transistor mobilities can be related to the misorientation of the substrate surface. For example, an n-channel mobility of 596 cm²/V-s and a p-channel mobility of 285 cm²/V-s were observed in silicon deposited on a substrate misoriented by ≈2°. In silicon deposited on a substrate misoriented ≈ −1°, the corresponding n- and p-channel mobilities were 457 and 232 cm²/V-s, respectively. A positive misorientation conceptually involves rotation around the a-axis in the direction to bring the c-axis more parallel to the (1$\bar{1}$02) plane.

Film crystallinity: low-energy electron diffraction

The crystal structure of the heteroepitaxial silicon, as determined by low-energy electron diffraction, has been examined as a function of the substrate orientation and temperature and is given in Figure 2.26. The interpretation of LEED patterns is somewhat complex; details of the nature of the structures observed may be found in the literature [see, for instance, Chang (1969)]. Although the LEED studies have proved to be particularly useful in defining the influence of such parameters as substrate pretreatment, orientation, and temperature on the overall crystal structure, this method is less applicable to characterizing the defect structure of silicon than are other methods described in this section.

Optical absorption and reflection

The optical absorption of heteroepitaxial films near the absorption edge of silicon has been examined by a number of investigators. In relatively thin (0.35 to 1.7 μm) vacuum-deposited silicon, Salama and coworkers (1967) observed a sharpening of the absorption edge (i.e., an increase in slope of the absorption vs. wavelength plots) with increasing substrate temperature between temperatures of 880 and 1100°C. A somewhat surprising conclusion reached here was that mobility increased with decreasing substrate temperatures

(between 1100 and 950°C) while the crystal perfection, as inferred from the optical absorption measurements, decreased. Salama and coworkers interpreted the data on the basis of degradation of conduction in the grain boundaries and a shift in the doping level and size of the "grains" with deposition temperature. Other investigators have observed that within a narrow range of crystalline perfection, the carrier mobility is more dominated by scattering from aluminum complexes than by the overall crystallinity and that the mobility decreases with increasing temperature as the autodoping increases (see Section 2.8). Such an effect is characteristic of the situation in which the deposit reacts with the substrate.

The effects reported by Salama and coworkers appear to be characteristic of the specific vacuum-deposition conditions and temperature range studied by them. The optical absorption (near the band edge) of chemically vapor-deposited silicon has been measured as a function of film thickness and deposition temperature by Dumin and Robinson (1968B). These investigators found that the optical absorption increased, in 0.2-μm-thick films, with a decrease in temperature by Dumin and Robinson (1968B). These investigators found that the optical absorption increased, in 0.2-μm-thick films, with a decrease in temperature from 1050 to 1000°C. In this investigation the higher optical absorption could be associated with degraded carrier mobilities. The same trend was observed in thicker (8-μm) deposits. It was also observed that the optical absorption decreased with increasing film thickness for thicknesses between 0.2 and 30 μm. This observation is in agreement with other measurements indicating that the defect density decreases rapidly with distance from the film/substrate interface.

Druminski and coworkers (1976) and Kühl and coworkers (1976A) have demonstrated a relationship between an optical "absorption factor" (F_A) and channel mobility in transistors fabricated in silicon on sapphire. The absorption factor, derived from optical reflectance measurements, is proportional to the linear slope of the absorption edge and the absolute value of the absorption constant (Kühl et al., 1974) at a predetermined wavelength (this is discussed in more detail in Chapter 5). This "quality factor" is considered to be a measure of lattice perfection of the deposit.

The channel mobility as a function of the absorption factor is shown in Figure 2.33 (Druminski and Wieczorek, 1975). The absorption factor of the heteroepitaxial silicon was measured prior to processing the transistor structures. It is clear that the correlation between the absorption factor and the device mobilities, at least for electron mobilities, falls outside the limits of error of the mobility measurement. The absorption factor was also shown to correlate well with ion backscattering yields, and "fair" correlation was seen with

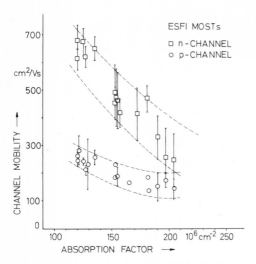

FIGURE 2.33   Channel mobility as a function of the optical absorption factor in silicon on sapphire. (Note that the Siemens investigators use the acronym ESFI®, epitaxial silicon films on insulators, for both silicon on sapphire and silicon on spinel.) After Druminski et al. (1976)

a "subjectively defined" factor derived from electron diffraction.

The Siemens* investigators (Druminski et al., 1976) also have measured the optical absorption as a function of thickness and related these values with mobility as a function of thickness. In Figure 2.34 the absorption constant and the mobility are plotted as a function of the thickness of a deposit of silicon on spinel. The absorption constant is shown both as a mean value measured in the deposit subsequent to successive thinning by etching and as a value obtained by computer differentiation of the absorption constant with respect to the thickness. The absorption constant at a particular wavelength is used here rather than the absorption factor to ease the mathematical manipulation. The mobility as a function of thickness was measured using the carrier-depletion method (Tihanyi, 1972). The high absorption and low mobilities near the silicon/spinel interface can be related to poor crystallinity.

It appears that correlations established with the optical reflectance measurements are specific to the method of preparation. Druminski and Wieczorek (1975) reported that the mobilities in silicon deposited on sapphire using dichlorosilane as the source exhibited optical absorption constants similar to those of silicon deposited from silane but that the electron mobilities in 2.8-μm-thick silicon deposited from dichlorosilane were only ~10 percent of bulk values.

More recently, attempts have been made to characterize the crystalline quality of the sapphire surface by

*Research Laboratories, Munich, Germany.

IR reflectance (Duffy et al., 1975, 1977; Zanzucchi et al., 1978). Changes in the optical refractive index (n) and the absorption index (k) were measured in the wavenumber region between 900 and 250 cm^{-1}. At 600 cm^{-1}, the reflectance is highly sensitive to lattice distortion in the sapphire. The requirement to derive absolute values is avoided by taking, as a quality factor, the ratio of the reflectance at a wavenumber sensitive to lattice distortion and the reflectance at a relatively insensitive wavenumber (~450 cm^{-1}). An interesting conclusion of these investigations is that changes are not seen in the semiconducting properties of the heteroepitaxial silicon unless the substrate surface is damaged to a degree considerably greater than originally anticipated.

The crystalline quality of the heteroepitaxial silicon was studied by the same group of investigators using absorption in the UV range. As compared to the work of Druminski and Kühl, described above, these latter investigators employed wavelengths in the region well above the absorption edge of silicon corresponding to a photon energy of 4.3 eV. In this region the refractive and absorption indexes are more sensitive to crystallographic changes in the silicon. Experience has shown that these two factors are generally interrelated. Combining the results from the IR measurements on the sapphire and the UV measurements on the silicon, these investigators conclude that the current state-of-the-art substrate surface finish is adequate, and that further improvement in the crystalline quality of the heteroepitaxial silicon will come from a study of prep-

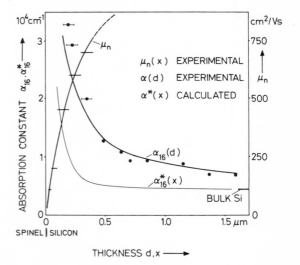

FIGURE 2.34   Optical absorption and carrier mobility as a function of the thickness of silicon on spinel. α is the absorption constant measured on films successively thinned by chemical etching; α^* is obtained by differentiating α with respect to thickness; and μ is obtained as a function of thickness by the carrier-depletion method. After Druminski et al. (1976)

arative parameters other than the substrate surface finish.

The measurement of film quality by optical absorption is particularly attractive because it is nondestructive and thus offers the possibility of being used in in-process quality control. These methods warrant further calibration and development. Druminski and coworkers (1974) reported that of a number of analytical techniques examined, only optical absorption, ion backscattering, and carrier mobility measurements were sufficiently sensitive to distinguish between small differences in the relatively good quality heteroepitaxial deposits.

The examination of the optical transmission and reflection of the silicon on spinel composite in the wavelength region away from the band edge provides more information on the thickness and uniformity of thickness than on the crystalline structure (Stein, 1972). An interesting conclusion of this work, however, is that excess reflectivity near a wavelength of 900 cm^{-1} suggests that the silicon growth alters the composition of the substrate adjacent to the silicon/substrate interface. As interest increases in understanding the microstructure of the material near the silicon/substrate interface, the significance of reflection measurements in this wavelength region warrants further attention.

Surface microscopy

In the early work on heteroepitaxial silicon, during the period when the emphasis was on the preparation of the thicker films ($> 5 \mu$m) from halide source materials, surface features were readily observed by optical microscopy. In the presence of the halides, during deposition an equilibrium is set up between deposition on and erosion of the silicon surface. This has the effect of bringing out crystallographic surface features, which become more evident as the film grows thicker.

As the less-corrosive silicon source materials were sought, replication microscopy was used to examine the differences in the films deposited from $SiCl_4$ and SiH_4 on sapphire (Manasevit et al., 1965; Nolder et al., 1965), from $SiHCl_3$ and SiH_4 on sapphire (Bicknell et al., 1966), from $SiCl_4$ and $SiHCl_3$ on spinel (Heywang, 1968), and from $SiCl_4$ and SiH_4 on spinel (Zaminer, 1968). It became evident that the crystal structure of the heteroepitaxial silicon improved with the reduction of the halide content of the growth atmosphere and with increasing growth rates. It was also noted that the substrate surface was covered more rapidly when SiH_4 was used as the source than when $SiCl_4$ was the source (Manasevit et al., 1965). Subsequent to these investigations the pyrolysis of SiH_4 has been generally employed as the silicon source for deposition of sapphire and spinel. Replication microscopy

has been employed to examine the thicker silicon deposited from silane on sapphire (Robinson and Mueller, 1966) and spinel (Wang et al., 1969). It is fair to say at this point, however, that with the exception of the examination of the heteroepitaxial deposits prior to complete coverage of the surface, there is little information to be derived from the examination of the "as-deposited" surface structure of less than 1-μm-thick silicon deposited from silane (under optimum deposition conditions). We have observed that even the effect of the discrete island formation during the early growth is lost immediately with the complete coverage of the surface (see the micrographs in Section 2.7). Schlötterer (1976), however, reports that surface roughness was observed at thicknesses about twice those of surface coverage.

The microscopic examination of chemically etched surfaces has been widely employed to evaluate the defect density of both bulk silicon [see, for instance, Kane and Larrabee (1970)] and the substrate single crystals (see Section 2.4). In the examination of the microstructure of etched bulk silicon, faceted etch pits are developed by exposing—typically for a period of minutes—the surface to standard etchants. A similar treatment of the heteroepitaxial silicon results in the complete removal of the thinner deposits ($<1 \mu$m) and the development of deep nonfaceted furrows in the thicker deposits. Structure with obvious directional relationships may be brought out by etching the heteroepitaxial films for very brief periods (i.e., seconds). After the brief etching, little is to be seen with optical microscopy, and the higher magnification available with scanning or replication electron microscopy must be employed. As the density of defects increases in the material near the silicon/substrate interface, characterization by microscopic observation of the etched surfaces becomes more difficult. At thicknesses of interest for device application ($<1 \mu$m), the etched-out structure may have directional features but generally is not faceted.

In deposits greater than 2μm in thickness, dislocation counts of $1-2 \times 10^8$ lines/cm² have been derived from the microscopic examination of Dash-etched (111) silicon on (111) spinel [Dumin and Robinson, 1968A; Hasegawa et al., 1969 (evaporated silicon); and Krause, 1970]. Yasuda and Ohmura (1969) observed evidence of stacking fault lamellae, dislocation pits associated with low-angle grain boundaries, and random dislocation pits by replication microscopy of Dash-etched evaporated silicon on sapphire.

The thinner deposits are more difficult to analyze by microscopic examination of etched surfaces. The defect density as a function of thickness was clearly demonstrated in silicon deposited on spinel from silicon tetrachloride at 1200°C (Schlötterer and Zaminer, 1966). Deposits were examined between thicknesses of 1 and 30 μm. After Sirtl etching (Sirtl and Adler,

1961), the surface features exhibit crystallographic relationships, as observed by replication electron microscopy. In conjunction with correlating the results of the ion-channeling investigations with the more conventional analytical methods, Picraux and Thomas (1973) examined etched surfaces of silicon on spinel. The surfaces were Sirtl etched for periods between 1.5 and 4 s. In (100) silicon on spinel, lines aligned in the $\langle 110 \rangle$ direction (as determined by x-ray diffraction) were observed. These investigators point out that the $\langle 110 \rangle$ direction is consistent with the intersection of (111) fault planes with the (100) surface. The defect density as a function of thickness in chemically thinned samples was examined. Defect densities of 3×10^7, 6×10^7, and $20 \times 10^7/cm^2$ were measured in films 2.2, 1.3, and 0.6 μm thick. The etch features on the (111) silicon surface were triangular, consistent with the intersection of the three (111) planes, which are inclined to the (111) surface. At a thickness of 2.1 μm, the defect density observed on the (111) surface was typically $2 \times 10^8/cm^2$. From TEM studies, however, the (111) silicon on spinel appeared to contain less defects than the (100) silicon on spinel. It should also be recalled that the mobilities measured in (111) silicon have always been higher than in the (100) silicon on spinel. Picraux and Thomas explain the apparent discrepancy between the etched surface and the TEM data on the basis of a greater etching efficiency of fault planes in the (111) silicon. The fault planes are inclined at a greater angle to the surface in the (111) silicon than in the (100) deposit. It was also observed that whereas consistent results were obtained on the examination of various sections of a single sample, the defect densities might change by a factor of 2 when different samples were examined. It is clear from these results that the derivation of quantitative data from the observation of etched heteroepitaxial surfaces must be done with caution.

Ion channeling/backscattering

Ion channeling/backscattering have been used to evaluate the crystalline quality of silicon on both sapphire and spinel (Picraux, 1972, 1973; Picraux and Thomas, 1973). The sensitivity to the defect content is limited, and the establishment of quantitative correlations is complicated by the somewhat difficult analysis required. Nonetheless, the method has proven to be useful for comparative analysis of silicon on sapphire and spinel deposited under various deposition conditions and for an evaluation of the defect density as a function of the film thickness. Picraux (1973) has characterized the method as filling in an "evaluation gap" between the realization of silicon heteroepitaxy and the growth of high-quality films and suggests that electrical and optical measurements are more appropriately used in the final phases of evaluating films for application in device structures. The influence of different types of crystalline defects cannot be isolated by ion channeling; in the analysis one must assume that all defects are equivalent. Picraux and Thomas (1973) have studied the relationship between ion channeling and scattering data and the characterization of the heteroepitaxial silicon by surface etching and transmission electron microscopy.

Protons (from 450 to 1340 keV) directed in specific crystallographic directions of silicon (for instance, along the $\langle 100 \rangle$ and $\langle 111 \rangle$ axes) undergo scattering essentially only from atoms displaced from their normal lattice positions. Thus, the comparison of the number of back-scattered ions from two samples of the same crystal type and thickness provides a qualitative measure of the relative crystalline quality. In order to derive the crystalline perfection as a function of thickness, the back-scattering yield is measured as a function of the back-scattered ion energy. An analysis of the depth profile is complicated by the fact that the back-scattered beam represents ions that have been scattered off of displaced atoms as well as dechanneled atoms deflected from the atoms in their normal lattice position. The analysis of the depth profiles in the heteroepitaxial thin films has been verified by carrying out measurements as a function of the sample temperature and thickness and the incident beam energy (Picraux, 1973).

A plot of the heteroepitaxial film defect density as a function of the distance from the silicon/substrate interface derived from ion-channeling data is shown in Figure 2.35. It is quite clear that the film improves rapidly with distance from the interface. The defect density near the interface, and the rate of decrease with distance from the interface, is a function of both the nature of the substrate material and the deposition (or substrate surface finishing) parameters. Since the samples in this study were obtained from commercial suppliers (Union Carbide, San Diego, CA and Inselek Corp., Princeton, NJ) the differences in deposition conditions were not reported. It should be kept in mind, however, that the stoichiometric Czochralski spinel substrates used in this study had proven to be relatively chemically reactive and that unless the surfaces were treated in a manner specific to the chemical nature of the material, the electrical properties were found to be degraded (see Sections 2.6 and 2.7). Although the data collected by ion scattering would be useful (at least on a comparative basis) for the development of preparative and substrate variables, they are not likely to be employed to a great extent because of the rather specialized equipment required.

Druminski and associates (1974, 1976) have also carried out ion backscattering investigations in silicon deposited on sapphire and spinel. They have observed

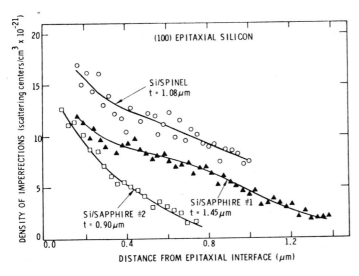

FIGURE 2.35   The density of imperfections as a function of distance from the silicon/substrate interface. The data were derived from ion-channeling measurements. The samples were obtained from commercial suppliers (see text). After Picraux and Thomas (1973)

good correlation between the backscattering results and optical absorption measurements previously discussed.

Defect structure inferred from electrical properties

In general, the increase in the carrier mobility with increasing distance from the silicon/substrate interface (see Figure 2.31) can be related to a decrease in the defect density with film thickness. A number of investigators have attempted to draw more specific correlations between the defect density and the electrical properties. It is a reasonable assumption that the degradation in the carrier mobilities and lifetimes in the heteroepitaxial silicon are related to defect-associated charged trapping centers (Schlötterer and Zaminer, 1966; Schlötterer, 1967, 1968; Dumin and Robinson, 1968A,B). Dumin and Robinson (1968B) suggested that the defect density was proportional to the difference between the electron and donor concentrations in the silicon. The "electrically active" defect density was estimated to be between 10^{15} and $10^{16}/cm^3$. Dumin (1971) has pointed out that the density of the electrically active defects could be changed by as much as two orders of magnitude by heating the films in a HCl-O_2 ambient. This suggests a rather complex interaction between the carriers and the defect structure which is still not completely understood and also points out the pitfalls involved in drawing quantitative relationships between the electrical properties and the physical defect structure.

The differences between the bulk and heteroepitaxial silicon carrier mobility as a function of temperature (Schlötterer, 1968; Dumin and Ross, 1970; Stein, 1972; see also Section 2.12) also suggest that the mobilities in the heteroepitaxial silicon are limited more by defects than by lattice scattering. Here again, how-

ever, it is difficult to develop quantitative relationships between the nature of the defects and the electrical properties.

Transmission electron microscopy

As attempts were made to improve the properties of the heteroepitaxial silicon by the examination of the microstructure as a function of the deposition conditions, several investigators in the field turned to the use of transmission electron microscopy. It is interesting to note that electron microscopic methods were employed very early in the heteroepitaxial silicon technology to examine the influence of the use of the various silicon source materials, the influence of the chemical nature of the substrate on the film properties, and the influence of growth rate. After this early activity, there appeared to have been a hiatus in the electron microscopic investigations. Historically this was a period in which the deposition and substrate variables were examined in detail, numerous samples were generated over short periods of time, and the semiconducting properties were in general used as a rapid measure of film quality and of the influence of the variables investigated. It would have been difficult to keep up with the progress using the more laborious and time-consuming transmission electron microscopic techniques. A factor that has seriously impeded the use of TEM has been the difficulty involved in thinning the sample. Now that preparative reproducibility has been considerably improved and the number of variables decreased, it has become worthwhile to examine "characteristic" samples in detail by electron microscopy, and the interest in this area has been rekindled.

In the early work in which TEM was used to examine the heteroepitaxial silicon it was established that the halide-bearing silicon source gases reacted with the substrate and led to formation of highly defected films. The information content was increased by examining the deposit prior to complete coverage of the substrate surface (Tamura and Nomura, 1966). Preparation of the silicon-on-insulator composite for TEM was expedited by the use of ion bombardment to thin the samples (Schlötterer and Zaminer, 1966). Both of these techniques continue to be useful in characterizing heteroepitaxial silicon. Using the ion-thinning technique, it was possible to select material for examination at various distances from the silicon/substrate interface. Attempts were made to relate the semiconducting properties with the various defects observed (for instance, primary and secondary growth twins). The results of these pioneering investigations are not discussed in more detail here because the material studied was relatively thick silicon deposited at 1200°C from halide sources. The nature and distribution of the defect structure in these deposits is quite different from that in silicon deposited using current preparative and substrate conditioning technology.

The microstructure of vacuum-evaporated silicon on sapphire has been examined by TEM in some detail by Linnington (1971). The films studied were deposited in an ultrahigh vacuum on annealed (1500°C in damp H_2) substrates at a substrate temperature of 950°C and at a deposition rate of 0.2 μm/min. More than 1 μm from the silicon/substrate interface, regions of orientation differing from that of the silicon matrix and twinned regions were observed. Essentially no free dislocations were detected. As the silicon/substrate interface was approached, the planar defects increased sharply. The microtwins were observed to terminate by intersection with other faults, with an associated increase in the density of free dislocations. Weak-beam electron microscopy (Howie and Basinski, 1968) was used in an attempt to resolve the defect structure near (<1 μm) to the silicon/substrate interface. Pyramids of stacking faults which originated from regions less than 20 Å in diameter were observed. Since it proved to be impossible to resolve the details of the structure of the silicon directly adjacent to the interface, it was concluded that the structure in this region was highly disoriented. Estimates were made from the TEM studies of the (planar) defect density as a function of the distance from the silicon/substrate interface. On a plot of the log of the defect density as a function of the distance from the interface, the data fall on a straight line between a density value of $\sim 1.5 \times 10^9$/cm² at the interface to $\sim 1 \times 10^7$/cm² at a distance of 1 μm from the interface. Linnington concluded that the origin of the defects in the silicon near the silicon/substrate interface resulted from etching of the substrate by silicon during the deposition. It is interesting to note that no evidence of

slip (commonly associated with strain relief) was observed in the silicon. The conclusion was drawn, therefore, that the stress resulting from cooling from the deposition temperature had been relieved by "reconfiguration" of the interface rather than by propagation of defects in the "bulk" of the film and that the reconfiguration was operable at temperatures below which defects in the silicon are immobile.

Cullis (1972), also using TEM, examined the microstructure of silicon vacuum deposited on the virgin surface of (0001) platelets of sapphire. The platelets were grown by chemical vapor deposition [White, 1972 (as cited by Cullis, 1972)]. The silicon was deposited in an ultrahigh vacuum at substrate temperatures of 900 and 1000°C. In the films grown at 900°C, Cullis also failed to observe strain-relieving defects. Evidence for reaction with the substrate was observed, which was more evident at a growth rate of 0.5 Å/s than at 20 Å/s. The early silicon growth centers were found to be composed of approximately equal proportions of two 60° rotational twins. Also observed were stacking faults and microtwins on inclined planes. In general the films less than 1 μm in thickness were described as being polycrystalline in nature. It should be recalled that, in general (Section 2.12), it is more difficult to achieve usable semiconducting properties in the (111) silicon on the (0001) sapphire than in the (100) silicon on the ($1\bar{1}02$) sapphire. Cullis investigated only the former orientation because of the availability of the naturally occurring basal plane facet on the CVD sapphire platelets.

Picraux and Thomas (1973) attempted to establish correlations between the microstructure observed by TEM (1 MeV), ion channeling, and the scanning microscopic examination of etched surfaces. The (111) and (100) silicon deposited by chemical vapor deposition on both sapphire and spinel were examined. The ion channeling and SEM results have been previously discussed. Using TEM these investigators observed, as have the previous workers already discussed, that the most prevalent defects in the silicon were microtwin lamellae and stacking faults and that the fault density decreased rapidly with distance from the silicon/substrate interface. They point out that it has been previously observed (Booker and Stickler, 1962) that the intersection of two stacking faults can result in the partial or complete cancellation of the faults and propose that this is one mechanism by which the fault density decreases with distance from the substrate.

Using conventional methods to prepare samples for transmission electron microscopy, the composite structure is thinned, and the observations are carried out in a direction normal to the plane of the substrate surface. Using this technique it is extremely laborious to derive the defect density of the heteroepitaxial silicon as a function of distance from the silicon/sub-

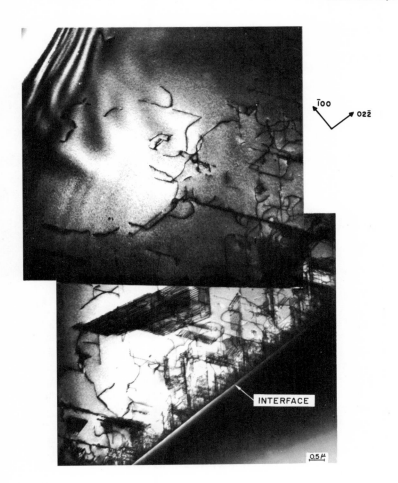

$\bar{1}00$

$02\bar{2}$

INTERFACE

0.5μ

FIGURE 2.36   A bright-field transmission electron micrograph of the cross section of (100) silicon on ($1\bar{1}02$) sapphire. From Abrahams and Buiocchi (1975)

strate interface. Each datum point on the density vs. distance plot is collected from a sample carefully thinned to include the silicon at the location of interest. If the density vs. distance from the interface relationship is to be examined for any number of preparative or substrate variables, the task becomes so tedious and laborious as to be unfeasible. To alleviate this problem, Abrahams and Buiocchi (1974) developed a technique to thin epitaxial deposits in such a way that the sample constitutes a cross section of the deposit that includes the substrate/deposit interface. The preparation of the cross-sectional samples was originally applied to the examination of chemically vapor-deposited III-V materials. Although the method is simple in concept, it is difficult in execution and takes a good deal of patience and skill. It is clear, however, that the information content of the cross-section sample is very large indeed, and in the long run it is a less laborious method to study the crystallographic characteristics of the deposit as a function of distance from the deposit/substrate interface.

A transmission electron micrograph of a cross sec-

tion of chemically vapor-deposited (100) silicon on ($1\bar{1}02$) sapphire is shown in Figure 2.36. The silicon was deposited from silane at a rate of 2 μm/min at 1000°C on hydrogen-annealed substrates (Cullen and Corboy, 1974). Included in the micrograph is the silicon/sapphire interface and about 6 μm of the silicon. The sapphire is featureless. Dislocations, microtwins, and isolated stacking faults are observed in the silicon deposit. It is clear that the fault density in the silicon deposit is a strong function of the distance from the silicon/sapphire interface. A log-log plot of the fault density as a function of the distance from the interface is shown in Figure 2.37. Only edge-on faults were counted. The fault density decreases from a value of 7 × 10⁵/cm at 0.044 μm to 2 × 10³/cm at a distance of 4.3 μm from the interface. At greater distances from the interface the fault density is ≤ 10²/cm (the limit of detection). There appears to be a break in the plot (using a least-squares fit) at ~0.2 μm from the interface. The fault density presented in the work of Abrahams and Buiocchi is approximately an order of magnitude greater than the values reported by Linnington,

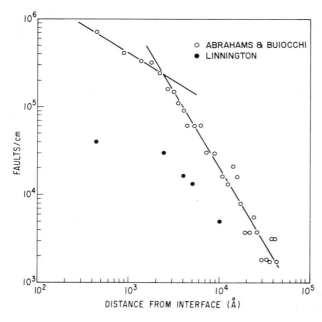

FIGURE 2.37   The fault density in chemically vapor-deposited (100) silicon on (1$\bar{1}$02) sapphire as a function of distance from the silicon/sapphire interface. After Abrahams and Buiocchi (1975)

but the shapes of the plots are similar. The origin of the break in the curve is not understood at this time. The TEM examination of the cross-sectional samples of the silicon on sapphire provides a means of establishing a physical relationship between the preparative variables (substrate and deposit) and the semiconducting properties of the heteroepitaxial deposit.

Abrahams and coworkers (1976) have also examined the formation and coalescence of the early growth islands. In this case, observations were made normal to the substrate surface. In order to realize control over deposition of the silicon prior to complete coverage of the substrate surface, the silicon was deposited at a rate of 0.4 μm/min. The morphology of the growth was examined after deposition periods of 0.5, 1.0, 2.5 and 3.5 s of growth (at 1000°C). It was observed that {110} domains nucleate independently of the (100) domains on the (1$\bar{1}$02) sapphire surface. The two domains grow at the same rate, but eventually the {110} domains are entrapped by the (100) domains. The volume percentage of the {110} domains prior to surface coverage differ from sample to sample. Attempts are in progress to relate the appearance of the {110}-domains with preparative variables. It is possible that this spurious orientation is the origin of the different physical and electrical properties previously observed in the silicon/substrate "interface layer." Blanc and Abrahams (1976) have developed a model for the early growth of silicon on sapphire based on their TEM examinations. An interesting conclusion is that under the high supersaturation commonly employed for heteroepitaxial growth of silicon on the insulators, the growth and coales-

cence are not dominated by surface energies. These investigators speculate that nucleation on silicon on sapphire occurs by a homogeneous surface mechanism and that debris and crystallographic defects are not involved in the nucleation process.

2.12. The semiconducting properties of silicon on sapphire and spinel

Electrical properties of interest for device application

For the fabrication of MOS transistor structures in either bulk or heteroepitaxial silicon, the magnitude of the carrier mobility (at the desired carrier concentration) is of primary concern. The transistor transconductance (i.e., the time required for the carrier to travel from the source to the drain) is proportional to the mobility. Thus, every effort is made to optimize the magnitude of the mobility.

For the fabrication of bipolar transistors, the magnitude of the carrier lifetime is the electrical property of most concern. In bipolar structures, the minority carriers injected from the emitter into the base region must live sufficiently long to be swept across the base to the collector. Some minimum, but not well defined, value of lifetime is required to fabricate bipolar transistors in the heteroepitaxial silicon.

Because of the current emphasis on the application of heteroepitaxial silicon in MOS transistor structures, the mobility—as a function of the preparative parameters and film thickness—is treated in more detail than are considerations involving the lifetime. Also, a considerably greater body of information has been developed on mobility than on lifetime for heteroepitaxial silicon.

Mobility measurements

Mobilities are measured by the Hall method (see Chapters 5 and 8) in heteroepitaxial silicon thin films in an effort to evaluate the applicability of the films in MOS transistor structures. Measuring carrier mobility this way is mechanically relatively simple and therefore provides rapid feedback to materials scientists involved in the development of silicon-on-insulator growth technology. There are factors, however, in the measurement of the Hall mobility, and in the relationship between Hall and MOS transistor mobilities, that warrant some comment in regard to the significance of mobility values. The Hall measurement is complicated by the fact that in the heteroepitaxial deposits the local mobility and carrier concentration are a strong function of the distance from the silicon/substrate interface, particularly for thicknesses less than 1 μm. The values derived from Hall measurements in thin films may also be influenced by depletion of low carrier concentration silicon ($<1 \times 10^{16}$/cm^3) by charge trapped at the silicon/silicon dioxide and the silicon/substrate interfaces. Thus, particular caution must be exercised in the analysis of Hall measurements made in thin, low-carrier-concentration films.

The influence of these factors has been brought out in attempts to explain the decrease in Hall mobility with decreasing carrier concentrations for concentrations less than about 1 × 10^{16}/cm^3. Schlötterer (1968) had attributed this effect to scattering by space charge regions around crystalline imperfections in the silicon. This was suggested by analogy to a model previously offered by Weisberg (1962) to account for a similar effect observed in epitaxial III-V materials. It was proposed that, as the carrier concentration decreases, the effective space charge region increases, resulting in degradation of the carrier mobility. In addition, Ham (1972) has suggested that the effect is associated with carrier depletion in the thin lightly doped films by surface charges at (or near) the silicon/silicon dioxide interface. If the films are partially depleted by surface charge, one cannot make the assumption that the current is evenly distributed throughout the thickness of the sample. For any fixed value of surface charge, as the carrier concentration decreases the thickness of the layer depleted from the silicon/silicon dioxide interface increases and the thickness of the layer in which the Hall measurement is made (i.e., the undepleted region) effectively decreases. While the average Hall mobility is independent of measured film thickness, in silicon on insulating substrates the mobility is degraded near the silicon/substrate interface

(Ipri, 1972A,B) by the relatively high impurity content and poor crystallinity. The net result is that the Hall mobility decreases with decreasing carrier concentration as the undepleted region is narrowed to material near the silicon/substrate interface.

The analysis of measured carrier concentration in lightly doped films is complicated not only by the possibility that the local carrier concentration is a function of the distance from the silicon/substrate interface but also by the limitation (by the space charge) of current to only a portion of the film, even though the entire film thickness is used in the calculation of the carrier concentration. Note that (unlike the mobility) the film thickness must be known in the calculation of the carrier concentration. Ham (1972) has demonstrated the influence of the surface charge by measuring the Hall voltage in a gated structure in which the film can be biased to flat band. This provides "bulk" values (i.e., values unaffected by surface charges) wherein it is assumed that the current flow is evenly distributed throughout the thickness of the sample.

If the mobility is measured in MOS transistor structures, only carriers remote from the silicon/substrate interface participate. This is true in both the enhancement mode and deep-depletion transistors, although for different reasons. In the conventional (ungated) Hall measurement, the properties near the silicon/substrate interface are weighted in thin ($<1 \mu$m) lightly doped ($<1 \times 10^{16}$/cm^3) films. In the gated-Hall-bar measurement (when the film is biased to flat band) the characteristics derived represent an integrated value of properties throughout the thickness of the film. Therefore, significantly different values for the transport properties may be derived using the different measurement techniques.

For the reasons outlined above, the Hall mobility vs. carrier concentration (unbiased) relationships cannot be used on an absolute scale to compare thin, lightly doped films with bulk silicon properties. On the other hand, if the surface charge is fixed immediately prior to the measurement by stripping the oxide in an aqueous HF solution (Ham, 1972), the Hall values are reproducible and characteristic of the specific deposition conditions, film thickness, and type of substrate. Thus, on a relative scale, the unbiased Hall values are useful in the development of deposition conditions and in the evaluation of the effect of film thickness and substrate orientation and composition.

The oxide/semiconductor surface charge is commonly minimized and held relatively constant in device processing by employing "clean" methods to grow a thermal oxide. Such a procedure could be used prior to making an ungated Hall measurement, but this obviates the possibility of making an "as-deposited" mobility measurement. Useful information on the

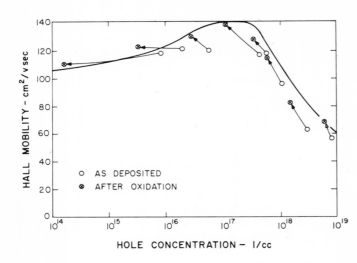

FIGURE 2.38   The hole mobility in 2-μm-thick (100) silicon on sapphire as a function of the hole concentration and thermal oxidation. After Dumin and Robinson (1968A)

effect of deposition conditions is derived by comparing Hall mobility measurements made before (in the "as-deposited" film) and after a thermal oxidation. For this evaluation, the oxidation conditions are designed to simulate the thermal treatment commonly employed during the processing of MOS transistor structures. The before- and after-oxidation Hall measurement can be carried out within hours of the completion of a deposition run, and it therefore provides much more rapid feedback than is possible if gated-Hall-bar or MOS transistor mobilities are measured.

Many of the problems in making meaningful Hall mobility measurements can be avoided if only films thicker than about 1.0 μm with carrier concentrations greater than about $2 \times 10^{16}/cm^3$ are used in the development of the deposition parameters. The optimization of thin low carrier concentration films has been important, however, for application in deep-depletion MOS transistor geometries, and it is desirable to fabricate devices in deposits less than 1 μm in thickness for a number of reasons discussed in Section 2.13 (the current standard thickness for device fabrication is 0.6 μm). The properties of the films become more sensitive to the deposition parameters with decreasing film thickness and carrier concentration. It is important to note that the properties of the after-oxidation films are very much more sensitive to deposition parameters than are the properties of the as-deposited silicon.

Despite these considerations, which cloud the relationship between Hall and device mobilities in the heteroepitaxial silicon films, good device characteristics have been realized in films developed by optimizing the Hall mobilities, and therein lies the justification for making Hall mobility measurements in the heteroepitaxial thin films.

The basic method used to measure Hall mobilities is described in Chapter 5, and factors influencing Hall mobilities, briefly outlined above, are discussed in more detail in Chapter 6. It is important to emphasize

that a mobility value is meaningful only if the film thickness, the carrier concentration, and the post-deposition thermal treatment of the sample is explicitly stated. Since one finds that all of these factors are often not specified in the literature, it has been difficult to compare the electrical properties of heteroepitaxial deposits prepared under different deposition conditions within various organizations.

The change in properties on thermal oxidation

In 1968 Dumin and Robinson reported on the carrier mobilities in 2 μm thick (100) silicon on (1$\bar{1}$02) sapphire as a function of the carrier concentration and thermal oxidation. The hole mobilities reported by these authors are shown in Figure 2.38, and the electron mobilities are shown in Figure 2.39. This work brought out a major problem that had to be solved in the preparation of the heteroepitaxial films for practical device application: on exposure to thermal processing similar to that used in device fabrication (at that time 1 hr in dry oxygen at 1200°C) the hole concentration decreased by as much as an order of magnitude (depending on the initial "as-deposited" value), and the electron mobilities increased by almost 100 percent with accompanying increases in the active electron concentration. At a carrier concentration of about $1 \times 10^{16}/cm^3$, the after-oxidation hole mobility of 120 cm²/V-s in the 2-μm-thick heteroepitaxial film represented about 30 percent of the bulk hole mobility (for a similar carrier concentration) while the after-oxidation electron mobility of 550 cm²/V-s represented approximately 50 percent of the bulk value. Although less specific in regard to the deposit thickness and thermal oxidation, Hart and coworkers (1967) reported similar hole mobilities at about the same time. Dumin and Robinson (1968A) referred to the thermal oxidation as "aluminum removal" on the assumption that aluminum (in an undefined oxidation state) was introduced into the deposit as the result of the reaction between silicon and the substrate and subsequently segregated into the silicon oxide during thermal oxidation.

In the following sections the change in electrical properties during and after thermal oxidation is used as an important quality factor in the optimization of the preparative procedures [see also Ross and Warfield (1969)].

The effect of temperature and rate of deposition on the carrier mobility

The changes in electrical properties during thermal oxidation have been suppressed by a combination of lowering the deposition temperature and increasing the deposition rate (Dumin et al., 1970; Cullen et al., 1970B). The chemical considerations involved in adjusting these deposition parameters have been discussed in a previous section. The influence of deposition temperature on the hole mobilities and change in hole mobilities caused by thermal oxidation in 1.5-μm-thick silicon on sapphire is shown in Table 2.7. These data were collected during the development of the "dual-rate" deposition technique, whereby a high growth rate was initially employed to rapidly cover the surface subsequently lowered during the latter part of growth of the film. It is noted that a good deal of scatter is observed in the characteristics of the heteroepitaxial silicon films deposited at this state of development in the technology. This is observed in the reports of efforts to optimize both the temperature and rate of silicon deposition. Nonetheless, sufficient trends were observed to indicate the most favorable growth conditions. In hindsight, the scatter can be associated largely with variation in the quality of the substrate surface finishing. The resistivities of the films deposited at temperatures below 975°C were too high to make meaningful Hall mobility measurements,

presumably because of poor crystallinity. The after-oxidation mobility value of the film deposited at 975°C represents approximately 34 percent of the bulk mobility. The after-oxidation mobilities for films deposited between 1000 and 1050°C were typically 55–65 percent of bulk. Above 1050°C the effect of autodoping was observed, and the after-oxidation mobilities were similar to the values observed in films deposited at 975°C. It is not unusual for the after-oxidation mobility values to be more reproducible than the as-deposited values. The as-deposited mobility value of 265 cm²/V-s for one of the films deposited at 1000°C was unusually high, while the after-oxidation value was within the range expected. In such a case, the absolute value of the after-oxidation value is more meaningful than the magnitude of the change during oxidation. In general, the after-oxidation mobilities are the highest and the change in both the mobility and carrier concentration the lowest for films deposited at 1000°C. The relatively large change in both carrier concentration and mobility observed for the film deposited at 1090°C can be associated with contamination of the film by substrate constituents at the relatively high deposition temperatures. Therefore, a deposition temperature of ~1000°C had been adopted as the optimum temperature for the deposition of (100) silicon on the (1$\bar{1}$02) sapphire.

Using similar procedures, a temperature of 1100°C was adopted as the optimum temperature for the growth of silicon on the flame fusion alumina-rich spinel (Cullen et al., 1969). That a higher growth temperature could be used for deposition on the flame fusion spinel substrates than for deposition on sapphire may be associated with the observation that the flame fusion spinel is less reactive (on exposure to the silicon deposition ambient) than the sapphire (Robinson and Dumin, 1968). In the spinel substrate system, the chemical reactivity of the surface proved to be a function of the method by which the substrate was

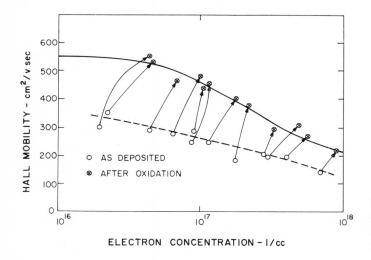

FIGURE 2.39   The electron mobility in 2-μm-thick (100) silicon on sapphire as a function of the electron concentration and thermal oxidation. From Dumin and Robinson (1968A)

TABLE 2.7   Effect of Deposition Temperature on the Electrical Properties on Silicon on Sapphire

Deposition Temperature (°C)	Thermal Treatment	Hole Concentration, n_a ($1/cm^3$)	Mobility, μ ($cm^2/V\text{-}s$)	Percentage of Bulk Mobility	Decrease in Percentage of Bulk Mobility on Oxidation
935	As deposited				
940	As deposited	Resistivity too high to measure Hall Constants			
965	As deposited				
975	As deposited	8.3×10^{15}	160	45	
	oxidized	1.5×10^{16}	115	34	11
1000	As deposited	1.5×10^{16}	265	78	
	oxidized	1.5×10^{16}	195	57	21
1000	As deposited	1.7×10^{16}	225	67	
	oxidized	2.1×10^{16}	220	66	1
1050	As deposited	1.3×10^{16}	190	55	
	oxidized	1.9×10^{16}	195	58	+3
1090	As deposited	3.8×10^{16}	195	62	
	oxidized	6.8×10^{15}	110	31	31

Source: Gottlieb and Corboy (1972).
Substrate: $(1\bar{1}02)$ Czochralski sapphire.
Film thickness: 1.5 μm.
Deposition rate: 2 μm/min for 30 s, then 0.3 μm/min to thickness.
Oxidation: 1 hr in dry 0_2 at 1000°C.

grown (Cullen and Wang, 1971), and it has therefore been necessary to take into consideration the specific chemical nature of the various substrates in the development of the deposition conditions (Cullen and Dougherty, 1972).

The influence of the growth rate on the hole mobility in 1.5-μm-thick $(100)Si/(1\bar{1}02)Al_2O_3$ is shown in Table 2.8. The rate shown is the rate used for the first 30 s of growth of the dual-rate method. As was observed on examination of the mobility vs. deposi-

tion temperature data, the significant value to examine is the after-oxidation hole mobility. Although the absolute mobility values realized in this series are lower than those observed in Table 2.7, it can be seen that the after-oxidation mobilities peaked and were reproducible at a value of 45–46 percent of bulk using a deposition rate of 2 μm/min.

Similar hole mobility vs. growth rate data are shown for 1.5-μm-thick $(111)Si/(111)$spinel in Table 2.9. In this case the rates shown were employed for the depo-

TABLE 2.8   Effect of Growth Rate on the Electrical Properties of Silicon on Sapphire

Growth Rate (μm/min)	Thermal Treatment	Hole Concentration, n_a ($1/cm^3$)	Mobility, μ ($cm^2/V\text{-}s$)	Percentage of Bulk Mobility	Decrease in Percentage of Bulk Mobility on Oxidation
0.5	As deposited	1.6×10^{16}	115	33	
	oxidized	1.3×10^{16}	60	17	16
1.5	As deposited	4.9×10^{15}	125	34	
	oxidized	8.2×10^{15}	110	31	3
2.0	As deposited	1.2×10^{16}	220	64	
	oxidized	1.9×10^{16}	150	46	18
2.0	As deposited	5.7×10^{15}	185	51	
	oxidized	1.0×10^{16}	160	46	5
2.0	As deposited	2.2×10^{15}	210	56	
	oxidized	5.2×10^{15}	165	45	11
3.0	As deposited	1.3×10^{16}	175	51	
	oxidized	2.9×10^{16}	115	35	16

Source: Gottlieb and Corboy (1972).
Substrate: $(1\bar{1}02)$ Czochralski sapphire.
Film thickness: 1.5 μm.
Deposition temperature: 1000°C.
Deposition rate: as stated for 30 s, then 0.3 μm/min to thickness.

TABLE 2.9   The Effect of Growth Rate on Stability During Thermal Oxidation in Silicon on Spinel with Similar As-Deposited Properties

Growth Rate (μm/min)	Thermal Treatment	Hole Concentration, n_a ($1/cm^3$)	Mobility, μ (cm^2/V-s)	Percentage of Bulk Mobility	Decrease in Percentage of Bulk Mobility on Oxidation
0.4	As-deposited	4.5×10^{15}	350	95	
	oxidized	1.5×10^{15}	55	13	82
0.8	As-deposited	5.6×10^{15}	335	92	
	oxidized	4.2×10^{15}	225	60	32
2.0	As-deposited	4.9×10^{15}	330	91	
	oxidized	4.0×10^{15}	300	82	9

Source: Cullen et al. (1969).
Substrate: (111) flame fusion spinel.
Deposit thickness: 1.5 μm.
Deposition temperature: 1100°C.
Oxidation: 1 hr in dry 0_2 at 1100°C.

sition of the entire deposit. It should be noted that the as-deposited hole concentrations and mobilities were very similar and that the deposition rate had a striking influence on the change in the mobility during thermal oxidation. Using a growth rate of 0.4 μm/min, the mobility decreased from a value of 350 to 55 cm^2/V-s on oxidation; while using a growth rate of 2.0 μm/min, the hole mobility decreased only from 330 to 300 $cm^2/$V-s on oxidation. It has been demonstrated (Section 2.7) that the thickness at which the surface is covered (~200 Å) is similar between growth rates of 0.4 and 2.0 μm/min, and therefore the decrease in the degradation of mobility with increasing growth rate must be related primarily to the duration of exposure of the partially covered substrate. Whereas the mobilities of silicon on Al_2O_3 appeared to have peaked at a rate of 2.0 μm/min, in silicon on spinel the after-oxidation

mobilities were not degraded, and similar values are observed for rates as high as 5.4 μm/min (Cullen, 1971).

The effect of the silicon thickness

The effect of the heteroepitaxial silicon thickness on hole mobility is summarized in Figure 2.40 for data collected prior to 1971. Although there has been further optimization of the electrical properties of the films less than 1 μm in thickness since that time, there has been very little data reported since 1971 for the thicker films, presumably because of the inapplicability of the thicker films in MOS transistor structures. The carrier concentration of the n- and p-type (100)Si/(1$\bar{1}$02)sapphire and (111)Si/(111)spinel films shown in Figure 2.30 were all between 1 and 5 $\times$ $10^{16}/cm^3$. It is

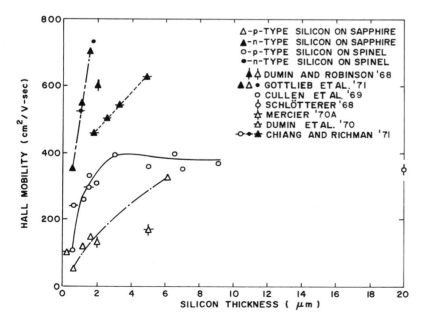

FIGURE 2.40   The hole and electron mobility in as-deposited (100) silicon on sapphire and (111) silicon on spinel as a function of the film thickness. The carrier concentrations are between 1 and $5\times10^{16}/cm^3$. Figure is after Cullen (1971); data sources are as shown

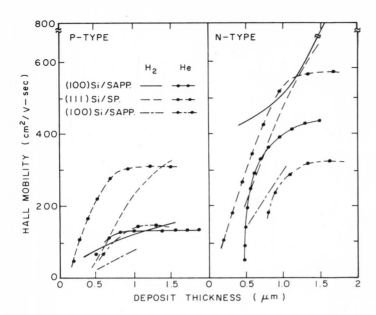

FIGURE 2.41   The Hall mobility as a function of thickness in silicon deposited in hydrogen and helium ambients. He ambient data: Chiang and Looney (1973); H₂ ambient data: Cullen and Corboy (1974). H₂ ambient data are extended to 2 μm thickness from data included in Cullen et al. (1969) and Gottlieb and Corboy (1972)

interesting to note that the hole mobilities in silicon on sapphire increase to values similar to bulk values at a thickness ~6 μm, whereas the silicon-on-spinel hole mobilities are similar to bulk at thickness of ~3 μm. In this case the (111) orientation of silicon on spinel was compared to the (100) orientation of silicon on sapphire because the highest mobility values were observed in these orientations on the two substrate materials.

More recent data for the Hall mobility in deposits less than 2 μm in thickness on both sapphire and spinel are shown in Figure 2.41. Silicon was deposited on hydrogen-annealed spinel in a helium ambient (Chiang and Looney, 1973) and on air-annealed spinel in a hydrogen ambient (Cullen and Corboy, 1974). In both studies the sapphire substrates were hydrogen annealed. The carrier concentration of the silicon deposited in helium was 1–8 × 10¹⁶/cm³, and in hydrogen ~1 × 10¹⁶/cm³. For these levels it is assumed that the influence of the surface charge (Ham, 1972) is negligible even for thickness less than 1 μm. It is evident that the Hall mobility is a strong function of the thickness in all the deposits for thicknesses less than 1 μm. No advantage was seen in depositing silicon on sapphire in a helium rather than a hydrogen atmosphere. This is consistent with the observation, previously discussed, that the chemical erosion of sapphire can be assigned to the reducing action of silicon rather than hydrogen. The hole mobilities of silicon deposited on (111) spinel were significantly higher for films deposited in helium than in hydrogen. More modest increases were seen in the hole mobilities of silicon deposited in helium on the (100) spinel and electron mobilities on the (111) spinel.

A summary of the mobilities observed in 0.5- and 1.0-μm-thick silicon on sapphire and spinel as a function of carrier concentration

The Hall mobility values as a function of carrier concentration for 1.0- and 0.5-μm p- and n-type silicon on sapphire and spinel are summarized in Figures 2.42 through 2.47. A growth rate of 2 μm/min was employed in the development of these data. Although it had proven to be advantageous to employ the dual-rate method (Gottlieb and Corboy, 1972) for films 1.5 μm thick, this procedure is difficult to execute for films 1 μm and less in thickness (i.e., the growth duration is 30 s for the 1-μm-thick films and 15 s for the 0.5-μm-thick films). A growth temperature of 1000°C was employed for the deposition of silicon on sapphire and 1100°C for the deposition of silicon on spinel. The mechanically polished sapphire substrates were annealed in a hydrogen ambient for 30 min at 1300°C, whereas the flame fusion spinel substrates are annealed for the same period in hydrogen at 1150°C. Spinel substrates grown by the Czochralski method were also included in the collection of these data. The highest mobilities observed on these substrates were similar to those observed on the flame fusion spinel if the Czochralski spinels are annealed in air for 5 hr at 1500–1600°C.

Before- and after-oxidation mobility values have not been presented in the figures because, under the deposition conditions described, the change in mobility on thermal oxidation has been suppressed to levels acceptable for device fabrication. The changes can be summarized as follows: in 1.0-μm silicon on both sapphire and spinel, for carrier concentrations greater

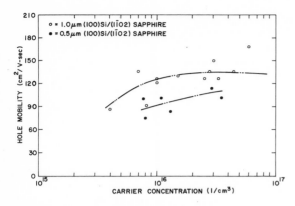

FIGURE 2.42   Hall mobility as a function of hole concentration in 0.5- and 1.0-μm-thick (100)Si(1$\bar{1}$02) sapphire. After Cullen and Corboy (1974)

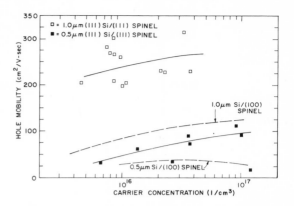

FIGURE 2.44   Hall mobility as a function of hole concentration in 0.5- and 1.0-μ-thick (111)Si/(111)spinel. Curves for (100)Si/(100)spinel are included (without data points) for reference. After Cullen and Corboy (1974)

than about 2×10^{16}/cm³, the carrier mobility typically changes by less than a factor of 1.5 on thermal oxidation. In films less than 1 μm in thickness, and for carrier concentrations less than 2×10^{16}/cm³, the magnitude of the changes on thermal oxidation is scattered and difficult to use as a "material quality" factor. Clear conclusions in this respect must be derived from device data.

Typical hole mobilities, as a function of the hole concentration, are shown for 0.5- and 1.0-μm-thick (100) silicon on (1$\bar{1}$02) sapphire (Figure 2.42), (100) silicon on (100) spinel (Figure 2.43), and (111) sillicon on (111) spinel (Figure 2.44). In order to facilitate comparison of these data, the curves presented in Figure 2.42 are reproduced (without data points) in Figure 2.43, and the curves of Figure 2.43 are reproduced in Figure 2.44.

Typical electron mobilities, as a function of the electron concentration, are shown for 0.5- and 1.0-

μm-thick (100) silicon on (1$\bar{1}$02) sapphire (Figure 2.45), (100) silicon on (100) spinel (Figure 2.46), and (111) silicon on (111) spinel (Figure 2.47). As with the p-type data, the curves of Figure 2.45 are superimposed on Figure 2.46, and the curves of Figure 2.46 are included in Figure 2.47. At the time the data were taken, comparable mobility values had not been reported in (111) silicon on sapphire.

A comparison of Hall and MOS transistor mobilities

The Hall mobilities are summarized and compared with effective (MOS transistor) mobilities in Table 2.11. Also included are the percentage of bulk mobility values. Hall mobilities of 120 cm²/V-s are typically observed in 1-μm-thick p-type (100)Si/(1$\bar{1}$02)sapphire. On the other hand, the effective mobilities of MOS fabricated in films of the same thickness are typically 210 cm²/V-s. Although significantly lower Hall mobili-

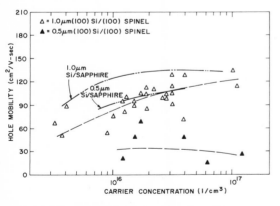

FIGURE 2.43   Hall mobility as a function of hole concentration in 0.5- and 1.0-μm-thick (100)Si/(100)spinel. Curves for (100)Si/(1$\bar{1}$02)sapphire are included (without data points) for reference. After Cullen and Corboy (1974)

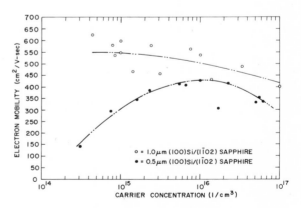

FIGURE 2.45   Hall mobility as a function of electron concentration in 0.5- and 1.0-μm-thick (100)Si/ (1$\bar{1}$02)sapphire. After Cullen and Carboy (1974)

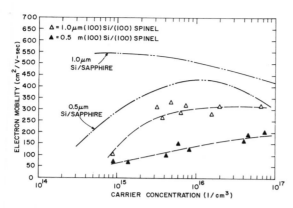

FIGURE 2.46   Hall mobility as a function of electron concentration in (100)Si/(100)spinel. Curves for (100)Si/(100)sapphire are included (without data points) for reference. After Cullen and Corboy (1974)

ties of 80 cm²/V-s are observed in 1-μm-thick (100)Si/(100)spinel, the effective mobilities of MOS transistors fabricated in 1-μm-thick silicon on the two substrates are similar. The mobilities in (100) silicon on spinel degrade more rapidly with decreasing film thickness than do the mobilities in (100) silicon on sapphire. In 0.5-μm-thick films, hole mobilities are typically 90 cm²/V-s in (100)Si/(1$\bar{1}$02)sapphire and 30 cm²/V-s in (100)Si/(100) spinel. Insufficient data have been collected in p-channel MOS transistors fabricated in the 0.5-μm-thick (n-type) films to make meaningful comparisons between the effective device and Hall mobilities in these films.

Relatively high (Hall) hole mobilities of 240 cm²/V-s are typically observed in (111)Si/(111)spinel, but as in the case of the (100)Si/(100)spinel, the electrical properties are a strong function of the film thickness, and the mobilities drop to 50 cm²/V-s in 0.5-μm-thick (111)Si/(111)spinel.

As in the case of the hole mobilities, the effective electron mobilities in the 1-μm-thick films of 420 cm²/V-s in (100)Si/(1$\bar{1}$02)sapphire and 350 cm²/V-s in (100)Si/(100)spinel are more similar than the Hall electron mobilities of 500 cm²/V-s and 300 cm²/V-s, respectively. The effective n-channel mobilities are lower than the Hall electron mobilities in (100)Si/(1$\bar{1}$02)sapphire but are slightly higher in (100)Si/(100)spinel. This same trend in electron mobilities is observed in the 0.5-μm-thick silicon, except that with the degradation of the Hall electron mobilities in the thinner (0.5-μm) (100)Si/(100)spinel, the effective p-channel value of 250 cm²/V-s is 1.8 times the Hall electron mobility of 140 cm²/V-s. The corresponding values in the 0.5-μm-thick (100)Si/(1$\bar{1}$02)sapphire are p-channel mobility of 340 cm²/V-s and a Hall electron mobility of 430 cm²/V-s.

Although the Hall hole mobilities in 1.0-μm-thick (111)Si/(111)spinel were double the values observed in

(100)Si/(1$\bar{1}$02)sapphire, this relationship does not hold in the n-type films. The Hall electron mobility in 1.0-μm-thick (111)Si/(111)spinel of 480 cm²/V-s is very similar to the values of 500 cm²/V-s typically observed in (100)Si/(1$\bar{1}$02)sapphire. As was previously observed, the Hall mobilities in the silicon on spinel are a strong function of the thickness, and the Hall electron mobility of 200 cm²/V-s in 0.5-μm (111)Si/(111)spinel is less than half that observed in the 1.0-μm-thick film.

Since the unbiased Hall mobility measurement is weighted by the properties of the material near the silicon/substrate interface, and the (enhancement mode) device measurement is characteristic of the carrier mobilities only within about 500 Å of the silicon surface most remote from the silicon/substrate interface, it is not unexpected that the device mobilities are significantly higher than the Hall mobilities for holes in silicon on both sapphire and spinel and for electrons in silicon on spinel. It is more difficult to explain, however, why the device electron mobilities in the silicon on sapphire are lower than the Hall electron mobilities measured in equivalent films. In general, processing results in the degradation in the carrier mobility even in bulk silicon devices. Expressed in terms of the percentage of equivalent values observed in bulk silicon, the percentage of effective device mobilities is in every case considerably in excess of the percentage of the Hall mobilities.

It is interesting to note that the Hall mobilities in 0.5- and 1.0-μm-thick films are more similar in silicon on sapphire than in silicon on spinel. It seems reasonable to relate this to the nature of surface coverage, discussed earlier (Section 2.7). The thickness of the heavily contaminated silicon formed prior to complete surface coverage is ~300 Å in the silicon on sapphire and ~1000Å in the silicon on spinel. The spinel substrates used in this study were annealed at 1500–1600°C for 5 hr in order to minimize the differences

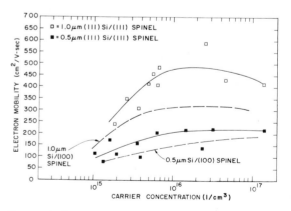

FIGURE 2.47   Hall mobility as a function of electron concentration in (111)Si/(111)spinel. Curves for (100)Si/(100)spinel are included (without data points) for reference. After Cullen and Corboy (1974)

TABLE 2.10   Typical Carrier Mobility Values for 1.0- and 0.5-μm Silicon on Sapphire and Silicon on Spinel for a Carrier Concentration of $\sim 1 \times 10^{16}$/cm^3

Film Type	Film Thickness (μm)	Substrate Composition and Orientation	Hall Mobility (cm^2/V-s)	Percentage of Bulk	Effective Mobility (cm^2/V-s)	Percentage of Bulk
p-type		Bulk silicon	350		200	
p-type	1.0	(100)Si/($1\bar{1}02$)sapphire	120	34	210	100
p-type	0.5	(100)Si/($1\bar{1}02$)sapphire	90	26	—	
p-type	1.0	(111)Si/(111)spinel	240	69	—	
p-type	0.5	(111)Si/(111)spinel	50	14	—	
p-type	1.0	(100)Si/(100)spinel	80	23	210	100
p-type	0.5	(100)Si/(100)spinel	30	9	—	
n-type		Bulk silicon	1100		500	
n-type	1.0	(100)Si/($1\bar{1}02$)sapphire	500	45	420	84
n-type	0.5	(100)Si/($1\bar{1}02$)sapphire	430	39	340	68
n-type	1.0	(111)Si/(111)spinel	480	44	—	
n-type	0.5	(111)Si/(111)spinel	200	18	—	
n-type	1.0	(100)Si/(100)spinel	300	27	350	70
n-type	0.5	(100)Si/(100)spinel	140	13	250	50

Source: Cullen and Corboy (1974).

between the silicon deposited on the flame fusion and on Czochralski-grown materials. It should be mentioned that the thickness at which the spinel is completely covered with silicon is a function of the annealing treatment.

Carrier mobilities in silicon deposited on sapphire orientations other than ($1\bar{1}02$)

The (100)Si/($1\bar{1}02$)Al$_2$O$_3$ composite orientation has, to date, been the most widely investigated and utilized in MOS device structures. As has been discussed in previous sections, relatively high carrier mobilities and low autodoping levels have been realized in thin ($\leq$ 1 μm) silicon deposited on the ($1\bar{1}02$) sapphire. Of the various other sapphire orientations on which single crystal silicon has been observed (see Section 2.2), definitive electrical properties have been reported only for (111) silicon deposited on (0001), ($11\bar{2}0$), and ($10\bar{1}4$) sapphire. It is evident from the work of Dumin (1967B) and Manasevit and coworkers (1968, 1974) that the deposition temperature at which the optimum electrical properties are achieved in silicon on sapphire is a function of the sapphire orientation. Hole mobilities of 200 cm^2/V-s and electron mobilities of 300 cm^2/V-s have been observed in thick ($>$4 μm) (111) silicon deposited on the (0001) sapphire at temperatures $\sim$1200°C (Dumin, 1967B). More recent efforts to achieve useful semiconducting properties in the thinner ($\sim$1 μm) silicon films deposited on the sapphire basal plane have not proven to be fruitful; the high growth temperatures required for deposition on this substrate orientation have been associated with heavy autodoping. Although Manasevit and coworkers (1968) identified single crystal (111) silicon on a

number of sapphire orientations some years ago, the electrical properties of the (111) silicon on ($10\bar{1}4$) and ($11\bar{2}0$) sapphire have only recently been reported by Manasevit and his coworkers (Ruth et al., 1973; Manasevit et al., 1974; Manasevit et al., 1976). In 2-μm-thick (111) silicon deposited at 1075°C, hole mobilities of $\sim$290 and 260 cm^2/V-s were observed, respectively, on ($11\bar{2}0$) and ($10\bar{1}4$) sapphire (the hole concentrations were in the 10^{16}/cm^3 range after a 30-min exposure to O$_2$ and 120-min exposure to N$_2$ at 1100°C). These hole mobilities are the highest reported to date in silicon, of equivalent thickness, on sapphire (compare with Figure 2.31). In the n-type films, the magnitude of mobilities decreases with substrate orientation: ($11\bar{2}0$) $>$ ($1\bar{1}02$) $>$ ($10\bar{1}4$).

Considering the background of information developed on the effect of autodoping at the higher growth temperatures, it is interesting to note that Manasevit and coworkers (1976) have claimed that the electrical properties of 2-μm-thick (100) silicon deposited on the ($1\bar{1}02$) sapphire at temperatures as high as 1100°C are as good as the best properties realized in films deposited at 1050°C if they are exposed to the O$_2$-N$_2$ anneal. It was also noted that the mobilities of both the (100)Si/($1\bar{1}02$)Al$_2$O$_3$ and (111)Si/($11\bar{2}0$)Al$_2$O$_c$ were nearly independent of growth temperature between temperatures of 1040 and 1100°C. It would be interesting to examine whether or not these relationships hold true for deposits less than 1 μm in thickness.

In films less than 1 μm thick there has been no report to date of carrier mobilities in (111) silicon on the ($11\bar{2}0$) or ($10\bar{1}4$) orientations significantly in excess of mobilities observed in the (100) silicon on the ($1\bar{1}02$) sapphire, Nonetheless, the preliminary work on the ($11\bar{2}0$) and ($10\bar{1}4$) orientations is promising, and the possibility of achieving thin (111) silicon on sapphire with useful semiconducting properties warrants further investigation. This orientation could be particularly useful in providing flexibility in the integration of silicon with other heteroepitaxial materials on sapphire substrates. In a number of applications of the heteroepitaxial III–V and II–VI materials, the device characteristics are more a function of the deposit orientations than is the case for MOS transistors (see Chapters 3 and 4). It should be noted, however, that the density of states at the silicon/silicon oxide (gate oxide) interface is a function of the silicon orientation and decreases in the following order: (111) > (110) > (100) [see Grove (1967)].

Gated-Hall measurement

In order to determine the effect of deposit thickness on the electrical properties, mobility and carrier concentrations have commonly been measured in heteroepitaxial silicon deposited to various thicknesses. The assumption is made that the properties of a 0.5-μm-thick film are the same as the properties within 0.5 μm of the interface of a thicker film and that reproducibility is achievable from one deposition run to another. On the other hand, in order to avoid the possibility that the properties are influenced by these variables, and to reduce the labor involved, attempts have been made to derive electrical properties as a function of thickness by thinning the heteroepitaxial silicon by chemical and mechanical methods or by sputter erosion. It has proven to be very difficult to thin the highly defected silicon films and maintain thickness uniformity, surface smoothness, and freedom from surface work damage, and the use of these procedures has therefore resulted in the introduction of more variables than have been avoided. Recently, methods have been developed to electrically control the thickness of the current-carrying portion of the deposit without physically altering the structure. The gated-Hall-bar structure (Ham, 1972) has previously been discussed for the measurement of "bulk" mobilities (i.e., mobilities uninfluenced by surface charges) in the thin silicon deposit. Using this same structure, the silicon below the gate can be inverted to controlled depths by the application of gate voltage (Ipri, 1972A,B; Elliot and Anderson, 1972). This procedure is, of course, similar to the operation of one type of MOS transistor, except that the silicon is depleted to controlled depths rather than completely to the silicon/substrate interface. The relation between the gate bias and the depletion depth is determined by capacitance measurement techniques. The average electron mobility of the silicon under the inverted layer, for various thicknesses of the inversion layer, is shown in Figure 2.48. Data for four 1-μm-thick samples, doped to 2–4 $\times$ 10^{16}/cm^3, are shown in the figure. Under the conditions used, and for this doping level, the silicon could be inverted only to a depth of ~0.3 μm. It can be seen that as the 0.3-μm layer of material most remote from the silicon/sapphire interface is excluded from carrying current, the average mobility decreases by more than 50 percent. From these data the average carrier concentration in the uninverted portion of the silicon can be calculated. The average net donor concentration, normalized to the thickness of the conducting portion of the deposit, is shown as a function of the thickness of the inverted region in Figure 2.49. Similar plots have been developed for the hole mobility and concentration.

In an effort to develop an understanding of the origins of the change in electrical properties with distance from the silicon/substrate interface, Ipri and Zemel (1973) measured the carrier concentration and mobility as a function of both the depth of the inverted region and temperature. The normalized (in relation to

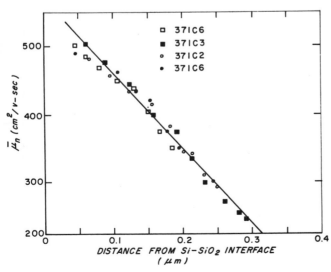

FIGURE 2.48   Average Hall mobility for electrons as a function of the depth of the inverted layer. The numbers refer to four similar 1-μm-thick samples doped to a level 2–4 $\times$ 10^{16}/cm^3. After Ipri (1972A)

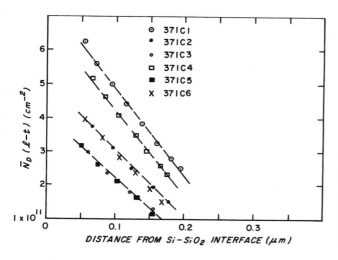

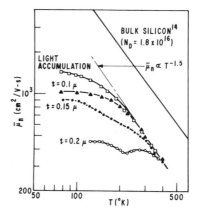

FIGURE 2.49   "N" type doping concentration × the conducting film thickness as a function of distance from the film surface. The product of the donor doping level and the thickness of the conducting portion of the film as a function of the depth of the inverted layer. The numbers refer to 1-μm-thick samples doped to an average level of 2–4 × 10^{16}/cm^3. After Ipri (1972A)

thickness) electron concentration as a function of the reciprocal temperature, for various depletion layer widths, is shown in Figure 2.50. It is interesting to note that, as the depleted region is forced deeper into the silicon (by applying potential to the gate), the curves become characteristic of the presence of two ionization levels. Ipri and Zemel associate the shallow level ($\epsilon_i = 0.052$ eV) with the intentionally added arsenic, whereas the deeper ionization state ($\epsilon_i = 0.1$ eV) is associated with a silicon–oxygen complex. It is reasonable to assume that this complex, which increases in density with depth into the silicon layer, is associated with constituents introduced into the film

(prior to complete coverage of the surface) as the result of the reaction between silicon and sapphire. The presence of aluminum–oxygen complexes has also been proposed by Dumin and Ross (1970). The presence of the silicon–oxygen complexes is invoked by these investigators to explain the leveling off of mobility at lower temperatures. Here again, the effect is a function of thickness in the film (see Figure 2.51). It is suggested that at the lower temperature the complex is not ionized and acts as a neutral scattering center impurity.

The normalized hole concentration was also measured as a function of the reciprocal temperature. In this case, there is no evidence of an acceptor level other than that characteristic of the intentionally added boron. As in the case of n-type films, however, it was observed that the hole mobilities level off with decreasing temperature, and that the extent of the leveling is a function of the depth of the depletion

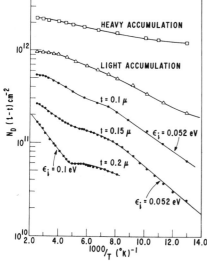

FIGURE 2.50   The product of the electron concentration and conducting film thickness as a function of reciprocal temperature. The depth of the inverted layer is indicated on each curve. After Ipri and Zemel (1973)

FIGURE 2.51   The average electron mobility as a function of temperature. The depth of the inverted layer is indicated on each curve. After Ipri and Zemel (1973)

layer. Ipri and Zemel attribute this effect to the presence of a boron–SiO_y^+ complex, which is expected to influence the mobility but not significantly alter the hole concentration.

Similar mobility vs. temperature characteristics for films of various thicknesses (depletion of the films was not employed) have been observed by Dumin and Ross (1970).

Tihanyi (1972, 1973) has also studied the electrical properties of the heteroepitaxial silicon deposits as a function of film thickness, using inversion of a portion of the layer to control the thickness of the current-carrying region. In the MOS transistor test structure employed, Tihanyi was able to extend the depth of the inverted region by making electrical contact to the inversion layer and applying additional voltage in the direction normal to the substrate surface. Using this technique he has measured the mobility and carrier concentrations completely to the silicon/substrate interface. He has concluded that the donor concentration of a 0.9-μm-thick deposit is homogeneous throughout the thickness at a level of 4×10^{15}/cm³ but that the mobility is at a minimum in the silicon near the interface and increases to values similar to bulk MOS mobilities in the silicon just below the gate electrode. On the issue of the doping profile, there is some disagreement between the work of Ipri and Tihanyi that cannot be easily reconciled on the basis of the differences in the measurement techniques.

Hynecek (1974B) also used MOS Hall measurements wherein the sample is forced into inversion or accumulation. He analyzes the results in terms of two carriers (i.e., A1 from autodoping and the intentionally added As or P). Making the assumption that both n and p mobilities are independent of the carrier concentration, Hynecek calculates an average Hall mobility as a function of the sheet resistivities. The "true" electron mobility (extrapolated to sheet resistivity ≈ 0) is 450 cm²/V-s.

Stress effects; degradation, enhancement, and anisotropy in Hall mobilities

Because of the differences in the thermal contraction of the silicon and the substrate, stress is induced in the silicon film when composite is cooled from the silicon deposition temperature to room temperature. At the lower temperatures, stress cannot be relieved by the formation of dislocation networks (see Chapter 8) and thus can be relieved only by bending of the composite. A number of investigators have measured the residual room temperature stress in silicon on the insulating substrates. Ang and Manasevit (1965) derived the stress from the deformation of a composite made up of silicon deposited on a basal plane sapphire whisker. The silicon was deposited by the hydrogen reduction of silicon tetrachloride. Dumin (1965) calculated the

stress in silicon deposited by the pyrolysis of silane on ($1\bar{1}02$) sapphire by measuring the circular composite's degree of bowing. Schlötterer (1968) calculated the stress from the bowing of silicon deposited from silane on both (100) and (111) magnesium aluminate spinel. Kamins and Meieran (1973) calculated the stress in silicon deposited from silane on ($1\bar{1}02$) sapphire from deviation of the lattice parameter of the heteroepitaxial silicon from the bulk value. Considering the differences in method of deposition, the nature of the substrate, and method of measurement, unusually good agreement was obtained by all of these investigators; all values reported fall within the range 5–11.5 × 10⁹ dyne/cm². It is quite clear that the crystallographic natures of both the silicon and the substrate examined by these investigators are markedly different. This demonstrates that the overall residual room temperature stress in the silicon is not strongly dependent on the crystalline defect structure of the silicon deposit.

Ang and Manasevit (1965) point out that the theoretical yield strength of silicon is ~2.5 times the stress values measured [reference is made to Menter and Pashley (1959)]. They also point out that the tensile yield strength of silicon (8 × 10⁹ dyne/cm²) is similar to the stress values measured in the silicon on insulators [reference is made to Sylwestrowicz (1962)]. Fortunately, the coefficient of thermal expansion of both sapphire and spinel is larger than that of silicon, and therefore the thin films are under compression and are crack free.

Schlötterer (1968) calculated that the piezoresistive effect is expected to lead to a degradation of Hall mobility in n-type (100) silicon and enhancement of the mobility in the p-type (111) silicon. He shows agreement, above carrier concentrations $\approx 10^{17}$/cm³, between calculated strained silicon values and experimentally measured (by the van der Pauw method) in 20-μm-thick n-type (100) silicon and p-type (111) silicon on spinel. Although the establishment of (111) hole mobility values in heteroepitaxial silicon significantly higher than bulk silicon values seems somewhat tenuous, it seems reasonable that the piezoelectric effect may compensate (to some extent) for degradation by grown-in defects and contribute to relatively high experimental hole mobilities reported by Schlötterer and later observed in thinner (1.5-μm) heteroepitaxial deposits (Cullen et al., 1969). Ipri and Zemel (1973) also analyzed the degradation of the Hall electron mobility in (100) silicon on ($1\bar{1}02$) sapphire in terms of the piezoelectric effect but pointed out that theoretically the compressive stress should have no effect on the Hall mobility of p-type films of the same orientation. This is in agreement with the calculations of Schlötterer (1968). Ipri also suggested that whereas it is generally assumed that the thin silicon films are free of shear stress, shear stress should influence the Hall voltage and, if present, could contribute to the differences observed (Boleky, 1970) in the Hall mobilities (measured in a magnetic field) and the MOS transistor mobilities (measured with no magnetic field imposed).

Thorsen and Hughes (1972) pointed out that within a specific substrate plane the thermal contraction of sapphire may be anisotropic and thus lead to differences in carrier mobilities depending on the orientation of the direction of current flow within the plane. They constructed Hall patterns in the silicon deposit in such a manner that the mobility could be measured at 18° intervals around a circle in the plane of the substrate. In (100) silicon on ($1\bar{1}02$) sapphire, differences of approximately 90 percent were observed between the maximum and minimum electron mobilities measured as a function of angle (Hughes and Thorsen, 1973). In (221) silicon on ($11\bar{2}2$) sapphire the electron mobility anisotropy proved to be about 40 percent (Thorsen and Hughes, 1972). The silicon films investigated were deposited at 1100C, using a rate of 2 μm/min, and subsequently annealed at 1100°C in O_2 for 30 min and N_2 for 2 hr. The net donor concentration was in the range $2-7 \times 10^{16}$/cm^3. The maximum stressed electron mobilities in the (221) silicon of ~700 cm^2/V-s are greater than the unstressed mobilities and also significantly greater than the maximum electron mobilities of ~450 cm^2/V-s observed in the (100) silicon. These investigators proposed, therefore, that optimization of the orientation might lead to significantly enhanced carrier mobilities in the heteroepitaxial silicon device structures. The maximum mobilities observed in these 1.9- to 2.5-μm-thick films are not, however, in excess of mobilities measured in (100) silicon (of equivalent thicknesses) by other investigators (see Figure 2.31). Hughes and Thorsen have demonstrated excellent agreement between theory and experiment and thus can account for the anisotropy in the electron mobility on the basis of thermally induced stresses only. They point out, however, that the influence of crystalline defects must be invoked to account for the observation that the unstressed mobilities are considerably lower than bulk silicon values. Hughes (1975) has proposed that the room temperature stress in the heteroepitaxial silicon is predominantly the result of thermal contraction of the substrate rather than stresses induced by crystallographic lattice mismatches or other growth effects. He has extended the model, which deals with mobility anisotropy within a specific plane, to calculations of mobilities in specific crystallographic orientations not yet investigated. Hughes cites specific orientation regimes that warrant further investigation in an effort toward maximizing the mobility in heteroepitaxial silicon. In n-type deposits, the mobility may be either enhanced or degraded by stress effects. The ratio of stressed to unstressed electron mobility falls within the range between 0.4 and 1.3. In p-type silicon the mobility is always enhanced, and the ratio between the unstressed and stressed hole mobility may theoretically be as great as 2.4. It is quite clear that this work has considerably contributed to a basic understanding of at least one important facet of the influence of orientation on the electrical properties of heteroepitaxial silicon. In a practical sense, the problem (and the challenge to the experimentalist) is to realize good crystallinity in the silicon deposits on the theoretically desirable substrate orientations. It has commonly been observed that the crystalline quality of the deposits becomes highly defected with deviation from a limited number of specific substrate orientations (see Section 2.2).

Hynecek (1974A) has also investigated the elastoresistance effect in silicon on sapphire. Strips with the long axes oriented in the $\langle 010 \rangle$, $\langle 100 \rangle$, and $\langle 110 \rangle$ directions were cut in n-type (100) silicon on ($1\bar{1}02$) sapphire wafers. MOS Hall bridges were fabricated in the silicon, and the strips were bent in a specially constructed jig. The Hall voltage was plotted as a function of the gate bias for the three crystallographic directions. Large differences in the Hall voltage were observed, and elastoresistance coefficients (different from bulk values) were calculated. Hynecek shows that, because of lateral stress, the energy band minima in the direction normal to the surface increase, "causing almost all the carriers to be transferred out to remaining energy surfaces." This is an intrinsic effect and not specific to silicon on sapphire, however, that this effect can account for the degradation in the electron mobility in silicon on sapphire and suggests that if schemes could be devised to reduce the stress in silicon on sapphire, the electrical properties could be significantly improved.

Carrier lifetime

The carrier lifetime is more sensitive to crystalline perfection and impurity content of the silicon than is the carrier mobility. Crystalline defects and unintentionally added impurities are associated with energy levels introduced into the forbidden gap, which act as recombination centers and thus lead to the degradation of the carrier lifetime [see, for instance, Grove (1967)]. Therefore, the lifetimes in silicon on sapphire and spinel deviate from bulk silicon values considerably more than do the mobilities.

Lifetime is difficult to measure and results are difficult to interpret in thin silicon films on sapphire and spinel because of the relatively low values involved, the influence of the two close interfaces as sources or sinks of carriers, the series resistance of the thin silicon films, and the nonuniformity of lifetime and carrier concentration throughout the thickness of the deposit. The lifetime in most of the measurement methods employed (derivation from reverse-bias diode characteristics is the exception) is the product of an observed response time and a parameter which may not be accurately known, such as the carrier concentration. Under these circumstances, relative values of response time measured by one technique are more meaningful than absolute values measured by different methods. In considering the validity of a lifetime value, the measurement technique and the variables involved must be taken into consideration. Dumin and Silver (1968) derived lifetimes from diffused diode characteristics and from pulsed charge–injection measurements on a single set of samples. The values agreed only within a factor of 3, which they considered to be within experimental error.

Heiman (1967), using the pulsed-MOS capacitor technique (capacitance is measured as a function of time) measured a 4.5-ns lifetime in 5-μm-thick silicon on sapphire. About the same time, Hart et al. (1967), working with silicon on sapphire, and Schlötterer (1968), working with silicon on spinel, estimated lifetimes ~1 ns from the characteristics of diodes fabricated in thick (>4 μm) films. Although higher values have been reported subsequent to these early investigations, Norris (1972) recently derived lifetimes of about 0.2–0.3 ns, in 4-μm-thick silicon on spinel films, from the observation of the transient conductance subsequent to exposure of the silicon to electron irradiation. Mueller (1970), Dumin (1970), and Robinson and Heiman (1971) found that the minority carrier lifetime in silicon on sapphire can be increased by an order of magnitude by impurity gettering techniques. Lifetimes of about 50 ns were reported. Cullen et al. (1969) observed a 40-ns lifetime in a 4-μm-thick gettered silicon deposit on spinel. Cullen and coworkers (1973A) also observed that the lifetime in thinner silicon on sapphire films can be increased by more than an order of magnitude by rapid coverage of the sapphire surface (the "burst" growth method has been described earlier in this chapter). Lifetimes of 1.7 to 3.1 ns were observed in 1-μm-thick films; these lifetimes were determined by measuring the source drain current (of a deep-depletion MOS transistor) as a function of time. Schroder and Rai-Choudhury (1973) reported minority carrier lifetimes as high as 1.5 μs in 2-μm-thick silicon films deposited on a 1-μm-thick heavily phosphorus-doped (~10^{20}/cm^3) layer of silicon on sapphire. Rai-Choudhury and coworkers (1974B) observed that the aluminum contamination in silicon "piles up" in the heavily phosphorus-doped layer adjacent to the silicon/sapphire interface, leaving the second-deposited n-layer (in the 10^{16}/cm^3) doping range) relatively free of aluminum. McGreivy and Viswanathan (1974) have observed an improvement in lifetime by about a factor of 10 in silicon deposited on a 3-μm-thick arsenic-doped (~10^{18}/cm^3) layer of silicon on sapphire. These investigators pointed out that it is desirable to employ the lower level of arsenic

doping in order to minimize the diffusion of the dopant away from the interface region during subsequent device processing. It was also observed that the lifetime increased exponentially with the silicon film thickness. The lifetime increased from a value of ~6×10^{-9} s in a 4.3-μm-thick composite structure to ~2×10^{-6} s in an 18-μm-thick composite structure (of Si, $N_D = 5 \times 10^{16}$/cm^3 deposited on the 3-μm Si; $N_D = 1 \times 10^{18}$/cm^3, arsenic doped). The pulsed MOS capacitance technique was employed to measure the lifetime.

In most of the minority carrier lifetime measurements the lifetime and the carrier concentration (usually involved in the lifetime calculation) are integrated over the thickness of the films. This contributes to the inaccuracy of the measurement and makes it difficult to compare the characteristics of silicon deposited under different conditions with different lifetime and impurity profiles. Dumin and Silver (1968) measured the lifetime profiles from the characteristics of diodes fabricated in mechanically thinned silicon on sapphire. The lifetimes increased from ~0.1 ns in ~0.5-μm-thick films to ~2.0 ns in 10-μm-thick films. In this work the lifetime was derived from the reverse-biased characteristics of diodes in which the diffusion did not reach the silicon/sapphire interface. It was noted that the lifetime measured in diodes in which the diffusion did not reach the substrate was a factor of 3 to 10 greater than in the "through-diffused" diode structures. This emphasized the role of the silicon/substrate interface in recombination processes. A good deal of progress has been made in improving the physicochemical nature of heteroepitaxial silicon since this work was done, and although the earlier results may not be quantitatively applicable to current technology, the overall trends observed in lifetime as a function of the film thickness are still qualitatively relevant.

Tihanyi (1972, 1973) calculated lifetimes at various distances from the silicon/sapphire interface with the same test structure employed to derive the carrier concentration and mobility as a function of the film thickness (i.e., contact was made to the depletion region under the gate in an MOS transistor structure). The conductance transient (between the source and the drain) is measured after application of a negative voltage step to the gate. The generation rates and the related lifetimes are calculated from these data. Tihanyi concluded that the lifetime in a 0.9-μm-thick deposit ($N_d = 4 \times 10^{15}$/cm^3) is ~5 ns within 0.1 μm of the silicon/sapphire interface, and greater than 30 ns in the remaining portion of the film more remote from the highly defected interface region.

Kranzer (1974) has also dealt with the problem of the variation of lifetime and carrier concentration with distance from the silicon/substrate interface. He points out that measurements of reverse-biased diode

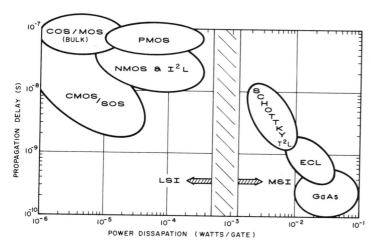

FIGURE 2.52   A comparison of the speed and power dissipation of CMOS transistors fabricated in silicon on sapphire with other technologies. The technologies are categorized according to their applicability in large scale integration (LSI) and medium scale integration (MSI). After Scott and Ipri, (1977)

characteristics yield lifetime values integrated over the "bulk" (of the thin film) depletion region and the silicon/sapphire interface (in the case where the silicon thickness is >4 μm), whereas the MOS capacitor technique yields lifetime values integrated over the "bulk" of the silicon and the silicon/SiO$_2$ interface. Kranzer electrically (i.e., without physically thinning the silicon) determined the lifetime profile by measuring the gate capacitance and reverse current of gate-controlled diodes fabricated in the heteroepitaxial films. Within the region of the 1-μm-thick films measured, it was observed that the lifetime decreased monotonically with distance from the silicon/SiO$_2$ interface. For example, in a film with an electron concentration of 2.5×10^{15}/cm^3, the lifetime decreased from a value of ~20 ns at a depth of 0.1 μm to ~8 ns at a depth of 0.2 μm. The profile, as well as the absolute values, proved to be dependent on the carrier concentration. In a film with a carrier concentration of 1×10^{16}/cm^3, the lifetime decreased from ~2 ns at 0.1 μm to 0.3 ns at 0.25 μm depth.

The carrier lifetime is a particularly sensitive measure of the crystallinity and impurity content of the heteroepitaxial films. As the bipolar transistor technology in films is advanced (probably with impetus from ion implantation), it appears very likely that lifetime measurements will be further understood, standardized, and employed. It would be very useful to develop clear-cut lifetime profiles as a function of deposition and processing parameters.

2.13. Device application

The applicability of heteroepitaxial silicon in CMOS/SOS structures

As has been discussed previously, the reduction of parasitic capacitance by use of an insulating substrate leads to an increase in the operational speed and

reduction of power dissipation in MOS transistor structures. A comparison is made in Figure 2.52 between the propagation delay time and the power dissipation of various technologies as applied to digital devices. It can be seen that the power consumption of the bulk CMOS transistor structures is several orders of magnitude less than that observed in bulk silicon bipolar device technologies such as transistor-transistor logic (T^{2}L) and emitter coupled logic (ECL). The p-channel MOS (PMOS), n-channel MOS (NMOS) and integrated injection logic (I^{2}L) are intermediate in regard to power consumption. By fabrication of the complementary pair MOS transistor structures in heteroepitaxial silicon, the speed associated with the bulk complementary pair technology is maintained, while the power dissipation is considerably reduced.

The support of the basic MOS transistor structure by the insulating substrate also offers stability in hostile radiation environments. Radiation effects are associated with processes occurring in the depleted region within a diffusion length of the junction (the photocurrent collection region). In bulk CMOS structures, the region in the bulk silicon substrate adjacent to the bottom of the source and drain regions acts as photocurrent collection areas. With the elimination of the bulk silicon support, the photocurrent collection area is limited to the silicon adjacent to the vertical junctions. Radiation effects at the silicon/substrate interface are a subject of current investigation.

Factors such as ease of production, yield, and reliability are clearly considerations in the choice of the various technologies for specific applications. Trade-offs must be made, for instance, between speed and cost. In this regard, the CMOS/SOS technology complements rather than supplements the various bulk silicon technologies. In considering the ease of processing and yield, it is interesting to note that processing errors (such as local mask failures and diffusion spikes) which would short out portions of the device to the semiconducting substrate in bulk silicon, do not lead to failure of the device when an insulating substrate is employed. It is also interesting to note that should a failure occur in processing which degrades an unacceptably large percentage of devices on a substrate, the silicon can be removed from the sapphire and the substrate can be re-used.

One of the earliest applications of CMOS/SOS circuitry was in electronic wrist-watches. Here the low power consumption is desirable in a compact battery operated device, and the high frequency capability permits the use of small oscillator crystals. Early interest in military applications has been in portable field communications and radiation resistant electronics. An emerging field of application is in very dense high-speed LSI components. The density of components on a chip is now sufficiently high that dissipation of the power in bulk silicon circuitry is a very real problem. For this reason it is anticipated that CMOS/SOS components will prove to have an important impact on the microprocessor and memory technology. As the functional advantage of the SOS technologies is realized, efforts are increased to minimize the substrate costs by optimization of the crystal growth techniques. In conjunction with developments in packing density and yield, SOS circuitry will become cost competitive with bulk silicon technology.

The fabrication of MOS circuitry in silicon on sapphire

In order to realize the speed inherently in the SOS devices, the effective channel mobilities and the source-drain spacings must be similar to those currently achievable in bulk silicon. These factors are clearly related to the crystalline quality of the hetero-epitaxial silicon. It is fortunate that the characteristics of MOS transistor structures are primarily related to the semiconducting properties of the silicon within < 1000 Å of the gate dielectric; this is the silicon most remote from the silicon/substrate interface in the heteroepitaxial structure. There is ample evidence, from examination of the film physical and electrical properties, that the crystalline quality of the silicon improves very rapidly with distance from the substrate surface. In fact, at the current state of development we are very near to the point where the channel mobility of MOS transistors fabricated in silicon films ($> 0.6 \ \mu$m in thickness) is more dominated by effects of charge at (or near) the silicon/silicon dioxide interface than by the crystalline perfection of the silicon in and near to the surface inversion layer. A number of years ago effective (channel) mobilities greater than 80% of the mobilities observed in equivalent devices fabricated in bulk silicon had been observed in NMOS transistors fabricated in 1.0 μm thick silicon on sapphire (Boleky, 1970). In regard to the capability to fabricate source-drain spacings (channel lengths) in silicon on sapphire comparable to the spacing in bulk device structures, it

should be noted that while the crystallinity of the entire film plays a role, the crystallinity at the surface is again weighed more heavily because it is only the spacing within the 1000 Å of the silicon/gate oxide interface that determines the speed of the device. The spacing may vary in the "bulk" of the film as long as the source and drain are not shorted. Early in the development of the technology, when relatively high deposition temperatures and low deposition rates were employed, anomalously rapid diffusion of dopants near the silicon/substrate interface resulted in closure of the channel. This mode of failure is not common, however, in the more recent heteroepitaxial silicon.

Problems related to processing MOS transistors in silicon on sapphire

Control of the characteristics of the starting material is frequently a problem in the development of a new technology. Since the physical and electrical properties of the heteroepitaxial silicon film vary with distance from the silicon/substrate interface, the properties of the silicon are more difficult to specify and measure than is the case with bulk silicon. It is unfortunate that a complete characterization of the semiconducting properties of the heteroepitaxial silicon is possible only after the processing of the device structure has been completed, particularly when low carrier concentration thin films are employed (see discussions in Section 2. and Chapter VIII). In-line process controls, for the most part based on analytical techniques discussed in Chapter V, are under development. Much of the effort in the preparation of the silicon-substrate composite, which serves as the starting material for processing of devices, has been directed toward the maintenance of reproducibility. Shaping the substrate crystal by methods which minimize the introduction of work damage, and finishing the substrate surface by chemical-mechanical methods has contributed significantly to control of the substrate surface and reproducibility in the film properties. With low temperature growth of the silicon, the variability in the electrical properties associated with autodoping has been minimized to acceptable levels. While satisfactory control of the carrier concentration can be achieved by introduction of the dopant during the growth of the film, ion implantation offers even more precise control of doping. This proves to be particularly useful at the low carrier concentration levels required for fabrication of the deep-depletion MOS device structures.

The source-drain leakage currents observed in MOS/SOS devices are typically ≈0.1 to 10 nA/mil at 5 V with a source-drain spacing of 0.2 mil (Ham, 1973), which is more than an order of magnitude greater than is commonly observed in similar devices fabricated in bulk silicon. The origin of the leakage currents is not well understood, and therefore this is the subject of considerable research effort.

The leakage-currents characteristic of the transistor structures fabricated in the silicon on sapphire places certain limitations on the application of the heteroepitaxial silicon devices. For instance, due to the d.c. power consumption associated with the leakage currents, the CMOS/SOS devices would not be the most appropriate choice in an application where the circuit is to be held in a powered stand-by mode and addressed infrequently.

An additional problem characteristic of MOS/SOS structures fabricated in silicon on sapphire is a shift in the threshold voltage associated with effects at the ramped edge of the silicon islands. In the removal of the unwanted silicon from the substrate, preferential (orientation sensitive) etches are used so that the edges of the silicon are (111) planes at ~57° to the substrate surface. Angling of the edges minimizes discontinuities in the evaporated metal contacts. Discontinuities may occur if the edges of the silicon islands are vertical, or even worse, undercut at the substrate. The gate electrode (metal or poly silicon) runs across the oxidized (111) edge to the oxidized (100) surface. This, in effect, creates an MOS transistor in (100) silicon in parallel with a similar transistor in (111) silicon (Flatley and Ham, 1974). Due to the differences in the charges at (or near) the silicon-silicon dioxide interface adjacent to the (100) and (111) silicon orientations, the two transistors may switch at different threshold voltages. Instabilities associated with the edge effect can be minimized by preferential doping of the edges or by the use of an appropriate dielectric between the gate and the oxide (Flatley and Ham, 1974).

There is a problem associated with maintaining the continuity of metal (or poly silicon) strips which must run over the edges of the 0.6 μm thick silicon islands. Due to shadowing effects, these materials may be thin or discontinuous over the juncture of the epitaxial silicon and the substrate. To avoid this problem, the processed surface is maintained nearly planar by use of the local oxidation silicon-on-sapphire process (LOSOS) (Capell et al., 1977). Using this process, the areas between the final silicon islands are thinned by etching, and the remaining silicon is oxidized. The oxide fills the area between the islands to the level of the islands.

The problems discussed above are related specifically to the fabrication of MOS transistor structures in silicon on sapphire. The fabrication of bipolar devices in silicon on sapphire is complicated by the relatively low carrier lifetimes observed in the thin ($\geqslant 1.0$ μm) films (see Section 12.11.). There is not agreement, however, on the minimum lifetime required for the application of the SOS material in bipolar structures.

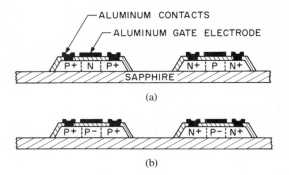

FIGURE 2.53   A cross-section schematic of CMOS transistors fabricated in silicon on sapphire. (a) The complementary pair fabricated in discrete n- and p-doped silicon deposited by the two-stage (double-epi) process, and in (b) lightly doped p-type silicon. The $p^+/p^-/p^+$ is a deep depletion device, the others are enhancement mode transistors

Mueller (1970) has proposed that bipolar structures can be fabricated in silicon on sapphire with lifetimes in excess of ~50 nsec, and indeed experimental operating bipolar transistors have been constructed in the heteroepitaxial silicon (Heiman and Robinson, 1968; Ronen and Robinson, 1971). Kokkas [in Cullen et al. (1973)] points out that bipolar performance may be more limited by the emitter efficiency than by the carrier lifetime. It should also be noted that the narrow junction spacings characteristic of the bipolar structures (typically 2000 Å) are controlled by diffusion or epitaxial growth, while the junction spacings characteristic of MOS transistor structures are determined by photolithographic techniques. Thus, the processing technology which would be brought to bear on the fabrication of bipolar structures in silicon on sapphire is considerably different than those already employed for the processing of MOS transistors. This is also an area in which ion implantation is expected to have a significant impact on the silicon on sapphire technology. Alternative dielectric isolation techniques oriented toward the fabrication of dielectrically isolated bipolar devices are discussed in Section 13.4.

CMOS transistor processing and the related materials requirements

The advantages offered by the silicon on sapphire composite are realized to the maximum extent in the complementary-pair MOS transistor configuration (see Figure 2.53). The complementary pair is made up of either two enhancement-mode transistors or an enhancement-mode transistor paired with a deep-depletion device.

The N-MOS enhancement mode transistor operates in the usual manner; the application of a positive gate-source voltage (V_{GS}) equal to or greater than the

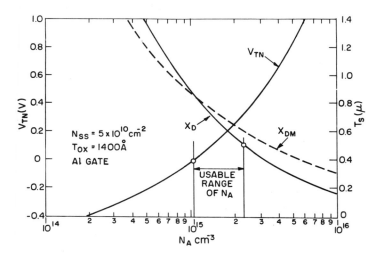

FIGURE 2.54   Selection of film thickness and doping level for the deep depletion process in p-type silicon films. ($N_{SS} = 5 \times 10^{10}/cm^2$). After Boleky (1970).

threshold voltage of the NMOS (V_{TN}) inverts the surface of the p-type silicon and establishes a conducting channel between n^+ source-and-drain regions. The operation of the deep-depletion mode transistor depends on the fact that the gate electrode (aluminum or poly-silicon) to silicon work function potential difference (V_{GSC}) is of such a polarity to deplete silicon near the surface of the film, creating a surface space-charge region composed entirely of immobile ionized atoms. It is necessary that the width of this space-charge region be at least equal to the film thickness (T_S) in order that negligible source-drain current flows when $V_{GS} = 0$. In the aluminum gate technology, the channel region is lightly boron doped silicon, and the application of small negative V_{GS} removes the gate electrode induced depletion region and accumulates the surface with holes. The PMOS transistor, therefore, operates in a manner similar to a conventional enhancement-mode PMOS with a very low threshold voltage. The requirement that the zero gate-bias width of the surface space charge region extend through the film determines the upper limits on the thickness and doping of thin p-type silicon film.

The epitaxial growth of silicon films for the deep-depletion CMOS technology constitutes a compromise between the film thickness and the carrier concentration. The carrier concentration which is acceptable is inversely proportional to the film thickness. The range of acceptable thicknesses and carrier concentrations for the simultaneous fabrication (using the Al gate technology) of NMOS transistor (enhancement-mode) and a PMOS transistor (deep-depletion mode) in a single p-type silicon film is shown in Figure 2.54 (Boleky, 1970). (Similar relationships have been developed for n-type films.) This figure is a plot of calculated values of:

X_D − the thickness of p-silicon depleted by the gate

electrode (aluminum)-silicon work function difference

X_{DM} − the maximum possible value of X_D

V_{TN} − the threshold voltage of an NMOS transistor

as a function of N_A, the acceptor concentration. For this plot a surface state concentration (N_{SS}) is assumed to be $5 \times 10^{10}/cm^2$. The lower limit of acceptable carrier concentration ($N_A \sim 10^{15}/cm^3$) is set by the condition that $V_{TN} \geq 0$. It is seen that for a film thickness of 0.5 μm, the upper limit of the carrier concentration is $\sim 2.4 \times 10^{15}/cm^3$. It is also evident from this plot that if films less than 0.5 μm in thickness can be used, higher (and more easily controlled) carrier concentrations are acceptable, whereas if thicker silicon deposits are employed, the range of acceptable carrier concentration narrows. The range appears to be exceedingly narrow, and difficult to realize. In practice, however, the usable range has not proven to be as narrow as illustrated in Figure 2.54. This is, at least in part, due to the fact that electrically the "apparent" film thickness is less than the measured physical thickness. Since the defect density increases with distance from the silicon/gate oxide interface, so does the resistivity of the film, with the result that the thickness of film that can be depleted is greater than calculated on the basis of the surface carrier concentration. This provides some latitude in the usable range of carrier concentration, so that emphasis can be placed on the control of the threshold voltage rather than on the depletion to the silicon/substrate interface. With the use of relatively low temperature deposition and processing, and closed-loop feedback electronic flowmeters to control the introduction of the dopant during deposition or ion implantation subsequent to deposition, it has been possible to achieve the required control of the carrier concentration in films as thin as 0.5 μm. The deep depletion transistor structures were originally fabricated in lightly doped p-type films using aluminum gates (Heiman, 1966; Hofstein, 1966). If polycrystalline silicon is used as the gate electrode, the deep depletion device can be formed in lightly doped ($\approx 1 \times 10^{15}/cm^3$) n-type films. Since the silicon adjacent to the silicon/substrate interface is more resistive in n-type than p-type films (Ipri, 1972B) (presumably due to compensation by the p-type aluminum autodoping), it was proposed (Sarace and Ipri, 1974) that the source-drain leakage currents could be minimized by using n-type silicon as the starting material for the deep-depletion technology. In this case the deep depletion n-channel device has N^+ source and drain areas, an N^- channel, and a P^+ poly silicon gate, and the enhancement mode transistor is a conventional p-channel device.

An outline of a typical sequence for the processing of the C-MOS/SOS deep depletion devices, using a p^+-poly gate, is shown in Figure 2.55 (Sarace and Ipri, 1974; Ipri and Sarace, 1976). The use of the poly silicon gate also provides "self-alignment" of the gate electrode with the edges of the channel, and thus eliminates the capacitance associated with the overlap of the (aluminum) gate and the source and drain regions. Self-alignment is also available with ion implantation, using a number of metals for the gate electrode.

Devices with relatively low threshold voltages may be fabricated using the deep depletion devices, and therefore this approach is used where low-voltage operation is required, such as in watch circuits.

When low threshold voltages are not required, the all-enhancement mode C-MOS structures are commonly employed. The control of processing is eased by the higher carrier concentrations employed in these structures. The early enhancement mode C-MOS devices were fabricated by the "two-stage" epitaxial process (Scott and Burns, 1970). Using this method, discrete n- and p-doped islands were formed. Silicon was deposited over the entire sapphire surface using *in situ* doping (from a gaseous source) of one type, typically starting with p-type in the low $10^{16}/cm^3$ range. The silicon to remain was masked, and the unwanted material etched from the substrate. The n-type material was then deposited over the entire structure, doped typically in the mid $10^{15}/cm^3$ range. While the n-islands were formed by conventional masking and etching, the previously formed p-type films were protected by an oxide layer. This process provided excellent control, but was subsequently replaced by processing employing the single epitaxial growth step. The enhancement mode devices are now formed using doped oxides as the dopant source. The processing scheme is similar to that shown in Figure 2.55, except that the channel regions are appropriately doped from the oxide sources. Clearly various combinations of

doping by diffusion from oxides and ion implantation can be employed (see, for instance, Ipri and Flatley, 1976).

Alternative dielectric isolation schemes

A number of schemes have been employed to thin bulk silicon and support it with a dielectrically isolated structure (see, for instance, Bean and Runyan, 1974). In most of these schemes the dielectric is grown silicon dioxide, and the mechanical support is polycrys-Otalline silicon. Conceptually, the dielectric isolation schemes are simple; it is in the execution that they are complex.

In a typical dielectric isolation sequence, isolation moats are formed in the surface of a bulk silicon wafer to provide separation between the individual islands. Selective etches are employed so that the depth of the moats is limited by the width of the opening in a protective oxide layer. The oxide is stripped, and the junction of the device is formed horizontal to the surface either by epitaxial growth or diffusion. Silicon oxide, which provides the dielectric isolation, is then grown on the surface. A thick layer of polycrystalline silicon, which acts as the mechanical support for the structure, is deposited on the oxide. The original silicon is mechanically abraded and chemically-mechanically polished to the desired thickness. The dimen-

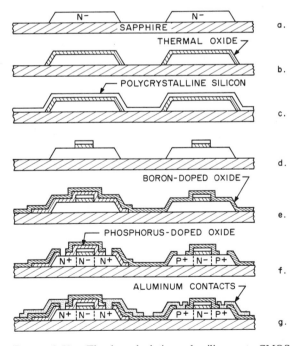

FIGURE 2.55   The deep-depletion poly-silicon gate CMOS processing sequence

sions of the isolation moats serve as a measure of the thickness of the bulk silicon. The removal of the unwanted silicon must be carried out with great precision, and in a manner to minimize the introduction of work damage into the remaining thin layer of silicon.

The precision lapping can be eliminated by the selective electrochemical removal of the silicon. The rate of the electrochemical etch is a function of the type and concentration of the doping, and thus the etching can be terminated at a p-n junction (van Dijk and de Jonge, 1970). To form the dielectrically isolated thin film structure by preferential etching (van Nielen and coworkers, 1970) a p-type silicon layer is epitaxially deposited on an n^+ substrate. The dielectric silicon oxide is then grown from the epi p-type layer, and the polycrystalline silicon support is deposited on the oxide. The unwanted n^+ silicon is then removed electrochemically back to the p-n junction. Under the proper conditions, the electrochemical etch stops at the junction, and thus the thickness of the isolated thin film is determined by the position of the junction. An epitaxial n layer grown to the desired thickness prior to thinning completes the structure, and devices are formed in the composite structure by conventional processing. In addition to avoiding precision lapping and polishing, the introduction of work damage characteristic of lapping is avoided by use of the selective electrochemical etch.

The principal advantage of the methods which involve thinning of bulk silicon is that the crystallographic and semiconducting properties of the isolated silicon are more similar to unprocessed bulk silicon than is heteroepitaxial silicon. Even when the semiconducting properties are degraded by the mechanical shaping, the properties of the silicon in the junction area are adequate for the formation of bipolar device structures. The disadvantage is that the processing schemes require many steps, and always include more than one vapor phase silicon growth operation. While considerable effort has been put forth in recent years in the development of the various dielectric isolation schemes, to date they have been used only in very specialized applications, such as radiation hardened military circuitry (Bean and Runyan, 1974).

2.14. Discussion: limiting factors

The identification of limiting factors

With the aim toward improving the semiconducting properties of thin heteroepitaxial silicon films, it is useful to identify the key limiting factors. It is also useful to define which limiting factors are inherent characteristics of heteroepitaxial deposits and which are amenable to further development. This is a difficult task for a number of reasons. To date there have

been very few experimental data in which direct correlation between the physical structure and the semiconducting properties has been established. Some of the evidence found in the literature that bears on limiting mechanisms is contradictory. It is difficult to compare the properties of films prepared at various locations, at least in part because the films may have been deposited under very different (and not always well defined) conditions. This latter factor impedes analysis because it has clearly been demonstrated that the influence of any one limiting factor is highly dependent on the preparative conditions.

As was mentioned earlier in this chapter, the factors commonly invoked as limiting the properties in heteroepitaxial silicon are (1) degradation of the thin-film crystallinity as the result of the spatial mismatch between the crystalline lattice of the substrate and the overgrowth, (2) contamination and degradation of the crystallinity of the film as the result of reaction between the substrate and the deposition constituents, and (3) stress in the film resulting from differential contraction of the film and substrate as the composite cools from the deposition temperature. The success in constructing fast, low-power MOS transistors in thin heteroepitaxial silicon can, to a large extent, be attributed to the effect of the first two limiting factors decreasing rapidly with distance from the silicon substrate interface: the density of defects decreases rapidly with distance from the interface, and the impurities introduced primarily prior to complete coverage of the substrate surface are concentrated near the interface. Therefore, in MOS transistor structures fabricated in silicon as thin as 0.6 μm, the circuit characteristics for some applications are more limited by junction leakage, which is associated with the silicon near the substrate surface, than by the channel mobility, which is associated with the silicon within about 200 Å of the silicon/silicon dioxide interface.

Limiting factors at the extremes of the deposition conditions

In silicon films deposited at the extremes of the preparative conditions, the limiting factors can be readily identified. The critical preparative parameters are the deposition temperature and rate, the substrate preparation, and the purity of the ambient. At deposition temperatures in excess of ~1150°C the films are heavily contaminated by substrate reaction products. At chemical vapor-deposition temperatures less than ~950°C, the crystalline quality is poor as the result of low surface mobility of the ad-atom. In the upper range of temperatures used for heteroepitaxial silicon growth (>1050°C), low rates (<1.0 μm/min) lead to degradation of the semiconducting properties by contamination with substrate reaction products. The deposition rate at which the crystallinity is degraded by too rapid growth does not appear to have been unambiguously identified. The presence of oxygen in the deposition atmosphere leads to degradation of the crystalline structure. The crystalline nature of the substrate surface will limit the crystallinity of the deposit if the subsurface work damage is not eliminated by chemical or thermal processes. In the preparation of heteroepitaxial silicon by vacuum evaporation, the factors that limit the properties at the extremes of the preparative conditions are similar but fall within different ranges of rate, temperature, and ambient purity. This has been briefly discussed in Section 2.9.

Limiting factors within the extremes of the deposition conditions

In recent years a number of investigators have adopted deposition conditions that provide usable semiconducting properties in thin (≤ 1.0 μm) heteroepitaxial silicon films. These properties do not, however, equal the properties of bulk (or homoepitaxial) silicon. It is within this range of conditions that the key limiting factors are more difficult to identify. Although many of the characteristics observed in silicon on sapphire can be related to a number of limiting factors, for purposes of discussion it is useful to attempt to categorize some of the more salient characteristics that have been observed.

Characteristics relating to crystallinity: lattice mismatch, orientation, and substrate surface preparation.

- Usable semiconducting properties in thin (≤ 1 μm) heteroepitaxial silicon have been realized only on sapphire and spinel, but these substrates are not of the same crystalline structure (La Chapelle et al., 1967; Filby and Nielson, 1967; Cullen 1971).
- Within fairly wide limits, the semiconducting properties of heteroepitaxial silicon have not been associated with the grown-in defect density of the substrate (Hart et al., 1967; Druminski and Kühl, 1974; Peel and Barry, 1976).
- The semiconducting properties of heteroepitaxial silicon are closely associated with the presence of subsurface work damage in the substrate (Naber and O'Neal, 1968; Cullen, 1971; Gassmann et al., 1972; Maurits, 1974, 1976).
- The structure at the surface of an annealed substrate may not be the same as in the bulk of the crystal (Chang, 1971).
- The highest reported carrier mobilities are similar for several silicon/sapphire orientations in relatively thick (4 μm) vacuum-evaporated silicon (Chang, 1971).
- The magnitude of the MOS transistor leakage current, which is associated with the crystallo-

graphically imperfect silicon adjacent to the substrate surface, is a strong function of chemical-thermal conditions employed during device processing (Ham, 1975).
- It has been observed by transmission electron microscopy that the silicon adjacent to the substrate surface is highly defected and that the density of the defects decreases rapidly with distance from the substrate (Picraux and Thomas, 1973; Linnington, 1974; Abrahams and Buiocchi, 1975).
- The mobility vs. carrier concentration relationship characteristic of silicon on the insulators has been theoretically explained on the basis of carrier scattering by the defect structure characteristic of silicon on the insulators (Schlötterer, 1968; Hasegawa et al., 1969; Pödör, 1970; Schlötterer, 1970; Eernisse and Norris, 1972).

Characteristics that relate to contamination of the deposit by substrate constituents.

- The after-oxidation (after device processing) semiconducting properties of heteroepitaxial silicon are much more sensitive to deposition conditions than are the as-deposited properties (Cullen et al., 1969).
- The hole concentration (associated in part with aluminum autodoping) may decrease or increase following thermal oxidation, depending on the deposition rate (Gottlieb and Corboy, 1972).
- Carrier mobilities are degraded by thermal oxidation; the amount of degradation is a strong function of the deposition conditions and the carrier concentration [for instance, Dumin and Robinson (1968A), Ross and Warfield (1969), Cullen et al. (1969), Gottlieb and Corboy (1972)].
- The level of the net carrier concentration (in the absence of intentional doping) is a function of the chemical reactivity of the substrate material (Robinson and Dumin, 1968).
- It has been reported that mobilities have increased by decreasing the (vacuum) deposition temperature even at some sacrifice to the heteroepitaxial film crystalline quality (Salama et al., 1967).
- Carrier lifetimes can be improved (in some, but not all deposits) by "gettering" oxidized heteroepitaxial silicon in an HCl atmosphere (Dumin, 1970).
- Carrier lifetimes are a strong function of the distance from the silicon/substrate interface and may be improved by heavily doping the silicon adjacent to the substrate interface with phosphorus (Schroder and Rai-Choudhury, 1973; Rai-Choudhury et al., 1974B) or arsenic (McGreivy and Viswanathan, 1974).

- The semiconducting properties of very thin heteroepitaxial silicon (≤ 0.5 μm) can be related to the thickness at which the substrate is completely covered with silicon (Cullen and Corboy, 1974).
- It should be noted that the characteristics described in the earlier literature apply to the entire range of deposition conditions investigated and may be less applicable to the current optimized low-temperature high-rate conditions.

Characteristics that relate to the influence of strain in the heteroepitaxial silicon.

- The stress derived from the experimentally measured values of strain in silicon on sapphire is about 0.4 of the theoretical yield strength of silicon (Ang and Manasevit, 1965).
- The experimentally measured strain levels in silicon on sapphire appear to be independent of the crystalline perfection of the deposits (see Section 2.11).
- The unusually high hole mobilities (higher than bulk) observed in 20-μm (111) silicon on (111) spinel are in agreement with values calculated on the basis of a stress-induced piezoresistance effect (Schlötterer, 1968).
- The anisotropy in mobilities (100) silicon on ($1\bar{1}02$) sapphire and (221) silicon on ($01\bar{1}2$) sapphire have been explained completely on the basis of anisotropies in the thermal stress induced from the trigonal substrate (Thorsen and Hughes, 1972; Hughes and Thorsen, 1973; Hughes, 1975).

Contribution by the various limiting factors: an overview

Early in the development of heteroepitaxial thin-film technology, the degradation of deposit properties was assigned primarily to the crystallographic lattice mismatch between the substrate and the silicon. One finds in the early literature a variety of lattice-matching schemes. Much of the degradation assigned at that time to lattice mismatch, at least in silicon on sapphire and spinel, was probably caused more by the presence of work damage introduced during fabrication of the substrate and by the reaction between silicon and substrate.

As methods to remove the substrate surface work damage improved, and reactions between the substrate and the deposit were suppressed, the issue of lattice mismatch became more clouded. Equivalent electrical properties were observed in composite structures with very dissimilar lattice matches, and it was appreciated that what was presumed to be a good lattice match might not be due to reconstruction of the crystal structure at the termination of the bulk lattice

(Chang, 1971). Simply because the properties of homoepitaxial silicon are superior to those of heteroepitaxial silicon even under optimized preparative conditions, it is clear that lattice mismatch is an important limiting factor in the heteroepitaxial system; but the confidence in understanding and predicting lattice-matching effects has faded. This is discussed in more detail in Chapter 7. It appears that the lattice mismatch obscures effects related to the propagation of grown-in defects in the substrate into the heteroepitaxial deposit. With the exception of discrete grain boundaries in the substrate, no direct correlation has been established between the defect density of the substrate and the deposit (within the range of defect densities commonly observed in the currently available substrate crystals). In regard to sapphire substrates, the challenge is more in the area of maintaining control of the surface-finishing process than in improving the crystallographic quality of the "bulk" of the substrate. Crystal quality considerations are new to the commercial fabrications of refractory oxide materials.

With the observation of the strong relationship between deposition rate and temperature and film properties came the realization that reaction between the substrate and the deposition constituents was a limiting factor at least equal in importance to the lattice mismatch. After silane was adopted as the source gas in preference to the more reactive (in relation to the substrate) chlorinated silanes, it became clear that, of the various constituents of the deposition ambient, deposited elemental silicon was primarily responsible for the erosion of both the sapphire and spinel substrates (Cullen and Dougherty, 1972). Elemental silicon at the commonly used deposition temperatures is a strong chemical reducing agent. The refractory oxide substrates, with the cationic constituent in the maximum oxidation state, are susceptible to reduction by silicon. It also became evident that the oxidation/reduction reaction between silicon and the oxide substrate takes place primarily prior to the complete coverage of the substrate surface during the time period in which gaseous intermediate oxidation state oxides of both aluminum and silicon can escape from the site of the reaction (Cullen and Corboy, 1974).

Reaction between silicon and substrate influences the deposited silicon properties as the result of (1) microroughening and faceting of the substrate surface and the associated growth of defected and off-oriented silicon; (2) distortion of the crystalline structure of silicon by incorporation of reaction products, particularly in the deposit adjacent to the substrate grown prior to complete coverage of the substrate surface; and (3) the introduction of reaction products into the "bulk" of the silicon deposit, some of which are electrically active (autodoping).

After having established the conditions in which single crystal silicon could be deposited, the most obvious effect on the film properties resulting from reaction between the substrate and the silicon was autodoping. This led to lack of control of the carrier concentration and to large and unacceptable changes in both the carrier concentration and mobility during device processing. In the heavily autodoped silicon deposited at relatively high temperatures and low rates, the change in the electrical properties during thermal oxidation was associated primarily with transfer of the aluminum-bearing species from the silicon into the silicon oxide (Dumin and Robinson, 1968A; Ross and Warfield, 1969). It is this author's view that the changes in the electrical properties of the more lightly doped silicon deposited at the lower temperature and higher rates are associated more with the change of the charge of the aluminum-bearing species trapped in the lattice than with segregation of the aluminum from the silicon into the silicon oxide (Cullen and Corboy, 1974). The change in the charge of the complex can be associated with a change in the oxidation state of the aluminum. In the analysis of the deactivation or removal of autodopants, caution must be exercised in drawing analogies between impurities incorporated during the growth of the silicon and impurities diffused into the silicon subsequent to the deposition (for a treatment of the latter, see for instance Rai-Choudhury et al., 1974A). From a chemical point of view, it is very likely that the nature of the dopants (for instance, oxidation state) and the manner in which they interact with the defect structure are strong functions of the manner in which they were introduced into the film. This point is brought up here because it appears to be neglected by many investigators involved in the analysis of the semiconducting properties of heteroepitaxial silicon.

During the early development of the silicon-on-insulator device technology, reproducibility in device fabrication was seriously impeded by autodoping. It is of historic interest that the development of methods to suppress autodoping was significantly delayed until it was appreciated that the after-oxidation (after-device processing) semiconducting properties of heteroepitaxial silicon are very much more sensitive to the preparative variables than are the as-deposited properties (Cullen et al., 1969). This observation also led to the conclusion that, during that stage in development of the heteroepitaxial silicon technology, the properties were more limited by autodoping than by crystallographic defects per se. Through the development of preparative techniques, postdeposition treatments and modification of device-processing procedures, good control of the doping level has been achieved. At least for applications of heteroepitaxial silicon in field effect devices, autodoping as related to the control of the carrier concentration is not now considered to be a serious limiting effect. The interaction of both electrically active and inactive species with the defect structure continues to be a factor in the optimization of

mobilities in thinner (<0.6 μm) deposits and in developing an understanding of the properties of heteroepitaxial silicon.

The incorporation of electrically active impurities may, however, still be a factor that seriously limits the application of heteroepitaxial silicon in minority carrier device structures. It has been shown (Rai-Choudhury et al., 1974B; McGreivy and Viswanathan, 1974) that the increase in lifetime with silicon film thickness can be significantly altered by heavy doping of the silicon near the silicon/substrate interface. The implication here is that the lifetime is more limited by the impurities from the substrate than by the crystallographic defects (or that the defects are less active in the absence of the impurities). Rai-Choudhury and coworkers explain their results on the basis of a "gettering" effect of phosphorus. There is some similarity here to the improvement of lifetime as the result of HCl gettering of the silicon on sapphire (Mueller, 1970; Dumin, 1970; Robinson and Heiman, 1971). Interpretations of the modification of the aluminum (in unspecified oxidation states) distribution by phosphorus near the interface based on secondary ion mass spectrometric (SIMS) data may be open to some question unless the influence of the presence of oxygen is accounted for. Kokkas, on the basis of lifetime measurements made in the presence of a potential applied from the back of the substrate [cited in Cullen et al. (1973A)], and Redfield (1975), on the basis of the enhancement in lifetime by dopants not known to act as getters, have suggested that the effect of doping the silicon near the interface may be attributed to shielding the material more remote from the interface from generation-recombination centers in the heavily defected silicon near the interface.

The effects of incorporation of reaction products in the silicon near the silicon/substrate interface are more subtle and difficult to identify. The influence of this region on the device mobilities is indirect, because the carriers in the MOS transistor structures are active only near the surface most remote from the interface. Incorporation of foreign species prior to complete substrate coverage can be related to the observation that the increase in mobility with distance from the substrate within 0.6 μm of the substrate is significantly greater in (111) silicon on spinel than in any orientation of silicon on sapphire, whereas at the same time the hole mobilities (within the same thickness) in the (111) silicon on the (111) spinel are lower than in silicon on sapphire (see Figure 2.41). The relatively low mobility at 0.5 μm in thickness in the (111)Si/(111)spinel can be attributed to the observation that surface coverage is realized at 1000–2000 Å (depending on the substrate annealing treatment) in silicon on spinel and at 300–500Å in silicon on sapphire (see Section 2.7). The observation that a relatively good crystalline structure is propagated beyond the 0.5-μm highly defected layer can be qualitatively explained by assuming that the reaction products of the substrate (oxides of aluminum) are trapped during surface coverage, and for a brief period thereafter as the species are depleted from the gas phase, and that they do not redistribute throughout the film to any great extent during subsequent heat treatment. It is apparent that there is an inherent driving force leading to a rapid improvement of the crystallinity as the deposit grows away from the material degraded by the lattice mismatch and the trapped substrate reaction products.

Leakage currents and radiation effects in MOS transistors fabricated in heteroepitaxial silicon are also associated with the nature of the silicon directly adjacent to the silicon/substrate interface. In the heavily defected silicon near the interface, it appears that the characteristics observed are related to a complex (and as yet undefined) interaction between the autodoping species and the defect structure rather than to the effect of the defect structure per se (i.e., the properties of the heteroepitaxial films cannot be analyzed on the basis of the effect of known crystallographic effects in bulk silicon). This is supported by the observation (Ham, 1975) that the level of leakage currents in MOS transistor structures is strongly related to device processing variables (more specifically, with the oxidation processes).

There has been a good deal of effort to develop models to bring together the effects of the physical defect structure and the complex interaction with impurities to explain the degradation mechanisms in the heteroepitaxial silicon. Both Schlötterer (1968) and Cullen et al. (1969) invoked defect-related space charge scattering to explain the anomalous decrease in both the hole and electron mobility with decreasing hole and electron concentrations below ~10^{16} carriers/cm³. This model had been originally proposed by Weisberg (1962) to explain a similar effect in III–V compounds. It is interesting to note here that Redfield (1975) has proposed that the low carrier concentration mobilities are more adequately explained by a compensation effect (Redfield, 1974) than by space charge scattering. Hasegawa and coworkers (1969) have considered defect-related current-limiting mechanisms in some detail. These investigators have calculated mobilities on the assumption that the mobility is limited by space charge scattering above 200°K and dislocation scattering below 100°K. On the basis of their model, space charge scattering should suppress the electron mobilities more than the hole mobilities. This is in agreement with experimental observations in both silicon on sapphire and silicon on spinel. In a related analysis, Pödör (1970) attributed the degradation in mobility in the heteroepitaxial silicon to scattering by a potential field associated with stress around dislocations and space charges around dislocations resulting from trapping of carriers on dislocations. He has proposed that the first mechanism predominates at

carrier concentrations above ~10^{16}/cm³ whereas the latter mechanism is operative below this carrier concentration level. From this analysis of the electrical data, defect densities were extracted of 2–4 × 10^8/cm² in silicon on spinel and 2–5 × 10^9/cm² in silicon on sapphire (thickness unspecified).

Schlötterer (1970) has developed an overview of several of the models that have been proposed to explain the dependence of the mobility in the heteroepitaxial silicon on the temperature and carrier concentration. He assumed the carrier scattering by defects to be independent of impurity and lattice scattering and used the general approximation

$$\frac{1}{\mu_f} = \frac{1}{\mu_D} + \frac{1}{\mu_B}$$

where μ_f is the measured mobility, μ_D is the mobility associated with the defects, and μ_B is the bulk mobility. Schlötterer points out that the various models have led to a similar formulation:

$$\mu_D \approx \frac{n^b T^c}{N_D}$$

where N_D is the observed dislocation density, T is the temperature, n is the carrier concentration and the exponents b and c depend on the specific model: for neutral dislocations (Dexter and Seitz, 1952) $b = 0$, $c = 1$; for space charge scattering (Weisberg, 1962; Hasegawa et al., 1969), $b = \frac{1}{3}$, $c = -\frac{1}{2}$; for scattering by charge cylinders (Pödör, 1966, 1970), $b = \frac{1}{2}$, $c = 1$.

These models do not, however, provide a good quantitative fit to the experimental data. Schlötterer (1970) has pointed out that one source of the inaccuracy is that the defects are assumed to be homogeneously distributed, which is not the case in the heteroepitaxial silicon. For low carrier concentrations the distances between twin boundaries or rows of dislocations observed in the silicon on insulators may be less than the radius of the space charge region, and therefore the interaction between adjacent defects must be taken into consideration.

Eernisse and Norris (1972) have also invoked scattering by space charge regions in silicon on spinel but have proposed that the space charge regions are associated with defects agglomerated into clusters, rather than uniformly distributed. Using this model, the calculations become insensitive to the defect density. Ham (1975) has also avoided invoking homogeneously distributed defects in the film. He has analyzed the degradation in the carrier mobility in heteroepitaxial silicon on the basis of transport across energetic barriers associated with planar (rather than line) defects in the heteroepitaxial films. Planar defects (stacking faults and twins) have been observed in the heteroepitaxial films by transmission electron microscopy (see Section 2.13). In this model, analogies are made to the transport properties that have been observed in polycrystalline silicon. This point of view, and the relationship to other models, is discussed in Chapter 6.

Some comment is appropriate in this overview on the issue of stress specifically resulting from the difference in thermal contraction between the substrate and the deposit. The influence of the stress-induced piezoresistive effect is more obvious in the enhancement than in the degradation of mobilities. It is difficult to account for the unusually high hole mobilities observed in the (111) silicon on (111) spinel by other than the piezoresistive effect (Schlötterer, 1968). On the other hand, the degradation of electrical properties is so complicated by the complex structural/impurity effects discussed earlier that it is not possible in any real sense to separate out the piezoresistive effect, except in the case of the anisotropy in the carrier mobilities observed within one crystallographic plane (Thorsen and Hughes, 1972; Hughes and Thorsen, 1973; Hughes, 1975). In this case the anisotropy can be accounted for by the piezoresistive effect alone, while degradation in the zero stress value within the plane can be associated with the defect/impurity effects. In the (100) silicon on the ($1\bar{1}02$) sapphire, which is of most interest at present for device applications, the mobility anisotropy associated with the piezoelectric effect is small. It seems reasonable to conclude at this point, therefore, that the thermally induced stress is not a major limiting factor in heteroepitaxial silicon. In extending the theory of the piezoelectric effect from anisotropy within a plane to a comparison of the maximum mobilities possible in silicon deposited on any sapphire orientation (Hughes, 1975), it is tempting to anticipate that major improvements will be realized by the choice of the proper substrate orientation. On the other hand, there is a problem in matching the theoretically desirable orientations with the experimental realization of good crystal structure. After a considerable amount of effort throughout the years, acceptable crystallinity in heteroepitaxial silicon has been achieved only on a limited number of substrate orientations. It seems unlikely, therefore, that significant enhancement of the mobilities will be achieved by the piezoresistive effect.

Limiting factors: summary

With the data collected to date on the characteristics of heteroepitaxial silicon, it is not possible to assign in a quantitative manner the contributions of the various limiting factors discussed above. In general, however, one can conclude that with the development of the technology the basic limiting factor in heteroepitaxial films has shifted from autodoping to crystal structure. This is true at least for application of the films in MOS transistor structures, wherein the transport properties in the silicon most remote from the substrate dominate the device characteristics. As the applications are broadened, the properties of the silicon near the silicon/substrate become of more concern. The challenge of the future is to understand, gain further control of, and learn to live with the gradients (normal to the substrate) in the defect structure and semiconducting characteristics, particularly in the silicon within less than 0.5 μm of the silicon/substrate surface.

It is not very speculative to predict that improved semiconducting properties would be realized on substrates with a significantly better lattice match to silicon, less reactivity in the presence of silicon, and better thermal contraction match to silicon than is observed with sapphire and spinel. It seems unlikely, however, considering the number of single crystal materials that have been examined for possible use as heteroepitaxial substrates, that a known (available) material will prove to be a better substrate. Nor is it likely that the properties of the known materials can be sufficiently modified with respect to lattice spacing, reactivity, or thermal expansion without unacceptable distortion of the crystalline lattice. For the foreseeable future, these factors must be accepted as inherent properties of the heteroepitaxial silicon/refractory oxide materials system.

This does not lead, however, to pessimism in regard to the possibility of further improving the semiconducting properties of heteroepitaxial silicon. Considering the highly defected silicon at the silicon/substrate interface, it is encouraging that the electrical properties of the silicon within a micron of the interface are as good as they are. It has been demonstrated that MOS transistors with field effect mobilities 80 percent of the values commonly observed in processed bulk silicon have been fabricated in heteroepitaxial silicon less than 1 μm in thickness. These mobilities are adequate to realize the advantages offered by the use of the insulating substrate. Bipolar structures have been prepared in heteroepitaxial silicon ~4 μm in thickness. The use of ion implantation offers a flexibility in the fabrication of bipolar devices that was not anticipated in the early development of heteroepitaxial silicon technology.

Rather than ask, "How similar to bulk silicon can we make the heteroepitaxial silicon?" it may be more appropriate at this point in the development of the technology to ask, "How do we prepare heteroepitaxial silicon in which the electrical properties improve more rapidly with distance from the substrate than is the case at present?" There appears to be real latitude for an advance in the technology here. It is this author's view that such an improvement will come from a more detailed understanding of the processes involved in the initial nucleation of the silicon, of the manner in which the silicon grows during the period prior to and immediately after the complete coverage of the substrate, and how the nature of the early growth relates to the physical and electrical properties of the silicon directly adjacent to (within 0.2 μm of) the silicon/substrate interface. During the period in the development of the technology when the semiconducting properties were dominated by autodoping and the properties were highly irreproducible, the factors involved in the improvement of the film with distance from the substrate were elusive. Now autodoping has been suppressed to acceptable levels, and excellent reproducibility in film properties has been achieved. Some of the classical tools for examination of the chemical, physical, and semiconducting properties have been refined and are more readily applied to the detailed examination of submicron heteroepitaxial deposits. Information is currently being gathered which will contribute not only to improvement in the semiconducting properties of the thin heteroepitaxial silicon but which will also contribute to a more fundamental understanding of the heteroepitaxial process.

References

Abrahams, M. S. and C. J. Buiocchi, *J. Appl. Phys. 45:* 3315 (1974).

Abrahams, M. S. and C. J. Buiocchi, *Appl. Phys. Letters 27:* 325 (1975).

Abrahams, M. S. and C. J. Buiocchi, private communication, RCA Laboratories, Princeton, NJ (1976).

Abrahams, M. S., C. J. Buiocchi, R. T. Smith, J. F. Corboy, Jr., J. Blanc, and G. W. Cullen, *Appl. Phys. 47,* No 12, 5139 (1976).

Adamski, J. A., *J. Appl. Phys. 36:* 1784 (1965).

Aeschlimann, R., F. Gassmann, and T. P. Woodman, *Mat. Res. Bull. 5:* 167 (1970).

Allison, J. F., D. J. Dumin, F. P. Heiman, C. W. Mueller, and P. H. Robinson, *Proc. IEEE 57:* 1490 (1969).

Alper, A. M., R. N. McNally, P. H. Ribbee, and R. C. Doman, *J. Am. Ceram. Soc. 45:* 263 (1962).

Ang, C. Y. and H. M. Manasevit, *Solid State Electronics 8:* 994 (1965).

Austerman, S. B., *J. Nucl. Mat. 14:* 225 (1964).

Austerman, S. B., D. K. Smith, and H. W. Newkirk, *Crystal Growth* (H. S. Peiser, Ed.), London: Pergamon Press (1967), p. 813.

Austerman, S. B., M. F. Ehman, and T. B. Johnson, Contract Report No. AFML-TR-72-175, Air Force Materials Laboratory, Wright-Patterson AFB, Ohio (1972).

Avigal, Y. and M. Schieber, *J. Crystal Growth 9:* 127 (1971).

Avigal, Y., Y. David, M. Schieber, and R. Levin, *J. Crystal Growth 24/25:* 188 (1974).

Bardsley, W., G. W. Green, C. H. Holliday, and D. T. J. Hurle, *J. Crystal Growth 16:* 277 (1972).

Barker, A. S., Jr., *Phys. Rev. 132:* 1474 (1963).

Bean, K. E. and W. R. Runyan, *J. Electrochem. Soc. 121:* 284C, Abstract No. 138 (1974).

Benzing, W. C., private communication, Applied Materials, Inc., Santa Clara, CA (1974).

Benzing, W. C., A. E. Ozias, and I. M. Helmer, Internal Report, Applied Materials, Inc., Santa Clara, CA (1974).

Berg, H. M. and E. L. Hall, Paper presented at Electrochemical Society Meeting, Toronto, Canada, 1975; Abstract: *J. Electrochem. Soc. 122:* 82C (1975); Extended Abstract 75-1, No. 155 (1975).

Berman, P. A., *Jet Propulsion Lab., Quarterly Tech. Review 2,* No. 4 (January 1973).

Bicknell, R. W., J. M., Charig, B. A. Joyce, and D. J. Stirland, *Phil. Mag. 9:* 965 (1964).

Bicknell, R. W., B. A. Joyce, J. H. Neave, and G. V. Smith, *Phil. Mag. 14:* 31 (1966).

Blanc, J. and M. S. Abrahams, *J. Appl. Phys.* 47, 5151 (1976).

Blank, J. M. and V. A. Russell, *Trans. AIME 236:* 291 (1966).

Boleky, E. J., *RCA Review 31:* 372 (1970).

Boleky, E. J., J. R. Burns, J. E. Meyer, and J. H. Scott, *Electronics,* July 1970, p. 82.

Booker, G. R. and R. Stickler, *J. Appl. Phys. 33:* 3281 (1962).

Brimm, E. and H. Humphreys, *J. Phys. Chem. 61:* 829 (1957).

Burns, J. and J. Scott, *Proc. AFIPS Conf., Fall Joint Computer Conf. 35:* 469 (1969).

Capell, A., D. Knoblock, L. Mather, and L. Lopp, *Electronics,* May 26, 1977, p. 99.

Chalmers, B., H. E. LaBelle, and A. I. Mlavsky, *J. Crystal Growth 13/14:* 84 (1972).

Chang, C. C., Proceedings of Fourth International Materials Symposium, University of California, Berkeley, June 17–21, 1968A.

Chang, C. C., *J. Appl. Phys. 39:* 5570 (1968B).

Chang, C. C., *The Structure and Chemistry of Solid Surfaces* (G. A. Somoyai, Ed.) New York: John Wiley and Sons, Inc. (1969).

Chang, C. C., *J. Vac. Sci. Technol. 8:* 500 (1971).

Charvat, F. R., J. C. Smith, and O. H. Nestor, "Crystal Growth," in *Procedings of an International Conference on Crystal Growth, Boston, 1966,* New York: Pergamon Press (1967), p. 45.

Chiang, Y. S., *Semiconductor Silicon,* Princeton, NJ: The Electrochemical Society Inc. (1973), p. 285.

Chiang, Y. S. and G. W. Looney, *J. Electrochem. Soc. 120:* 550 (1973).

Chiang, Y. S., and D. Richman, *Metallurgical Trans. 2,* 743 (1971).

Chopra, K. I., *Thin Film Phenomena,* New York: McGraw-Hill Book Co. (1969).

Chu, T. L., M. H. Francombe, G. A. Gruber, J. J. Oberly, and R. L. Tallman, Westinghouse Research Laboratory, Rpt. No. AFCRL-65-574, AD61992 (1965).

Cockayne, B. and M. Chesswas, *J. Mat. Sci. 2:* 498 (1967).

Cockayne, B., M. Chesswas, P. J. Born, and J. D. Filby, *J. Mat. Sci. 4:* 236 (1969).

Cockayne, B., M. Chesswas, and D. B. Gasson, *J. Mat. Sci. 2,* 7 (1967).

Cocks, P. H. and J. T. A. Pollock, *J. Appl. Phys. 43,* 3878 (1972).

Corboy, J. F., private communication, RCA Laboratories, Princeton, NJ (1974).

Corboy, J. F., private communication (1976).

Crossley, P. A., and W. E. Ham, J. Electronic Materials *2,* 465 (1973).

Cullen, G. W., *J. Crystal Growth 9:* 107 (1971).

Cullen, G. W., Paper presented at the IEEE Workshop on Silicon on Sapphire, Vail, Colorado, October 1977.

Cullen, G. W. and J. F. Corboy, *J. Electrochem. Soc. 121:* 1345 (1974).

Cullen, G. W. and F. C. Dougherty, *J. Crystal Growth 17:* 230 (1972).

Cullen, G. W. and J. H. Scott, unpublished, RCA Laboratories, Princeton, NJ (1972).

Cullen, G. W. and C. C. Wang, *J. Electrochem. Soc. 118:* 640 (1971).

Cullen, G. W., G. E. Gottlieb, C. C. Wang, and K. H. Zaininger, *J. Electrochem. Soc. 116:* 1444 (1969).

Cullen, G. W., G. E. Gottlieb, and D. Leibowitz, Final Report of Contract No. N00039-69-C-0549, Naval Electronics Systems Command, Dept. of the Navy, Washington, DC, July, 1970A.

Cullen, G. W., G. E. Gottlieb, and C. C. Wang, *RCA Review 31:* 355 (1970B).

Cullen, G. W., J. F. Corboy, and A. G. Kokkas, Technical Rpt. AFAL-TR-73-200, June, 1973A.

Cullen, G. W., S. R. Bolin, A. D. Morrison, and C. C. Wang, U.S. Patent 3,736,158, May, 1973B.

Cullen, G. W., J. F. Corboy, and R. T. Smith, *J. Crystal Growth 31,* 274 (1975A).

Cullen, G. W., S. R. Bolin, J. Creamer, J. F. Corboy, and A. J. Wasielewski, U.S. Patent 3,883,313, May 13, 1975B.

Cullis, A. G., Ph.D. Thesis, University of Oxford, 1972.

Cullis, A. G. and G. R. Booker, *Thin Solid Films. 31* 53 (1976).

Cutter, I. A. and R. McPherson, *J. Am. Ceram. Soc. 56:* 266 (1973).

Deines, J. L. and A. Spiro, Paper presented at Electrochemical Society Meeting, San Francisco, California (1974); Abstract: *J. Electrochem. Soc. 121:* 91C (1974); Extended Abstract 74-1, No. 161 (1974).

Dell'Oca, C. J., and H. Nonnenmacher, Paper presented at Electrochemical Society Meeting, New York, 1974; Abstract: *J. Electrochem. Soc. 121:* 285C (1974); Extended Abstract 74-2, No. 146 (1974).

Dexter, D. L. and F. Seitz, Phys. Rev. *86,* 964 (1952).

Dixon, M. T. and J. P. Willis, AEI Central Res. Lab. Rpt. No. 6, Nov. 1967, Rugby, England.

Djevahirdjian, M. V., Swiss Pat. 354,428, July 15, 1961.

Dobrovinskaya, E. R., A. N. Galagurya, and L. A. Litvinov, *Sov. Phys.—Solid State 12:* 1460 (1970).

Druminski, M., Paper presented at the Electrochemical Society Meeting, Las Vegas, Nevada, October, 1976.

Druminski, M. and Ch. Kühl, Paper presented at Fundamentals of Epitaxy Conference, Erlangen, Germany, 1974.

Druminski, M. and H. Schlötterer, *J. Crystal Growth 17:* 249 (1972).

Druminski, M. and Cl. Wieczorek, Paper presented at Electrochemical Society Meeting, Toronto, Canada, 1975; Abstract: *J. Electrochem. Soc. 122:* 82C (1975); Extended Abstract 75-1, No. 154 (1975).

Druminski, M., Ch., Kühl, E. Preuss, F. Schwidefsky, J. Tihanyi, and K. Schmid, *J. Electrochem. Soc. 121:* 284C, Abstract No. 139 (1974).

Druminski, M., Ch. Kühl, E. Preuss, F. Schwidefsky, H. Splittgerber, and D. Takacs, Proceedings of the 7th Conference on Solid State Devices, Tokyo, Japan, 1975; *Japanese J. Appl. Phys. 15:* 217 (1976).

Duffy, M. T. and P. J. Zanzucchi, Paper presented at Electrochemical Society Meeting, Las Vegas, NA, 1976.

Duffy, M. T., P. J. Zanzucchi, and G. W. Cullen, Contract N. 5-35915, National Bureau of Standards, Washington, D.C., Report No. 2, October, 1975.

Duffy, M. T., P. J. Zanzucchi, and G. W. Cullen, Reports of Contract No. 5-35915, National Bureau of Standards, Washington, DC 20234, 1975–1977.

Dumin, D. J., *J. Appl. Phys. 36:* 2700 (1965).

Dumin, D. J., *Rev. Sci. Inst. 38:* 1107 (1967A).

Dumin, D. J., *J. Appl. Phys. 38:* 1909 (1967B).

Dumin, D. J., *Solid State Electronics 13:* 415 (1970).

Dumin D. J., *J. Vac. Sci. Technol. 8:* 235 (1971).

Dumin, D. J., Paper presented at Electrochemical Society Meeting, New York, 1974; Abstract: *J. Electrochem. Soc. 121:* 284C (1974); Extended Abstract 74-2, No. 137 (1974).

Dumin, D. J. and P. H. Robinson, *J. Appl. Phys. 39:* 2759 (1968A).

Dumin, D. J. and P. H. Robinson, *J. Crystal Growth 3/4:* 214 (1968B).

Dumin, D. J. and E. C. Ross, *J. Appl. Phys. 41:* 3139 (1970).

Dumin, D. J. and R. S. Silver, *Solid State Electronics 11:* 353 (1968).

Dumin, D. J., P. H. Robinson, G. W. Cullen, and G. E. Gottlieb, *RCA Review 31:* 620 (1970).

Eernisse, E. P. and C. B. Norris, *Solid State Electronics 16:* 315 (1972).

Ehman, M. F., Paper presented at Electrochemical Society Meeting, Boston, MA, 1973; Abstract: *J. Electrochem. Soc. 120:* 244C (1973); Extended Abstract 73-2, No. 198 (1973).

Elliot, A. B. M. and J. C. Anderson, *Solid State Electronics 15:* 531 (1972).

Estrup, P. J. and E. G. McRae, *Surface Science 25:* 1 (1971).

Eversteijn, F. C., *Philips Res. Reports 26:* 134 (1971).

Eversteijn, F. C., and H. L. Peek, *Philips Res. Reports 25:* 472 (1970).

Eversteijn, F. C., P. J. W. Severin, C. H. J. v.d. Brekel, and H. L. Peek, *J. Electrochem. Soc. 117:* 925 (1970).

Faktor, M. M., D. G. Fiddyment, and G. R. Newns, *J. Electrochem. Soc. 114:* 356 (1967).

Falckenberg, R., *J. Crystal Growth 13,14:* 718 (1972).

Falckenberg, R., *J. Crystal Growth 29:* 195 (1975).

Falckenberg, R., *J. Electrochem. Soc. 123:* 63 (1976).

Farrington, D., *J. Electrochem. Soc. 121:* 285C, Abstract No. 145 (1974).

Faust, J. W., Jr., *Surface Science 13:* 60 (1969).

Filby, J. D., *J. Electrochem. Soc. 113:* 1085 (1966).

Filby, J. D., *Modern Oxide Materials* (B. Cockayne and D. W. Jones, Eds.), London and New York: Academic Press, (1972), p. 203.

Filby, J. D. and S. Nielsen, *Brit. J. Appl. Phys. 18:* 1357 (1967).

Flatley, D. W. and W. E. Ham, Paper presented at Electrochemical Society Meeting, New York, 1974; Abstract: *J. Electrochem. Soc. 121:* 290C (1974); Extended Abstract 74-2, No. 198 (1974).

Fraimbault, J. L., I. Gyomlai, R. Montmory, and J. Vuillod, *Proceedings of International Symposium of the Basic Problem Thin Film Physics,* Goettingen, Germany (1965) p. 638.

Frenkel, D. R. and J. A. Venables, *Advances in Physics 19:* 409 (1970).

Garcia, M., J. F. Bresse, and B. Blanchard, Paper presented at the Third European Devices Research Conference, Munich, Germany. Abstract A4.1, p. 828, 1973.

Gassmann, F., A. Della Casa, and R. Aeschlimann, *Mat. Res. Bull. 6:* 817 (1971).

Gassmann, F., R. Aeschlimann, and U. Bänziger, *Mat. Res. Bull. 7:* 1493 (1972).

Gottlieb, G. E., Ph.D. Thesis, Rutgers University, 1971.

Gottlieb, G. E., *J. Crystal Growth 12:* 327 (1972).

Gottlieb, G. E. and J. F. Corboy, *J. Crystal Growth 17:* 261 (1972).

Gottlieb, G. E., J. F. Corboy, G. W. Cullen, and J. H. Scott, *Metallurgical Trans. 2:* 653 (1971).

Grabmaier, J. G. and C. Watson, *Phys. Stat. Sol. 25:* K7 (1968A).

Grabmaier, J. G. and C. Watson, *Z. Angew. Phys. 24:* 108 (1968B).

Grabmaier, J. G. and C. Watson, *J. Am. Ceram. Soc. 51:* 355 (1968C).

Green, J. M., *J. Electrochem. Soc. 119:* 1765 (1972).

Green, A. K., J. Dancy, and E. Bauer, *J. Vac. Sci. Technol. 7:* 159 (1970).

Gross, U. and R. Kersten, *J. Crystal Growth 15:* 85 (1972).

Grove, A. S., *Physics and Technology of Semiconductor Devices,* New York: John Wiley and Sons, Inc. (1967), p. 300.

von Grube, G., A. Schneider, U. Esch, and M. Flad, *Z. Anorg. Chem. 260:* 120 (1949).

Gunn, S. R., *J. Phys. Chem. 65:* 779 (1961).

Gupta, D. C., *Solid State Electronics 13:* 543 (1970).

Haidinger, W., B. Zega, H. Steiner, and M. Pongratz, International Conference on Physics. Chemistry of Semiconductor Heterojunctions and Layer Structures; Budapest, Hungary, October, 1970.

Ham. W. E., *Appl. Phys. Letters 21:* 440 (1972).

Ham, W. E., Electron Device Conference, Washington, DC, 1973.

Ham, W. E., Paper presented at Electrochemical Society Meeting, New York, 1974; Abstract: *J. Electrochem. Soc. 121:* 285C (1974); Extended Abstract 74-2, 144 (1974).

Ham, W. E., private communication, RCA Laboratories, Princeton, NJ (1975).

Hammond, M. R., Recent newspaper presented at Electrochemical Society Meeting, Detroit, Mich., 1969; Abstract: *J. Electrochem. Soc. 116:* 376C (1969).

Harrington, W., C. W. Magee, G. W. Cullen, and J. F. Corboy, Proceedings of the 24th Annual Conference on Mass Spectrometry and Allied Topics, San Diego, California, May, 1976.

Hart, P. B., P. J. Etter, B. W. Jervis, and J. M. Flanders, *Brit. J. Appl. Phys. 18:* 1389 (1967).

Hasegawa, S., N. Kaminaka, T. Nakamura, and T. Itoh, *J. Appl. Phys. 40:* 4620 (1969).

Heiman, F. P., *IEEE Trans. Electron Devices ED-13:* 855 (1966).

Heiman, F. P., *IEEE Trans. Electron Devices ED-14:* 781 (1967).

Heiman, F. P. and P. H. Robinson, *Solid State Electronics 11:* 411 (1968).

Heywang, W., *Mat. Res. Bull. 3:* 315 (1968).

Higson, G. R., *J. Sci. Instr. 41:* 405 (1964).

Hockey, B. J., *The Science of Ceramic Machining and Surface Finishing* (S. J. Schneider, Jr. and R. W. Rice, Eds.) NBS Special Publication 348 (1972A), p. 119.

Hockey, B. J., Proc. No. 20, Publ. of the Brit. Ceramic Soc. (1972B), p. 95.

Hofstein, S. R., IEEE Trans. Electron Devices *ED-13,* 846 (1966).

Horiuche, S., *Solid State Electronics 18:* 1111 (1975).

Howie, A. and Z. S. Basinski, *Phil. Mag. 17:* 1039 (1968).

Huff, H. R. and R. R. Burgess, Eds., *Semiconductor Silicon 1973,* 2nd International Symposium on Silicon Materials Science and Technology, May 13–18, 1973, Electrochemical Society Meeting, Chicago.

Hughes, A. J., *J. Appl. Phys. 46:* 2849 (1975).

Hughes, A. J. and A. C. Thorsen, *J. Appl. Phys. 44:* 2304 (1973).

Hynecek, J., *J. Appl. Phys. 45:* 2631 (1974A).

Hynecek, J., *J. Appl. Phys. 45:* 2806 (1974B).

Ipri, A. C., *Appl. Phys. Letters 20:* 1 (1972A).

Ipri, A. C., *J. Appl. Phys. 43:* 2770 (1972B).

Ipri, A. C. and D. W. Flatley, IEEE Trans. Electron Devices, *September,* 1110 (1976).

Ipri, A. C. and J. C. Sarace, IEEE International Solid State Circuits Conference Digest of Technical Papers, Paper No. WPM6.3, p. 58 (1974).

Ipri, A. C. and J. C. Sarace, *IEEE J. Solid State Circuits* SC-11, 329 (1976).

Ipri, A. C. and J. N. Zemel, *J. Appl. Phys. 44:* 744 (1973).

Itoh, T., S. Hasegawa, and N. Kaminaka, *J. Appl. Phys. 40:* 2597 (1969).

Jona, F., *Appl. Phys. Letters 9:* 253 (1966).

Joyce, B. A., *J. Crystal Growth 3/4:* 43 (1968).

Joyce, B. A., *Rep. Prog. Phys. 37:* 363 (1974).

Joyce, B. A. and R. R. Bradley, *J. Electrochem. Soc. 110:* 1235 (1963).

Joyce, B. A., R. J. Bennett, R. W. Bicknell, and P. J. Etter, *Trans. AIME 233:* 556 (1965).

Kamins, T. I. and T. R. Cass, *Thin Solid Films 16:* 147 (1973).

Kamins, T. I. and E. S. Meieran, *J. Appl. Phys. 44:* 5064 (1973).

Kane, P. F. and G. B. Larrabee, *Characterization of Semiconductor Materials,* Texas Institute Electronics Series, New York: McGraw-Hill Book Co. (1970), p. 182.

Koepke, B. G., *The Science of Ceramic Machining and Surface Finishing* (S. J. Schneider, Jr. and R. W. Rice, Eds.) NBS Special Publication 348 (1972), p. 119.

Kranzer, D., *Appl. Phys. Letters 25:* 103 (1974).

Krause, H., *Phys. Stat. Sol. (A) 1:* K53 (1970).

Kroko, L. J. and G. L. Shaw, *Semiconductor Silicon* (R. R. Haberecht and E. L. Kern, Eds.) New York: The Electrochemical Society (1969) p. 316.

Kühl, Ch., H. Schlötterer, and F. Schwidefsky, *J. Electrochem. Soc. 121:* 1496 (1974).

Kühl, Ch., M. Druminski, and K. Wittmaack, *Thin Solid Films. 37,* 317 (1976A).

Kühl, Ch., H. Schlötterer, and F. Schwidefsky, *J. Electrochem. Soc. 123:* 97 (1976B).

Kuhn, G. L. and C. J. Rhee, *J. Electrochem. Soc. 120:* 1563 (1973).

Kyle, T. R. and G. Zydzik, *Mat. Res. Bull. 8:* 443 (1973).

LaBelle, H. E., Jr., U.S. Patent 3,591,348 (1971A).

LaBelle, H. E., Jr., *Mat. Res. Bull. 6:* 581 (1971B).

LaBelle, H. E., Jr. and A. I. Mlavsky, U.K. Patent 1,205,-544, June 9, 1969.

LaBelle, H. E., Jr. and A. I. Mlavsky, *Mat. Res. Bull. 6:* 571 (1971).

LaBelle, H. E., Jr., G. F. Hurley, and A. D. Morrison, Reports under Contract No. DAAB05-72-C-5841, U.S. Army Electronics Command, Philadelphia, PA (1973/74).

LaChapelle, T. J., A. Miller, and F. L. Morritz, *Progress in Solid State Chemistry* (H. Riess, Ed.), New York: Pergamon Press (1967), p 1.

Larssen, P. A., *Acta Cryst. 20:* 599 (1966).

Latanision, R. M., Paper presented at the International Conference on Surface Technology, May, 1973.

Lawson, R. W. and D. M. Jefkins, *J. Physics D: Appl. Physics 3:* 1627 (1970).

Lee, Chia-Chuan and K. Hu, Paper presented at Electrochemical Society Meeting, Toronto, Canada, 1975; Abstract: *J. Electrochem. Soc. 122:* 85C (1975); Extended Abstract 75-1, No. 177 (1975).

Lindstrom, G. A., U.S. Patent 3,321,872, May 30, 1967.

Linnington, P. F., Proceedings 7th International Conference Electron Microscopy, Grenoble, France, Vol. 2, p. 447 (1970).

Linnington, P. F., Procedings of the 25th Anniversary Meeting, EMAG, Institute of Physics, Cambridge, England, (1971) p. 182.

Linnington, P. F., Ph.D. Thesis, Cavendish Laboratory, Cambridge, England (1974).

Lucarini, V. J., R. L. Bratter, and J. E. Basso, Paper presented at Electrochemical Society Meeting, San Francisco, CA, 1974; Abstract: *J. Electrochem. Soc. 121:* 91C (1974); Extended Abstract 74-1, 164 (1974).

MacKenzie, J. M. W., *Minerals Sci. Engng. 3:* 25 (1971).

Manasevit, H. M., U.S. Patent 3,414,434, December 3, 1968A.

Manasevit, H. M., *J. Electrochem. Soc. 115:* 434 (1968B).

Manasevit, H. M., *J. Crystal Growth 22:* 125 (1974).

Manasevit, H. M., Paper presented at Electrochemical Society Meeting, Toronto, Canada (1975); Abstract: *J. Electrochem. Soc. 122:* 82C (1975); Extended Abstract 75-1, No. 153 (1975).

Manasevit, H. M. and D. H. Forbes, *J. Appl. Phys. 37:* 734 (1966).

Manasevit, H. M. and F. L. Morritz, *J. Electrochem. Soc. 114:* 204 (1967).

Manasevit, H. M. and W. I. Simpson, *J. Appl. Phys. 35:* 1349 (1964).

Manasevit, H. M. and A. C. Thorsen, Heteroepitaxial GaAs on Aluminum Oxide. I—Nucleation Studies, NASA Contract No. NAS12-2010.

Manasevit, H. M. and W. I. Simpson, Recentnews paper, American Physical Society, Edmonton, Canada (1963).

Manasevit, H. M., A. Miller, F. L. Morritz, and R. Nolder, *Trans. AIME 233:* 540 (1965).

Manasevit, H. M., D. H. Forbes, and I. B. Cadoff, Trans. AIME *236:* 275 (1966).

Manasevit, H. M., R. L. Nolder, and L. A. Moudy, *Trans. AIME 242:* 465 (1968).

Manasevit, H. M., F. M. Erdmann, and A. C. Thorsen, *J. Electrochem. Soc. 121:* 284C, Abstract No. 140 (1974).

Manasevit, H. M., F. M. Erdmann, and A. C. Thorsen, *J. Electrochem. Soc. 123:* 52 (1976).

Matthews, J. W. (Ed.), *Epitaxial Grwoth,* Parts A and B, New York: Academic Press (1975).

Maurits, J. E. A., *J. Electrochem. Soc. 121:* 284C, Abstract No. 136, (1974).

Maurits, J. E. A., Paper presented at the Electrochemical Society Meeting, Las Vegas, Nevada, October, 1976.

Maurits, J. E. A., and A. M. Hawley, *Advances in X-Ray Analysis* (Heinrich, Barrett, Newkirk, and Rund, Eds.), Vol. 15, New York: Plenum Publish. Corp. (1971), p. 516.

Mayer, S. E. and D. E. Shea, *J. Electrochem. Soc. 111:* 550 (1964).

McGreivy, P. J. and C. R. Viswanathan, *Appl. Phys. Letters 25:* 505 (1974).

Mendelson, S., *J. Appl. Phys. 39:* 2477 (1967).

Menter, J. W. and D. W. Pashley, *The Structure and Properties of Thin Films* New York: John Wiley and Sons, Inc. (1959).

Mercier, J., *Rev. Physique Appliquee 3:* 127 (1968).

Mercier, J., *J. Eleetrochem. Soc. 117:* 666 (1970A).

Mercier, J., *J. Electrochem. Soc. 117:* 812 (1970B).

Mercier, J., *J. Electrochem. Soc. 118:* 962 (1971).

Mercier, J., Internal Report, Grenoble, France.

Messineo, P. J., private communication, RCA Laboratories, Princeton, NJ (1975).

Miller, A., *Electronics 40:* 171 (1967).

Miller, A. and H. M. Manasevit, J. Vac. Sci. Technol. *3,* 68 (1966).

Moss, H. I., private communication, RCA Laboratories, Princeton, NJ (1973).

Mueller, C. W., *J. Vac. Sci. Technol. 7:* 147 (1970).

Mueller, C. W. and P. H. Robinson, *Proc. IRE 52:* 1487 (1964).

Mustachi, A., private communication, RCA Laboratories, Princeton, NJ (1974).

Naber, C. T. and J. E. O'Neal, *Trans. AIME 242:* 470 (1968).

Namba, S., A. Kawazu, and T. Maruyama, *Sci. Pap. Inst. Phys. Chem. Res. (Tokyo) 61:* 45 (1967A).

Namba, S., A. Kawazu, and T. Maruyama, *Procedings of the Second Colloquium on Thin Films,* Budapest, Hungary, 1967B, p. 213.

Nolder, R. L. and I. B. Cadoff, *Trans. AIME 233:* 549 (1965).

Nolder, R. L., D. J. Klein, and D. H. Forbes, *J. Appl. Phys. 36:* 3444 (1965).

Norris, C. B., *Appl. Phys. Letters 20:* 187 (1972).

Novak, R. E., (RCA Laboratories, Princeton, NJ), private communication (1976).

Oh, H. L., K. P. L. Oh, S. Vaidyanathan, and I. Finnie, *The Science of Ceramic Machining and Surface Finishing* (S. J. Schneider, Jr., and R. W. Rice, Eds.) NBS Special Publication 348 (1972) p. 119.

Paladino, A. E. and B. D. Roiter, *J. Am. Ceram. Soc. 47:* 465 (1964).

Pashley, D. W., *Advances in Physics 14:* 327 (1965).

Patterson, D. L., private communication, RCA Laboratories, Princeton, NJ (1974).

Peel, J. L. and M. D. Barry, Paper presented at the 18th Annual Electronic Materials Conference on the Preparation and Properties of Electronic Materials, Metallurgical Society of the AIME, Salt Lake City, Utah, June, 1976.

Picraux, S. T., *Appl. Phys. Letters 20:* 91 (1972).

Picraux, S. T., *J. Appl. Phys. 44:* 587 (1973).

Picraux, S. T. and G. J. Thomas, *J. Appl. Phys. 44:* 594 (1973).

Pödör, B., phys. stat. sol. (a) *16,* K167 (1966).

Podor, B., *Phys. Stat. Sol. (A)2:* K197 (1970).

Porter, J. L. and R. G. Wolfson, *J. Appl. Phys. 36:* 2746 (1965).

Powell, C. F., J. H. Oxley, and J. M. Blocher, Jr., *Vapor Deposition,* New York: John Wiley and Sons, Inc. (1966).

Rabinowicz, E., *Scientific American* (June): 91 (1968).

Raetzel, C. and J. Tihanyi, Siemens R&D Reports 1, No. 3, 307 (1972).

Raetzel, C., S. Schild, and H. Schlotterer, *J. Electrochem. Soc. 121:* 284C, Abstract No. 142 (1974).

Rai-Choudhury, P., Y. C. Kao, and G. G. Sweeney, *J. Electrochem. Soc. 121:* 1509 (1974A).

Rai-Choudhury, P., D. K. Schroder, T. M. S. Heng, and G. G. Sweeney, Paper presented at Electrochemical Society Meeting, New York, NY, 1974, Abstract: *J. Electrochem. Soc. 121:* 284C, No. 143 (1974); Extended Abstract 74-2, 339 (1974B).

Redfield, D., *IEEE Trans. Parts, Hybrids, and Packaging PHP-10:* 239 (1974).

Redfield, D., private communication, RCA Laboratories, Princeton, NJ (1975).

Reisman, A., M. Berkenblit, J. Cuomo, and S. A. Chan, *J. Electrochem. Soc. 118:* 1653 (1971).

Reynolds, F. H. and A. B. M. Elliott, *Phil. Mag. 13:* 1073 (1966).

Reynolds, F. H. and A. B. M. Elliott, *Solid State Electronics 10:* 1093 (1967).

Ricard, J., private communication, Ugine Kuhlmann, Grenoble, France (1975).

Ricard, J. and A. Cisccolaini, *J. Crystal Growth 13,14:* 718 (1972).

Richman, D. and R. Arlett, *J. Electrochem. Soc. 116:* 872 (1969).

Richman, D., Y. S. Chiang, and P. H. Robinson, *RCA Review 31:* 613 (1970).

Robinson, P. H. and D. J. Dumin, *J. Electrochem. Soc. 115:* 75 (1968).

Robinson, P. H. and F. P. Heiman, *J. Electrochem. Soc. 118:* 141 (1971).

Robinson, P. H. and C. W. Mueller, *Trans. AIME 236:* 268 (1966).

Robinson, P. H. and R. O. Wance, *RCA Review 34:* 616 (1973).

Ronen, R. and P. H. Robinson, *Proc. IEEE 59:* 1506 (1971).

Ronen, R. and P. H. Robinson, *J. Electrochem. Soc. 119:* 747 (1972).

Rosenblatt, A., *Electronics,* April 10, 1972, p. 77.

Ross, E. C., and G. Warfield, *J. Appl. Phys. 40:* 2339 (1969).

Rubin, J. J. and L. G. Van Uitert, *Mat. Res. Bull. 1:* 211 (1966).

Ruth, R. P., A. J. Hughes, J. L. Kenty, H. M. Manasevit, D. Medellin, A. C. Thorsen, Y. T. Chan, C. R. Viswanathan, and M. A. Ring, Final Report under Contract No. DAAH01-70-C-1311, ARPA Order No. 1585 (1973).

Safdar, M., G. H. Frischat, and H. Salge, *J. Am. Ceram. Soc. 57:* 106 (1974).

Sahagian, C. S., Air Force Cambridge Research Laboratories Rpt. No. 66-659, Physical Sciences Research Paper No. 268, September 1966.

Sahagian, C. S. and M. Schieber, Procdings of Symposium on Crystal Growth, 7th International Crystallography Congress, Moscow, July, 1966.

Sahagian, C. S. and M. Schieber, *Growth of Crystals,* Vol. VII (N. N. Sheftal', Ed.) New York: Consultants Bureau (1969) p. 183.

Salama, C. A. T., T. W. Tucker, and L. Young, *Solid State Electronics 10:* 339 (1967).

Sarace, J. C. and A. C. Ipri, Paper presented at Electrochemical Society Meeting, New York, 1974; Abstract: *J. Electrochem Soc. 121:* 290C (1974); Extended abstract 74-2, No. 197 (1974).

Scheuplein, R. and P. Gibbs, *J. Am. Ceram. Soc. 43:* 458 (1960).

Schlötterer, H., IEEE European Meeting on Semiconductor Device Research, Bad Nauheim, Germany, April, 1967.

Schlötterer, H., *Solid State Electronics 11:* 947 (1968).

Schlötterer, H., *Electronics,* October 27, 1969, p. 113.

Schlötterer, H., Procdings of the International Conference on Physical Chemistry of Semiconductor Heterojunctions and Layered Structures, Budapest, Hungary, 1970, p. IV-5.

Schlötterer, H., Paper presented at the American Vacuum Society Meeting, Philadelphia, PA, 1975; *J. Vac. Sci. and Technol. 13:* 29 (1976).

Schlötterer, H. and Ch. Zaminer, *Phys. Stat. Sol. 15:* 399 (1966).

Schmid, F., *J. Crystal Growth 26:* 162 (1974).

Schmid, F., Paper presented at Electrochemical Society Meeting, Toronto, Canada, 1975; Abstract: *J. Electrochem. Soc. 122:* 90C (1975); Extended Abstract 75-1, No. 222 (1975).

Schmid, F. and D. J. Viechnicki, *J. Am. Ceram. Soc. 53:* 528 (1970).

Schmid, F. and D. J. Viechnicki, *Solid State Tech.,* Sept., XXX (1973).

Schmidt, W. A. and J. E. Davey, *The Science of Ceramic Machining and Surface Finishing* (S. J. Schneider, Jr. and R. W. Rice, Eds.) NBS Special Publication 348 (1972), p. 259.

Schneider, H. G. and V. Ruth, *Advances in Epitaxy and Endotaxy,* Leipzig: VEB Deutscher Verlag fur Grundstoffindustrie (1971).

Schneider, S. J., Jr. and R. W. Rice (Eds.), *The Science of Ceramic Machining and Surface Finishing,* NBS Special Publication 348 (1972).

Schroder, D. K. and P. Rai-Choudhury, *Appl. Phys. Letters 22:* 445 (1973).

Scott, J. H. and J. R. Burns, Paper presented at the Electrochemical Society Meeting, Los Angeles, CA, 1970; Abstract: *J. Electrochem. Soc. 117:* 101C (1970); Extended Abstract 70-1, No. 138 (1970).

Scott, J. H. and A. C. Ipri, private communication (1977).

Seiter, H. and Ch. Zaminer, *Z. Angew. Phys. 20:* 158 (1965).

Seiter, H. and Ch. Zaminer, U.S. Patent 3,424,955, January, 1969.

Seiter, H., E. Sirtl, and Ch. Zaminer, Procedings 128th Electrochemical Society Meeting, Buffalo, 1965.

Shakh-Budagov, A. L. and A. V. Stepanov, Summary of the 1950 report of the Leningrad Physico-Technical Institute, Acad. Sci. USSR.

Sirtl, E. and A. Adler, *Z. Metallkunde 52:* 529 (1961).

Sladek, K. J., *J. Electrochem. Soc. 118:* 654 (1971).

Stadnik, A. V., V. E. Kosenko, V. I. Poludin, and N. M. Torchun, *Poluprov. Tekh. Mikroelektron.,* No. 10: 74 (1972).

Stein, H. J., *Solid State Electronics 15:* 1209 (1972).

Stepanov, A. V., Summary of the 1941 report of the Leningrad Physico-Technical Institute, Acad. Sci. USSR.

Stowell, M. J., *Thin Films 1:* 55 (1968).

Strepkoff, S. J., *LeVide* No. 2, Special A.V.I. SEM, p. 52, Oct. (1966).

Swain, M. V., R. M. Latanision, and A. R. C. Westwood, *J. Am. Ceram. Soc. 9:* 372 (1975).

Sylwestrowicz, W. D., *Phil. Mag. 7:* 1825 (1962).

Tallman, R. L., T. L. Chu, G. A. Gruber, J. J. Oberly, and E. D. Wolley, *J. Appl. Phys. 37:* 1588 (1966).

Tamura, M. and M. Nomura, *Appl. Phys. Letters 11:* 196 (1966).

Tanikawa, E., O. Takayama, and K. Maeda, Paper presented at Electrochemical Society Meeting, Boston, MA, 1973; Abstract: *J. Electrochem. Soc. 120:* 247C (1973); Extended Abstract 73-2, 569 (1973).

Thorsen, A. C. and A. J. Hughes, *Appl. Phys. Letters 21:* 579 (1972).

Tihanyi, J., *Siemens Forsch. und Entwickl. Ber. 1,* Nr. 3/72: 263 (1972).

Tihanyi, J., Paper presented at the European Solid State Device Conference, Munich, Germany, 1973.

Trilhe, J., H. Legal, and G. Rolland, Paper presented at the Third International Conference on Vapour Growth and Epitaxy, Amsterdam, The Netherlands, 1975.

Tsivinskii, S. V., *Sov. Phys.—Crystallogr. 12:* 119 (1967).

Tsivinskii, S. V. and A. V. Stepanov, *Sov. Phys.—Solid State 7:* 148 (1965).

van der Pauw, L. J., *Philips Res. Reports 13:* 1 (1958).

van Dijk, H. J. and J. de Jonge, *J. Electrochem. Soc. 117:* 533 (1970).

van Nielen, J. A., M. J. J. Theunissen, and J. A. Appels, *Philips Tech. Rev. 31:* 271 (1970).

Via, G. G. and R. E. Thun, Procedings American Vacuum Society Meeting, New York, 1962.

Viechnicki, D. J. and F. Schmid, *J. Crystal Growth 11:* 345 (1971).

Viechnicki, D. J., F. Schmid, and J. W. McCauley, *J. Appl. Phys. 43:* 4508 (1972).

Viechnicki, D. J., F. Schmid, and J. W. McCauley, Technical Report No. AMMRC TR 73-3, February, 1973.

Walsh, R. J. and A. H. Herzog, U.S. Patent 3,170,273, February, 1965.

Wang, C. C., *J. Appl. Phys. 40:* 3433 (1969).

Wang, C. C. and G. W. Cullen, *J. Electrochem. Soc. 121;* Abstract No. 141 (1974).

Wang, C. C., G. E. Gottlieb, G. W. Cullen, S. H. McFarlane, and K. H. Zaininger, *Trans. AIME 245:* 441 (1969).

Weisberg, L. R., *J. Appl. Phys. 33:* 1817 (1962).

Weisberg, L. R. and E. A. Miller, *Trans. AIME 242:* 479 (1968).

Weitzel, C. W. and R. T. Smith, Paper presented at Electrochemical Society Meeting, Las Vegas, Nevada (1976).

Westwood, A. R. C. and D. L. Goldheim, *J. Am. Ceram. Soc. 53:* 142 (1970).

Westwood, A. R. C. and R. M. Latanision, *The Science of Ceramic Machining and Surface Finishing* (S. J. Schneider, Jr., and R. W. Rice, Eds.) NBS Special Publication 348 (1972) p. 119.

Westwood, A. R. C., N. H. MacMillan, and R. S. Kalyoncu, *J. Am. Ceram. Soc. 56:* 258 (1973).

Yasuda, Y., *Japanese J. Appl. Phys. 7:* 1171 (1968).

Yasuda, Y., *Japanese J. Appl. Phys. 10:* 45 (1971).

Yasuda, Y. and Y. Ohmura, *Japanese J. Appl. Phys. 8:* 1098 (1969).

Zaminer, Ch., *Z. Angew, Phys. 24,* No. 4 (1968).

Zeveke, T. A., L. N. Kornev, and V. A. Tolomasov, *Sov. Phys.—Crystallogr. 13* (3): 493 (1968) [in English].

Zanzucchi, P. J., M. T. Duffy, and R. C. Alig, J. Electrochem. Soc., *125,* 299 (1978).

Zinnes, A. E., B. E. Nevis, and C. D. Brandle, *J. Crystal Growth 19:* 187 (1973).

Chapter 3

Heteroepitaxial Growth and Characterization of Compound Semiconductors

C. C. Wang

3.1. Introduction

The achievement of single crystal growth of large-area semiconductor films on substrates of different materials is of technological importance to many solid state electronic devices. The heteroepitaxial composite structure is also of scientific interest because the epitaxy is determined by the spatial relationship between the atomic arrangement in the substrate and that of the atoms in the appropriate crystallographic plane of the semiconductor. The degree of crystalline perfection of the semiconductor deposits further depends largely on the physicochemical nature of the substrate surfaces.

Extensive studies on the epitaxy of silicon on sapphire (α-Al_2O_3) and spinel ($MgAl_2O_4$) substrates have been reported in the past decade (see Chapter 2). This work has contributed to a basic understanding of such composite material systems and to the commercial development of the silicon-on-sapphire technology. The trend of development has extended to compound semiconductors, notably the III-V compounds. However, because of the increasing complexity in material growth and device fabrication, developments on the III-V film/oxide substrate composite systems are still at an early stage. Research studies on epitaxial growth by new techniques and on defect characterization of the composite materials have been increasing in many leading laboratories. Several device structures have been successfully fabricated in heteroepitaxial systems with interesting results. These include GaAs/ sapphire transmission-mode photocathodes; GaN/sapphire, GaAs/spinel, and GaP/spinel electroluminescent diodes; GaAs/sapphire Gunn oscillators; GaAs/ BeO microwave transistors; and AlN/sapphire surface acoustic wave devices.

In this chapter, a status report is presented on recent and current developments in the field of heteroepitaxial compound semiconductor films grown on dielectric oxide substrates by chemical vapor-deposition techniques. Certain areas of this field have been covered by review articles in literature. They are cited in this chapter where appropriate, but no effort is made to discuss the topics previously presented in detail. Emphasis is placed in areas that have not been comprehensively and critically reviewed and on subjects where up-to-date supplements need to be added to the previous reviews. In Section 3.2, a brief discussion on the oxide substrates is presented; Sections 3.3 and 3.4 review epitaxial growth and the characterization of heteroepitaxial compound semiconductor films; potential device applications are presented in Section 3.5; and conclusions based on experimental results and the trend of future development are discussed in the last section, Section 3.6.

3.2. Substrate preparation

Materials requirements. The quality of the epitaxial compound semiconductor films depends largely on (1) the crystallographic relationship at the interface between the film and substrate and (2) the physicochemical nature of the substrate surface under the growth conditions. Therefore, the substrate materials and surface preparation play an important role in the characteristics of the film–substrate composites. The general substrate requirements and their implications for the epitaxial growth of compound semiconductors are summarized in Table 3.1. In addition to these considerations, specific applications require that the substrates exhibit certain physical and chemical properties. These requirements will be discussed through-out the chapter.

A number of dielectric oxides have been considered or investigated for use as substrates for epitaxy of semiconductor films [for a review see Filby (1972)]. However, most of them do not adequately fulfill the general requirements described in Table 3.1. From a practical standpoint, sapphire (α-Al_2O_3) and spinel ($MgAl_2O_4$) have best met the requirements as sub-strates for the compound semiconductors. They have been the most studied to date.

A comprehensive discussion has been presented in Chapter 2 on the single crystal growth and properties of sapphire and spinel, on the fabrication of substrate wafers, and on the preparation of substrate surfaces for epitaxial growth.

In addition to sapphire and spinel, beryllia (BeO) is a potentially useful substrate material for the epitaxy of compound semiconductors. This material exhibits high thermal conductivity, and therefore an improve-ment in power dissipation is anticipated when BeO is used as substrate for thin-film devices. However, BeO single crystals available to date have been grown only by a flux method (Austerman, 1964). In comparison with melt-grown sapphire and spinel crystals, flux-grown BeO crystals are smaller in size and less pure, with probable flux inclusions. Epitaxial GaAs films grown on the BeO substrates have exhibited inferior electrical characteristics as compared to GaAs grown on sapphire and spinel substrates (see Section 3.4). The practical use of BeO as substrate relies on the further advancement of the state of the art of crystal growth of this material.

Experimental results obtained to date have shown that the growth and electrical characteristics of the epitaxial films depend on substrate orientation. This dependence is much more critical for the compound semiconductors (such as GaAs) than for the elemental semiconductors (such as Si), imposing a further limita-

tion on the availability of useful substrate surfaces. Details on the substrate orientation effects are pre-sented in other parts of this chapter.

Surface preparation

Surface preparation of the substrates prior to epitaxial growth is an important aspect of the heteroepitaxial growth process. Because the general subject has been fully discussed in Chapter 2, the discussion in the present section will be limited to the aspects of surface preparation pertinent to compound semiconductor epitaxial growth.

The mechanical polishing process is the traditional method of producing a smooth and flat surface for epitaxial growth. After the completion of mechanical polishing, scratches, mounds, adsorbed layers, and impurity aggregates are generally present on the sub-strate surfaces. Surface damage obviously influences the film nucleating process and leads to the introduc-tion of defects in the deposited films (see Section 3.4). To obtain surface finishes with a minimum amount of damage, procedures must be optimized for the specific material and orientation of interest. The amount and rate of removal of materials at various polishing stages is a function of the type and orientation of the substrate.

The chemical removal of the work damage intro-duced by the polishing operation has been the subject of many investigations. Molten salt etches, acid etch-es, and high-temperature vapor-phase etches have been employed (references to these studies may be found in Chapter 2). A common problem in the use of these materials has been that chemical etching of sap-phire and spineal surfaces to remove work damage often produces difficult-to-remove reaction products and etch pits. In addition to chemical etching, high-temperature annealing has been used as an effective means to reduce the surface damage in sapphire and

TABLE 3.1   General Substrate Requirements and Considerations

Requirement	Purpose of Requirement
Single crystal of good crystalline perfection and with minimum degree of lattice mismatch to the film material	To nucleate the growth of single crystal films with high degree of perfection
Chemical, mechanical, and thermal stability	To withstand, without structural or chemical degradation, the process-ing and thermal cycling necessary for device fabrication
Thermal expansion coefficient similar to film	To minimize film stress, deforma-tion, and dislocations
Chemical inertness to the growth conditions	To avoid or minimize contamination and autodoping effects
Single crystal can be cut and polished with minimum surface defects	To provide surface conditions neces-sary for epitaxial film growth

spinel. The annealed surfaces can be effectively polished, leaving essentially featureless surfaces, by wet chemicals.

Experimental results (see Section 3.3) to date indicate that compound semiconductor films grown on substrates with treatment other than mechanical polishing followed by standard wet solvent cleaning are generally not superior in quality and that the growth characteristics are generally not reproducible. These results may not necessarily be fundamental to the heteroepitaxial growth of the compound semiconductors but represent the present state of the art. The formation of heteroepitaxial compound semiconductor films using the organometallic growth process on unannealed and unetched substrates may be attributed to the exposure of the single crystal substrate as the result of certain chemical reactions between the constituents in the vapor phase and the substrate surface at the growth temperature (see Section 3.3). Improved substrate surface preparation techniques remain an important objective to be accomplished in the area of heteroepitaxy.

3.3. Epitaxial growth

Growth process

Preparation of epitaxial semiconductors for practical uses in advanced electronic devices has become a realistic technology through the development of chemical vapor-deposition processes and techniques during the last two decades. Many chemical systems have been investigated or used (or both) in the epitaxial growth, and they may be classified, according to the nature of the reactants, into two general processes: (1) chemical vapor transport and (2) organometallic. Because the III-V compounds, in particular GaAs, have received most study to date, GaAs is mainly used here to illustrate the growth processes.

Chemical transport process.   The epitaxial growth of GaAs and many other compound semiconductors by chemical transport reactions has formed the subject of several comprehensive review articles (Schwartz, 1966; Joyce, 1966; Francombe and Johnson, 1969; Feist et al., 1969; Bradshaw and Knappett, 1970; Jain and Sharma, 1970; Pashley, 1972; Dugue et al., 1972; Tietjen et al., 1972; Chu and Smeltzer, 1973; Joyce, 1974; Shaw, 1974.) For complete and detailed discussions on this subject, the reader is referred to these review papers and to the annotated bibliographic surveys of Turnbull (1967), Neuberger (1971), and Connelly (1972).

Among the many chemical transport systems used for the growth of epitaxial GaAs (and many other semiconductors), the most successful, reproducible, and widely used one to date is the open-tube chloride-transport process, which is discussed in detail by Jain and Sharma (1970) using slightly different chemical systems. In principle, all the systems require the formation of volatile GaCl which is made to react with arsenic vapor to form GaAs. For example, the system described by Tietjen and coworkers (1972) used Ga, HCl, AsH_3, and H_2 as the source materials. The volatile GaCl was formed through the reaction of Ga and HCl. The chloride was mixed with arsine at a temperature high enough to prevent reaction, and the gas mixture was then passed over a substrate at a lower temperature where deposition takes place. Hydrogen was used for the carrier gas. The principal reactions leading to the deposition of GaAs are

$$2AsH_3 \rightarrow As_2 + 3H_2 \qquad (3.1)$$
$$2HCl + 2Ga \rightarrow 2GaCl + H_2 \qquad (3.2)$$
$$2GaCl + As_2 + H_2 \rightarrow 2GaAs + 2HCl. \qquad (3.3)$$

Other systems use source materials of $AsCl_3$-Ga-H_2 (Knight et al., 1965) and of As-Ga-HCl-H_2 (Conard et al., 1967), and the final reaction leading to the formation of GaAs is the same as described in Equation (3.3). During the last decade, the chloride process has been widely used for the synthesis of a broad spectrum of III-V compounds and their alloys [see Tietjen et al. (1972)] with widely differing electrical and optical properties and in a variety of geometries for various devices. These compounds include both homoepitaxial materials and heteroepitaxial films grown on foreign III-V substrates (mostly GaAs and GaP).

In contrast to the extensive literature available on the growth of III-V compound semiconductors by the chloride process on substrates of the same composition or of similar chemical and structural properties, little is known of the heteroepitaxial growth of compound semiconductors on oxide insulators such as sapphire and spinel. A literature survey reveals that, among the III-V compounds, only the nitrides have been reported forming epitaxial films on oxide substrates by the chloride growth process. These include GaN on sapphire (Maruska and Tietjen, 1969), AlN on sapphire (Yim et al., 1973), and ScN on sapphire (Dismukes et al., 1972). Attempts to grow other III-V compounds, such as GaAs, on oxides have been found either unsuccessful or not reproducible. For example, Gutierrez et al. (1970) studied the growth of GaAs on sapphire and spinel substrates using the chloride process. They reported that it was extremely difficult to grow continuous layers less than 2 μm thick and that the epitaxial yield was only about 30 percent.

The ability to prepare single crystal films directly on sapphire or spinel using the halide process has been questioned because of the large quantities of HCl present in the growth system. It was felt that the HCl might react with the insulator surface and thereby prevent or hinder the epitaxy. However, since the nitrides can be successfully grown, it is unlikely that the HCl/substrate reaction plays a key role in preventing the epitaxy. Experimental results (Gutierrez et al., 1970) have shown that it is the surface preparation of the substrate, rather than the presence of HCl, that predominantly controls the ability to obtain epitaxy. For reasons yet unknown, the halide deposition process is much more sensitive to the condition of the substrate surface than the organometallic process. For a given group III element this sensitivity seems to increase in the order: nitride < arsenide < phosphide. To date the technology of oxide substrate surface preparation for epitaxy has not been established and the heteroepitaxial growth of compound semiconductors on oxides by the chloride process has not been popularly used. The only one material system, grown by the chloride process, that exhibits interesting properties for potential electroluminescent applications (Pankove, 1973) is GaN on sapphire.

In addition to the III-V compounds, II-VI compounds have been grown heteroepitaxially on oxide substrates using chemical vapor-phase reactions between the group II elements (transported to the reaction zone by H_2) and the group VI hydrides (Yim and Stofko, 1972, 1974). Epitaxy has been achieved on the following systems: ZnSe/sapphire, CdS/sapphire, and CdSe/sapphire. In addition, ZnO films have been reported grown (Galli and Coker, 1970) epitaxially on sapphire using HCl/H_2 as the transporting agents. The film/substrate epitaxial relationships and the film properties are discussed on the subsequent sections in this chapter.

Organometallic process. Although the synthesis of compound semiconductors using chemical reactions involving organometallics was studied some 15 years ago (Didchenko et al., 1960), it was only recently that the organometallics were successfully and advantageously used to grow compound semiconductors on a variety of substrates, in particular the insulating oxides. The successful heteroepitaxial growth of GaAs on sapphire, spinel, beryllia, and thoria was first reported by Manasevit (1968) using reactions between $(CH_3)_3Ga$ and AsH_3. The organometallic process since then has received increasing attention, and many film/substrate composite systems have been investigated in a relatively short period. The epitaxial growth of compound semiconductor films by the organometallic process has been reviewed by Manasevit (1972, 1974).

In the organometallic process, semiconductor films of III-V and II-VI compounds and alloys are produced by decomposing appropriate group III and group II organometallic compounds in the presence of the appropriate group V and group VI hydrides or alkyls.

In addition, the group II organometallics and group VI hydrides are commonly used as dopants for the III-V compounds. The overall reaction leading to the growth of GaAs, for example, is

$$(CH_3)_3Ga + AsH_3 = GaAs + 3CH_4.$$

H_2S (or H_2Se) and $(C_2H_5)_2Zn$ have been used as dopants for, respectively, n and p epitaxial GaAs films. The pertinent properties of several commonly used organometallics are listed in Table 3.2.

There are several distinct advantages of the organometallic process over others. The process requires only one controlled hot-temperature zone for the in situ formation and growth of the compound semiconductors on the heated substrates and occurs in an atmosphere free of halide-containing species which may etch undesirably certain film/substrate systems. The epitaxial growth can be achieved at relatively low temperatures (for example 650°C for GaAs), and films can be grown readily on oxide substrates. However, there are alos drawbacks. The control of impurities in the organometallic source materials is difficult, and therefore, at low carrier densities ($< 10^{16}/cm^3$), the control of doping is not easy to achieve. Improvements of the source material purity and handling techniques are necessary to further advance the state of the art of the organometallic growth processes.

The growth apparatus is rather simple in construction and easy to operate. The schematic of a typical system used to grow GaAs (Wang et al., 1974A) is shown in Figure 3.1, illustrating the growth techniques. A variety of compound semiconductors can be grown from the same system with appropriate modifications using appropriate source materials. The growth system consists of a water-cooled vertical quartz reactor (6.25 cm diameter, 40 cm long) equipped with an adjustable quartz pedestal, a trimethyl gallium source reservoir (kept at 0°C) and dispenser, an arsine supply (10 percent by volume in H_2), all stainless steel gas lines and valves, and precision gas-flow controls. Hydrogen purified by a Pd-diffuser has been used as the carrier gas. The growth of doped GaAs films can be achieved using $(C_2H_5)_2Zn$ and H_2Se (also H_2S) as the p- and n-dopants, respectively. An RF induction heating system (with precision electronic control unit) and pyrolytic graphite or SiC-coated graphite susceptors have been employed for the deposition. The substrate temperature has been measured by an infrared pyrometer.

In a typical growth run, the reactor is first thoroughly flushed by H_2 to purge the system of air. Predetermined and equilibrated AsH_3 and $(CH_3)_3Ga$ gas flows are then sequentially admitted to the reactor in which the substrate is heated to the growth temperature. The crystal growth is carried out for the desired length of time after which the $(CH_3)_3Ga$ carrier gas is turned to

TABLE 3.2 Properties of Organometallics Commonly Used For Epitaxial Growth

Compound	Formula	Molecular Weight	Melting Point (C)	Boiling Point (C)	Density (gm/cm³)	Vapor Pressure (mm Hg)	Chemical Reactivity with Air and Water	Storage Stability	Remarks
Trimethyl boron	$(CH_3)_3B$	25.86	−153	−21.8	0.625 (−100°C)	—	Ignites spontaneously upon exposure to air; does not react with water in the absence of O_2.	Stable at room temperature; decomposes thermally only above 200°C.	Releases unusual pungent odors.
Triethyl boron	$(C_2H_5)_3B$	98.10	−92.9	95	0.696 (23°C)	12.5 (0°C) 42.6 (20°C) 108 (40°C)	Ignites spontaneously upon exposure to air; does not react with water in the absence of O_2.	Stable at room temperature; decomposes thermally only above 200°C.	Releases unusual pungent odors.
Trimethyl aluminum	$(CH_3)_3Al$	72.08	15.4	126	0.752 (20°C)	8.4 (20°C) 68.5 (60°C) 332 (100°C)	Highly pyrophoric; ignites spontaneously upon exposure to air: reacts violently with water; Al-alkyls are the most reactive of all the group III alkyls.	Stable at room temperature.	Liquid exhibits high viscosity; molecules in dimeric form.
Triethyl aluminum	$(C_2H_5)_3Al$	114.16	−52.5	207	0.837 (20°C)	0.8 (60°C) 13 (110°C) 110 (140°C)	Highly pyrophoric; ignites spontaneously upon exposure to air; reacts violently with water.	Stable at room temperature.	Liquid exhibits high viscosity; molecules in dimeric form.
Trimethyl gallium	$(CH_3)_3Ga$	114.82	−15.8	55.7	1.151 (15°C)	8.9 (−33.5°C) 64.5 (0°C) 221.8 (25°C)	Pyrophoric; ignites spontaneously upon exposure to air; reacts violently with water.	Stable at room temperature.	Exhibits higher vapor pressure than $(C_2H_5)_3Ga$: vapor can be conveniently generated at or below room temperature.
Triethyl gallium	$(C_2H_5)_3Ga$	156.91	−82.3	142	1.0583 (30°C)	16 (43°C) 62 (72°C) 72 (118°C)	Pyrophoric; ignites spontaneously upon exposure to air; reacts violently with water.	Stable at room temperature.	See above for $(CH_3)_3Ga$.
Trimethyl indium	$(CH_3)_3In$	159.93	88.4	135.8	1.568 (10°C)	7.2 (30°C) 72 (70°C)	Reactive with both air and water; In-alkyls are less reactive than the corresponding compounds of Ga.	Stable at room temperature.	In solid form at room temperature; preferable $(C_2H_5)_3In$ which is in liquid form but less stable
Triethyl indium	$(C_2H_5)_3In$	202.40	−32	144	1.538 (20°C)	3 (54°C) 12 (83°C) 38 (118°C)	Reactive with both air and water.	Gradually decomposes by light at room temperature.	See above for $(CH_3)_3In$.
Trimethyl arsine	$(CH_3)_3As$	120.02	−87.3	51	1.124 (22°C)	—	Moderately pyrophoric; ignites in air; not sensitive to water, hydrolysis proceeds with difficulty.	Stable at room temperature; should avoid storage near its boiling point.	High toxicity containing As.
Triethyl arsine	$(C_2H_5)_3As$	162.11	—	140	1.079 (15°C)	15.5 (37°C) 93 (73°C)	Moderately pyrophoric; ignites in air; not sensitive to water, hydrolysis proceeds with difficulty.	Stable at room temperature.	High toxicity containing As.

Compound	Formula	MW	mp	bp	density	(°C)	v.p.	(°C)	Reactivity	Thermal stability	Other
Trimethyl stibine	(CH₃)₃Sb	166.86	−87.6	80.6	1.528	(15°C)	—		Moderately pyrophoric; not sensitive to water.	Stable at room temperature.	Releases characteristic unpleasant odor.
Triethyl stibine	(C₂H₅)₃Sb	208.94	−98.0	160	1.324	(16°C)	—		Moderately pyrophoric; not sensitive to water.	Stable at room temperature.	Releases characteristic unpleasant odor.
Dimethyl cadmium	(CH₃)₂Cd	142.88	−4.5	105.5	—		—		Pyrophoric; ignites spontaneously upon exposure to air; decomposes in water.	Stable at room temperature.	Releases characteristic unpleasant odor.
Dimethyl zinc	(CH₃)₂Zn	95.45	−42	46	1.386	(10°C)	124	(0°C)	Pyrophoric; ignites spontaneously or detonates upon exposure to air; decomposes violently in water.	Stable at room temperature; decomposes rapidly near its boiling point	Some photodecomposition may occur in strong light over long periods.
Diethyl zinc	(C₂H₅)₂Zn	123.50	−28	118	1.182	(18°C)	15 / 27 / 91	(20°C) / (30°C) / (60°C)	Pyrophoric; ignites spontaneously or detonates upon exposure to air; decomposes violently in water.	Stable at room temperature.	Some photodecomposition may occur in strong light over long periods.

Sources: E. G. Rochow, D. T. Hurd, and R. N. Lewis, *The Chemistry of Organometallic Compounds*, New York: John Wiley and Sons (1957). Technical Brochure, Organometallics for Electronics, Alfa Inorganics, Inc., Beverly, MA. (1970) Technical Brochure, Alkyl-Metals for Semiconductors, Sumitoma Chemical Co., Osaka, Japan (1974). Research and Development Product Data Sheets, Texas Alkyls Inc.., Houston, TX (1971).

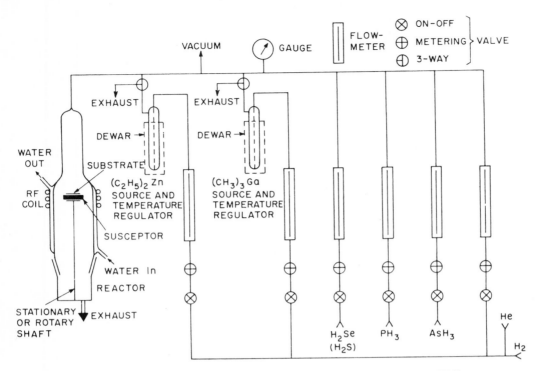

FIGURE 3.1   Epitaxial growth system using organometallic process. After Wang et al. (1974A)

the exhaust. The AsH$_3$ gas valve is then turned off, and the substrate is slowly cooled at an average rate of about 50°C/min in the H$_2$ ambient.

Growth parameters

The important growth parameters affecting heteroepitaxial film characteristics are substrate orientation, substrate crystallinity and surface preparation, growth temperature, gas flows, reactor geometry, source material quality, and growth procedure. The effect of the various parameters on the layer properties depends on the growth process and may vary also from one material system to another. In the discussion that follows, major emphasis is placed on the growth of GaAs on various oxides by the organometallic process in order to illustrate the effect of the growth parameters.

Substrate orientation. Substrate orientation plays a major role in the ease with which the epitaxial compound semiconductor layers grow. In the GaAs/spinel system, it was found (Wang et al., 1974A) that the epitaxial growth of GaAs on (111) spinel is less sensitive to substrate surface quality, and also depends less critically on the growth temperature and other variables, than is the growth of GaAs on other major cubic spinel orientations. Scanning electron micrographs of GaAs films simultaneously grown at about 700°C on the three major cubic spinel orientations are presented

in Figure 3.2 showing the surface crystalline structures. It is seen that the surface crystalline quality improves with spinel orientation in the order of (100) < (110) < (111). The oriented single crystal growth of (111) and (100) GaAs films was obtained on respectively, (111) and (110) spinel surfaces, but not on (100) orientation. The (100) surface is so sensitive to surface preparation and other growth variables that only polycrystalline films with isolated regions of single crystal (100) GaAs were grown. It was also found that smooth single crystal GaAs films (Figure 3.2d) could be grown on (100) spinel with only slight misorientation of 1° or greater. The GaAs grown on the misoriented (100) spinel exhibits an unexpected (111) orientation. For the (111) and (110) substrates, (111) and (100) films, respectively, with improved surface crystallinity have been grown on slightly off-oriented (1–3 degree) substrate surfaces.

The effect of substrate orientation on the film growth characteristics was also studied in the GaP/spinel system (Wang et al., 1974B). This system is similar to the GaAs/spinel system in that the ease of growth increases with spinel orientation in the order (100) < (110) < (111). However, it differs from the GaAs/spinel system in that the GaP epitaxial films grown on (110) spinel substrates are composed of both (110) and (100) grains, instead of (100) orientation only as found in GaAs/spinel epitaxy. The results indicate that two orientations of the GaP overgrowth fulfill the condition for a single substrate orientation. Therefore, the GaP film is not a single crystal but is still epitaxial. Similar situations are encountered in other film/substrate systems as listed in tables in Section 3.4. The GaP/spinel system also differs from the GaAs/spinel system in that the GaP films exhibit no significant difference in growth characteristics (surface orientation or crystallinity) as a function of the degree of substrate off-orientation.

The strong dependence of GaAs film crystalline quality on substrate orientation has also been observed in the GaAs/sapphire and GaP/sapphire systems (Manasevit and Simpson, 1969; Wang and McFarlane, 1972). Film growth on the (0001) sapphire basal plane is least sensitive to surface preparation and to growth conditions. The characteristics of (111) GaAs films grown on (0001) sapphire have been the most extensively studied (Manasevit and Thorsen, 1970; Thorsen and Manasevit, 1971). In contrast to the (0001) orientation, it is relatively difficult to grow single crystal GaAs and GaP of good quality on other sapphire orientations. In the GaAs/BeO system, among nine substrate orientations studied, it was found that the $(10\bar{1}1)$ and $(11\bar{2}2)$ BeO surfaces tend to yield better quality of, respectively, (111) and (100) GaAs films (Thorsen et al., 1974).

It is interesting to note that the dependence of the ease of heteroepitaxial semiconductor film growth on substrate orientation is much less critical for the elemental semiconductors (such as Si) than for the compound semiconductors (such as GaAs). For example,

single crystal silicon films can be conveniently grown simultaneously on the major cubic spinel surfaces or on several common sapphire orientations. Another interesting aspect is that the film/substrate epitaxial orientation relationships (Section 3.4) are different between the Si/oxides and GaAs/oxides systems despite the similar crystal structure and lattice parameters of Si and GaAs. This suggests that the film/substrate interfacial bonding is quite different between the Si/oxide and GaAs/oxide composite systems.

Substrate crystallinity and surface preparation. As a general rule in heteroepitaxy, the maximum crystalline perfection which may be achieved in the film is of the same order of magnitude as the substrate. For example, epitaxial GaAs, under certain growth conditions, may be prepared approaching the perfection of Ge substrate. However, the heteroepitaxial materials grown on oxide substrates with large structural differences generally exhibit far less crystalline perfection than the substrates.

The dependence of the film crystalline perfection on the substrate bulk crystallinity has been studied for the GaAs and GaP heteroepitaxial films grown on oxide substrates (McFarlane and Wang, 1972; Wang and McFarlane, 1972; Wang et al., 1974A and B). The results revealed that the general orientation of the epitaxial layers follows the orientation of large grains in the substrate. Films grown on substrate oxides prepared by different growth methods (Czochralski,

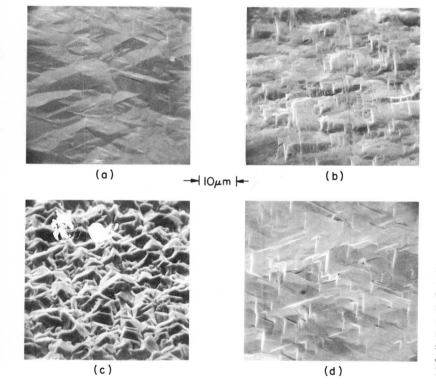

(a)   →| 10μm |←   (b)

(c)   (d)

FIGURE 3.2
Scanning electron micrographs of GaAs films (~20 μm thick); beam-sample angle 45°; (a) on a (111) spinel substrate; (b) on a (110) spinel substrate; (c) on a (100) spinel substrate; (d) on 1° off-oriented (100) spinel substrate. From Wang et al. (1974A)

TABLE 3.3   Departure of (111) GaAs from Exact Parallel Epitaxy

Substrate	Misorientation from Exact Parallel Epitaxy
Flux spinel	0.15°
Czochralski spinel	0.32°
Flame fusion spinel	0.43°

flux, and Verneuil) exhibit differences in physical appearance. GaAs grown on Verneuil spinel exhibits the subgrain structures originating from the substrates, whereas films grown on Czochralski (or flux) substrates do not show the decorating subgrain structures.

GaAs films grown on spinel substrates prepared by different methods also exhibit a difference in the departure from exact parallel epitaxy with respect to the substrates. Experimental results from x-ray diffraction studies are shown in Table 3.3. It is interesting to note that the flux spinel exhibits high crystalline perfection without lattice distortion and that the GaAs grown on the flux spinel exhibits the least misorientation from exact parellel epitaxy.

Despite the differences in crystalline properties described above, there does not appear to be a significant trend of critical dependence of film electrical characteristics on substrate bulk crystalline perfection. This is because the bulk imperfections in films grown on various substrates are similar (see Section 3.4). Consequently, factors other than the substrate crystallinity (such as surface quality and impurity contamination) may play a dominant role on the film electrical characteristics. As shown in Table 3.4, the electrical characteristics of GaAs grown on Verneuil spinel are comparable to those of GaAs grown on Czochralski spinel. Films grown on flux spinel substrates exhibit generally inferior electrical properties.

Contamination introduced into the substrate crystals from the flux (Wang and McFarlane, 1968) is released at the film growth temperatures.

Substrate surface preparation is one of the most important growth parameters. The quality of the mechanical polishing has a direct effect on the film growth and electrical characteristics. Surface scratches on the substrates generally provide preferential nucleation sites. Absorbed layers and impurity aggregates on the substrate surfaces can cause various defects in the epitaxial films. In an effort to improve the substrate surface quality, the effect of various treatments of substrate prior to growth has been investigated for the growth of GaAs/oxide and GaP/oxide systems using the organometallic processes. The treatments include thermal annealing (Cullen, 1971; Wang, 1969) and chemical polishing (Reisman et al., 1971) of the substrates. However, experimental results to date indicate that films grown on substrates with treatments other than just mechanical polishing followed by standard wet-solvent cleaning are generally not superior in quality and that the growth characteristics are generally not reproducible. These results may not necessarily be fundamental but only represent the present state of the art. Improved substrate surface preparation techniques remain the important objective to be accomplished in the area of heteroepitaxy. Heteroepitaxy using the organometallic growth processes in the III-V compound/oxide systems has been achieved on substrates that had not been annealed or etched after the final mechanical polish. This may be attributed to chemical reactions between the constituents in the vapor phase and the substrate surfaces at the growth temperatures, which lead to the exposure of single crystalline substrate surfaces. For example, it was found (Wang et al., 1974A) that AsH_3 etches spinel at a temperature of about 700°C, and the etched single crystal surfaces promote the epitaxial growth of GaAs on spinel.

TABLE 3.4   Electrical Properties of (111) Epitaxial GaAs Grown on Spinel[a]

Substrate	Film Thickness (μm)	Conductivity Type	Resistivity (Ω-cm)	Carrier Concentration (cm^{-3})	Hall Mobility (cm^2/V-s)
Flux $MgAl_2O_3$	30	n	9	9.4×10^{14}	740
Verneuil $MgO \cdot 1.7Al_2O_3$	50	n	0.40	4.1×10^{15}	3820
Czochralski $MgAl_2O_4$	23	n	0.31	4.7×10^{15}	4200
Czochralski $MgAl_2O_4$	26.4	p	0.58	3.4×10^{16}	320
Verneuil $MgO \; 2.0Al_2O_3$	39	p	4.5	4.5×10^{15}	316

[a]Unintentionally doped films.

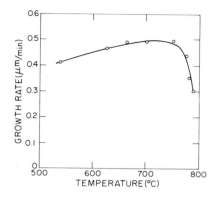

FIGURE 3.3   Growth rate as a function of growth temperature of (111) GaAs. After Wang et al. (1974A)

Growth temperature and gas flows. The optimum growth temperature range for a growth process and film/substrate system depends, to a large extent, on the substrate orientation. The effect of growth temperature on the crystalline quality of heteroepitaxial GaAs on spinel was studied (Wang et al., 1974A) in the temperature range 500 to 800°C. The general trends observed were (1) at low temperatures (<600C), the GaAs films are polycrystalline; (2) at high temperatures (>800°C), the gases in the reactor are highly turbulent and the films tend to be inhomogeneous with high densities of growth defects; and (3) within a limited temperature range [680 to 720°C for (111) spinel and 630 to 700°C for (110) spinel], highly reflective GaAs of good crystallinity could be grown. A similar effect was observed (Manasevit and Simpson, 1969) on the epitaxial growth of GaAs on (0001) sapphire in the temperature range 600 to 800°C. Films grown at 800°C were found less perfect than those grown at 675°C.

The dependence of growth rate on the growth temperature was studied for both GaAs/spinel (Wang et al., 1974A) and GaAs/sapphire (Manasevit and Simpson, 1969) systems. In both systems, the growth rate was found (see Figure 3.3 for GaAs/spinel systems) to be essentially constant over a wide temperature range corresponding to the optimum range for single crystal growth. The growth rate and the film quality depend largely on the gas flows. The flow conditions that yield the best results for a material system can be experimentally determined from the particular apparatus geometry. Generally, higher flow rates than optimum favor heavy deposition at the center of the substrate, whereas slower rates favor deposition at the periphery.

The growth rate of (111) GaAs on (0001) sapphire by the reaction of $(CH_3)_3Ga$ and AsH_3 was found (Manasevit and Simpson, 1969) to be essentially linear with $(CH_3)_3Ga$ concentration when it is decomposed in an atmosphere containing As and at least a tenfold excess of AsH_3 over $(CH_3)_3Ga$ in the gas stream entering the reactor. The dependence of growth rate and surface crystallinity on the reactant ratio of $AsH_3/(CH_3)_3Ga$ is shown in Figure 3.4 for the growth of (111) GaAs on (111) spinel. An optimum range of source-gas ratio is shown to be required for the growth of high-quality films.

Thorsen and Manasevit (1971) have shown that the electrical characteristics of GaAs on sapphire depend largely on the $AsH_3/(CH_3)_3Ga$ flow-rate ratios. The carrier concentration of as-grown n-type films has been found to be dependent on the AsH_3 flow rate for a given growth temperature and a fixed $(CH_3)_3Ga$ flow rate. For the higher AsH_3 flow rates, the carrier concentration saturates at some value characteristic of the gases used. As the AsH_3 flow is reduced, the net donor concentration also tends to decrease. With continued decrease in AsH_3, the films eventually become p-type. The origin of the acceptor state responsible for the p-type conductivity is probably related to a defect state, possibly As vacancy. By changing the flow rates of the reactants, p-n junctions have been grown in heteroepitaxial GaAs films (Thorsen et al., 1974).

Source materials and doping techniques. Experimental results obtained from growth studies have indicated that the purity of the source material is of critical importance in determining the electrical properties of the heteroepitaxial films. High-purity AsH_3 is commercially available. The purity of the AsH_3 gas used for the GaAs growth may be evaluated in a homoepitaxial GaAs growth system employing the reaction between gallium chloride and arsine. However, the quality of the $(CH_3)_3Ga$ varies from lot to lot. The impurities generally found, by emission spectroscopy, in typical lots of $(CH_3)_3Ga$ include Cu, Fe, Zn, Al, Si, and Mg on the order of 10^{-1}–10^2 ppm

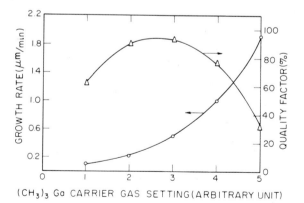

FIGURE 3.4   Growth rate and surface crystallinity as a function of $(CH_3)_3Ga$ feed rate (quality factor = reflectivity measured/reflectivity calculated). From Wang et al. (1974A).

(by weight). These impurities may cause significant unintentional doping of the GaAs films. Analysis of $(CH_3)_3Ga$ samples by infrared spectroscopy also revealed that there are hydrocarbons (degradation products) present in some samples with widely varying concentrations from sample to sample.

To date $(CH_3)_3Ga$ and other organometallics of controlled quality for epitaxial growth of compound semiconductors are not commercially available. The nominal high-purity materials are generally tested for acceptance by growing the epitaxial films and then measuring the electrical characteristics. GaAs films grown from defective $(CH_3)_3Ga$ exhibit very high resistivity ($>10^3$ Ω-cm) independent of thickness. The films are highly compensated and contain localized p-n junctions. Further discussions of the effect of source materials on film electrical properties are in Section 3.4.

The growth of doped heteroepitaxial GaAs and GaP has been reported using $(C_2H_5)_2Zn$ [also $(C_2H_5)Cd$] and H_2S (or H_2Se) as the p- and n-dopants, respectively (Manasevit and Thorsen, 1972; Wang et al., 1974A,B). The techniques of handling the organometallic dopants are the same as that used for $(CH_3)_3Ga$ as discussed before with the flow rates adjusted to yield the desired doping levels. For n-type doping, the control of the degree of doping was found more difficult for H_2Se than for H_2S because H_2Se is not chemically stable in storage.

Growth procedure. As described above, the ease of film growth depends to a large extent on the substrate orientation, using the normal growth procedure in which the growth is initiated and continued at a constant rate for the desired length of growth. For example, in the GaAs/spinel system the quality of the (111)

GaAs grown on (111) spinel is far superior to that of the (100) GaAs grown on (110) spinel. The (100) GaAs so grown exhibits fairly well defined band or strip structures (Figure 3.5) across the crystal with a high degree of crystallite misorientation.

In order to achieve the growth of high-quality (100) GaAs on spinel, which is desired for certain devices, a new growth procedure has been developed. It consists of a three-step nucleation, annealing, and steady state growth process. In performing a growth run by this technique, the initial growth is the same as in the one-step procedure. After the predetermined initial nucleation stage (typically 40 s to 1 min), the $(CH_3)_3Ga$ feed is isolated from the system, and the thin GaAs grown on spinel is annealed at the growth temperature ($\sim700°C$) for a period of time (typically 10 min) under the AsH_3-H_2 atmosphere. After the annealing, film growth is continued by reintroducing the $(CH_3)_3Ga$ source into the system. (100) GaAs films of good quality have been consistently prepared by this three-stage process. The band structure generally associated with the (100) GaAs films has not observed in films prepared by this new method. Moreover, (100) GaAs films of fairly good quality can be grown by this technique even on (110) spinel substrate of relatively poor surface quality (high scratch density or poorly polished surfaces).

It is probable that for certain film/substrate orientations the annealing promotes the atomic rearrangement or interactions (or both) at the film/substrate interface so that oriented nuclei are formed for further single crystal growth. Beside the GaAs/spinel system, the three-step growth procedure was also applied to the GaP/spinel system with positive effects as evidenced by the GaP electroluminescent properties described in Section 3.5.

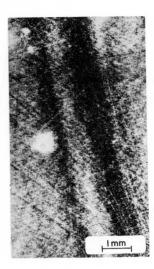
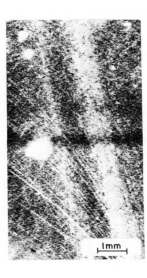

FIGURE 3.5
Lang x-ray topographs of a (100) GaAs layer (~20 μm thick) grown on a (110) spinel substrate [silver radiation, (022) reflection]. The crystal was rotated 0.4° between topographs. From McFarlane and Wang (1972)

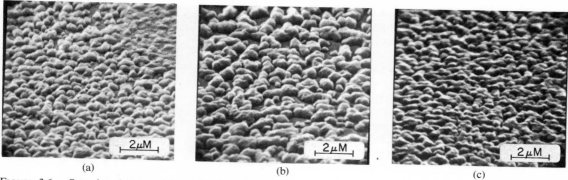

FIGURE 3.6   Scanning electron micrograph of GaAs grown on (111) spinel showing growth islands just before complete coverage; growth rate of about 0.8 μm/min for 8 s; beam-sample angle, 60°. (a) Flux spinel substrate; (b) flame fusion spinel substrate; (c) Czochralski spinel substrate. From Wang et al. (1974A)

Early growth stage studies

The nature of the initial film growth of GaAs on sapphire has been examined by means of electrical measurements and electron diffraction and electron microscopy techniques (Manasevit and Thorsen, 1970). The early stage of growth, under conditions consistent with epitaxial film formation, appears to be by the formation of many discrete nuclei which coalesce to form large islands and eventually produce complete surface coverage. A similar growth mechanism was found also in the GaAs-on-spinel system (Wang et al., 1974A). The scanning electron micrographs in Figure 3.6 show the growth islands of (111) GaAs on (111) spinel just before complete coverage of the substrate. A slight difference is seen in the percentage of coverage for substrates grown by different methods. This may be due to differences in surface perfection.

Electron diffraction studies revealed evidence of a high density of defects near the film/substrate interface with a considerable improvement in crystal quality as the film thickness increases. Films at the interface exhibit p-type conductivity which may be related to the defect structures near the film/substrate interface

Growth of III-V compounds

During the last decade, a large number of III-V semiconductors and their alloys have been successfully grown on several insulating oxide substrates by chemical vapor growth processes. The film and substrate materials, their orientation relationships, and the growth processes are listed in Tables 3.5 to 3.7. Electrical characteristics and potential device applications of the films are presented in Sections 3.4 and 3.5.

The epitaxy of group III-nitrides on oxide substrates, as described earlier in this section, can be achieved by both the chloride transport process and the organometallic growth process. In certain material systems, films grown by the two processes exhibit a significant difference in properties. For example, in the AlN-on-sapphire system films grown by the chloride process on (01$\bar{1}$2) sapphire exhibit at least three orientations: (11$\bar{2}$0), (30$\bar{3}$5), and (11$\bar{2}$7) (Yim et al., 1973). AlN grown on (01$\bar{1}$2) sapphire by the organometallic process exhibits only (11$\bar{2}$0) orientation (Manasevit et al., 1971; Duffy et al., 1973). (11$\bar{2}$0) AlN grown on (01$\bar{1}$2) sapphire (organometallic process) and (11$\bar{2}$0) GaN grown on (01$\bar{1}$2) sapphire (chloride process) have been extensively studied for potential applications in, respectively, surface acoustic wave and electroluminescent devices (see Section 3.5).

For the phosphides, aresenides, and antimonides, epitaxy on oxide substrates has been reported only for films grown by the organometallic process. Among the group III–phosphides, GaP has received the most study because of its well known electroluminescent properties. Epitaxial GaP has been successfully grown (Wang and McFarlane, 1972) on sapphire, spinel, and Si on sapphire using the organometallic process [(CH$_3$)$_3$Ga + PH$_3$]. A vapor/liquid two-stage process of growing (111) GaP on (111) spinel has been successfully developed (Wang et al., 1974A) with improved results (see Section 3.5). AlP can be grown by the reaction between (CH$_3$)$_3$Al and PH$_3$ on the oxide substrates (Manasevit, 1972; Wang and Dougherty, 1974), but the films are unstable (hydrolyze) in air. The stability improves upon forming solid solutions with GaP, and Al$_{1-x}$Ga$_x$P (x = 0.3–0.6) films have been grown on sapphire and spinel substrates (Wang and Dougherty, 1974). InP has been grown on sapphire by the reaction between (C$_2$H$_5$)$_3$In and phosphine (Manasevit and Simpson, 1971A,B).

Epitaxial growth and properties of GaAs films grown on oxides have been most extensively studied, as described throughout the whole chapter. High-quality films can be obtained using the organometallic process [(CH$_3$)$_3$Ga + AsH$_3$] with the following film/

substrate combinations: (111)GaAs/(0001)sapphire; (111)GaAs/(111)spinel; (111)GaAs/(100)off-oriented spinel; (100)GaAs/(110)spinel; (100)GaAs/(11$\bar{2}$2)beryllia; (111)GaAs/(10$\bar{1}$1)beryllia. These results have been summarized from the work of Manasevit and Simpson (1969), Wang and coworkers (1974A), and Thorsen and coworkers (1974). Films grown on the oxides exhibit electron mobility up to 5000 cm²/V-s. p-n junctions in the films can be formed either by the organometallic process or by the vapor/liquid two-stage epitaxial process (Ladany and Wang, 1972). The heteroepitaxial GaAs films are potentially important for applications in microwave integrated circuits, transmission-mode photoemitters, and electroluminescent devices.

Heteroepitaxial GaAs alloys can be formed readily by appropriate reactions: $Ga_xAl_{1-x}As[(CH_3)_3Ga + (CH_3)_3Al + AsH_3]$; $Ga_xIn_{1-x}As[(CH_3)_3Ga + (C_2H_5)_2In + AsH_3]$; $GaAs_{1-x}P_x[(CH_3)_3Ga + AsH_3 + PH_3]$; $GaAs_{1-x}Sb_x [(CH_3)_3Ga + AsH_3 + SbH_3]$. InAs was reported (Manasevit and Simpson, 1971B) grown on sapphire by the reaction between $(C_2H_5)_3In$ and AsH_3.

Growth of II-VI compounds

The interest in the heteroepitaxy of compound semiconductors on oxides has extended from the III-V compounds to the II-VI compounds in the last few years. The achievement of heteroepitaxial growth of several II-VI compounds on insulating oxide substrates has been reported recently. Yim and Stofko (1972, 1974) reported the expitaxial growth on (0001)

FIGURE 3.8   X-ray Laue back-reflection pattern of epitaxial (111) ZnSe grown on (111) spinel. From Wang and Blumentritt (1970)

sapphire of cubic (111) ZnSe, hexagonal (0001) CdS, and hexagonal (0001) CdSe films. The growth has been carried out by the vapor-phase reactions between the group II elements (transported by H_2) and the group VI hydrides. Galli and Coker (1970) reported the epitaxial growth of (11$\bar{2}$0) ZnO on (01$\bar{1}$2) sapphire by a close-spaced transport (ZnO transported by HCl + H_2). The ZnO films are piezoelectric and potentially useful in surface acoustic wave and integrated optic devices.

In addition to the chemical-transport processes, the organometallic process has been successfully used to grow epitaxial II-VI compounds on oxide substrates. These include cubic (111) ZnSe on several sapphire, spinel, and beryllia orientations (Manasevit and Simpson, 1971A); cubic (110) ZnSe on (110) spinel (Wang and Blumentritt, 1970); cubic (111) ZnTe on (0001) sapphire (Manasevit and Simpson, 1971A); and cubic (111) CdTe on (11$\bar{2}$6) sapphire (Manasevit and Simpson, 1971A). The reactions employed for the growth are $[(C_2H_5)_2Zn + H_2Se]$ for ZnSe growth; $[(C_2H_5)_2Zn + (CH_3)_2Te]$ for ZnTe growth; $[(CH_3)_2Cd + (CH_3)_2Te]$ for CdTe growth.

Using the organometallic growth processes the control of growth is generally more difficult for the II-VI compound than for the III-V compounds. Consequently, the film growth characteristics are less reproducible. In many cases, vapor-phase chemical reactions take place at room temperature forming condensed phases long before the vapor streams reach the heated substrates. Single crystal films, therefore, are not formed on the substrates. For example, in the growth of ZnSe, $(C_2H_5)_2Zn$ and H_2Se react at room temperature forming an amorphous solid. In order to prevent the undesirable side reactions, the gas streams are admitted to the growth system independently and are allowed to mix and react at a critical distance from

FIGURE 3.7   An optical micrograph of epitaxial (111) ZnSe (~20 μm thick) grown on (111) spinel showing the surface structure. From Wang and Blumentritt (1970)

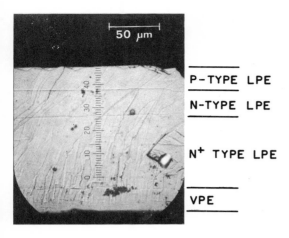

FIGURE 3.9   Cross section of GaP/spinel composite showing junction structures. From Wang et al. (1974B)

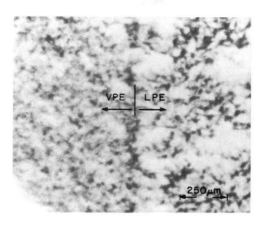

FIGURE 3.11   X-ray reflection topograph of OVP/LPE (111) GaP grown on spinel; Cu radiation, diffracting from (440) planes, showing the grain structures in OVP and LPE layers. From Wang et al. (1974B)

the heated substrate forming the single crystal growth. The surface structures and a Laue x-ray back-reflection pattern of a single crystal ZnSe film grown on (111) spinel are shown in Figures 3.7 and 3.8, respectively. The II-VI heteroepitaxial films reported in literature to date generally exhibit high resistivities (see Table 3.11), and little is known about their electrical characteristics and potential device applications.

Two-stage epitaxial growth

A 2-stage process has been used to form grown-in p-n junction electroluminescent diodes in both GaAs and GaP (the two conventional LED materials) on sapphire and spinel substrates (Ladany and Wang, 1972; Wang et al., 1974B). The process consists of the organometallic vapor-phase (OVP) deposition of layers onto the oxide substrates, followed by liquid-phase epitaxial (LPE) growth (Nelson, 1963) of two appropriately doped layers to produce the junction. This approach was chosen because the OVP technique has the distinct advantage over the liquid-phase technique of easier nucleation and initial growth. However, the control of stoichiometry and doping are difficult. This

can be complemented by the LPE method, which offers the advantages of automatic control of stoichiometry, gettering of impurities, and convenient incorporation of dopants. Furthermore, the LPE layers often show improvement in crystal perfection (Ladany et al., 1969) over the substrates on which they are grown. A cross section of a GaP-on-spinel composite junction structure is shown in Figure 3.9.

Studies on the crystallinity and growth characteristics of the III-V films grown by the 2-stage method have shown that the LPE layers generally exhibit better surface quality (higher reflectivity and uniformity) and bulk perfection (larger grain size) than the parent OVP layers as shown, respectively, in Figures 3.10 and 3.11. The film perfection of LPE III-V layers on various substrates increases in the order (111)spinel > (0001)sapphire > (110)spinel.

The heteroepitaxial growth of GaP was found to be more difficult to control than that of GaAs. The GaP epitaxy is more sensitive to oxide substrate surface preparation and to the gas-flow conditions. Although OVP GaAs of good quality may be routinely prepared by the conventional one-step growth procedure (no change or discontinuity during the growth) for further

(a)

(b)

FIGURE 3.10
Scanning electron micrographs of GaAs on spinel. (a) OVP GaAs; (b) LPE GaAs. Beam-sample angle, 45°. From Ladany and Wang (1972)

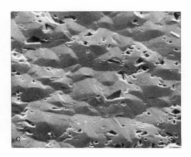
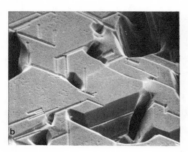
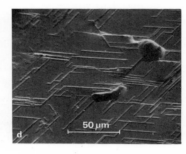

FIGURE 3.12
Scanning electron micrographs of epitaxial (111) GaP films grown on spinel substrates; beam-sample angle, 45°. (a) One-step VPE layer; (b) LPE layer grown on (a); (c) 3-step VPE layer; (d) LPE layer grown on (c). From Wang et al. (1974B).

LPE growth, the OVP GaP prepared by this way is not reproducible in quality. To improve the OVP GaP, the conventional one-step growth procedure was modified to a 3-step nucleation, annealing, and growth process (Wang et al., 1974B).

To perform a growth run by the modified procedure, the $(CH_3)_3Ga$ feed into the growth system is interrupted after a predetermined initial nucleation stage (typically 40 s to 1 min), and the thin GaP is annealed at the growth temperature ($\sim$800°C) for a period of time (typically 10 min) in the PH_3/H_2 atmosphere. After the annealing, film growth is continued by reintroducing the $(CH_3)_3Ga$ source into the system. (111) GaP films of improved surface and bulk perfections were consistently prepared by this 3-step procedure. It is probable that the annealing promotes the atomic rearrangement, interactions, or both at the film/substrate interface, facilitating the formation of oriented nuclei for further single crystal growth. Scanning electron micrographs of (111) GaP films by the two different procedures are presented in Figure 3.12; also shown are the LPE overgrowths. The surface appearance is improved by the modified 3-step VPE procedure over the conventional one and by the LPE process over the VPE one. LPE growth on 3-step VPE (111)GaP/ (111)spinel has yielded the most successful results. Electrical and optical characteristics of the GaAs and GaP electroluminescent diodes are discussed in Section 3.5.

In addition to the LPE layers, second epitaxial layers have been grown by the chloride vapor-phase method on the OVP first layers grown on oxide substrates forming microwave device structures. Charac-

teristics of the composites are discussed in Section 3.4.

3.4. Physical and electrical characteristics

Crystalline perfection

A general and preliminary assessment of the crystalline perfection of heteroepitaxial films grown on oxide substrates may be conveniently made by conventional methods such as Laue x-ray back reflection, low-angle electron diffraction, and various microscopic techniques. In addition, the techniques of x-ray diffraction topography are particularly useful in obtaining information about the nature of defect structures in the bulk of the films and, sometimes, about the effect of substrate defects on film imperfections. Several epitaxial III-V film/oxide substrate composites have been characterized by the topographic methods with interesting results (McFarlane and Wang, 1972).

The use of Lang transmission topography (Lang, 1959) to characterize heteroepitaxial systems usually has been confined to the examination of layers after removal from the substrate. However, for the III-V compounds (such as GaAs and GaP) grown on sapphire and spinel substrates, the topographic studies can be made on the as-grown sample and allows separate imaging of the layer and substrate. This is due to the high linear x-ray absorption coefficient (μ) of the relatively thin GaP or GaAs layer and the low linear absorption coefficient of the thick substrate. Thus, the product of the linear absorption coefficient and thick-

(a)

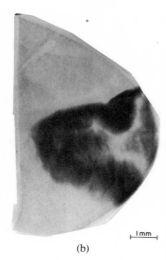

(b)

FIGURE 3.13
Lang x-ray topographs of a (0001) sapphire substrate [silver radiation, $(1\bar{2}10)$ reflection]. The crystal was rotated $0.1°$ between the two topographs. From Wang and McFarlane (1972)

ness (μt) is less than one for both layer and substrate and meets the criterion of $\mu t < 1$ for Lang topography. For example, the μt values for 20-μm-thick GaAs and GaP layers are 0.36 and 0.18, respectively, and for 0.5-mm-thick spinel and sapphire substrates, μt is 0.30 and 0.35, respectively. Because the two crystalline components of the heteroepitaxial system have different lattice parameters, it is possible to obtain separate topographs of the layer and substrate by appropriate orientation of the composite, first at the Bragg angle for a set of diffracting planes in the substrate, then at the Bragg angle for a set in the layer.

Examination by x-ray topography reveals that the III-V heteroepitaxial layers shown to be overall monocrystalline by Laue methods are composed of small crystallites (~10 μm large) and most are misoriented in a range ±0.1° from the nominal orientation of the layer. Further, the general orientation of the layers follows the orientation of large grains in the substrate. Figures 3.13 and 3.14 are topographs taken with silver

radiation of a (111) GaP film (20 μm thick) grown on a (0001) sapphire substrate. The film was grown by the organometallic process (Section 3.3). The two topographs of the sapphire substrate shown in Figure 3.13 were taken at crystal settings 0.1° apart and clearly show the presence of at least two subgrains in the crystal. Figure 3.14 shows topographs obtained from the GaP film, again at a crystal setting difference of 0.1° between the two topographs. It is evident that the GaP layers in Figures 3.14a and 3.14b follow the orientation of the sapphire grains shown in Figures 3.13a and 3.13b, respectively. Note also in Figure 3.13 the clear distinction between the areas where the sapphire grain is present or absent. However, in Figure 3.14, the GaP film orientation is much less sharply defined. That is, in the top part of Figure 3.13a, a sapphire grain exhibits strong diffraction contrast, and there is no diffraction from this same region in Figure 3.13b. The GaP film in the top part of Figure 3.14a also exhibits strong diffracted contrast, and the image

(a)

(b)

FIGURE 3.14
Lang x-ray topograph of a (111) GaP layer (20 μm thick) grown on the substrate shown in Figure 3.13 [silver radiation, $(02\bar{2})$ reflection]. The crystal was rotated 0.1° between the two topographs. From Wang and McFarlane (1972)

appears to be due to a high density of small crystallites. In Figure 3.14b, this same region still exhibits diffracted intensity from a small number of crystallites. Thus, the x-ray topographs show that the GaP film is composed of small crystallites which generally follow the orientation of the grains in the substrate but are misoriented from each other by ±0.1°.

Another example is the topographic examination of (100) GaP grown (by the organometallic process) on a (100) Si/($1\bar{1}02$) sapphire composite. The Si is 0.3 μm thick. Figure 3.15a is a topograph of a ($\bar{1}012$) sapphire substrate, and Figures 3.15b and 3.15c are topographs of the (100) GaP layer (20 μm thick) taken at a crystal setting difference of 0.1°. Since the film/substrate composite is bent, the crystal does not remain in proper alignment for diffraction from the $K_{\alpha 1}$ component of the incident x-rays as the crystal is translated in the beam. Thus, the two vertical bands of strongly diffracted intensity in Figure 3.15a are caused by diffraction from both the $K_{\alpha 1}$ and $K_{\alpha 2}$ components of the incident beam. No grains were observed in topographs taken of this wafer, but the general dislocation density is too high to resolve individual dislocations. As in the case of (111) GaP, the topographs of the (100 GaP layer in Figure 3.15b and 3.15c show contrast caused by small crystallites. This (100) layer appears, however, to have less spread in the range of misorientation than was observed on the (111) layer.

Topographs of the substrates usually show polishing scratches and grown-in defects. The polishing scratches are also visible in topographs of the layers as lines of no diffracted intensity, indicating that crystallites grown on the scratches are misoriented from the surrounding layer. Topographs of a (111) GaAs/(111) MgAl₂O₄ composites are shown in Figure 3.16. The GaAs (3.8 μm thick) was grown by the organometallic process. The topograph of the spinel substrate is shown in Figure 3.16a. Several polishing scratches are visible as lines of darker diffracted intensity. Note that there is a uniform background of dislocations but the density is too high (>10⁶ lines/cm²) to resolve individual dislocations. The topograph of the GaAs film is shown in Figure 3.16b. The diffracted image over the entire area of the layer is evidence that the layer is

monocrystalline. However, the dark image is not uniform, and the small whiter areas over the entire topograph are small regions that are misoriented from the nominal crystal orientation. The white lines in the GaAs topographs can be seen to correspond to scratches visible in the topograph of the substrate. This indicates that the GaAs grown on the scratches is misoriented from the rest of the layer. That many of the lines visible on the GaAs topographs are not evident on the substrate topograph shows that the GaAs growth is influenced by inperfections (scratches) in the surface of the substrate that do not have large enough strain fields to be imaged by x-ray topography.

Heteroepitaxial GaN films grown on sapphire by the chloride process (see Section 3.3) were examined by both reflection and transmission x-ray topography and found (McFarlane, 1974) to exhibit the same kinds of defect structures as seen in the GaAs and GaP films discussed above. Figure 3.17 shows a transmission topograph of a ($11\bar{2}0$) GaN film grown on a ($1\bar{1}02$) sapphire substrate. The film is composed of small grains and is strikingly similar to the structure of the other heteroepitaxial films.

Deformation and stress

The heteroepitaxial films grown on oxide substrates are strained and generally under compressive stress. This is because of the difference in thermal expansion behavior between the film and substrate and the elevated growth temperatures. The residual stress (σ_t) may be estimated, to a first order approximation, by the expression (Budo and Priest, 1963)

$$\sigma_t = Y_f(\alpha_s - \alpha_f)\Delta T, \qquad (3.4)$$

where α_s and α_f are, respectively, the coefficients of thermal expansion of the substrate and the film, Y_f is the elastic modulus of the film, and ΔT is the temperature change. The thermal expansion coefficients of most III-V and II-VI semiconductors are less than those of the commonly used oxide substrates such as sapphire, spinel, and beryllia, and therefore the contraction of the substrate upon cooling forces the film into compression. The film/substrate composite is

(a)   (b)   (c)

FIGURE 3.15
Lang x-ray topographs (silver radiation) of a ($\bar{1}012$) sapphire substrate and a (100) GaP deposit. (a) Sapphire substrate, ($0\bar{1}1\bar{2}$) reflection; (b) and (c) GaP layer, 20 μm thick, ($02\bar{2}$) reflection. The crystal was rotated 0.1° between the two topographs. From Wang and McFarlane (1972)

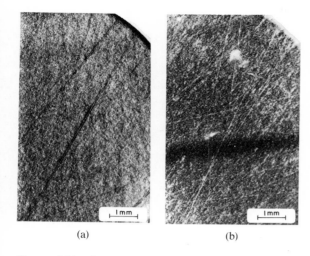

(a)   (b)

FIGURE 3.16   Lang x-ray topographs (silver radiation) of a Czochralski (111) spinel substrate and a (111) GaAs deposit. (a) Spinel substrate, ($4\bar{4}0$) reflection; (b) GaAs layer, 3.8 μm thick, ($2\bar{2}0$) reflection. From McFarlane and Wang (1972)

deformed with the film convexed outward. The deformation and stress may alter certain physical and electrical properties of a film with respect to the bulk crystalline properties of the same material.

The deformation of a film deposited on a single crystal substrate can be measured by using an x-ray technique (McFarlane and Wang, 1972) in which the angular setting (measured to a second of arc) of the substrate crystal, corresponding to the Bragg condition for reflection (for maximum $K_{\alpha 1}$ intensity) from selected planes, is determined as a function of substrate displacement. Because the structures are curved, the reflecting crystallographic planes are no longer parallel, and the deviation relative to the initial measurement at an edge of the substrate is a function of the horizontal translation of the substrate. The slope of the curve is inversely proportional to the radius of curvature of the crystal. If the curve is a straight line, the bending is uniform across the crystal. The bending is measured about a vertical axis which lies in the reflecting planes and is perpendicular to the direction of translation across the crystal. Thus, it is possible to determine if the deformation is isotropic by measuring the bending about various crystallographic directions in the substrate wafer.

The deformation (δ) and stress (σ) in an epitaxial film grown on a circular substrate material of radius r may be estimated by the following relations (Cahn and Hanneman, 1964; Denhartag, 1952; Dumin (1965):

$$\delta = \frac{r^2}{2R} \qquad (3.5)$$

and

$$\sigma = \frac{E\delta t_s^2}{3(1-\nu)r^2\, t_f}, \qquad (3.6)$$

where R is the radius of curvature, t_f is the film thickness, t_s is the substrate thickness, E is the elastic modulus, and ν is the Poisson ratio of the substrate material.

The deformation of GaAs grown on spinel (McFarlane and Wang, 1972) and of AlN grown on sapphire (Duffy et al., 1973) was determined by this x-ray method. The results are summarized in Figures 3.18 and 3.19, respectively. The experimentally measured deformation of GaAs on spinel is isotropic since the bending about all measured bending axes is the same. This is expected, because crystallographically the GaAs/spinel composite is a cubic-on-cubic system. For a given substrate thickness, the deformation increases with increasing film thickness. On the other hand, the film crystalline perfection improves as films become thicker. These two factors, the deformation and crystalline perfection, may determine the optimum film thickness at which the best film properties for devices can be achieved. GaAs films up to 70 μm grown on spinel and sapphire substrates (~0.02 inch thick) have been free of cracks. The stress in the film, as estimated from Equation (3.6), is approximately 1×10^9 dyne/cm^2 (McFarlane and Wang, 1972), which is nearly an order of magnitude less than the stress in Si grown on the oxides (Dumin, 1965; Ang and Manasevit, 1965; Wang et al., 1969).

Schlotterer (1968) analyzed theoretically the piezoresistance effect (change of resistivity caused by isotropic stress) of heteroepitaxial silicon grown on spinel

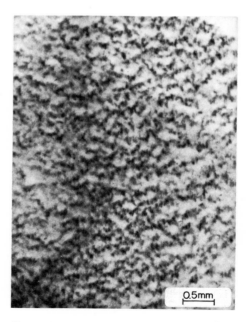

FIGURE 3.17   Lang x-ray topograph of a ($1\bar{1}20$) GaN film grown on a ($1\bar{1}02$) sapphire substrate [silver radiation, diffracting from the ($10\bar{1}0$) planes]

with results qualitatively in agreement with experimental observations of the carrier mobility dependence on the various crystallographic orientations of the films. The expected trend of dependence is that the effective mobility increases and decreases for, respectively, p-(111) and n-(100) silicon films. The piezoresistance effect is less pronounced in the case of GaAs on spinel. This is because the piezoresistance coefficients of GaAs and the film stress are such that the change of resistivity with respect to orientation is minimal. Experimental results (discussed below) indicate that the carrier mobility does not depend significantly on substrate orientation, as in the case of silicon on spinel.

The deformation of AlN on sapphire is more complicated. Figure 3.19 shows the deformation resulting from residual strain of two (11$\bar{2}$0) AlN/(1$\bar{1}$02) sapphire samples of different film thickness along two perpendicular crystallographic zones as indicated in the insert stereographic projection. It is clearly seen that the deformation is anisotropic as might be expected because of anisotropy of the elastic and thermal expansion properties of the rhombohedral-on-hexagonal system. The strain in AlN films becomes excessive

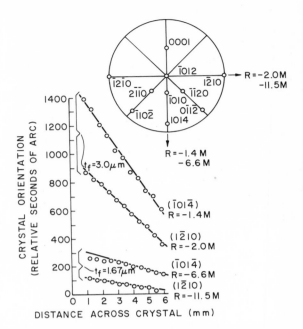

FIGURE 3.19   Plots of angular orientation of (1$\bar{1}$02) Al$_2$O$_3$ substrates (with AlN film), for x-ray diffraction from planes indicated, as a function of distance across substrate for two samples (measurements made on back of substrate). The radius of curvature along each of two zones is indicated in the inset stereographic projection of (1$\bar{1}$02) Al$_2$O$_3$. Film thickness is indicated in graph. From Duffy et al. (1973)

at thicknesses above about 4 μm. In some cases films shattered when subjected to a relatively small temperature gradient or mechanical stress for film thicknesses in the range 5–10 μm (Duffy et al., 1973).

Film/substrate orientation relationships

In Tables 3.5 to 3.8, the film/substrate orientation relationships of compound semiconductor/oxide composite systems are listed according to the classification of the film compounds. The classification is divided into group III–nitrides (Table 3.5), group III–phosphides and group III–arsenides (Table 3.6), III-V alloys (Table 3.7), and II-VI compounds (Table 3.8). The composite systems discussed here are limited to those in which the epitaxial growth is achieved by chemical vapor-phase growth processes (see Section 3.3), as specified in Section 3.1. The orientation relationships were experimentally determined principally by x-ray diffraction as described in the original work cited in the references. The growth processes used to achieve the epitaxy for each system are cited in the tables. Stereographic projections of several film/substrate systems are presented in Figures 3.20 to 3.23, illustrating the complete epitaxial relationships. The systems include (111) GaAs/(0001)

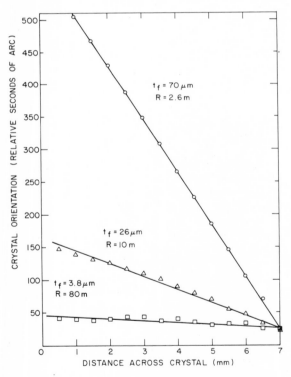

FIGURE 3.18   The change in angular setting of a spinel substrate (with GaAs film) for the maximum intensity of a particular reflection as a function of distance across the crystal. The radius of curvature and film thickness are shown in the graph. From McFarlane and Wang (1972)

TABLE 3.5   Film/Substrate Epitaxial Orientation Relationships III-V Compounds (Nitrides)/Oxides

Film	Substrate	Parallel Orientations Film/Substrate		Reference	Remarks
AlN	α-Al$_2$O$_3$	$(0001)/(0001)$;	$[\bar{1}210]/[\bar{1}100]$	(1),(2),(3)	(1) and (3) OM^a process; (2) chloride process
AlN	α-Al$_2$O$_3$	$(0001)/(11\bar{2}0)$;	$[\bar{1}2\bar{1}0]/[1\bar{1}00]$	(1)	OM process
AlN	α-Al$_2$O$_3$	$(11\bar{2}0)/(01\bar{1}2)$;	$[0001]/[01\bar{1}1]$	(1),(3)	OM process
AlN	α-Al$_2$O$_3$	$(11\bar{2}0),(30\bar{3}5),$ and $(11\bar{2}7)/$ $(01\bar{1}2)$;	— —	(2)	Chloride process; films exhibit several orientations
AlN	MgAl$_2$O$_4$	$(11\bar{2}0)/(110)$;	— —	(1)	OM process
GaN	α-Al$_2$O$_3$	$(0001)/(0001)$;	$[\bar{1}210]/[\bar{1}100]$	(1),(3),(4),(5),(6)	(1) and (3) OM process; (4), (5) and (6) chloride process
GaN	α-Al$_2$O$_3$	$(11\bar{2}0)/(01\bar{1}2)$;	$[0001]/[01\bar{1}1]$	(1),(3),(7)	(1) and (3) OM process; (7) chloride process
GaN	α-Al$_2$O$_3$	$(0001)/(10\bar{1}4)$;	$[\bar{1}2\bar{1}0]/10°$ off $[20\bar{2}\bar{1}]$	(1)	OM process
GaN	α-Al$_2$O$_3$	$(0001)/(11\bar{2}0)$;	— —	(4)	Chloride process
GaN	α-Al$_2$O$_3$	$(10\bar{1}5)/(10\bar{1}0)$;	— —	(4)	Chloride process
GaN	MgAl$_2$O$_4$	$(10\bar{1}1)/(100)$;	— —	(8)	OM process
ScN	α-Al$_2$O$_3$	$(111)/(0001)$;	— —	(9)	Chloride process
ScN	α-Al$_2$O$_3$	$(100)/(01\bar{1}2)$;	— —	(9)	Chloride process

^{a}OM: organometallic growth process

Sources: (1) Manasevit et al. (1971); (2) Yim et al. (1973); (3) Duffy et al. (1973); (4) Wickenden et al. (1971); (5) Ilegems (1972); (6) Maruska and Tietjen (1969); (7) Pankove et al. (1974); (8) Duffy and Wang (1974); (9) Dismukes et al. (1972).

sapphire, (100) GaAs/(100) spinel, (111) GaAs/(100) spinel, and (100) GaP/(100) Si/($1\bar{1}02$) sapphire.

It is seen that the orientation relationships are quite complicated. There are systems (such as GaAs on spinel) exhibiting unusual parallel orientation relationships unexpected from a general physicochemical point of view. Moreover, there are films in which the crystal structures and lattice parameters are similar but exhibit distinctively different parallel orientation relationships with respect to a common substrate material. For example, GaAs and GaP are well known to be closely related in structure (complete miscibility; belong to the same III-V class). GaAs and ZnSe are isoelectronic and exhibit similar lattice parameters. All the three compounds exhibit the same cubic structure. However, the observed orientation of the films grown on spinel substrates with the same (110) orientation are distinctively different:

$$(100)\text{GaAs} \| (110)\text{MgAl}_2\text{O}_4;$$
$$(100) \text{ and } (110)\text{GaP} \| (110)\text{MgAl}_2\text{O}_4;$$

and

$$(110)\text{ZnSe} \| (110)\text{MgAl}_2\text{O}_4.$$

The GaAs/spinel orientation relationships (Wang et al., 1974A) are described partially in Section 3.3. The epitaxy is unusual from the point of view of crystal structure. Since both GaAs and MgAl$_2$O$_4$ are cubic and belong to the same space group, one would expect

that a unique parallel epitaxial relationship exists between the film and substrate as in the case of Si on spinel (see Chapter 2). However, the experimentally determined relationships are:

$$(111)\text{GaAs} \| (111)\text{MgAl}_2\text{O}_4;$$
$$(100)\text{GaAs} \| (100)\text{MgAl}_2\text{O}_4;$$

and

$$(100)\text{GaAs} \| (110)\text{MgAl}_2\text{O}_4.$$

An apparent $(111)\text{GaAs} \| (100)\text{MgAl}_2\text{O}_4$ relationship has also been observed when the (100)MgAl$_2$O$_4$ substrate is off oriented by 1° or more.

Concurrently, it was found that the spinel substrate orientation plays a critical role in the ease with which the GaAs epitaxial layers can be grown (see Section 3.3). This is, again, in contrast to the Si-on-spinel growth case in which simultaneous epitaxial growth of Si can be achieved in spinel substrates of different orientations (see Chapter 2). The ease of GaAs growth increases with spinel orientation in the order (111) > (110) > (100). The epitaxy of (100)GaAs on (110)spinel is more sensitive to substrate surface quality and to growth temperature than that of (111)GaAs on (111)spinel. The growth on (100)spinel is most difficult in that polycrystalline films [with isolated (100)GaAs single crystal areas] are generally formed under various growth conditions. However, with a substrate off-orientation as small as one degree, the growth is enhanced remarkably with (111)GaAs formed on the

TABLE 3.6   Film/Substrate Epitaxial Orientation Relationships III-V Compounds (Phosphides and Arsenides)/Oxides

Film	Substrate	Parallel Orientations Film/Substrate		Reference	Remarks[a]
AlP	αAl_2O_3	$(111)/(0001)$	; — —	(1)	Films unstable (hydrolyze) in air
AlP	$MgAl_2O_4$	$(111)/(111)$	; — —	(1)	Films unstable (hydrolyze) in air
AlAs	$\alpha\text{-}Al_2O_3$	$(111)/(0001)$	; — —	(1),(2)	Films unstable (hydrolyze) in air
AlAs	$MgAl_2O_4$	$(111)/(111)$	; — —	(1),(2)	Films unstable (hydrolyze) in air
GaAs	$\alpha\text{-}Al_2O_3$	$(111)/(0001)$	; $[1\bar{1}0]/[11\bar{2}0]$	(3),(4),(5),(6)	(0001) substrate yields best quality GaAs among all sapphire orientations
GaAs	$\alpha\text{-}Al_2O_3$	$(111)/(11\bar{2}3)$	; — —	(4)	
GaAs	$\alpha\text{-}Al_2O_3$	$(111)/(11\bar{2}5)$	; — —	(4)	
GaAs	$\alpha\text{-}Al{;}_2O_3$	$(111)/(11\bar{2}6)$	; — —	(4)	
GaAs	$\alpha\text{-}Al_2O_3$	$(111)/(01\bar{1}2)$	; — —	(4)	
GaAs	$MgAl_2O_4$	$(111)/(111)$	; $[01\bar{1}]/[01\bar{1}]$	(3),(7)	Details of GaAs/spinel epitaxy in text
GaAs	$MgAl_2O_4$	$(100)/(110)$	; $[011]/[\bar{1}10]$	(3),(7)	
GaAs	$MgAl_2O_4$	$(100)/(100)$	; $[010]/[010]$	(3),(6),(7)	
GaAs	$MgAl_2O_4$	$(111)/(100)$	; $[01\bar{1}]/[\bar{1}10]$	(7)	
GaAs	BeO	$(100)/(10\bar{1}0)$	; — —	(8)	
GaAs	BeO	$(100)/(11\bar{2}2)$	; — —	(8)	$(11\bar{2}2)$ BeO yields the best quality (100) GaAs
GaAs	BeO	$(111)/(10\bar{1}1)$	; — —	(8)	$(10\bar{1}1)$ BeO yields the best quality (111) GaAs
GaAs	BeO	$(111)/(10\bar{1}\bar{1})$	; — —	(8)	
GaAs	BeO	$(111)/(0001)$	; — —	(8)	
GaAs	BeO	$(11)/(000\bar{1})$	; — —	(8)	
GaAs	BeO	$(111)/(11\bar{2}1)$	; — —	(8)	
GaAs	BeO	$(111)/(11\bar{2}1)$	; — —	(8)	
GaAs	BeO	(100) and $(111)/(11\bar{2}0)$	; — —	(8)	Films exhibit both (111) and (100) orientations
GaAs	ThO_2	$(100)/(100)$	; — —	(4)	
GaP	$\alpha\text{-}Al_2O_3$	$(111)/(0001)$	; $[1\bar{1}0]/[11\bar{2}0]$	(4),(9)	
GaP	$\alpha\text{-}Al_2O_3$	$(111)/(01\bar{1}2)$	; — —	(9)	
GaP	$Si/\alpha\text{-}Al_2O_3$	$(100)/(100)/(01\bar{1}2)$	; — —	(9)	(100) Si ($< 1\ \mu m$) grown on $(01\bar{1}2)$ sapphire yields (100) GaP
GaP	$MgAl_2O_4$	$(111)/(111)$	; $[01\bar{1}]/[01\bar{1}]$	(9),(10)	
GaP	$MgAl_2O_4$	$(100)/(100)$	; $[1\bar{1}0]/[010]$	(11)	
GaP	$MgAl_2O_4$	(100) and $(110)/(110)$	; — —	(10)	Films exhibit both (100) and (110) orientations
InP	$\alpha\text{-}Al_2O_3$	$(111)/(0001)$	; — —	(12)	
InAs	$\alpha\text{-}Al_2O_3$	$(111)/(0001)$	; — —	(12)	

Sources: (1) Wang (1973); (2) Manasevit (1971); (3) Manasevit and Simpson (1969); (4) Manasevit (1972); (5) Rai-Choudhury (1969); (6) Gutierrez et al. (1970); (7) Wang et al. (1974A); (8) Thorsen et al. (1974); (9) Wang and McFarlane (1972); (10) Wang et al. (1974B); (11) Wang (1974); (12) Manasevit and Simpson (1971B).

[a]In this table, all films were grown using organometallic processes, except in the cases (Gutierrez et al., 1970) of $((111)GaAs/(0001)\alpha\text{-}Al_2O_3$ and $(100)GaAs/(100)MgAl_2O_4$ using the chloride process.

TABLE 3.7   Film/Substrate Epitaxial Orientation Relationships III-V Compound Alloys/Oxides

Film	Substrate	Parallel Plane Orientations Film/Substrate	Reference	Remarks [a]
$Al_{1-x}Ga_xP$	α-Al_2O_3	$(111)/(0001)$	(1)	$x = 0.3$ to 0.6
$Al_{1-x}Ga_xP$	$MgAl_2O_4$	$(111)/(111)$	(1)	$x = 0.3$ to 0.6
$Al_{1-x}Ga_xAs$	α-Al_2O_3	$(111)/(0001)$	(2)	$x = 0.8$ to 0.1
$Al_{1-x}Ga_xAs$	$MgAl_2O_4$	$(111)/(111)$	(1)	$x = 0.8$ to 0.1
$Al_{1-x}Ga_xAs$	$MgAl_2O_4$	$(100)/(110)$	(1)	$x = 0.8$ to 0.1
$Ga_{1-x}In_xAs$	α-Al_2O_3	$(111)/(0001)$	(3)	—
$Ga_{1-x}In_xAs$	$MgAl_2O_4$	$(111)/(111)$	(1)	$x = 0.1$ to 0.3
$Ga_{1-x}In_xAs$	$MgAl_2O_4$	$(100)/(110)$	(1)	$x = 0.1$ to 0.3
$GaAs_{1-x}P_x$	α-Al_2O_3	$(111)/(0001)$	(4)	$x = 0.1$ to 0.6
$GaAs_{1-x}P_x$	$MgAl_2O_4$	$(111)/(111)$	(1),(4)	$x = 0.1$ to 0.6
$GaAs_{1-x}Sb_x$	α-Al_2O_3	$(111)/(0001)$	(4)	$x = 0.1$ to 0.3

Sources: (1) Wang (1973); (2) Manasevit (1971); (3) Manasevit and Simpson (1971B); (4) Manasevit and Simpson (1969).
[a]All films in this table were grown using organometallic process.

TABLE 3.8   Film/Substrate Orientation Relationships II-VI Compounds/Oxides

Film	Substrate	Parallel Plane Orientations Film/Substrate	Reference	Remarks
AnO	α-Al_2O_3	$(11\bar{2}0)/(01\bar{1}2)$	(1)	CVD^a transport process (ZnO transported by $H_2 + HCl$)
ZnSe	α-Al_2O_3	$(111)/(0001)$	(2),(3)	(2) CVD transport process, Zn (transported by $H_2 + H_2Se$; (3) OM^b process; ZnSe cubic form
	α-Al_2O_3	$(111)/(11\bar{2}3)$	(3)	OM process; ZnSe cubic form
		$(111)/(10\bar{1}4)$	(3)	OM process; ZnSe cubic form
		$(111)/(01\bar{1}2)$	(3)	OM process; ZnSe cubic form
	$MgAl_2O_4$	$(111)/(111)$	(3)	OM process; ZnSe cubic form
	$MgAl_2O_4$	$(110)/(110)$	(4)	OM process; ZnSe cubic form
	BeO	$(111)/(10\bar{1}1)$	(3)	OM process; ZnSe cubic form
	BeO	$(111)/(10\bar{1}0)$	(3)	OM process; ZnSe cubic form
	BeO	$(111)/(0001)$	(3)	OM process; ZnSe cubic form
ZnTe	α-Al_2O_3	$(111)/(0001)$	(3)	OM process; ZnTe Cubic form
CdS	α-Al_2O_3	$(0001)/(0001)$	(2)	CVD transport process, Cd (transported by H_2) + H_2S; CdS hexagonal form
CdSe	α-Al_2O_3	$(0001)/(0001)$	(5)	CVD transport process, Cd (transported by H_2) + H_2Se; CdSe hexagonal form
CdTe	α-Al_2O_3	$(111)/(11\bar{2}6)$	(3)	OM process; CdTe cubic form

Sources: (1) Galli and Coker (1970); (2) Yim and Stofko (1972); (3) Manasevit and Simpson (1971A); (4) Wang and Blumentritt (1970); (5) Yim and Stofko (1974).
[a]CVD: chemical vapor deposition.
[b]OM: organometallic process.

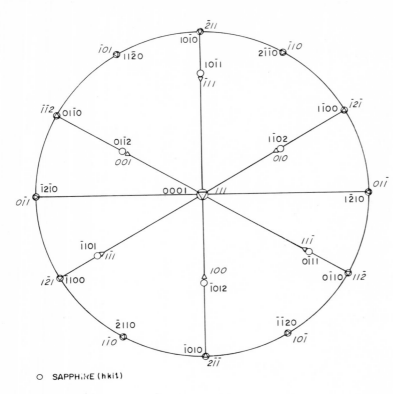

○ SAPPHIRE (hkil)

△ GaAs(hkl)

FIGURE 3.20
Stereographic projection of (111) GaAs on (0001) sapphire. From Wang and McFarlane (1972)

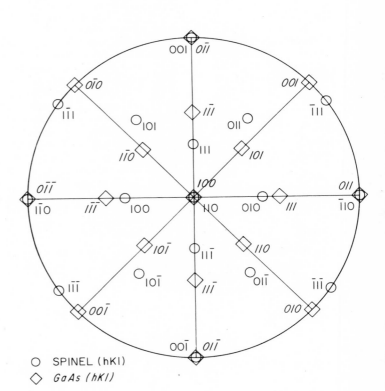

FIGURE 3.21
Stereographic projection of (100) GaAs on (110) spinel

○ SPINEL (hKl)

◇ GaAs (hKl)

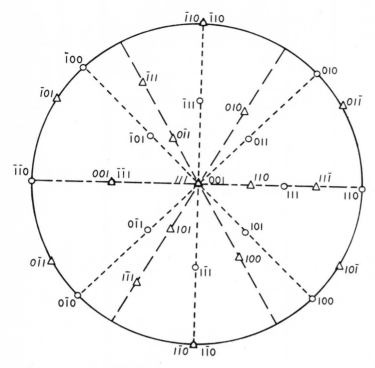

O — SPINEL (hkl)
△ — GaAs (hkl)

FIGURE 3.22
Stereographic projection of (111) GaAs on (100) spinel

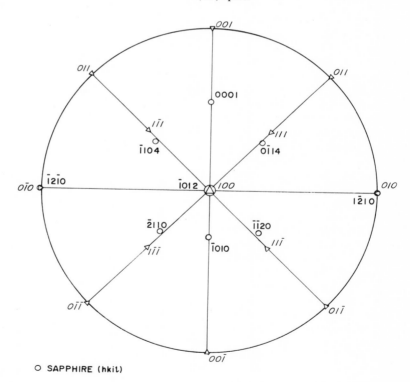

FIGURE 3.23
Stereographic projection of (100) GaP on (100)Si/(1$\bar{1}$02)sapphire

O SAPPHIRE (hkil)

△ GaP (hkl)

off-oriented (100)spinel. The ease of growth of the (111)GaAs on off-oriented (100)spinel is comparable to that of the (111)GaAs on (111)spinel.

Manasevit and Simpson (1969) have also reported that different epitaxial relationships exist in Si-on-sapphire and GaAs-on-sapphire systems. They suggested that the bonding at the GaAs-Al_2O_3 interface involves an As bridge between metal ions at the substrate surface and the succeeding Ga layer in the film. This mechanism is different from that of Si on sapphire, in which the bonding was proposed (Manasevit et al., 1968) to be related to a filling-in of metal ion sites with bonding to the oxygen ions.

Experimental results obtained from infrared specular reflection measurements of several compound-semiconductor/oxide systems (see also Chapter 5) indicate that the nature of interfacial bonding in a composite system may vary from one mechanism to another depending on the film/substrate orientations.

In spite of the unusual orientation relationships found in some systems, a few generalized trends can be derived from the compound-semiconductor/oxide heteroepitaxial relationships observed. First, for films with cubic crystal structures, the (111)film∥(0001)sapphire and (111)film∥(111)spinel relationships have generally been found. In fact, for many film materials, the (111) epitaxial films can be grown more easily than any other orientations under a wide range of growth conditions. Second, of the compound semiconductor films investigated to date, more epitaxial films have been successfully grown on the (0001)sapphire orientation (the basal plane) than on any other substrate orientations. In some material systems, such as GaAs on sapphire, the (0001) surface is the only orientation that yields films of good quality. In others, for example the AlN-on-sapphire system, more than one orientation [(0001) and (01$\bar{1}$2)] can be used to yield films of good quality. The (0001) sapphire has been most popularly used to date for many film materials and conceivably is the most logical first choice for the study of film growth of new heteroepitaxial materials.

The film/substrate orientation relationships have practical significance in the consideration of the feasibility of specific device structures. From the crystallinity point of view, a substrate orientation should be chosen so that the best film epitaxial growth may be achieved. On the other hand, there are devices in which a certain orientation is desirable (see Section 3.5), and therefore, films with the best crystalline quality may not exhibit the most desirable orientation for devices. In most cases, an alternate approach may be made to fulfill both the crystallinity and the orientation requirements. For example, (100) thin (~1 μ) GaAs grown on a transparent oxide substrate (such as sapphire) is desired for photocathode applications (see Section 3.5). However, it has not been possible to

achieve the (100)GaAs epitaxy on any sapphire orientations investigated. A further consideration is that a layer as thin as 1 μm grown on a structurally different substrate material may exhibit poor crystallinity. A third buffer layer such as GaP grown in between the GaAs film and sapphire substrate may be feasible since GaP is optically transparent and structurally similar to GaAs. The (100)GaP/sapphire epitaxy is achieved (Wang and McFarlane, 1972) through an intermediate thin Si (~2000 Å), and thus the desired multilayer composite substrate is (100)GaP/(100)Si/(1$\bar{1}$02)sapphire. (100)GaAs epitaxy has been achieved (Wang et al., 1974A) on this composite substrate.

Optical properties

The heteroepitaxial III-V and II-VI compounds in general have large refractive indices, strong lattice vibrational absorption bands, and band gap absorptions (direct and indirect) that occur in a wide UV to IR range. Optical properties of the films and a comparison between the film and bulk optical properties are of scientific interest as well as practical importance for consideration in device applications. Published information in this area, however, is scanty. Materials that have been investigated and reported in literature for some detailed aspects of the optical properties include GaAs on spinel (Wang et al., 1974A), AlN on sapphire

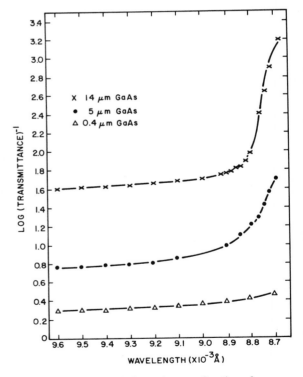

FIGURE 3.24   Optical absorption as a function of wavelength of GaAs films. From Wang et al. (1974A)

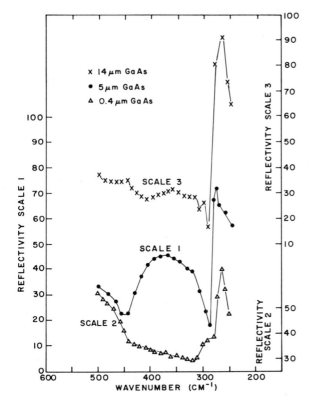

FIGURE 3.25 Reflection characteristics of GaAs films in the lattice band regions. From Zanzucchi and Wang (1974)

(Yim et al., 1973), and GaN on sapphire (Pankove, 1973).

Transmission characteristics near the band gap region of heteroepitaxial GaAs films (on spinel) of various thickness are shown in Figure 3.24. The sharpness of the absorption edge increases with increasing thickness, indicating that the crystallinity improves with increasing film thickness. The apparent absorption observed below the band gap is caused, in part, by reflection losses (accounting for an optical density of about 0.25) from the interfaces in the composite structure. Reflection measurements indicate that the apparent absorption is caused in large measure by light scattering from the as-grown surface structure (see for instance Figure 3.2), which become rougher with increasing film thickness (Zanzucchi and Wang, 1974). The same behavior has been also observed in heteroepitaxial GaP films (Zanzucchi and Wang, 1974) and may well be a fundamental characteristic of the as-grown heteroepitaxial compound semiconductor films.

Reflection characteristics of the GaAs films in the lattice band region are shown in Figure 3.25. The films all exhibit a sharp transition in reflectivity near 300 cm^{-1} consistent with the characteristics for bulk mate-

rial (Spitzer and Whelan, 1959; Jones and Hilton, 1965). The spectra suggest that for a thickness of 4000 Å the heteroepitaxial film does not diverge grossly in stoichiometry or crystal structure from the bulk.

Ellipsometric measurements were made at a wavelength of 5461 Å on epitaxial GaAs grown on spinel. The optical constants (the real and imaginary parts of the refractive index $n - ik$) and the dielectric constants ϵ_1 and ϵ_2 (the real and imaginary parts of the complex dielectric constant $\epsilon_1 - i\epsilon_2$) were derived from the measurements. Typical values of n, k, ϵ_1, and ϵ_2 of the films are compared with those obtained on epitaxial GaAs grown on bulk GaAs, as shown in Table 3.9. It is apparent that the optical and dielectric constants of the heteroepitaxial GaAs are quite close to those of the homoepitaxial films.

Optical absorption (in the band gap region) and infrared specular reflection (in the wavelength region 10–30 μm) spectra of heteroepitaxial AlN grown on sapphire are shown, respectively, in Figures 3.26 and 3.27. In Figure 3.26 data (Pastnak and Roskovcova, 1968) obtained from single crystal bulk material are included for comparison. From these data, Yim and coworkers (1973) concluded that AlN is a direct band gap material with a value of about 6.2 eV at room temperature.

Optical specular reflection measurements in the infrared region (see Zanzucchi, Chapter 5) of thin (<1 μm) semiconductor films grown on oxide substrates may provide information on the nature of film/substrate interfacial bonding and on the defect and disorder of the substrate surfaces. In the composite structure, reflections from both the air/semiconductor and the semiconductor/substrate interface contribute to the spectrum. For AlN on sapphire, AlN has a restrahlen reflection band beginning at about 11 μm and extending to about 15 μm where the other strong modes of sapphire occur. Consequently, if the AlN epitaxial layer is very thin (<1 μm in the present case) and thus very weakly absorbing and reflecting in the wavelength region greater than 15 μm, then the reflection spectrum obtained through the AlN layer at those wavelengths represents almost exclusively the contribution from the first few hundred angstrom thicknesses of the sapphire substrate at the AlN–sapphire interface. Comparison of the reflection spectrum in the wavelength region beyond about 15 μm of the sap-

TABLE 3.9   Optical and Dielectric Constants of Epitaxial GaAs Films at $\lambda = 5461$ Å

(111) GaAs Films	ϵ_1	ϵ_2	n	k
Film grown on spinel substrate Film thickness = 25 μm	15.195	3.602	3.925	0.459
Film grown on GaAs substrate Film thickness = 2 μm	15.636	3.674	3.981	0.461

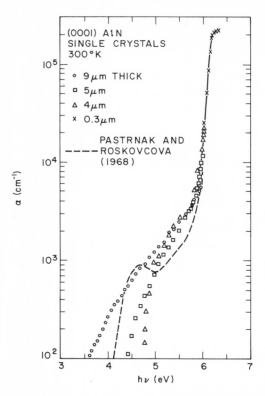

FIGURE 3.26   Dependence of absorption coefficient α upon photon energy $h\nu$ near the fundamental edge of (0001) AlN on (0001) sapphire at room temperature. In the data of Pastrnak and Roskovcova (dotted curve) the bump near 4.5–4.8 eV was attributed to absorption bands of oxygen in the AlN samples used. From Yim et al. (1973)

phire substrate with and without an AlN epitaxial layer, therefore, provides information on the nature of the epitaxy interface.

The spectra shown in Figure 3.27 were obtained from a (0001) sapphire substrate (undeposited back surface; 250 μm thick) and from the same substrate but through a thin (less than 1 μm) (0001)AlN epitaxial layer. It can be seen in the two infrared spectra shown that the number of sapphire lattice modes has increased upon epitaxy of AlN, with the appearance of new bands in the region of 650–550 cm^{-1} and 500–400 cm^{-1}. These new bands are quite similar to those reported (Barker, 1963) for sapphire with a work-damaged surface on which, in addition to the normal lattice vibration modes, forbidden modes were also observed. The overall features of the AlN–sapphire infrared spectrum thus clearly indicate that the epitaxy results in a large distortion of the sapphire structure in the film/substrate interface region, with the result that this region is highly strained.

Optical properties of GaN have been the most extensively studied [for a review, see Pankove (1973)] of the heteroepitaxial semiconductor films, primarily because of the interesting luminescent properties and the potential device applications (see Section 3.5). Measurements were mostly made on (11$\bar{2}$0) GaN grown on (1$\bar{1}$02) sapphire. GaN was found to have a direct bandgap of 3.50 eV. The index of refraction varies from 2.33 at 1 eV to 2.67 at 3.38 eV. The absorption spectrum, which is illustrated schematically in Figure 3.28, consists of four regions:

1. For photon energy less than about 1.5 eV, the dominant mechanism involves free carrier absorption which varies as $h\nu^{-n}$. n is in the range of 2.5–3.9, depending on the doping.

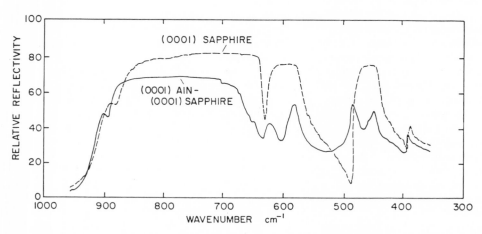

FIGURE 3.27   Infrared reflection spectra at room temperature of (0001)AlN-(0001)sapphire and (0001)sapphire. From Yim et al. (1973)

2. In the region where $1.7 < h\nu < 3.0\,\text{eV}$, the absorption coefficient varies as $(h\nu-1.7)^2$, corresponding to a transition involving a deep center in the middle of the gap.
3. Between 3.0 and 3.3 eV, the absorption coefficient varies exponentially with $h\nu$.
4. Between 3.3 and 3.5 eV, a steeper exponential dependence is observed.

The band structure of GaN was determined by Bloom (1971). Photoluminescence measurements made by Pankove and coworkers (1970) have shown that the near-gap emission occurs at 3.47 eV with approximately 100 percent internal quantum efficiency. The electroluminescence of GaN on sapphire is discussed in Section 3.5.

Electrical properties

Although the heteroepitaxy of a large number of compound semiconductors has been achieved (see Section 3.3), many of the films have been prepared only in preliminary experiments, and the film electrical properties have not been studied in detail. It is anticipated that more electrical data will be reported as more films grown under improved or optimum conditions become available. In the following discussion, the film electrical properties known to date are presented. Emphasis is placed on GaAs heteroepitaxial films, which have been the most extensively investigated of the compound semiconductor films.

GaAs. UNDOPED FILMS. The carrier-transport properties of heteroepitaxial GaAs grown (by organometallic process) on sapphire and on beryllia have been studied in some detail (Thorsen and Manasevit, 1971; Thorsen et al., 1974). Properties were reported for films grown on substrate planes that yield the best quality GaAs overgrowths. These include (111) GaAs on (0001) sapphire, (111) GaAs on (10$\bar{1}$1) beryllia, and (100) GaAs on (11$\bar{2}$2) beryllia.

Films grown without intentional doping on the two substrate materials were found to exhibit, qualitatively, a similar behavior in electrical properties. The first few microns of GaAs grown tend to have high resistivity and exhibit p-type conductivity. With subsequent growth, the outer layers of the film convert to n-type, with donor carrier concentrations ranging from 10^{15} to $10^{18}/\text{cm}^3$, depending largely on the purity of the reactant source materials. The thickness of the underlying high-resistivity layer was found to depend primarily on the donor carrier concentrations and be independent of the total film thickness.

Once the film has converted to n-type, the carrier concentration rises and the carrier mobility improves as the film becomes thicker, usually saturating at a film

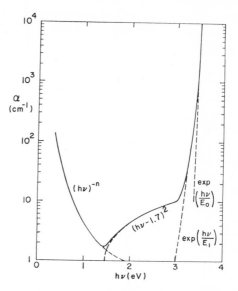

FIGURE 3.28   Schematized optical absorption of GaN. From Pankove (1973)

thickness of about 15 to 20 μm. The exact origins of the acceptor states near the GaAs/oxide interfaces are not known, and several possibilities may exist (Manasevit and Thorsen, 1970). These include (1) defects generated as a result of lattice mismatch, thermal contraction differences upon cooling from the growth temperatures, or irregularities in the substrate surface topology and (2) acceptor impurities either diffusing outward from the substrate or present on the surface at the initiation of growth.

The variation of the interface p-layer thickness as a function of net donor concentration is shown in Figure 3.29 for GaAs/α-Al$_2$O$_3$ and GaAs/BeO. The carrier mobility and carrier concentration vs. film thickness plots are shown in Figures 3.30 and 3.31, for, respectively, GaAs/α-Al$_2$O$_3$ and GaAs/BeO. It is seen from Figure 3.29 that for a given donor concentration the p-layer thickness in the GaAs/BeO system is substantially higher than that in the GaAs/α-Al$_2$O$_3$ system. The improvement of carrier mobility with increasing film thickness is shown in Figures 3.30 and 3.31. The maximum mobilities measured for undoped thick films are about 5000 cm^2/V-s for GaAs/α-Al$_2$O$_3$ and 3000 cm^2/V-s for GaAs/BeO at carrier concentrations in the mid-10^{16}/cm^3 range. For comparison, homoepitaxial GaAs films grown by the same process were reported to exhibit mobility values of 6900 cm^2/V-s (Ito et al., 1973). In an earlier paper, mobility values of 2480 cm^2/V-s (at 6.5×10^{17}/cm^3 carrier concentration) and of 1740–2550 cm^2/V-s (at 2.6-9 $\times 10^{17}$/cm^3 carrier concentration range) were reported (Rai-Choudhury, 1969) for, respectively, GaAs films grown on (0001) sapphire and on (100) semiinsulating GaAs substrates.

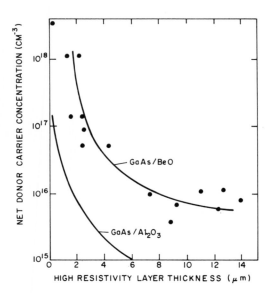

FIGURE 3.29   Carrier concentration vs. interface layer thickness for n-GaAs/Al$_2$O$_3$ and n-GaAs/BeO. From Thorsen et al. (1974)

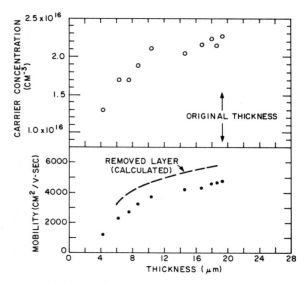

FIGURE 3.30   Variation of carrier concentration and mobility with thickness in thick n-GaAs film on (0001) α-Al$_2$O$_3$ substrate. From Thorsen and Manasevit (1971)

FIGURE 3.31
Variation of carrier concentration and mobility with thickness in thick GaAs film on (11$\bar{2}$2) BeO substrate. From Thorsen et al. (1974)

The Hall mobility vs. carrier concentration plots of (100)GaAs/(11$\bar{2}$2)BeO and (111)GaAs/(10$\bar{1}$1)BeO thick films are shown in Figure 3.32. The data suggest that the (100)GaAs/(11$\bar{2}$2)BeO films are electrically superior to (111)GaAs/(10$\bar{1}$1) films. A general trend of mobility drop with decreasing film thickness is seen for both films. This same trend was also observed in (111)GaAs/(0001)sapphire but with a less pronounced effect.

In addition to the carrier mobility properties, some characteristics of p-n junctions in GaAs films grown on beryllia (Thorsen et al., 1974) and on sapphire (Behrudt, 1971) were evaluated. The growth (organometallic process) of the junctions was accomplished by reducing the AsH$_3$ flow during growth to produce p-type material on an initial n-type layer. Preliminary measurements on the current–voltage relations indicated that junction properties approaching those obtained in homoepitaxial GaAs may be achieved.

Carrier transport properties of GaAs on spinel have been studied (Wang et al., 1974A). It was found that, in many respects, the properties of undoped GaAs grown on spinel are similar to those of films grown on sapphire and beryllia. Categorically, three kinds of films were obtained under the optimum growth conditions, depending on the impurities present in the growth system. The first class of films exhibits n-type conductivity. A high-resistivity ($>10^8$ Ω-cm) p-layer is often present during the initial growth. With subsequent growth, the material becomes n-type, the carrier concentration increases, and the carrier mobility improves to a certain level. This kind of film exhibits transport properties similar to those of GaAs/α-Al$_2$O$_3$ and GaAs/BeO films. The highest electron mobility measured was about 5000 cm^2/V-s at a carrier concen-

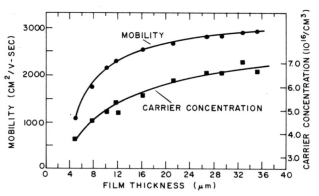

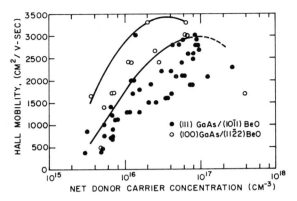

FIGURE 3.32 Mobility vs. carrier concentration for n-GaAs/BeO films. From Thorsen et al. (1974)

tration of mid-10^{16}/cm³. The electrical properties vs. thickness plots of a typical film is shown in Figure 3.33.

The second class of films exhibit p-type conductivity. Thick (>10 μm) films with mobilities (~ 300 cm²/V-s) close to the bulk value have been prepared in the carrier concentration range of mid-10^{15}/cm³ to mid-10^{17}/cm³. Since many lots of $(CH_3)_3Ga$ contain appreciable amounts of Cu and Zn, which are well known to be effective acceptors in GaAs, these p-films are most probably Cu or Zn doped (or both). The variation of electrical properties as a function of film thickness of a typical p-film is shown in Figure 3.34.

The third class of films exhibit very high resistivity ($>10^3$ Ω-cm) independent of thickness. The films were highly compensated and contain localized p-n junctions.

DOPED FILMS. Electrical characteristics of p-GaAs heteroepitaxial films grown with intentional doping, on oxides, have been reported [films grown by the chloride transport process: Gutierrez et al. (1970); films grown by organometallic process: Manasevit and Thorsen (1972); Wang et al. (1974A)]. Emphasis has been placed on the evaluation of properties of thin (~ 1 μm) degenerate GaAs films. The thin epitaxial GaAs grown on transparent oxide substrates is of interest for use as a negative electron affinity (NEA) photocathode (see Section 3.5). Typical results of thin-film studies obtained from various investigarors are summarized in Table 3.10. These include the film/substrate orientations, growth processes, acceptor concentrations, and hole mobilities. The thin films generally exhibit hole mobilities slightly less than those of the thick films. Zn-doped (100) GaAs films grown on (110) spinel were essentially equivalent electrically to the (111) films grown on (0001) sapphire. Results obtained from GaAs grown by the chloride-transport process indicate that this growth process is not feasible for the GaAs growth on oxides, as the epitaxy yield (~ 30 percent) is low and the films are of poor quality.

Both $(C_2H_5)_2Zn$ and $(CH_3)_2Cd$ have been shown to be effective sources of dopant for the growth of GaAs by the organometallic process. Thick films were found to have hole mobilities typical of the best bulk material for acceptor concentrations from 3×10^{16} to mid-10^{19}/cm³.

Electrical measurements on doped n-type GaAs have been reported for films grown on sapphire (Manasevit and Simpson, 1969) and on spinel (Wang et al., 1974A). Mobilities of Se-doped (from H_2Se) GaAs films grown on (0001) α-Al_2O_3 were found to be dependent on film thickness. Films less than about 8 μ thick

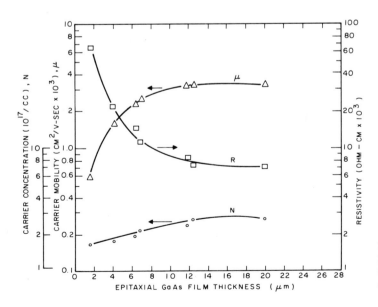

FIGURE 3.33
Electrical properties as a function of film thickness of an n-type (111) GaAs film grown on spinel

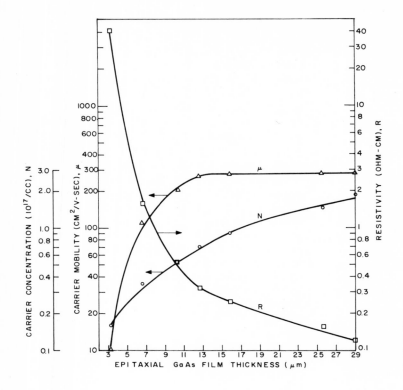

FIGURE 3.34
Electrical properties as a function of film thickness of a p-type (111) GaAs film grown on spinel

TABLE 3.10   Electrical Properties of p$^+$ GaAs Heteroepitaxial Films

Film/Substrate	Film Thickness (μm)	Growth Process	Dopant	Carrier Concentration ($1/cm^3$)	Hole Mobility (cm^2/V-s)	Reference	Remarks
(111)GaAs/ (0001)α-Al$_2$O$_3$	1.0	Organometallic [(CH$_3$)$_3$Ga + AsH$_3$]	(C$_2$H$_5$)$_2$Zn	8.6×10^{18}	91	(1)	GaAs films were grown simultaneously on (0001)α-Al$_2$O$_3$ and (110)MgAl$_2$O$_4$. Results of GaAs on MgAl$_2$O$_4$ are shown below.
(100)GaAs/ (110)α-Al$_2$O$_3$	1.0	Organometallic [(CH$_3$)$_3$Ga + AsH$_3$]	(C$_2$H$_5$)$_2$Zn	7.4×10^{18}	84	(1)	GaAs films were grown simultaneously on (0001)α-Al$_2$O$_3$ and (110)MgAl$_2$O$_4$. Results of GaAs on α-Al$_2$O$_3$ are shown above.
(111)GaAs/ (0001)α-Al$_2$O$_3$	0.9–1.5	Organometallic [(CH$_3$)$_3$Ga + AsH$_3$]	(C$_2$H$_5$)$_2$Zn	2×10^{17} 4×10^{18}	95 105	(1)	—
(111)GaAs/ (0001)α-Al$_2$O$_3$	< 1.5	Organometallic [(CH$_3$)$_3$Ga + AsH$_3$]	(CH$_3$)$_2$Cd	7×10^{16} 3×10^{17}	80 120	(1)	—
(111)GaAs/ (111)MgAl$_2$O$_4$	$\approx$ 1.0	Organometallic [(CH$_3$)$_3$Ga + AsH$_3$]	(C$_2$H$_5$)$_2$Zn	8×10^{18} 5×10^{19}	50 90	(2)	—
(100)GaAs/ (110)MgAl$_2$O$_4$	$\approx$ 1.0	Organometallic [(CH$_3$)$_3$Ga + AsH$_3$]	(C$_2$H$_5$)$_2$Zn	8×10^{18} 5×10^{19}	50 90	(2)	—
(111)GaAs/ (0001)α-Al$_2$O$_3$ (100)GaAs/ (100)MgAl$_2$O$_4$	< 2.0	Chloride transport [GaCl+AsH$_3$]	Zn	—	—	(3)	Films less than 2 μm thick are not continuous. Epitaxy achieved at 30 percent of growths. Thick degenerate films exhibit mobility values up to 55 cm^2/V-s.

Sources: (1) Manasevit and Thorsen (1972); (2) Wang et al. (1974A); (3) Gutierrez et al. (1970).

exhibited average mobilities usually less than 2000 cm^2/V-s, whereas films about 8–25 μm thick had average mobilities of about 4000 cm^2/V-s for carrier concentrations of about $4 \times 10^{16}/cm^3$. n-type GaAs films were grown on both (111) and (110) spinel using H_2Se and H_2S as the dopants. The control of the degree of doping was found more difficult for H_2Se than for H_2S because H_2Se was not chemically stable in storage. Thick films (> 10 μm) with carrier concentration in the range 2×10^{17} to $1 \times 10^{19}/cm^3$ were grown with mobilities ranging from 50 to 90 percent of the bulk values.

Other compound semiconductors In Table 3.11 the electrical properties (known to date from literature) of several III-V and II-VI heteroepitaxial semiconductor films grown on oxide substrates are summarized. As described before, many of these films have not been characterized in detail, primarily because of the lack of well defined epitaxial films suitable for electrical measurements at this time. It is anticipated that new experimental data will supplement the literature as films of suitable quality become increasingly available in this growing field.

3.5. Potential device applications

Transmission-mode photocathodes

The achievement of single crystal growth of III-V compounds on transparent oxide substrates is desirable for the realization of negative electron affinity (NEA) photocathodes that can operate in the transmission mode* in the wavelength ranges desired for applications in photomultipliers, image tubes, and camera tubes. The principle of negative electron affinity and its development and use in III-V compound electron emitters have been described in several

*NEA photocathodes are of two basic types; reflection-mode cathodes, for which the light is incident on the cathode–vacuum surface, and transmission-mode cathodes, for which the light is incident on the substrate or back of the cathode.

TABLE 3.11   Electrical Properties of Heteroepitaxial Compound Semiconductor Films

Film/Substrate	Film Thickness (μm)	Growth Process	Carrier Concentration ($1/cm^3$)	Conductivity Type	Carrier Mobility (cm^2/V-s)	Reference	Remarks
$(11\bar{2}0)$GaN[a]/ $(01\bar{1}2)\alpha$-Al_2O_3	5	Chloride transport [GaCl + NH_3]	$2 \pm 1 \times 10^{18}$	n	130 ± 20	(1),(2)	Undoped film; can be compensated by Zn, Mg, and Be doping.
$(11\bar{2}0)$GaN[a]/ $(01\bar{1}2)\alpha$-Al_2O_3	4.8	Organometallic [$(CH_3)_3Ga + NH_3$]	3×10^{19}	n	75	(3)	Undoped film; can be made insulating by compensation with Zn or Li doping.
GaN[a]/α-Al_2O_3	6	Organometallic [$(CH_3)_3Ga + NH_3$]	10^{19}	n	~60	(4)	Undoped film; orientation not reported.
(111)GaP[b] $(0001)\alpha$-Al_2O_3	6	Organometallic [$(CH_3)_3Ga + PH_3$]	1.4×10^{18}	n	70	(5)	As grown film; mobility 63 percent bulk value.
(111)GaP[b]/ (111)$MgAl_2O_4$	10	Organometallic [$(CH_3)_3Ga + PH_3$]	$10^{16} - 10^{18}$	p	80	(6)	As grown film; mobility 70 percent bulk value.
SnN/ $(1\bar{1}02)\alpha$-Al_2O_3	4.8	Chloride transport [Sc-chloride + NH_3]	8.6×10^{19}	n	158	(7)	Undoped film; film orientation is (100) or (321).
$(11\bar{2}0)$ZnO[c]/ $(01\bar{1}2)\alpha$-Al_2O_3	~10	Chemical vapor transport [ZnO + HCl]	—	n	—	(8)	Undoped films exhibit resistivity 1–10 Ω-cm; can be compensated by Li or Na doping.
(111)ZnSe/ $(0001)\alpha$-Al_2O_3	~10	Chemical vapor transport [Zn + H_2Se]	—	—	—	(9)	Films exhibit high resistivity ($>10^2$ Ω-cm).
(0001)CdS/ $(0001)\alpha$-Al_2O_3	~10	Chemical vapor transport [Cd + H_2S]	—	n	—	(9)	Films exhibit resistivity 10^{-2} to 10^2 Ω-cm range.
(0001)CdSe/ $(0001)\alpha$-Al_2O_3	~10	Chemical vapor transport [Cd + H_2Se]	—	—	—	(10)	Films exhibit high resistivity (5×10^6 to 1×10^7 Ω-cm range).

Sources: (1) Maruska and Tietjen (1969); (2) Pankove et al. (1974); (3) Duffy et al. (1973); (4) Manasevit et al. (1971); (5) Wang and McFarlane (1972); (6) Wang et al. (1974B); (7) Dismukes et al. (1972); (8) Galli and Coker (1970); (9) Yim and Stofko (1972); (10) Yim and Stofko (1974)
[a]GaN films exhibit piezoelectric and electroluminescent properties. These properties are described in other parts of this book.
[b]GaP exhibits transport properties with trends similar to that of GaAs. Depending on the system purities, films can be either highly compensated containing localized junctions, or semiconducting. Carrier mobilities of the semiconducting films improve with increasing film thickness, leveling off at about 10 μm. The properties of p-n junctions in GaP/spinel grown by a two-stage epitaxial process are described in Section 3.5.
[c]ZnO films are piezoelectric, and the properties are further described in other parts of this book.

recent publications including a monograph (Bell, 1973) and several review articles (Bell and Spicer 1970; Williams and Tietjen, 1971; Simon, 1972; Sommer, 1973; Martinelli and Fisher, 1974). The NEA concept was first proposed in the early sixties. From the advent of this concept, the materials and device technology of electron emission have received new impetus, and significant improvements have been achieved in many types of electron emission devices during the last few years.

The basic difference between conventional electron emitters† (such as alkali antimonide) and NEA emitters (such as GaAs) can be shown through a comparison of the photoemission process for each case. Figure 3.35 shows the schematic energy band diagrams of the two kinds of emitters. In the case of the conventional emitter, photoemission from the valence band occurs for photons have a minimum energy ϕ, but the

---

†For a comprehensive review of the conventional photoemissive materials, see Sommer (1968).

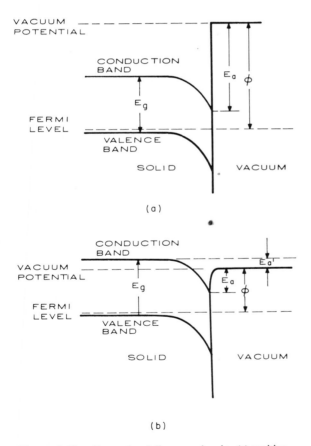

(a)

(b)

FIGURE 3.35   Energy band diagrams showing (a) positive electron affinity and (b) negative electron affinity conditions in relation to band gap energy E_g, work function ϕ, electron affinity E_a, and effective electron affinity E_a'

"hot" electrons excited to levels above the vacuum level lose energy quickly ($\sim 10^{-12}$ s) through interactions with phonons until they have been "thermalized" at the bottom of the conduction band. As a result, only those electrons can escape into vacuum that have been produced so close to the surface (a few hundred Angstroms) that their energy still exceeds the value ϕ. The lower the electron affinity (E_a) at the emitter surface, the larger the fraction of electrons with sufficient energy to be emitted.

By contrast, in NEA materials the vacuum level lies below the bottom of the conduction band minimum in the bulk material (i.e., the effective electron affinity (E_a') is negative). In this case, the thermalized electrons have sufficient energy to escape, and because they have long lifetimes ($\sim 10^{-9}$ s), they can diffuse from a considerable distance to the surface and still be emitted. Thus, the essential difference between the conventional (positive electron affinity) and the NEA emitters is that the latter have an escape depth several orders of magnitude greater, resulting in a significant increase in emission efficiency.

To obtain the NEA conditions, the work function of a p-type semiconductor is reduced by the adsorption of electropositive—and in some cases by a combination of electropositive and electronegative—elements on the surfaces. Cesium and oxygen are the elements most commonly used to date. Addition of cesium or cesium oxide to the surface causes a dipole potential to be introduced at the surface of the semiconductor, which reduces the barrier normally present there. The reduction occurs over atomic dimensions. The addition of cesium may also produce band bending. The complete NEA condition is achieved by combining materials such that the conduction band minimum in the bulk of the semiconductor lies above the vacuum potential and the semiconductor is sufficiently doped so that the band-bending distance is small compared with the hot-electron escape depth. If this condition is achieved, the energy lost by an electron as it passes through the band-bending region determines the fraction of electrons that will be emitted into vacuum. This fraction P is known as the surface escape probability.

As described above, the electron transport in the NEA materials is by minority carrier diffusion. Therefore, high-quality single crystalline materials with controlled doping are desired for good device performance, as imperfections in the crystal reduce the electron diffusion length and thus reduce the large escape depth which is the basis of the high efficiency of NEA emitters. The crystallite size must be larger than the escape depth. By contrast, conventional emitters can employ small crystallites because the escape depth is small compared with the size of the microcrystals.

The major effort in the development of NEA emitters has been directed toward III-V photoemitters with enhanced sensitivity in the infrared region of the spectrum. Among the III-V compounds, GaAs is the most widely used NEA photocathode. In addition, GaAs$_x$P$_{1-x}$ and In$_x$Ga$_{1-x}$As have found practical applications. In photomultipliers, GaAs generally exhibits white light response (to radiation from a 2854 K tungsten lamp) of about 800 μA/lumen. The highest reported value in literature (Sommer, 1974) is 1200 μA/lumen.

To date the III-V NEA photocathodes have been made almost exclusively for operation in the reflection mode. The NEA films have been grown by either chloride vapor-phase epitaxy or liquid-phase epitaxy on substrates of the same material (such as GaAs on GaAs) or of materials with chemical/structural similarities (such as GaAs$_x$P$_{1-x}$ on GaAs). The use of reflection-type cathodes is limited to only a few types of photomultipliers and photodiodes. For imaging devices (image tubes and TV camera tubes) it is more practical to employ photocathodes that operate in the transmission mode, whereby electrons are emitted from the side opposite that on which the radiation is incident. The heteroepitaxial III-V film/oxide substrate structures are particularly attractive for this application because of its high versatility, good mechanical and chemical stability, and high transparency in a wide wavelength range of the substrate oxides (such as sapphire and spinel); further advantages include large areas and reduced costs. Moreover, the substrate can serve as the tube window itself.

Although in principle the III-V film/oxide substrate composites are feasible for use as transmission-mode photocathodes, many technical problems remain to be solved before cathodes of high performance can be realized. Developments in this area (materials growth, activation procedure, and device fabrication) are still in an early stage, but much research is being devoted in many laboratories to the objective of obtaining high-quality devices employing heteroepitaxial III-V cathodes.

Theoretical analysis of the quantum efficiency of transmission mode photocathode has been made by Allen (1971), Antypas and coworkers (1970), Liu and coworkers (1970), and Syms (1969). For the GaAs-on-sapphire structure (Figure 3.36), the quantum yield in the transmission mode per incident photon is given by the expressions (Liu et al., 1970):

$$Y_t(\lambda)|_{S=0} = \frac{LAP}{1-(\alpha L)^2} \div \left[\alpha L(1-R_2)e^{-\alpha t} \right.$$
$$+ (1+R_2)e^{-\alpha t}\tanh\left(\frac{t}{L}\right)$$
$$\left. - \alpha L \operatorname{sech}\left(\frac{t}{L}\right)(1-R_2e^{-2\alpha t}) \right]$$

(3.7)

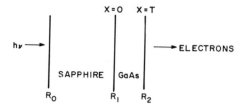

FIGURE 3.36   Geometry of GaAs-on-sapphire photocathode. From Liu et al. (1970).

and

$$Y_t(\lambda)|_{S=\infty} = \frac{LAP}{1-(\alpha L)^2} \div \left[\alpha L(1-R_2)e^{-\alpha t} \right.$$
$$+ (1+R_2)e^{-\alpha t}\operatorname{ctnh}\left(\frac{t}{L}\right)$$
$$\left. - (1+R_2e^{-2\alpha t})\operatorname{csch}\left(\frac{t}{L}\right) \right].$$

(3.8)

The parameter A is related to R by

$$A = \frac{\alpha(1-R_0)(1-R_1)}{1-R_0R_1 - [R_0R_2(1-R_1)^2 + R_2R_2]e^{-2\alpha t}},$$

(3.9)

where L is the electron diffusion length, P is the electron escape probability, α is the optical absorption coefficient of GaAs, R is the reflectivity at the interfaces, t is the GaAs film thickness, and S is surface recombination velocity at the GaAs/sapphire interface. The theoretical expression for transmission quantum efficiency $Y_t(\lambda)$ is complicated, and it is difficult to see the dependence of $Y_t(\lambda)$ on the parameters α, L, t, and S. Moreover, variation of nominally constant parameters with photon energy and with distance from the interface must be also considered in correlating theoretical and experimental results. Nonetheless, the theoretical analysis serves as a guide to the interpretation and understanding of the more complex behavior of real photocathodes.

In order to assess the relative importance of the parameters, Martinelli and Fisher (1974) have made analyses of several key parameters including the cathode thickness t, optical absorption constant α, diffusion length L, and surface recombination velocity S. Several conclusions were obtained: (1) For a given combination of L, S, and α, there is a value of t that maximizes $Y_t(\lambda)$. The cathode should be thick enough for light absorption but thin enough so that photoexcited electrons can reach the emitting surface. For values of α and L in the vicinity of 1 μm^{-1} (in GaAs, $\alpha = 1$ μm^{-1} at 0.85 μm) and 2 μm, respectively, the optimal value of t, t_m, is about 1 μm relatively independent of S. Furthermore, within a range of several tenths of a micrometer, around the optimum thickness, Y_t is relatively insensitive to t. (2) For values of $t \lesssim t_m$, $Y_t(\lambda)$ increases with increasing $\alpha(\lambda)$, owing to the increased number of photoexcited electrons, most of which can escape. For $t >> t_m$, $Y_t(\lambda)$ decreases with increasing $\alpha(\lambda)$. The reason for this latter effect is twofold: First, for small α, light absorption is approximately uniform throughout the cathode, which means that there are an appreciable number of electrons generated within a diffusion length of the NEA surface. As α increases, fewer electrons are generated near the NEA surface. Second, when α is small, there are multiple internal reflections, increasing the number of electrons generated near the emitting surface. As α increases, this enhancement decreases. (3) For any combination of α, S, and t, $Y_t(\lambda)$ is an increasing function of L. However, increases in L beyond a certain value produce no significant improvements in Y_t. For values of α near 1 μm^{-1} and t about 1 μm, this value of L is about 1.0 μm when $S >> 1.0$ and about 3.0 μm when $S \lesssim 1.0$. Values of L in the 1–3 μm range are realistic.

The minority diffusion length L is the parameter characterizing transport in the bulk of an NEA emitter. It is given by

$$L = (\mu k T \tau / e)^{1/2}, \qquad (3.10)$$

where μ is the electron mobility, τ is the carrier lifetime, e is the electronic charge, k is Boltzmann constant, and T is the temperature. Since μ and τ both depend on doping and crystalline perfection, L is strongly dependent on these two factors. L increases with improved crystalline perfection but decreases as doping increases. With optimum doping (mid-10^{18}/cm^3 to low-10^{19}/cm^3 for GaAs), the diffusion length is decisively determined by the crystalline perfection which, in turn, is largely affected by several factors including the cathode/substrate lattice and thermal 1 expansion matches, cathode/substrate orientation relationships (see Sections 3.3 and 3.4), cathode thickness, growth technique, and growth conditions.

In Equations (3.7) and (3.8) the electron escape probability P enters into the expression for $Y_t(\lambda)$ as a multiplicative factor. Recall that P is the probability that an electron, having reached the bent-band region, will traverse that region and be emitted into vacuum. It is determined, to a large extent by the band gap, doping density, and crystallinity of the cathode materials. For a given material, P is related to the band-bending effect (van Larr, 1973) such that it decreases as the doping density decreases. In addition, P depends on surface crystallographic orientation, and this dependence has a significant effect on the heteroepitaxial photocathodes. Studies by James and coworkers (1971) revealed that, for GaAs:Cs-O cathodes, the (111)A, (100), (110), and (111)B faces exhibit P values of 0.212, 0.317, 0.307, and 0.489, respectively, with corresponding luminous sensitivity values of 810, 1225, 1125, 1837 μA/lumen.

From the preceding discussion, the basic materials requirements for transmission photocathodes of high quantum efficiency* may be defined. For GaAs cathodes grown on oxide substrates (such as sapphire, spinel, and beryllia), there are the following requirements:

1. The cathode thickness and doping level should be, respectively, 1 μm and in the range of 5×10^{18} to 2×10^{19}/cm^3.
2. The GaAs films should exhibit a high degree of both surface and bulk crystalline perfection.
3. The GaAs films should exhibit, preferably, (111)b or (100) surface orientations.

The first requirement can be satisfactorily met by the presently established growth and doping techniques (see Section 3.3). However, the second and third requirements are difficult to meet, and they are the subject of current research in many laboratories. As described in early sections, the bulk crystalline perfection of heteroepitaxial III-V films grown on oxides improves with increasing thickness. In the thickness range up to several microns near the film/substrate interface, the epitaxial films exhibit a very high density of defects. In addition, experimental results have shown that the (111) III-V films grown on oxides exhibit the unwanted (111)A [rather than the desired (111)B] surface orientation, and that, for many film/substrate combinations, the (111) epitaxial films can be grown more easily or more perfectly than can any other orientations under a wide range of growth conditions. For instance, in the GaAs-on-sapphire system, the (0001) surface, also known as the basal plane, is the only orientation that yields GaAs films of good quality. The films exhibit (111)A orientation. It should be noted that the "poor crystallinity" effect is probably more dominant than the "wrong orientation" effect, as a moderate photosensitivity [about 40 percent of the value obtained from the (111)B surface] can be still obtained from the (111)A GaAs surface (James et al., 1971). The above discussions have revealed that it is inherently difficult to attempt the use of GaAs thin films, directly grown on sapphire or other oxides, as transmission-mode photocathodes with high performance. This conclusion has been verified from results reported in literature.

---

*In addition to transmission quantum efficiency, the resolution and speed of response of a transmission photocathode are important properties in imaging devices. Cathode parameters affecting resolution and response speed together with the optimization of the parameters are discussed in some detail by Bell (1973) and by Martinelli and Fisher (1974). No attempt is made to further discuss these parameters and properties in the present chapter. Qualitatively, resolution is determined by cathode thickness or diffusion length, whichever is smaller, and the response time is determined by the minority carrier lifetime.

The first GaAs-on-sapphire transmission photocathode was reported by Syms (1969). The films were polycrystalline and were grown by the Ga-AsCl$_3$-H$_2$ vapor-transport method. Information on the sapphire substrate orientation was not provided. The transmission luminous sensitivity was reported to be about 1 μA/lumen for a cathode with a layer thickness of 2.5 μm. Andrew and coworkers (1970) reported a transmission luminous sensitivity value of 70 μA/lumen from a GaAs layer 0.6 μm thick grown on sapphire. The layers were reported also to be polycrystalline and grown by the Ga-AsCl$_3$-H$_2$ vapor-transport method. The sapphire orientation was not given. Liu and coworkers (1970) reported a transmission quantum efficiency twice that of the conventional S-I photocathode at about 0.8-μm wavelength, for a GaAs single crystal film (0.85 μm thick) grown on sapphire. The photosensitivity is about the same as that reported by Andrew and coworkers (1970). The GaAs was reported to be grown by the organometallic process, and the sapphire used was presumably (0001) oriented.

In order to improve the crystallinity of the thin heteroepitaxial cathode layer, the film/substrate interface mismatches (structural and thermal) must be reduced. This may be achieved by introducing an intermediate buffer layer between the cathode thin layer and the oxide substrate. The buffer material should be structurally very close to the active layer and should be thick enough (generally 10–30 μm) so that defects originating from the buffer/oxide interface do not propagate to the buffer/cathode interface. Optically the buffer material should be compatible with the oxide substrate and the cathode layer. The short wavelength cutoff of the photocathode is limited by the minimum band gap in the buffer/oxide composite. For GaAs cathode, tertiary III-V alloys may be suitably used as the buffer layers. These include Ga(As,P), (In,Ga)P, and (Ga,Al)As. The later two are particularly attractive. At a composition of In$_{0.48}$Ga$_{0.52}$P, the buffer material matches GaAs in lattice dimension is transparent for wavelengths greater than 0.65 μm. In the (Ga,Al)As buffer system, all compositions match. As the Al/Ga ratio increases, the spectral window broadens. Curves showing the optical transmission of a Ga$_{0.2}$Al$_{0.8}$As/spinel and a In$_{0.48}$Ga$_{0.52}$P/GaP composite are presented in Figure 3.37. The film and substrate are, respectively, 10 and 500 μm thick.

The intermediate layer not only can improve the cathode crystalline perfection, but in certain systems it can modify the cathode surface orientation to a more desired one. This has been discussed in Section 3.4. For example, GaAs films grown on sapphire exhibit the undesirable (111)a orientation. When GaP is used as the "orientation modifying" layer, a (100) GaAs cathode layer can be realized on a sapphire-based composite substrate. The (100)GaP/sapphire epitaxy is

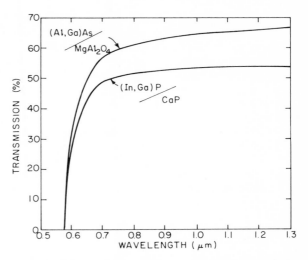

FIGURE 3.37   Optical transmission of (Ga,Al)As on spinel and (In,Ga)P on GaP

achieved (Wang and McFarlane, (1972) through another intermediate thin ($\sim$2000-Å)Si layer, resulting in the multilayered composite structure: (100)GaP/(100)Si/(1$\bar{1}$02)sapphire. (100) GaAs has been achieved (Wang et al., 1974A) on this composite substrate. There are several interesting aspects of the complicated composite structure which should be noted. First, the (100) GaP films grown on thin heteroepitaxial Si unexpectedly exhibit a better crystallinity than do (111) GaP films grown directly on (0001) sapphire, as revealed by x-ray diffraction topography studies (Wang and McFarlane, 1972). Second, the Si film has an undesirable light-absorption effect and should be made just thick enough ($\sim$2000 Å) to provide a growth of GaP with good quality. Without the thin Si, the GaP grown directly on (1$\bar{1}$02) sapphire exhibits a (111)a orientation. Attempts have also been made, but with poor results, to grow thin (1-μm) GaAs on the (100)Si/(1$\bar{1}$02)sapphire composite without the intermediate GaP layer.

GaAs photocathodes grown on GaP/Si/sapphire composite substrates exhibit transmission photoemission values (Williams et al. 1974) substantially higher than 150 μA/lumen. The results represent a considerable advance in the state of the art. In order to achieve enhanced photosensitivity, materials of improved quality are required. As discussed before, the maximum sensitivity is obtained when the cathode material has such a long electron-diffusion length that the thickness of the emitting layer required for total light absorption is less than the diffusion length. The cathode crystallinity may be improved by introducing buffer layers to reduce the film/substrate lattice and thermal expansion mismatches or to modify the surface orientations. In addition, the crystallinity of the cathode and the buffer layers may be improved by growing the materials by a combination of growth processes (see Section 3.3) including the organometallic vapor-phase method, the chloride vapor-phase method, and the liquid-phase epitaxial method. For a given film/substrate combination, one growth method may be more suited than others. Discussions presented in Sections 3.3 and 3.4 may be helpful as a general guide for the selection of the optimum methods.

In addition to sapphire and other oxides, GaP has been considered as a substrate for III-V transmission photocathode. To date GaP is the only III-V material available in large single crystal form and transparent to visible light (for $\lambda > 0.55$ μm). Developmental GaAs transmission-mode cathodes grown on GaP (with various III-V tertiary buffer layers) have been reported (Guttierrez and Pommerrenig, 1973; Fisher et al., 1974; Allenson et al., 1972; Hughes et al., 1974) in recent years with encouraging photoemission results (for details see the original references).

Light-emitting diodes

Light-emitting diodes (LED) have been fabricated in III-V compounds grown on oxide substrates (Ladany and Wang, 1972, 1974; Pankove, 1973). The interest in the heteroepitaxial composite systems for electroluminescence applications arises from new prospects regarding substrate size and cost, from the optical transparency of the substrate, and from the possible integration of more than one functional material on the same substrate.

GaAs and GaP electroluminescent diodes have been fabricated on sapphire and spinel substrates by a vapor/liquid two-stage epitaxial process as described in Section 3.3.

Emission spectra of typical GaAs and GaP film diodes are shown in Figure 3.38. The characteristics are very similar to those of standard bulk diodes. The diode current (I)–voltage (V) curves are shown in Figure 3.39. The curves exhibit the usual exponential voltage dependence caused by a combination of ther-

mal injection and space charge recombination. At low or reverse biases, there is some evidence of nonradiative tunneling. The reported external quantum efficiency is 0.1 percent for GaAs, 0.1 percent for GaP (red), and 0.01 percent for GaP (green). In comparison with bulk materials, the heteroepitaxial films exhibit lower light-emission efficiency. This is expected because of the strains generated at the film/substrate interface and the imperfections developed during the heteroepitaxial growth. Nevertheless, the efficiency reported is sufficiently high to make useful displays with the special features discussed earlier.

In addition to GaAs and GaP, GaN films grown on sapphire have been extensively studied in recent years for electroluminescent applications. A review including comprehensive literature references, made by Pankove (1973), deals with material synthesis, optical properties, and electroluminescent device performance of heteroepitaxial GaN.

The epitaxial growth of GaN on sapphire for electroluminescent devices has been most successfully carried out using the vapor-phase reaction of GaCl and NH$_3$ (Maruska and Tietjen, 1969). The $(1\bar{1}02)$ sapphire substrate is most frequently employed for GaN growth, and the film/substrate orientation relationship is $(11\bar{2}0)$GaN$\|(1\bar{1}02)$sapphire; $[0001]$GaN$\|[0\bar{1}11]$ sapphire. As grown, undoped GaN films are always n-type and highly conducting (carrier concentrations $\approx 10^{18}$/cm^3; electron mobility ≈ 130 cm^2/V-s). The film can be made insulating by adding such compensating dopants as zinc during growth. Optical measurements reveal that GaN has a direct energy gap of 3.5 eV. The wide band gap offers the possibility of generating light over a large range of wavelengths, including the visible and the near ultraviolet.

GaN electroluminescent diodes have been made with metal-insulating n-type (MIN) structures. An undoped n-type layer is first grown on sapphire. Then a thin Zn-doped insulating layer is grown on the n-type layer. Finally, a metal contact is formed on the top. The i-n diode passes a current when a sufficiently large

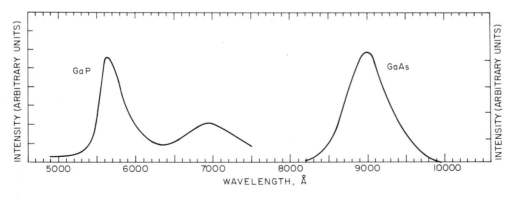

FIGURE 3.38   Emission spectra of heteroepitaxial GaAs and GaP diodes. From Ladany and Wang (1972, 1974)

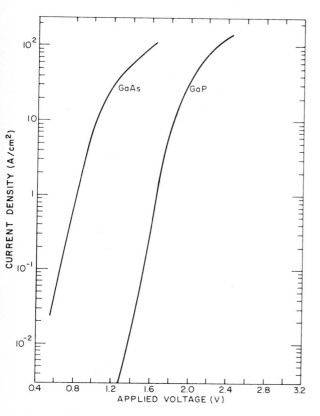

FIGURE 3.39   Current vs. voltage characteristics of heteroepitaxial GaAs and GaP diodes. From Ladany and Wang (1972, 1974)

respect to the n-GaN, an emission peak at 3850 Å is obtained. Another type of diode consists of a metal–insulator–semiconductor (MIS) structure using Si_3N_4 as the insulator (Pankove and Norris, 1972). The emission peak is at 3780 Å. The power efficiency and brightness of the GaN emitter are its most important aspects. Power efficiency up to 0.1 percent and light output of greater than 300 ft-lamberts have been achieved.

Surface acoustic wave devices

Certain heteroepitaxial III-V and II-VI compounds (such as Aln/sapphire, GaAs/spinel, and ZnO/sapphire) exhibit desirable piezoelectric or semiconducting properties (or both), and these composite systems are of potential importance in the evolution of new and improved surface acoustic wave (SAW) devices in the microwave acoustic technology [for a comprehensive review of the SAW technology, see Carr (1969), Stern (1969), Gandolfo (1969), White (1970)].

The surface wave acoustic technology evolved from the early ultrasonic work in bulk material media. Surface waves, like the bulk waves in solid media, propagate with a velocity smaller by five orders of magnitude than that of electromagnetic waves in free space and therefore offer the prospect of miniaturized microwave components. However, there exists an important difference between surface and bulk acoustic waves. In the bulk device the acoustic wave cannot be manipulated until it emerges from the output. The bulk systems are confined largely to fixed delays, and additional circuits must be employed to further process the signal. On the other hand, the surface waves are accessible and can be tapped, amplified, and manipulated along the entire propagation path. Surface wave devices thus lend themselves naturally to the implementation of all those signal-processing functions that depend upon the simultaneous sampling of a signal at many points. It is possible, therefore, to fabricate a new generation of sophisticated, miniature, low-power, high-speed signal-processing devices.

Although the principle and concept have been known for years, the advancement of SAW devices has been limited mainly by the associated material technology. The realization of sophisticated monolithic SAW devices requires improved material systems possessing compatible functional properties. Piezoelectric materials are required for transducers; low-loss materials are required for transmission and delay lines; semiconductor materials are required for amplification. For optimum generation, guidance, propagation, and amplification of surface waves, efficient, compatible, and integrated single crystalline functional materials are essential. Because no one single material is known possessing all the required properties, the availability of multifunctional epitaxial composite

field is induced in the insulating layer. Light is emitted through the layered device and out through the substrate.

The color of the light is determined by growth conditions such as the partial pressure of Zn during growth, the duration of the growth, and the thickness of the layers. The color of the emission also can depend on the polarity of the bias. Blue and red have always been obtained with a negative bias on the metal contact, whereas green and yellow can be realized with either polarity. As shown in the emission spectra (Figure 3.40), blue, green, yellow, or red emission can be obtained. The electroluminescence process, as described by Pankove and Lampert (1974), consists of impact excitation of electrons from deep centers with a subsequent radiative transition of a conduction band electron to the empty center.

UV electroluminescence has been generated in undoped GaN, as shown also in Figure 3.40. The diode consists of a surface barrier under colloidal carbon; when the carbon is biased positively with

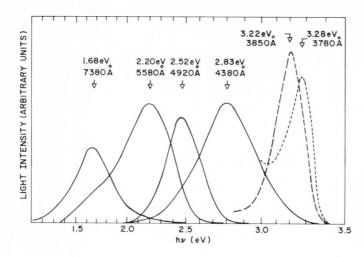

FIGURE 3.40

Emission spectra of various GaN LEDs. Solid lines are for i-n structures using Zn-doped material. Dashed and dotted lines are for diodes made of undoped GaN. From Pankove (1973)

material systems may provide the ultimate in subsystem miniaturization of SAW devices including amplifiers, waveguides, delay lines, and signal-processing devices. The heteroepitaxial materials for SAW devices have been reviewed by Collins and coworkers (1970). In the last few years, epitaxial growth and integration of SAW materials have been actively studied, and advancements have been made to the stage of fabricating active devices. For example, an acoustic amplifier device structure was recently reported (Manasevit et al., 1974) having heteroepitaxial AlN and silicon, both grown on a common sapphire substrate. A state-of-the-art discussion covering the subject of SAW materials and integration is in a separate chapter of this book.

Microwave integrated circuits

The availability of high-quality epitaxial GaAs grown on dielectric oxide substrates provides the possibility of obtaining new device structures useful in microwave integrated circuits. In the GaAs/oxide composite structures, the substrates afford an excellent dielectric isolation between components and make it possible to combine passive elements with monolithic devices on the same substrates. Furthermore, the heteroepitaxial composites may offer advantages in terms of substrate cost, thermal dissipation, film/substrate functional coupling, and subsystem integration.

The development of microwave solid state materials and devices has been reviewed in many articles [see, for example, Scrupski (1967); Mehal and Walker (1968), Minden (1969), Watson (1969), Bass (1970), Young and Sobol (1974)]. The state of the art of microwave integrated circuits has been discussed by Sobol and Caulton (1974). At present, because of its unique properties GaAs is the one semiconductor

material with the greatest scope and promise for use in microwave devices. These include a high carrier mobility ($\mu_e = 8000$ cm²/V-s), a wide energy gap (1.4 eV), a direct interband transition, a high-temperature capability, and a negative resistance effect (Gunn effect). In addition, GaAs can be made semiinsulating by Cr-doping with resistivities greater than 10^6 Ω-cm. This high-resistivity GaAs provides a host matrix for homoepitaxial growth of GaAs active elements and affords electrical isolation between components. The high-resistivity material is not sensitive to thermal treatments and can be etched to desired patterns by conventional techniques.

Microwave integrated circuits can be made in monolithic, quasimonolithic, and hybrid forms (Sobol and Caulton, 1974). In the monolithic circuit, active devices are grown on or in the semiconductor (semiinsulating) substrate, and passive circuitry is deposited on the substrate. The quasimonolithic circuit employs dielectric oxide substrates which provide for the heteroepitaxy of the active elements. In the hybrid form, active devices are attached to a ceramic, glass, or ferrite substrate containing the passive circuitry. Detailed analyses and assessments have been made (Sobol, 1974; Sobol and Caulton, 1974) of the materials and processing techniques, circuit functions, performance requirements, and the advantages and disadvantages of the monolithic and hybrid forms of microwave integrated circuits. The results have shown that the hybrid form of technology permits the use of many varieties of devices with fewer yield and performance problems and fewer processing difficulties than the monolithic approach. At present the hybrid circuits are used almost exclusively in the frequency range of 1 to 35 GHz. At high frequencies the monolithic approach is more desirable because of the reduction in parasitic inductance encountered in joining the circuits and devices. The one inherent disadvantage of the monolithic approach is that it permits the use of only those devices that can be grown on the substrate. Therefore, optimum combinations of materials (such as GaAs and Si) for the integration of microwave subsystems cannot be utilized.

The problem of integration of microwave subsystems may be overcome by using the quasimonolithic approach as, in principle, the heteroepitaxy of several materials (such as Si and GaAs) may be achieved on the same substrate for a variety of devices. Moreover, the quasimonolithic approach offers several other advantages. For example, when sapphire is used to replace the semiinsulating GaAs as the substrate material, a substantial reduction of substrate cost may be achieved. Another advantage is the possibility of obtaining improved power dissipation. In the epitaxial film/substrate device structures, an important property governing a device's power dissipation is the thermal conductivity of the substrate material. Thermal energy generated in the active region of a device can be removed by conduction through the substrate, by conduction through the electrodes on the device, and by radiation. Any heat loss by direct convection is extremely small. Page (1968) has shown that the heat loss caused by substrate conduction is directly proportional to substrate thermal conductivity and constitutes more than 90 percent of the total heat loss. Of the oxide substrate materials that yield single crystal growth of GaAs, beryllia (BeO) exhibits an extremely high thermal conductivity (7 times that of GaAs; about the same as Cu). Therefore, when beryllia is used as the substrate material for epitaxial GaAs, improved power dissipation (and, thus, device capability) is expected. A third potential advantage is the possibility of obtaining a functional coupling between film and substrate. For example, spinel ($MgAl_2O_4$) is a material exhibiting the lowest acoustic loss of any known material (Lewis and Patterson, 1968), and has a potential for microwave delay line applications. When spinel is used as the substrate for epitaxial GaAs, the intimate contact between the semiconductor and the substrate offers an opportunity for low-loss coupling between the microwave device and the acoustic circuit. One major area of system applications is in the development of transferred electron amplifier–delay line structures to replace the conventional traveling wave tube in amplifier–delay line loop systems.

Although in principle GaAs/oxide heteroepitaxial systems have characteristics desirable for applications in certain microwave devices, in practice the development is at a very early stage. Many technical problems in the areas of materials growth, device processing, and element integration need to be solved before major system applications can be anticipated. In the material aspect of heteroepitaxial GaAs, the principal requirements are that the epitaxial material in the active region be of good bulk perfection with high carrier mobility and well controlled doping and that the film be of high surface quality for device processing. The device structures employ either thick films ($>$ 20 μm), as required for Gunn oscillators, or thin films (submicron to several-micron range) grown on thick semiinsulating films, as needed for Schottky-barrier field effect transistors and transferred-electron devices. The inherent characteristics associated with heteroepitaxial films, including the interfacial p-region, high defect density near interfaces, and the preferential growth of $(111)a$ on many substrates, are not considered to be severe problems.

The first heteroepitaxial GaAs microwave Gunn effect devices were fabricated by Owens (1970) in GaAs grown on sapphire. The films were grown by the organometallic process. Presumably, the film and substrate were, respectively, (111) and (0001) oriented. The devices consist of undoped n-type films (25–34 μm) with Sn ohmic contacts. The best sample exhibited the following characteristics: film thickness, 34 μm; carrier concentration, $3 \times 10^{15}/cm^3$; carrier mobility, 5300 cm^2/V-s; oscillation frequency, 140 MHz; threshold field, 1900 V/cm. The oscillation in the better samples was reported to be comparable with that seen in bulk material with similar characteristics.

In addition to Gunn oscillators, Schottky-barrier field effect transistors (FET) have been fabricated in GaAs grown on BeO (Thorsen et al., 1974). The device structures, as shown in Figure 3.41, were fabricated on double layers of (100) GaAs grown by the organometallic process on either $(11\bar{2}2)$ or $(10\bar{1}0)$ BeO. The (100) films were used because they exhibit higher mobilities than the (111) films (Thorsen et al., 1974). The layer next to the BeO substrate was semiinsulating or of high sensitivity and about 15 μm thick. The top, active layer, about 0.25 μm thick, was n-type with a carrier concentration in the $10^{17}/cm^3$ range, in accordance with established design calculations (Statz and Munch, 1969) for operation at microwave frequencies. The source and drain contacts were made of evaporated Au–12 percent Ge thin films (1500 Å thick) and were rendered ohmic by sintering at 300–400°C in He. Schottky-barrier gate films were made by evaporation of 3000-Å-thick Al and were defined in the channel as shown in Figure 3.41. Pinch-off has been attained in the Schottky-barrier FET devices on a sample having an n-type layer 0.25 μm thick with a carrier concentration of $1.2 \times 10^{17}/cm^3$ and an electron mobility of 935 cm^2/V-s. The pinch-off voltage was near 2.5 V. The Schottky-barrier diode breakdowns for the devices ranged from -5 V to -7 V. A maximum transconductance of 5.6 millimhos was measured. For some devices, loops in the current–voltage curves were seen because of trapping effects. The results have demonstrated the possibility of fabricating Schottky-barrier FET devices in heteroepitaxial GaAs grown on BeO with properties approaching to those obtained in homoepitaxial material. However, further improvements in GaAs characteristics are required for practically useful devices.

A third kind of device, the transferred-electron oscillator (Perlman et al., 1971), was fabricated in heteroepitaxial GaAs grown on spinel (Narayan and Wang, 1974). The device structure is shown in Figure 3.42. The layer next to the (110) $MgAl_2O_4$ substrate was p-type of high resistivity, grown by the organometallic process and exhibiting (100) orientation. Subsequent growth of the n^+-n-n^+ structure was carried out by the chloride process with thicknesses (3–11–4 μm) and carrier concentrations ($10^{18}/cm^3$–2 × $10^{15}/cm^3$–$10^{18}/cm^3$), chosen to be typical of an X-band transferred-electron device. Ag-Sn ohmic contacts were deposited on the two n^+ layers using standard photolithographic techniques. The device structure was grown by the two-stage epitaxial process because the organometallic process offers the ease of nucleation in starting the growth, while the chloride process has been commonly used to grow the homoepitaxial devices with precise control of film thickness and doping. The (110) spinel was chosen because it yields the (100) GaAs which is an established conventional orientation for growth using the chloride process. The device structure was evaluated under DC bias, and the current voltage characteristics demonstrated the current saturation characteristic of the transferred-electron effect. When biased beyond the threshold voltage, vigorous bias circuit oscillations were observed. The results are preliminary in nature, yet they have demonstrated the feasibility of fabricating multilayered device structures using the best features of different growth processes.

3.6. Conclusions

The successful growth of epitaxial compound semiconductor films on oxide dielectric substrates provides a new class of composite material systems of technical interest for potential electronic device applications. The interest in these heteroepitaxial films arises from the prospects regarding substrate size and cost, dielectric isolation and improved thermal dissipation, possible film/substrate functional coupling, and subsystem integration.

Experimental results to date indicate that the epitaxial growth has been most successfully and reproducibly achieved by the organometallic growth process; other conventional growth methods may be advantageously used to complement the initial epitaxy achieved by this process. Epitaxial film growth characteristics are largely affected by several key parameters including the substrate orientation, substrate surface preparation, growth temperature, gas flow, and source material purity. These parameters may be adjusted for a specific material system to achieve the optimum film characteristics. Because of the complexity in the growth optimization, a large number of films have been prepared only in preliminary experiments, as cited in the preceding sections. These films do not exhibit optimum characteristics, and consequently, the in-depth characterizations of their properties remain to be accomplished. It is expected that more films grown under improved or optimum conditions will become available and that the films will be fully evaluated for potential device applications.

Because of the lattice and thermal expansion mismatches between the film and substrate, the heteroepitaxial films are deformed under generally compressive stress. Moreover, the films exhibit characteristic subgrain structures as revelaed by x-ray diffraction topography (see Section 3.4). These inherent defects are the main factors limiting consideration of the device uses of the films. The degree of the limitation depends mainly on the specific film/substrate combination and on the film growth conditions. Experimental results have shown that the electrical properties of the films are, as expected, inferior to the bulk material properties. For example, the highest electron mobility measured in GaAs films grown on sapphire is ~6000 cm^2/V-s, which is about 75 percent of the bulk value. This value represents a measure of the degree of limitation of heteroepitaxial GaAs films. GaAs of this quality is quite useful for a variety of device applications.

The development of practical devices using heteroepitaxial semiconductor films is still at a very early stage. Future progress depends largely on the advancement of the associated material technology. The trend of device development is expected to be in the areas of electroluminescence, photoemission, microwave integrated circuit, surface acoustic wave technology, and photovoltaic energy conversion.

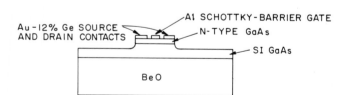

FIGURE 3.41
Schottky-barrier FET in GaAs on BeO. From Thorsen et al., (1974)

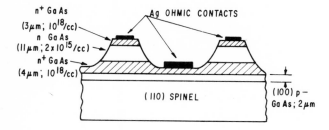

FIGURE 3.42
Transferred-electron device structure in GaAs on spinel

ACKNOWLEDGMENTS. The author is grateful to colleagues and associates at RCA Laboratories for assistance in preparing this manuscript. The critical review of, and suggestions relating to, this work by G. W. Cullen, S. F. Dierk, F. C. Dougherty, M. T. Duffy, I. Ladany, R. U. Martinelli, S. H. McFarlane, J. I. Pankove, D. Richman, and P. J. Zanzucchi are especially appreciated.

References

Allen, G. A., *J. Phys. D-4:* 308 (1971).

Allenson, M. B., P. G. R. King, M. C. Rowland, G. J. Steward, and C. H. A. Syms, *J. Phys. D-5:* 189 (1972).

Andrew, D., J. P. Gowers, J. A. Henderson, M. J. Plummer, B. J. Stocker, and A. A. Turnbull, *J. Phys. 3:* 320 (1970).

Ang, C. Y. and H. M. Manasevit, *Solid State Electronics 8:* 994 (1965).

Antypas, G. A., L. W. James, and J. U. Uebbing, *J. Appl. Phys. 41:* 2888 (1970).

Austerman, S. B., *J. Nucl. Mat. 14:* 225 (1964).

Barker, A. S., Jr., *Phys. Rev. 132:* 1474 (1963).

Bass, J. C., *Gallium Arsenide and Related Compounds* (K. Paulus, Ed.) London: The Institute of Physics (1971).

Behrudt, K. H., *Solid State Electronics 14:* 191 (1971).

Bell, R. L., *Negative Electron Affinity Devices,* Oxford: Clarendon Press (1973).

Bell, R. L. and W. E. Spicer, *Proc. IEEE 58:* 1788 (1970).

Bloom, S., *J. Phys. Chem. Solids 32:* 2027 (1971).

Bradshaw, A. and J. E. Knappett, *Solid State Tech. 13:* 45 (1970).

Budo, Y. and J. Priest, *Solid State Electronics 6:* 159 (1963).

Cahn, J. W. and R. E. Hanneman, *Surface Sci. 1:* 387 (1964).

Carr, P. H., *IEEE Trans. MTT-17:* 845 (1969).

Chu, T. L. and R. K. Smeltzer, *J. Vac. Sci. and Tech. 10:* 1 (1973).

Collins, J. H., P. J. Hagon, and G. R. Pulliam, *Ultrasonics 8:* 218 (1970).

Conard, R. W., R. A. Reynolds, and M. W. Jeffcoat, *Solid State Electronics 10:* 507 (1967).

Connolly, T. F., *Solid State Physics Literature Guides, Vol. 2: Semiconductors—Preparation, Crystal Growth, and Selected Properties,* New York: IFT/Plenum (1972).

Cullen, G. W., *J. Crystal Growth 9:* 107 (1971).

Denhartag, J. P., *Advanced Strength of Materials,* New York: McGraw Hill Book Co. (1952).

Didchenko, R., J. E. Alix, and R. H. Toeniskoetter, *J. Inorg. Nucl. Chem. 14:* 35 (1960).

Dismukes, J. P., W. M. Yim, and V. S. Ban, *J. Crystal Growth 13/14:* 365 (1972).

Duffy, M. T. and C. C. Wang, private communication, RCA Laboratories, Princeton, NJ (1974).

Duffy, M. T., C. C. Wang, G. D. O'Clock, Jr., S. H. McFarlane III, and P. J. Zanzucchi, *J. Electronic Materials 2:* 359 (1973).

Dugue, M., J. F. Goullin, P. Merenda, and M. Moulin, *Preparative Methods in Solid State Chemistry* (P. Hagenmuller, Ed.) New York: Academic Press (1972).

Dumin, D. J., . *Appl. Phys. 36:* 2700 (1965).

Feist, W. M., S. R. Steele, and D. W. Readey, *Physics of Thin Films,* Vol. 5 (G. Hass and R. E. Thun, Eds.) New York: Academic Press (1969).

Filby, J. D., *Modern Oxide Materials* (B. Cockayne and D. W. Jones, Eds.) New York: Academic Press (1972).

Fisher, D. G., R. E. Enstrom, J. S. Escher, H. F. Gossenberger, and J. R. Appert, *IEEE Trans. ED-21:* 641 (1974).

Francombe, M. H. and J. E. Johnson, *Physics of Thin Films,* Vol. 5 (G. Hass and R. E. Thun, Eds.) New York: Academic Press (1969).

Galli, G. and J. E. Coker, *Appl. Phys. Letters 16:* 439 (1970).

Gandolfo, D. A., *RCA Engineer 15:* 54 (1969).

Gutierrez, W. A. and H. D. Pommerrenig, Appl. Phys. Letters *22,* 292 (1973).

Gutierrez, W. A., H. D. Pommerrenig, and M. A. Jasper, *Solid State Electronics 13:* 1199 (1970).

Hughes, F. R., E. D. Savoye, and D. L. Thoman, *J. Elect. Materials 3:* 9 (1974).

Ilegems, M., *J. Crystal Growth 13/14:* 360 (1972).

Ito, S., T. Shinohara, and Y. Seki, *J. Electrochem. Soc. 120:* 1419 (1973).

Jain, W. K. and S. H. Sharma, *Solid State Electronics 13* 1145 (1970).

James, L. W., G. A. Antypass, J. Edgecumbe, R. L. Moon, and R. L. Bell, *J. Appl. Phys. 42:* 4976 (1971).

Jones, C. E. and A. R. Hilton, *J. Electrochem. Soc. 112:* 908 (1965).

Joyce, B. A., *The Use of Thin Films in Physical Investigation* (J. C. Anderson, Ed.) New York: Academic Press (1966).

Joyce, B. A., *Rep. Prog. Phys. 37:* 363 (1974).

Knight, J. R., D. Effer, and P. R. Evans, *Solid State Electronics 8:* 178 (1965).

Ladany, I., S. H. McFarlane III, and S. J. Bass, *J. Appl. Phys. 40:* 4984 (1969).

Ladany, I. and C. C. Wang, *J. Appl. Phys. 43:* 236 (1972).

Ladany, I. and C. C. Wang, *Solid State Electronics 17:* 573 (1974).

Lang, A. R., *J. Appl. Phys. 30:* 1748 (1959).

Lewis, M. F. and E. Patterson, *J. Appl. Phys. 39:* 3420 (1968).

Liu, Y. Z., J. L. Moll, and W. E. Spicer, *Appl. Phys. Letters 17:* 60 (1970).

Manasevit, H. M., *Appl. Phys. Letters 12:* 156 (1968).

Manasevit, H. M., *J. Electrochem. Soc. 118:* 647 (1971).

Manasevit, H. M., *J. Crystal Growth 13/14:* 306 (1972).

Manasevit, H. M., *J. Crystal Growth 22:* 125 (1974).

Manasevit, H. M. and W. I. Simpson, *J. Electrochem. Soc. 116:* 1725 (1969).

Manasevit, H. M. and W. I. Simpson, *J. Electrochem. Soc. 118:* 644 (1971A).

Manasevit, H. M. and W. I. Simpson, *J. Electrochem. Soc. 118:* 291c (1971B).

Manasevit, H. M. and A. C. Thorsen, *Met. Trans. 1:* 623 (1970).

Manasevit, H. M. and A. C. Thorsen, *J. Electrochem. Soc. 119: 99 (1972).*

Manasevit, H. M., R. L. Nolder, and L. A. Moudy, *Trans. Metall. Soc. AIME 242:* 465 (1968).

Manasevit, H. M., F. M. Erdmann, and W. I. Simpson, *J. Electrochem. Soc. 118:* 1864 (1971).

Manasevit, H. M., J. E. Coker, F. A. Pizzaello, and R. P. Ruth, Contract Report No. AFML-TR-73-295, Air Force Materials Laboratory, Wright-Patterson Air Force Base, Ohio (1974).

Martinelli, R. U. and D. G. Fisher, *Proc. IEEE 62:* 1339 (1974).

Maruska, H. P. and J. J. Tietjen, *Appl. Phys. Letter 15:* 327 (1969).

McFarlane, S. H., III, private communication, RCA Laboratories, Princeton, NJ (1974).

McFarlane, S. H., III and C. C. Wang, *J. Appl. Phys. 43:* 1724 (1972).

Mehal, E. W. and R. W. Walker, *IEEE Trans. ED-15:* 513 (1968).

Minden, H. T., *Solid State Tech. 12:* 25 (1969).

Narayan, S. Y. and C. C. Wang, unpublished, RCA Laboratories, Princeton, NJ (1972).

Nelson, H., *RCA Review 24:* 603 (1963).

Neuberger, M., *Handbook of Electronic Materials,* Vol. 2, New York: IFT/Plenum (1971).

Owens, J. M., *Proc. IEEE 59:* 930 (1971).

Page, D. J., *Solid State Electronics 11:* 87 (1968).

Pankove, J. I., *J. Luminescence 7:* 114 (1973).

Pankove, J. I. and M. A. Lampert, *Phys. Rev. Letters 33:* 361 (1974).

Pankove, J. I. and P. E. Norris, *RCA Review 33:* 377 (1972).

Pankove, J. I., J. E. Berkeyheiser, H. P. Maruska, and J. Wittke, *Solid State Commun. 8:* 1051 (1970).

Pankove, J. I., J. E. Berkeyheiser, and E. A. Miller, *J. Appl. Phys. 45:* 1280 (1974).

Pashley, D. W., *Prog. Surface Sci. 3:* 23 (1972).

Pastrnak, J. and L. Roskovcova, *Phys. Status Solidi 26:* 591 (1968).

Perlman, B. S., L. C. Upadhyayula, and W. W. Siekanowicz, *Proc. IEEE 59:* 1229 (1971).

Rai-Choudhury, P., *J. Electrochem. Soc. 116:* 1745 (1969).

Reisman, A., M. Berkenblit, J. Cuomo, and S. A. Chan, *J. Electrochem. Soc. 118:* 1653 (1971).

Schlotterer, H., Solid State Electronics 11:947 (1968)

Schwartz, B., *Vapor Deposition* (C. F. Powell, J. H. Oxley, and J. M. Blocher, Jr., Eds.) New York: John Wiley and Sons: (1966).

Scrupski, S. E., *Electronics 40:* 106 (1967).

Shaw, D. W., *Crystal Growth, Theory and Techniques,* Vol. 1 (C. H. L. Goodman, Ed.) New York: Plenum Press (1974).

Simon, R. E., *IEEE Spectrum 9:* 74 (1972).

Sobol, H., *Advances in Microwaves* (L. Young and H. Sobol, Eds.) New York: Academic Press (1974).

Sobol, H. and M. Caulton, *Advances in Microwaves* (L. Young and H. Sobol, New York: Academic Press (1974).

Sommer, A. H., *Photoemissive Materials,* New York: John Wiley and Sons (1968).

Sommer, A. H., *RCA Review 34:* 95 (1973).

Sommer, A. H., *J. Phys. (Paris) 34:* C6-51 (1974).

Spitzer, W. G. and J. M. Whelan, *Phys. Rev. 114:* 59 (1959).

Statz, H. and W. V. Munch, *Solid State Electronics 12:* 111 (1969).

Stern, E., *IEEE Trans. MTT-17:* 835 (1969).

Syms, C. H. A., *Adv. in Electronics and Electron Physics,* Vol. 28A (L. Morton, Ed.) New York: Academic Press (1969).

Thorsen, A. C. and H. M. Manasevit, *J. Appl. Phys. 42:* 2519 (1971).

Thorsen, A. C., H. M. Manasevit, and R. H. Harada, *Solid State Electronics 17:* 855 (1974).

Tietjen, J. J., R. E. Enstrom, V. S. Ban, and D. Richman, *Solid State Tech. 15:* 42 (1972).

Turnbull, W. R., *Semiconductor Thin Films,* AD 655 100, Washington, DC: U.S. Department of Commerce (1967).

van Laar, J., *Acta Electron. 16:* 215 (1973).

Wang, C. C., *J. Appl. Phys. 40:* 3433 (1969).

Wang, C. C., unpublished, RCA Laboratories, Princeton, N.J (1973).

Wang, C. C. (RCA Laboratories, Princeton, N.J.), unpublished, (1974).

Wang, C. C. and J. Blumentritt, unpublished RCA Laboratories, Princeton, NJ (1970).

Wang, C. C. and F. C. Dougherty, unpublished, RCA Laboratories, Princeton, NJ (1974).

Wang, C. C. and S. H. McFarlane III, *J. Crystal Growth 3/4:* 485 (1968).

Wang, C. C. and S. H. McFarlane III, *J. Crystal Growth 13/14:* 262 (1972).

Wang, C. C., G. E. Gottlieb, G. W. Cullen, S. H. McFarlane III, and K. H. Zaininger, *Trans. Met. Soc. AIME 245:* 441 (1969).

Wang, C. C., F. C. Dougherty, P. J. Zanzucchi, and S. H. McFarlane III, *J. Electrochem. Soc. 121:* 571 (1974A).

Wang, C. C., I. Ladany, S. H. McFarlane III, and F. C. Dougherty, *J. Crystal Growth 24/25:* 239 (1974B).

Watson, H. A., *Microwave Semiconductor Devices and Their Circuit Applications,* New York: McGraw-Hill (1969).

White, R. M., *Proc. IEEE 58:* 1238 (1970).

Wickenden, D. K., K. R. Faulkner, R. W. Brander, and B. J. Isherwood, *J. Crystal Growth 9:* 158 (1971).

Williams, B. F. and J. J. Tietjen, *Proc. IEEE 59:* 1489 (1971).

Williams, B. F., J. J. Tietjen, and C. C. Wang, private communication RCA Laboratories, Princeton, NJ (1974).

Yim, W. M. and E. J. Stofko, *J. Electrochem. Soc. 119:* 381 (1972).

Yim, W. M. and E. J. Stofko, *J. Electrochem. Soc. 121:* 965 (1974).

Yim, W. M., E. J. Stofko, P. J. Zanzucchi, J. I. Pankove, M. Ettenberg, and S. L. Gilbert, *J. Appl. Phys. 44:* 292 (1973).

Young, L. and H. Sobol, *Advances in Microwaves,* New York: Academic Press (1974).

Zanzucchi, P. J. and C. C. Wang, personal communication, RCA Laboratories, Princeton, NJ (1974).

Chapter 4

The Preparation and Properties of Heteroepitaxial III-V and II-VI Compounds for Surface Acoustic Wave and Electrooptic Devices

M. T. Duffy

4.1. Introduction

Recent advances in the heteroepitaxial growth of various electronic materials on insulating substrates have enhanced the possibility of attaining a wide range of signal-processing operations in monolithic structures. The ability to combine compatible materials with different functional properties on a common substrate also offers the prospect of developing low-cost, miniaturized, and high-speed communications systems. The rapid development of microsonics and integrated optics in recent years attests to the anticipated importance of these technologies. Both activities increasingly stress the importance of layered structures involving piezoelectric and electrooptic films on low-loss substrates for the generation, propagation, manipulation, and detection of surface acoustic waves and coherent light signals. In addition, the planar configuration makes it possible to incorporate passive components and interface with the highly developed and versatile MOS technology.

The information density carried by a transmission medium is related to the signal wavelength; the shorter the wavelength the greater this density. Ultrasonic and laser wavelengths may be several orders of magnitude shorter than the shortest electromagnetic microwaves, thus providing engineering incentive for use in communications and signal processing. Apart from the potential advantages of a higher information density, the thickness of the propagation medium can be as thin as one wavelength, or on the order of micrometers. In certain cases, such as ridge waveguides, the transverse dimensions can also be as little as a few wavelengths. This factor permits a substantial reduc-

tion in component size and makes possible the fabrication of both passive and active components in thin-film materials heteroepitaxially grown or deposited on suitable substrates. Likewise, the planar structure is compatible with MOS photolithographic processing techniques involving device fabrication and interconnections, thus further promoting the development of monolithic structures.

Interest in the use of microsonic components is related to the relatively low propagation velocity of sound waves, which travel about five orders of magnitude more slowly than electromagnetic waves. Thus, it is possible to obtain a few microseconds delay time in a 1-cm acoustic path length in a crystal, whereas a 1-km transmission line is required to accomplish a comparable delay in the case of electromagnetic waves. The periodic nature of sound waves permits the use of gratings and transducer pattern arrays of various configurations for the generation, manipulation, and detection of surface acoustic waves on piezoelectric materials. Such device functions as filtering, pulse compression, convolution and correlation, and dispersive and nondispersive delay lines are frequently cited in the literature. [For a review of applications and bibliography see, for example, White (1970), Carr (1969), Stern (1969), Holland and Claihorne (1974), Maines (1974).] Although electromagnetic microwave circuits can perform essentially all the functions of microsound components, the benefits of reduced size, loss, and cost of the latter offer a valuable complement to current microwave technology. The ability to operate acoustic wave delay lines and signal-processing devices at microwave frequencies without the necessity for frequency down-conversion and up-conversion—which result in increased insertion loss, greater complexity, and loss of phase information—provides an additional incentive. There is, however, a requirement for higher operating frequencies in surface wave microsonic devices than are usually reported. Transducer pattern linewidths and spacings are limited by the resolution of optical photolithographic fabrication techniques. The frequency range may be extended by employing higher acoustic velocity materials and by utilizing the higher resolution of x-rays and electron beams to define linewidths less than 1 μm in photoresist processing for generating transducer and other patterns.

The prospect of developing optical communications systems in which a large volume of information could be carried on a single light beam has fascinated engineers ever since the development of the laser. Proponents of integrated optics envision an optical analog of IC circuitry, producing similar basic functions optically. Recent advances in the development of low-loss optical fibers have provided an important transmission medium for the propagation of coherent light. Other components needed for a thin-film optical system are waveguides to link the optical circuit to the transmission medium, deflectors and modulators to impress the signal to be transmitted (such as a telephone conversation) onto the optical carrier, and detectors to convert the light signal to electronic output. Couplers may be needed to efficiently transfer energy between various parts of the circuit. Several materials have been studied in the search for active and passive components in both hybrid and monolithic approaches to integrated optics. In the hybrid approach the goal would be to combine the best components available, preferably in compatible processes through materials integration on a common substrate. The monolithic approach would endeavor to generate the various components: laser, modulator, waveguide, and detector in a single direct band gap semiconductor such as GaAs/Ga$_x$Al$_{1-x}$As system. [For a treatment of integrated optics and related topics, see, for example, Tamir (1975), Kaminow (1974), Barneski (1976), Balkanski and Lallemand (1973).]

This chapter reviews the preparation and properties of piezoelectric and electrooptic materials that may be grown heteroepitaxially on sapphire with a view to both surface wave and electrooptics applications. The discussion is confined to those crystallographic orientations compatible with both applications and with the state of the art in SOS technology in anticipation that silicon programmable functions might be combined on the same substrate. These restrictions confine the discussion to those few materials which received greatest attention in recent years. Among the III-V compounds heteroepitaxial AlN films have been most favored, while ZnO films have received greatest attention among the II-VI compounds.

4.2. Heteroepitaxial growth of piezoelectric and electrooptic materials on sapphire

Substrate requirements

It is generally accepted that the quality of the substrate surface finish is of paramount importance to the nucleation and growth of single crystal films. Work-damaged surfaces can give rise to poorly oriented or polycrystalline films which result in diminished piezoelectric and electrooptic activity and optical waveguide propagation losses too high to measure. It is interesting to note that, even in those cases in which conventional analytical methods such as x-ray analysis and electron diffraction indicate that the films are single crystalline in character, inferior waveguide properties may still be obtained. This indicates that the piezoelectric and electrooptic properties of heteroepitaxial films may be much more sensitive measures of crystalline perfection than the properties derived from more conventional analyses. Thus, the necessity for substrate surface perfection is stressed here as a first requirement for the heteroepitaxy of good waveguides. Surface irregularities and defects may become scattering sites in addition to affecting the crystal growth habit of the epitaxial layer. Acoustic and optical waveguides may be traversed throughout the depth and much of the area of the film by the signal. Consequently, both bulk and interfacial properties are important in determining propagation losses. The surface perfection of sapphire substrates can be determined prior to annealing by infrared specular reflectance measurements as discussed in Chapter 5.

The substrate orientation must also be specified, since this determines the direction of the crystallographic axes of the films and hence the desired direction of propagation for acoustic or light waves. Most attention has been given to (01$\bar{1}$2)-oriented sapphire as the substrate material for the growth of piezoelectric and electrooptic films. The purpose is twofold. This is the orientation of greatest interest to SOS technology. It allows for the integration of silicon programmable functions on the same substrate. In general, electroacoustic and electrooptic properties are dependent upon the plane and direction of propagation because of the asymmetric structure of piezoelectric and electrooptic films. It is necessary to arrange for propagation along predetermined crystallographic directions in order to obtain pure-mode propagation and to optimize the properties of the film/substrate composites for the intended appli cation. Several of the heteroepitaxial materials of interest are hexagonal in crystalline structure and, fortuitously, grow on the above orientation of sapphire with the crystallographic c axis in the plane of the film. The direction of the c axis possesses important piezoelectric and electrooptic coefficients, thus providing the possibility of fabricating surface acoustic wave devices or electrooptic and acoustooptic devices in such films.

Preparation and acoustic properties of heteroepitaxial AlN and GaN films

Growth process

Various methods have been reported for the preparation of aluminum nitride or gallium nitride including reactive sputtering (Noreika et al., 1969), plasma discharge (Pastrnak and Souckova, 1963; Cox et al., 1967; Kosicki and Kahng, 1969), sublimation (Taylor and Lenie, 1960; Cox et al., 1967), and chemical-vapor deposition (CVD). The CVD technique has been the most widely used method for the heteroepitaxial growth of these nitrides on various substrates. Aluminum nitride films have been grown with single crystal character on Si, α-Al$_2$O$_3$, MgAl$_2$O$_4$, and α-SiC (Cox et al., 1967; Chu et al., 1967; Noreika and Ing, 1968; Manasevit et al., 1971; Yim et al., 1973; Duffy et al., 1973; Lakin et al., 1974). Likewise, heteroepitaxial gallium nitride films have been grown on substrates such as α-Al$_2$O$_3$, GaAs, and α-SiC (Kosicki and Kahng, 1969; Maruska and Tietjen, 1969; Wickenden et al., 1971; Manasevit et al., 1971; Duffy et al., 1973). In most of these studies the deposition process involved reaction at elevated temperatures between either the metal chlorides (AlCl$_3$ or GaCl) and ammonia (NH$_3$) [see Ban (1972)], or between the metal–organic trimethyl compounds [(CH$_3$)$_3$Al or (CH$_3$)$_3$Ga] and ammonia. It has been reported by Yim and coworkers (1973) and later by Callaghan and coworkers (1974) that several orientations of AlN with varying degrees of crystalline perfection may be produced simultaneously on (01$\bar{1}$2)-oriented sapphire substrates by the chloride process. In contrast, AlN films grown by the metal–organic process are (11$\bar{2}$0)-oriented (Manasevit et al., 1971; Duffy et al., 1973; Lakin et al., 1974; Pizzarello and Coker, 1975; Liu et al., 1975). This has been the orientation of greatest interest to surface acoustic wave applications (Collins et al., 1970) and, thus, most acoustic measurements have been performed on films prepared by the latter pro-

cess. The preparation of (11$\bar{2}$0)-oriented GaN layers, on the other hand, is feasible by either growth method. Relatively little is known about the piezoelectric properties of these films. This may be caused by the low resistivity of the as-grown films and the difficulty in compensating them electrically to provide high-resistivity samples for piezoelectric measurements. The electrical properties of some typical GaN samples prepared by the metal–organic process are presented in Table 4.1. In some instances diethyl zinc [(C$_2$H$_5$)$_2$Zn] vapors were mixed with the reactants to compensate the GaN films with Zn. It can be seen from this table that the carrier concentration of uncompensated films is rather high ($\geqslant 10^{19}$/cm^3) and that high-resistivity films resulting from compensation with zinc are obtained only at the lower growth temperatures irrespective of the amount of diethyl zinc added during the growth process. By contrast, GaN films prepared by the chloride process may have carrier concentrations lower by two orders of magnitude than those noted above, with a corresponding increase in carrier mobility (Ilegems and Montgomery, 1973), and may be compensated with zinc to provide suitably high-resistivity films throughout the growth temperature range (Pankove et al., 1971A,B, 1973, 1974, 1975; Jacob et al., 1976). From these observations it appears that the chloride process is currently the preferred method for the preparation of high-resistivity epitaxial GaN films on sapphire. For a more detailed discussion on the preparation of group III nitrides, refer to Chapter 3 and the above references.

Single crystalline films of these materials grown on (01$\bar{1}$2)-oriented Al$_2$O$_3$ by the above preferred methods exhibit the following parallel and directional relationships, as shown for the case of AlN:

$$(11\bar{2}0)\text{AlN}\|(01\bar{1}2)\text{Al}_2\text{O}_3 \qquad [1\bar{1}00]\text{AlN}\|[21\bar{1}0]\text{Al}_2\text{O}_3.$$

A schematic representation of this configuration is depicted in Figure 4.1. The crystallographic c axis of the film lies in the plane of the film and coincides with the projection of the substrate c axis on the (01$\bar{1}$2) plane. One a axis of the film structure is normal to its surface. For surface acoustic wave applications, interdigitated transducer patterns are aligned on the film

TABLE 4.1   Electrical Properties of GaN Heteroepitaxial Layers [After Duffy et al. (1973)]

Composite	Growth Temperature (°C)	Film Thickness (μm)	Carrier Concentration (cm^{-3})	Resistivity, n-type (Ω-cm)	Hall Mobility (cm^2/V-s)
(11$\bar{2}$0)GaN/(01$\bar{1}$2)Al$_2$O$_3$	1050	2.8	1×10^{19}	0.017	55
(11$\bar{2}$0)GaN/(01$\bar{1}$2)Al$_2$O$_3$ (Zn-doped)	1050	2.2	4×10^{17}	0.306	51
(11$\bar{2}$0)GaN/(01$\bar{1}$2)Al$_2$O$_3$	850	4.8	3×10^{19}	0.003	75
(11$\bar{2}$0)GaN/01$\bar{1}$2)Al$_2$O$_3$ (Zn-doped)	850	5.0	—	Insulating	—

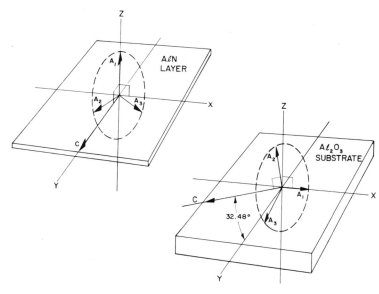

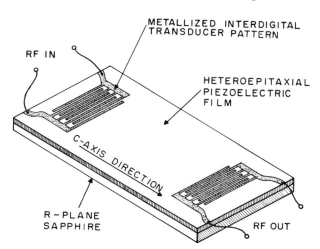

FIGURE 4.1
Diagram showing heteroepitaxial relationship of $(11\bar{2}0)\text{AlN}//(01\bar{1}2)\text{Al}_2\text{O}_3$ structure

surface so that acoustic wave propagation occurs parallel to the direction of the c axis, as illustrated in Figure 4.2. Before discussing acoustic properties, however, it is necessary to examine such film properties as crystallinity and surface topography which may influence the generation and propagation of surface waves. Not only are smooth surfaces required for photolithographic definition of transducer patterns, but surface roughness can contribute strongly to propagation loss (Maradudin and Mills, 1976). Film crystallinity as related to grain structure, misorientation of grains relative to the nominal orientation of the film, residual strain, and departure from parallel heteroepitaxy may also impair sample performance.

Crystallinity and surface topography. A scanning electron micrograph of an AlN film with the above heteroepitaxial configuration is shown in Figure 4.3

together with that of an AlN film grown on (111)-oriented silicon. In the latter case, the crystallographic c axis of the film is normal to the film surface; the heteroepitaxial parallel and directional relationships being

$$(0001)\text{AlN}\|(111)\text{Si} \qquad [\bar{1}2\bar{1}0]\text{AlN}\|[\bar{1}10]\text{Si}.$$

It appears from the micrographs and the heteroepitaxial relationships given above that the surface structure is comprised of crystallites aligned in parallel fashion with the principal axis parallel to the direction of the crystallographic c axis in the respective films. The micrographs also suggest a mosaic structure to the films. McFarlane and Wang (1972) have shown by Lang transmission x-ray topography that other heteroepitaxial III-V compounds, GaAs/spinel and GaP/sapphire, have a structure in which the grains are misoriented $\pm 0.1°$ from the nominal orientation of the

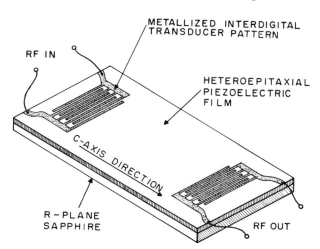

FIGURE 4.2
Schematic diagram of surface acoustic wave delay line or filter

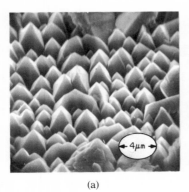
(a)

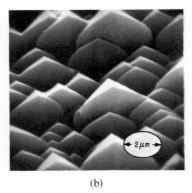
(b)

FIGURE 4.3
Scanning electron micrographs of AlN films (~10 μm) on (a) (111) silicon substrate; AlN c axis is perpendicular to plane on film; (b) $(01\bar{1}2)$ Al$_2$O$_3$ substrate; AlN c axis is in plane of film. From Duffy et al. (1973)

layer. A similar misorientation of grains relative to the nominal orientation of the film may be expected also for AlN films. The relatively low linear x-ray absorption coefficient, μ, of this material and the difficulty in growing films of suitable thickness, t, make similar measurements difficult in this case ($\mu t << 1$). Noreika and Ing (1968) have shown, by transmission electron microscopy, that AlN films on silicon are composed of mosaic-like arrays of small crystallites. The crystallite size depends strongly on the growth temperature as indicated by their plot of crystallite surface density vs. reciprocal growth temperature (Figure 4.4). Their measurements were made on films between 500 and 1000 Å thick, as estimated from growth rate data. They determined that the average linear dimensions of the crystallites formed at 900°C were about 10 times greater than those formed at 1350°C (i.e., 1600 and 150 Å, respectively). These authors point out that, unlike the growth of metal films, processes involving agglomeration and recrystallization of crystallites are not effective during the early growth stages of AlN films. They attribute this to the refractory nature of AlN and cite similar observations in earlier work on the growth of AlN on SiC (Chu et al., 1967). The similarity in the surface structure of films grown on silicon and on sapphire, as shown in Figure 4.3, suggests that the same consideration may hold for AlN films grown on sapphire substrates. These observations indicate that it is difficult to grow AlN films with a high degree of crystalline perfection and horizontal continuity. Since smooth optical-quality surfaces are required for the fabrication of transducer patterns and waveguide studies, it is necessary to polish films that are thicker than a few micrometers to remove surface faceting, especially for high-frequency operation. Several polishing agents may be used, but silica sol seems to give best results. Polished films may have a grain-like surface texture, depending on the initial thickness. An example is shown in Figure 4.5. When a film of this texture is stained with an insoluble dye and polished further, the dye remains until much of the original film thickness is removed, indicating considerable initial surface

porosity. The scanning electron micrograph of Figure 4.6 illustrates this point. A $(11\bar{2}0)$AlN/$(01\bar{1}2)$Al$_2$O$_3$ sample was broken along a direction parallel to the c axis of the film after scribing the back surface of the sapphire, and SEM micrographs were obtained of the broken edge. Figure 4.6 shows a micrograph taken at approximately 90° to the surface normal and viewed in a direction perpendicular to the c axis of the film. Two strata are apparent: a coherent layer of AlN about 5 μm thick, and a top porous structure of about equal thickness. The film was characterized as single crys-

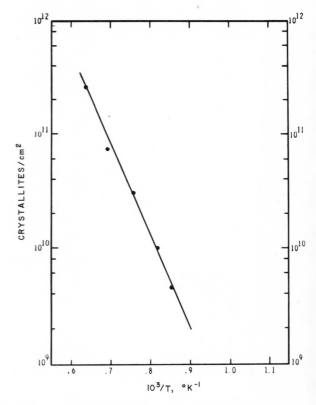

FIGURE 4.4   Variation of AlN crystallite size with deposition temperature on silicon. [Noreika and Ing (1968)]

talline by x-ray analysis. The picture was taken in a region where a portion of the top porous layer had broken away from the coherent layer. The overgrowth shows structural features running parallel with the direction of the c axis of the film. The surface topography of such a film is depicted in the SEM micrograph of Figure 4.7 showing overall surface texture. The morphology of the crystallites has been described by Lakin and coworkers (1974). The exposed facets correspond to the $(01\bar{1}2)$ faces of AlN and, as indicated previously, the c axis of each crystallite is aligned with the sapphire c-axis projection on the substrate surface. The morphology shown above is characteristic of films grown under "favorable" conditions. Under nonideal conditions, the surface structure may consist of misoriented material ranging from polycrystalline to preferentially oriented crystallites. An example is shown in Figure 4.8, which shows the general surface texture at high magnification, of the different crystallites for this particular example. One feature of concern relating to the morphology of piezoelectric or electrooptic films of this type is c-axis inversion, which cannot be detected by x-ray analysis or electron diffraction but diminishes the piezoelectric or electrooptic activity of the film. Evidence of c-axis inversion is apparent in the SEM micrograph shown in Figure 4.8. Pizzarello and Coker (1975) reported variation by a factor of three in the value of k^2 over a given film surface. They were unable to correlate this variation with the results of x-ray and electron diffraction analyses but were able to establish a correlation with topographs, obtained by replication electron microscopy, of the surface regions where transducer patterns had been arranged. They proposed a random variation defect structure that does not largely alter the single crystalline character of the films. Among such crystallographic imperfections as random inclusions of polycrystalline particles and grain boundaries, they postulated c-axis inversion as a likely cause of the degradation of k^2. Liu and coworkers (1975) have dealt more explicitly with this problem, showing evidence of c-axis inversion in SEM micrographs and demonstrated its influence on the piezoelectric properties of AlN films. They also indicated that c-axis inversion may or may not occur under apparently similar growth conditions.

It has also been demonstrated by x-ray analysis (Duffy and Smith, 1974) that AlN films may have a polycrystalline component occurring simultaneously with a single crystal component. Figure 4.9 is an x-ray diffractometer trace ($\theta{:}2\theta$ mode) showing the presence of a polycrystalline component in the angular range of 30° to 40° and a strong peak corresponding to $(11\bar{2}0)$-oriented material at $2\theta = 59.3°$ (Cu, K_{α_1}). When such a film is polished and reexamined, the intensity of diffracted radiation, corresponding to the polycrystalline

component, diminishes more rapidly than the intensity of the single crystalline component. It was found that films in this category invariably displayed much reduced piezoelectric activity or no activity at all. In contrast, films characterized by the absence of a polycrystalline component invariably displayed good piezoelectric activity. Accordingly, film examination by x-ray diffractometry for the presence of a polycrystalline component in the angular (2θ) region of 30° to 40° and at high sensitivity can also be an effective means of screening samples. A relatively strong peak corresponding to single crystal material is usually obtained even for completely inactive films. The polycrystalline component is more often present in films deposited at high growth rates and low growth temperatures. For growth rates of ≤ 3 μm/hr in the temperature range 1200 to 1300°C and with proper substrate preparation, the presence of a polycrystalline component is infrequent. At lower growth temperatures the growth rate must be slowed considerably to maintain good crystallinity. It was also observed in these measurements (pole-figure analysis) that an exact heteroepitaxial relationship, as depicted in Figure 4.1, did not exist between film and substrate in several samples examined. The extent of this deviation depends upon the initial misorientation of the sapphire surface and also on other factors which are not understood at this time. For example, a sapphire substrate that had been cut and polished with horizontal and vertical components of misorientation of $|\epsilon_H| \leq 0.01°$ and $|\epsilon_V| \leq 0.01°$ resulted in the growth of an AlN film which had a deviation from parallel epitaxy of 0.5° in the surface plane between the film c axis and the projection of the sapphire c axis, while the deviation in the vertical plane was not more than 0.01°. Other samples with larger components of substrate misorientation deviated from parallel heteroepitaxy in an unpredictable manner. Among the commercially available sapphire substrats used, orientation tolerances varied over a wide range: from $\pm 0.5°$ to $\pm 4°$ off-orientation. An x-ray analysis of the heteroepitaxial relationship between film and substrate showed a correspondingly wide deviation from parallel epitaxy. In general, the alignment of the c axis of the AlN film does not conform to that shown in Figure 4.1 but can diverge by two degrees or more from exact parallelism between film c axis and the projection of the sapphire c axis on the substrate surface. The problem is further complicated by the divergence having both normal and horizontal components with respect to the plane of the film. The safest procedure to follow in seeking parallel epitaxy is to start with substrates that are cut and polished to a tolerance of about $\pm 0.5°$.

Most investigators involved in the heteroepitaxial growth of AlN have witnessed film cracking caused by residual strain in the composite structures. Yim and

(a)

FIGURE 4.7   SEM micrograph of surface texture of AlN film. This sample is rotated 90° counterclockwise relative to the viewing direction in Figure 4.3b

(b)

FIGURE 4.5   Surface topography of a polished AlN sample at different thicknesses. (a) 5-μm-thick film after brief polishing step. (b) Same film after polishing to 3 μm in thickness

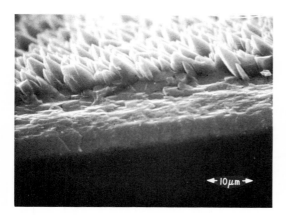

FIGURE 4.6   SEM micrograph of the edge of a broken AlN/Al$_2$O$_3$ sample showing porous surface structure

FIGURE 4.8   SEM micrograph of surface texture of AlN film showing spurious growth and evidence of c-axis inversion (crystallites to left and right in bottom micrograph)

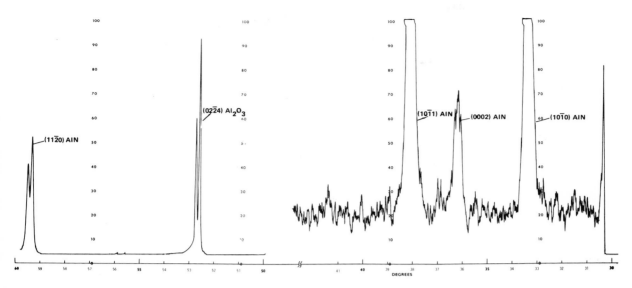

FIGURE 4.9   X-ray diffractometer trace (θ:2θ mode) of AlN/Al$_2$O$_3$ composite showing polycrystalline component

coworkers (1973, 1974) have compared the lattice constants and thermal expansion coefficients of AlN and
sapphire and attributed cracking to the large mismatch
in lattice constants and expansion coefficients
between the two materials. They indicated, however,
that cracking can be eliminated by slow cooling from
the growth temperature. The temperature variation of
lattic parameters and thermal expansion coefficients
of aluminum nitride, sapphire, and silicon, as reported
by Yim and coworkers, is presented in Table 4.2. The
curvature arising from residual strain in AlN/Al$_2$O$_3$
composites with different film thickness was discussed
by Duffy and coworkers (1973). Examples are given in
Figure 3.18. In this case the curvature increases quite
rapidly with film thickness. It has been demonstrated
by Liu and coworkers (1975) that the presence of
silicon as an impurity can give rise to polycrystalline
growth, spurious nucleation, and increased mechanical instability in the form of film cracking and peeling.
In their experiments, they introduced silicon as an
impurity from a SiH$_4$ source but indicated that silicon
is a major impurity in the growth apparatus, originating from exposed quartz surfaces, from the inductively heated silicon carbide–coated susceptor, and as the
major impurity in the trimethyl aluminum source
material.

Evidence of interfacial strain has also been obtained
from infrared reflectance measurements (see Chapter
5) as described by Yim and coworkers (1973) for
(0001)AlN/(0001)Al$_2$O$_3$ heteroepitaxial composites. A
similar evaluation was also performed by Duffy and
Zanzucchi (1973) on (11$\bar{2}$0)AlN/(01$\bar{1}$2)Al$_2$O$_3$ composites prepared at different growth temperatures. Spec-
tra corresponding to three growth temperatures
(1000°C, 1125°C, and 1260°C) are shown in Figure

4.10. The spectra were obtained in specular unpolarized infrared-reflectance measurements on both the
polished back surface of the substrate and the interfacial surface of the sapphire on which the AlN film (film
thickness, ~1.5 μm) was grown. As discussed in
Chapter 5, reflectance from the interfacial surface
occurs through the transparent AlN film. Residual
interfacial strain results in band splitting because of
perturbation in the normal vibrational modes of sapphire. Qualitatively, the interface spectra show band
splitting and distortion associated with sapphire lattice
strain corresponding to all three growth temperatures.
As the growth temperature is increased, there is a
noticeable increase in the observed distortion in the
wavenumber interval 600 to 900 cm^{-1}. This has been
tentatively ascribed to a combination of increased
strain and chemical reaction between film and substrate at the higher growth temperatures. It should be
noted that the spectra obtained from the back surfaces
of these substrates do not represent ideal surfaces
since a high-quality surface finish was not applied to
the back surfaces. If a high-quality finish had been
applied, the difference in the spectra would be even
greater. (An instrumental slit change occurs at 600
cm^{-1} in each case giving rise to a vertical shift in the
spectra at this wavenumber.) Absorption edge measurements were made on the same set of samples to
check for other differences in optical properties. Figure 4.11a shows curves of absorption coefficient
(obtained at room temperature) vs. photon energy for
AlN grown at the various temperatures. Similar data
have been published previously for (0001)-oriented
AlN films by Yim and coworkers (1973) and are
included here for completeness (Figure 4.12). An
interesting feature of each curve is the absorption

TABLE 4.2   Temperature Variation of Lattice Parameters and Mean Expansion Coefficients of AlN, Sapphire, and Silicon

Material	T (°C)	Measured lattice parameter (Å) Hexagonal a^a	Hexagonal c^b	Cubic a^c	Calculated Temperature Dependence of Lattice Parameters	rms Deviation ($\times 10^{-4}$ Å)	Mean Expansion Coefficient ($\times 10^{-6}$/°C) 20–800°C α_1	α_{11}	α
AlN	18	3.1115	4.9798		$a_T = 3.1113 + 1.3130 \times 10^{-5}T + 4.147 \times 10^{-9}T^2$	0.36			
	154	3.1134	4.9811				5.27	4.15	
	338	3.1162	4.9855		$c_T = 4.9793 + 1.4789 \times 10^{-5}T + 7.255 \times 10^{-9}T^2$	4.5			
	494	3.1187	4.9887						
	666	3.1219	4.9917						
	800	3.1244	4.9960						
α-Al$_2$O$_3$	25	4.7576	12.9834		$a_T = 4.7569 + 3.3726 \times 10^{-5}T + 1.037 \times 10^{-9}T^2$	1.8			
	156	4.7623	13.0000				7.28	8.11	
	350	4.7689	13.0205		$c_T = 12.9815 + 11.2939 \times 10^{-5}T - 9.321 \times 10^{-9}T^2$	13			
	496	4.7739	13.0356				7.71^d	8.56^d	
	650	4.7789	13.0485						
	803	4.7848	13.0675						
							8.17^e	9.06^e	
Si	25			5.4309	$a_T = 5.4304 + 1.8138 \times 10^{-5}T + 1.542 \times 10^{-9}T^2$	0.53			
	157			5.4332					
	356			5.4370					3.57
	496			5.4398					3.90^f
	651			5.4429					
	813			5.4461					3.59^g

Source: Yim and Paff (1974).

[a]Values shown here were calculated by using (205)–(220) lines for AlN and (054)–(330) lines for α-Al$_2$O$_3$.

[b]Values shown here were calculated by using (302)–(205) lines for AlN and (1,1,15)–(4,0,10) lines for α-Al$_2$O$_3$.

[c]Calculated by using $K_{\alpha1}$ and $K_{\alpha2}$ lines of (620), (533), and (444).

[d]Calculated from data of Wachtman et al. (1962).

[e]For the temperature range of 28–806°C; as reported in Campbell and Grain (1961).

[f]Calculated from data of Hall (1961).

[g]For the temperature range 25–800°C.

"tail," which is more pronounced at the higher growth temperatures. Similar measurements made at liquid-nitrogen temperature showed no appreciable change in the shape of the absorption tail (see Figure 4.11b), indicating no appreciable response from impurities. Though the nature of the absorption tail is not understood, it is interesting to note that infrared reflectance measurements made at the AlN/Al$_2$O$_3$ interface region show parallel trends dependent on deposition temperature. The above absorption behavior may be related to the existence of an interface compound. The results of Yim and coworkers show a similar variation in the absorption edge tail with film thickness; the greater the film thickness, the more pronounced this tail. This again may be related to an interface reaction because of the longer growth times required for thicker films. Evidence of chemical reaction is also apparent microscopically when AlN is etched from Al$_2$O$_3$ substrates and is particularly noticeable if the growth of heteroepitaxial silicon is attempted on the resulting surfaces. The residual strain present in AlN films may, thus, depend on the composition of the AlN/Al$_2$O$_3$ interface in addition to such other factors as lattice constant

mismatch, thermal expansion coefficient mismatch, impurity content, and rate of cooling. It has been suggested by Liu and coworkers (1975) that changes in the flow kinetics of the reaction gases during growth may also cause subtle changes in film strain. They also estimate that the depth to which strain in AlN films affects the acoustic properties is about 1 μm.

Another factor which can degrade the piezoelectric properties of AlN films is the defect density present in the form of pits and inclusions. Manasevit and coworkers (1971) have discussed the formation of AlN from (CH$_3$)$_3$Al and NH$_3$. These reagents react at room temperature to form the solid addition compound (CH$_3$)$_3$Al:NH$_3$. Dealkylation occurs in a stepwise manner with a mole of CH$_4$ being released at about 60°C, a second mole at about 160°C, and the final mole at some higher temperature. Vapor-phase reaction resulting in powder formation and incorporation of the latter in the film is possible unless the reactants are prevented from mixing until within the vicinity of the heated substrate. Incompletely reacted material from the vapor phase may accumulate on adjacent surfaces and become dislodged and incorporated in the film

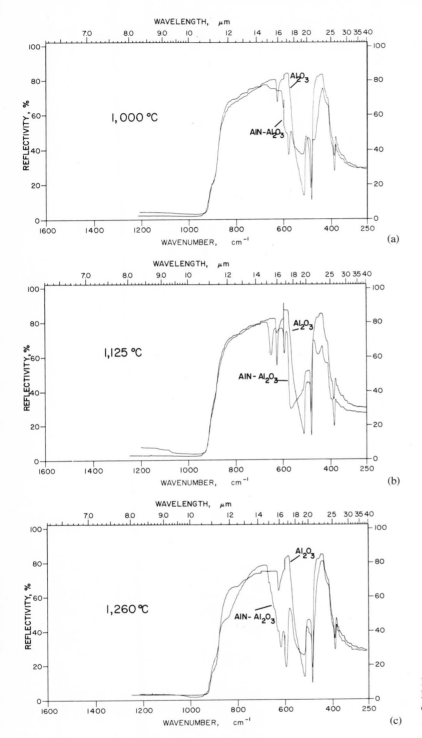

FIGURE 4.10
Reflectivity as a function of wavenumber on AlN/Al$_2$O$_3$ composites prepared at (a) 1000°C, (b) 1125°C, and (c) 1260°C

during growth. The problem is particularly serious at fast growth rates when the concentration of reactants is high.

Similar considerations apply to heteroepitaxial GaN films on sapphire. An SEM micrograph of a film sur-

face prepared by the chloride process is shown in Figure 4.13. It is evident from the micrograph that these films can be prepared with a greater degree of horizontal continuity than AlN films. Surface faceting can be removed by polishing with silica sol in a similar

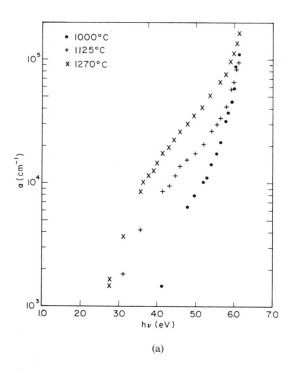

(a)

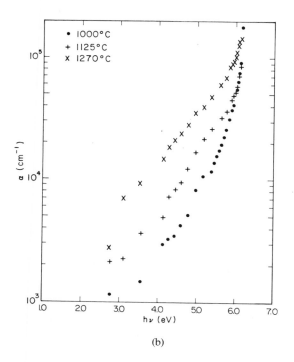

(b)

FIGURE 4.11   Plot of absorption coefficient vs. photon energy for AlN films prepared at different growth temperatures. (a) Measurements made at room temperature; (b) measurements made at liquid-nitrogen temperature

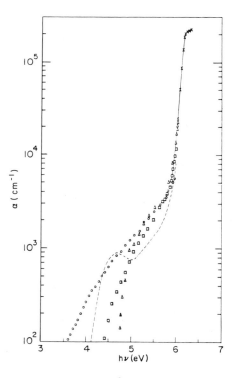

FIGURE 4.12   Plot of absorption coefficient vs. photon energy for different film thicknesses. After Yim et al. (1973), dashed curve; data from J. Pastrnak and L. Roskovcova (1968).

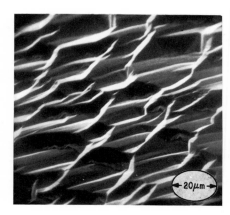

FIGURE 4.13   Scanning electron micrograph of (11$\bar{2}$0)GaN on (01$\bar{1}$2)Al$_2$O$_3$ substrate

manner to AlN and ZnO films, though the removal rate is much slower, probably because of the chemical inertness of the material. A granular structure and residual strain have also been recognized in GaN films and used to explain certain optical, luminescence, and electroluminescence properties of the films (Pankove et al., 1971A; Matsumoto and Aoki, 1974). Although a considerable amount of information is available on the optical and luminescence properties of GaN, relatively little is known about the piezoelectric properties of these films. Gallium nitride is not as refractory as AlN (Manasevit et al., 1971; Chu, 1971) and, therefore, is less compatible with semiconductor interfacing (Cullen et al., 1973). However, GaN has the hexagonal Wurtzite structure characteristic of AlN and ZnO films and, thus, of interest for the same applications. As indicated previously, GaN films are semiconducting unless compensated during the growth process. Preliminary results on the piezoelectric properties of compensated GaN films (O'Clock and Duffy, 1973) are disappointing because of the weak activity of the films. These films, prepared by the metal–organic process at the low growth temperatures at which compensation with Zn was possible, were highly strained and frequently displayed microcracks. More recent studies on films prepared by the chloride process (Onyshkevych and Duffy, 1975) indicate that more desirable properties are possible as a result of this process. Further research is necessary to establish the usefulness of GaN films in microsonic applications.

Piezoelectric properties. It is not surprising, based on the preceding discussion, that there can be considerable scatter in the results of piezoelectric measurements on these materials. The porosity of the top surface, if not removed by polishing, may be expected to lead to increased waveguide propagation and insertion losses depending on wavelength and other factors. Very likely this also means a deterioration of acoustic properties with increasing film thickness beyond some coherent film thickness value which, in turn, depends upon growth conditions. The apparent value of the electromechanical coupling coefficient, k^2, determined for a given thickness to wavelength ratio, t/λ, may thus depend upon whether measurements were made at high frequency on a thin film or at low frequency on a thick film. Consequently, some scatter may be expected in the results of measurements made on films thick enough to have porous surface structures. The presence of a polycrystalline component, c-axis inversion, residual strain, and pits and inclusions can diminish the value of k^2. A departure from parallel epitaxy can increase the uncertainty in the alignment of transducer patterns for acoustic propagation along the c axis of the film and lead to an apparently diminished k^2 value resulting from beam-

steering effects and variation in coupling efficiency with direction in the plane of the film. Lakin and coworkers (1974) and Penunuri and Lakin (1975) have also outlined other factors that can lead to low k^2 values even for high-quality films. They indicate that surface waves propagating at an angle to the projection of the sapphire c axis on the substrate surface plane may be attenuated by "leakage" into the substrate bulk. Thus, they predict diminished k^2 values, increased propagation loss, and a velocity sensitive to small changes in film thickness if the propagation direction deviates from the c-axis projection and the ratio t/λ is less than two. For propagation precisely parallel to the substrate c-axis projection no leakage occurs, but they also emphasize the practical difficulty in accomplishing this parallelism because of transducer fabrication alignment errors and beam diffraction effects. In addition, the transducer patterns used for the generation of surface waves may not be optimum in design for the given material and, because of film thickness variation and velocity dispersion for t/λ < 2, the output transducer may not be exactly tuned to the output signal, thus diminishing the value of k^2. It is evident that several factors influence the magnitude of the electromechanical coupling coefficient, not all of which are related to film morphology. Accordingly, it is reasonable to assume that only the maximum value of k^2, measured at a particular value of t/λ, approaches the intrinsic value of k^2 for high-quality single crystal films (Pizzarello and Coker, 1975).

The dependence of the electromechanical coupling coefficient of heteroepitaxial AlN on film thickness has been studied by several workers. A summary of recent data is presented in Figure 4.14 together with references. The broken lines give an outline of the envelope within which most data points lie. Differences in the data of independent investigators within this envelope are caused, in part, by differences in data-collecting methods. For example, in some cases k^2 is measured for different film thicknesses keeping the value of λ constant. In other cases, k^2 is measured for different values of λ keeping film thickness constant. This is accomplished by varying transducer line-widths and spacings and measuring k^2 at the corresponding resonant frequency. Still others may have compiled their data utilizing both methods. It is unlikely that these methods give equivalent results because of the porous surface texture of AlN films which becomes more pronounced with increasing film thickness. However, the results of the independent investigators show a more consistent trend than the scatter in results in Figure 4.14 would suggest. The actual shape of the k^2 vs. t/λ curve has been the subject of some speculation. In their publication, Pizzarello and Coker (1975) combined their results with those of O'Clock and Duffy (1973) and tentatively

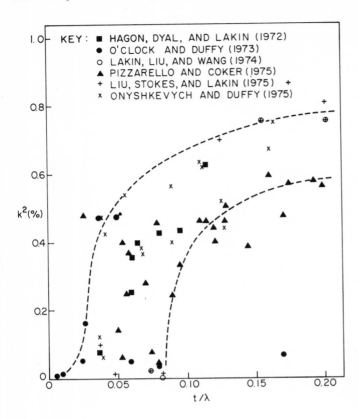

FIGURE 4.14
Piezoelectric coupling coefficient, k^2, vs. thickness to wavelength ratio, t/λ, as determined by several investigators

suggested the occurrence of an auxiliary maximum in the plot of k^2 vs. t/λ at $t/\lambda \approx 0.05$ followed by a minimum at $t/\lambda \approx 0.07$ before rising to a second broad maximum. They cited theoretical considerations (Smith, 1971; Lakin and Penunuri, 1971; Penunuri and Lakin, 1972) on the dispersion characteristics of layered systems predicting the occurrence of an auxiliary maximum in k^2 vs. t/λ data. Experimental work by Pizzarello (1972) on heteropeitaxial (11$\bar{2}$0)-oriented ZnO on sapphire, which is isomorphic with AlN, showed the presence of such an auxiliary maximum (see Figure 4.26). The elastic and piezoelectric constants for AlN have not been determined because of the difficulty in growing single crystals of AlN (Dugger, 1974). Thus, because of scatter in the results, any conclusions as to the dependence of k^2 are speculative at this time. The results reported by O'Clock and Duffy were obtained at relatively low measuring frequencies on relatively thick films (up to 10 μm). There is little doubt that surface porosity of the nature described earlier and possible polycrystalline components contributed to a deterioration of film properties at higher film thicknesses. This may have been responsible, at least in part, for some of the fall-off in k^2 for $t/\lambda > 0.06$ (i.e., $t > 3.6$ μm). Recent unpublished results by Onyshkevych and Duffy (1975) are included in Figure 4.14 and show no such fall-off in the value of k^2. In this work greater attention was given to

crystalline quality and surface continuity, and no deterioration in properties was observed to a film thickness of 6 μm. The results of Hagon and coworkers (1972) show a plateau at the position of the minimum discussed above. Lakin and coworkers (1974), and Liu and coworkers (1975) have made a detailed study of k^2 over a wider t/λ range up to a ratio of 0.75 and observed little piezoelectric activity below a ratio of 0.1. Their data show a sharp rise in k^2 above $t/\lambda \approx 0.1$, a slight dip near $t/\lambda \approx 0.2$ before rising to $k^2 \approx 0.8$ percent near $t/\lambda = 0.3$, followed by a gradual decrease at higher ratios (see Figures 4.14 and 4.15). Liu and coworkers attributed the lack of piezoelectric activity at low t/λ ratios (<0.1) to the presence of interfacial strain. They also point out that the theoretical work on the dispersion characteristics of layered systems does not take into account interfacial strain, which they suggest influences the piezoelectric properties of the first micrometer of film thickness. On this basis the shape of the k^2 vs. t/λ curve near the origin may be determined by the frequency of measurement since the same value of t/λ may be obtained at high measurement frequency on thin films or at low measurement frequency on relatively thick films. Thus, it is important to note the measurement frequency or acoustic wavelength, λ, when examining the shape of the curve near the origin. This may be the reason for some of the scatter in individual results near the origin in Figure 4.14. The data in Figures 4.14 and 4.15, however, give an empirical assessment of the dependence of k^2 over the frequency range 100–600 MHz and demonstrates useful piezoelectric properties even for thin dispersive films of AlN.

Velocity dispersion in AlN/Al_2O_3 composites was also studied by Lakin and coworkers, Liu and coworkers, and over a short t/λ range, by Onyshkevych and Duffy. The results are presented in Figure 4.16 and show reasonable agreement. Liu and coworkers also reported the temperature coefficient of delay for these composites. Their results show a negative temperature coefficient dropping almost linearly from about 47 ppm/°C at $t/\lambda = 0.15$ to about 25 ppm/°C at $t/\lambda = 0.57$.

The piezoelectric properties of GaN films will not be discussed at this time because of the scant information available.

Preparation and properties of heteroepitaxial ZnO films

Growth process. The literature contains several papers describing the crystal growth of II-VI compounds on various substrates. Yim and others (1970, 1972) reported on the heteroepitaxial growth of II-VI compounds on a variety of substrates such as GaAs, GaP, and sapphire by transporting the group II element (Zn or Cd) vapor in H_2 carrier gas passing over

the molten metal. The group VI element (S or Se) was introduced as the gaseous hydride (H_2S or H_2Se) in H_2 diluent. Manasevit and Simpson (1971) employed the metal–organic compounds (diethylzinc and dimethyl-cadmium) as the source materials for Zn and Cd in conjunction with the group VI hydrides (H_2S and H_2Se). Vapor transport was accomplished by passing a carrier gas through the liquid metal–organic compounds. Mixing of the vapors occurs in the vicinity of the heated substrate. These authors report on the growth of ZnS, ZnSe, CdS, and CdSe layers on Al_2O_3, $MgAl_2O_4$, and BeO by this method. Despite the development of these heteroepitaxial processes, relatively little information is available on the piezoelectric and electrooptic properties of thin films of these materials on sapphire substrates.

Of the II-VI compounds, greatest attention has been paid to ZnO films in this respect. In the early studies, zinc oxide films have been grown by such methods as reactive evaporation and sputtering. Films prepared by these methods were generally polycrystalline with only preferred orientation. There was also a strong tendency for film orientation to be independent of the orientation and nature of the substrate. Polarity inversion of adjacent crystallites parallel to the c axis was also observed, leading to diminished piezoelectric activity. Strictly speaking, films characterized in this manner cannot be classified as heteroepitaxial. Although heteroepitaxial ZnO films have been pre-

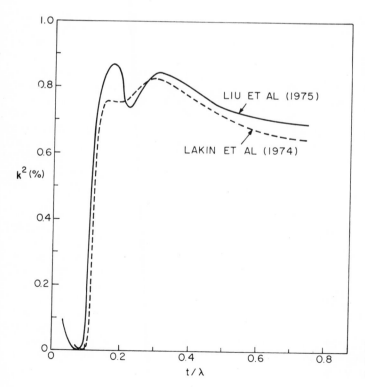

FIGURE 4.15   Piezoelectric coupling coefficient, k^2, of AlN vs. thickness to wavelength ratio, t/λ, over extended range

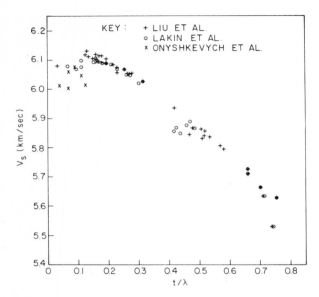

FIGURE 4.16   Acoustic surface wave velocity, V_s, of AlN/
Al_2O_3 composites vs. thickness to wavelength ratio, t/λ. Data
from Liu et al. (1975), Lakin et al. (1974), and Onyshkevich
and Duffy (1975)

pared on sapphire by sputtering when the substrate
temperature is raised to relatively high values, the
most widely used preparative technique for this purpose has been the close-spaced chemical-vapor transport method reported by Galli and Coker (1970). Most
results, pertaining to the piezoelectric and electrooptic
properties of ZnO films on sapphire, were obtained on
films prepared in this manner. This method uses ZnO
as the source material which is heated in a reducing
ambient such as H_2 or NH_3 in close proximity to a
heated substrate. The temperature of the substrate is
maintained at a lower temperature than that of the
source material. The ZnO source is reduced according
to the equilibrium reaction

$$ZnO(solid) + H_2(gas) \rightleftarrows Zn(vapor)$$
$$+ H_2O(vapor), \quad (4.1)$$

and the products are transported to the substrate
where recombination and film growth occurs. Other
species have been added to the reducing ambient such
as HCl and H_2O (vapor) (Shiloh and Gutman, 1971). A
typical temperature range for the source material in H_2
ambient is 800 to 850°C, and that of the substrate from
50 to 100°C lower. The addition of water vapor raises
the growth temperature as might be expected. Reisman and Landstein (1971) have carried out a thermodynamic analysis of the latter process in relation to the
reaction

$$Zn(vapor) + H_2O(vapor) \rightleftarrows ZnO(solid) + H_2(gas)$$
$$(4.2)$$

in He carrier gas. This reaction is the reverse of
Reaction (4.1). Galli and Coker have reported on the
heteroepitaxial growth of ZnO on sapphire at a substrate temperature of 775°C by means of Reaction (4.1)
in the presence of HCl gas. Reisman and coworkers
(1973) have reported on the growth of ZnO films on
sapphire by means of Reaction (4.2) in the substrate
temperature range 775 to 910°C. The different
approaches to the preparation of heteroepitaxial ZnO
are of interest because of the possibility of obtaining
films of different orientations on the same substrate
orientation (see above references). Single crystal films
of ZnO are semiconducting, n-type, with a resistivity
of less than 1 Ω-cm. This value is much too low for
efficient use of the films in devices which utilize the
piezoelectric and electrooptic properties of ZnO; a
value of $\rho \simeq 10^6$ Ω-cm is desirable. The semiconducting and surface state properties of ZnO have been
abundantly documented in the literature and are
beyond the scope of this discussion. The low resistivity has been attributed, at least in part, to a departure
from true stoichiometry of ZnO, with interstitial Zn
acting as a donor. One might hope that, for the applications of interest here, the different preparative
methods might offer the prospect of preparing high-
resistivity films by greater control of stoichiometry.
Unfortunately, this has not been the case, and resistivities have been consistently low in the case of single
crystal films irrespective of the preparative method.

During studies on the waveguiding properties of
heteroepitaxial ZnO on sapphire (Hammer et al., 1972,
1973, 1975; Channin et al., 1975), it was found that, at
the higher growth or processing temperatures, ZnO
reacts with sapphire resulting in increased waveguide
propagation losses. Consequently, these workers
found it desirable to grow heteroexpitaxial ZnO films
at a temperature of about 750°C by means of Reaction
(4.1) above. For exploratory purposes, they used a
simple growth apparatus as depicted schematically in
Figure 4.17. Sapphire substrates were supported at the
top of a vitreous carbon crucible by a quartz spacer as
indicated, while the ZnO (powder) source material
was contained in the bottom of the crucible. The
assembly was heated by RF induction. A temperature
differential of approximatley 100°C was maintained
between source material (at ~850°C) and substrate (at
~750°C) by positioning the top of the crucible a suitable distance above the induction coil. A graphite disc
placed over the substrate helped to maintain temperature uniformity. Substrate temperature determinations
were made by optical pyrometry on the graphite disc.

Like AlN and GaN, ZnO films have the hexagonal
Wurtzite structure, and the heteroepitaxial relation-
ship of greatest interest has similarly been $(11\bar{2}0)ZnO\|$
$(01\bar{1}2)Al_2O_3$ where the same directional relationships
apply as depicted in Figure 4.1. Films of this orientation ranging in thickness up to 100 μm have been

grown by the above technique. However, anisotropic residual strain in the films, of the nature described by Duffy and coworkers (1973) for the case of AlN, frequently results in cracking of the films at thicknesses of about 20 μm. This depends upon the cooling rate of the epitaxial composite and substrate thickness.

As-grown ZnO films exhibit varying degrees of surface smoothness depending upon growth conditions. Scanning electron micrographs of two samples of comparable thickness ($\sim$100 μm) are given in Figures 4.18 and 4.19. Surface topography is similar for both samples but surface smoothness differs widely. The smoother surface corresponds to a growth rate of about 1 μm/min, whereas the other surface corresponds to a growth rate of about 10 μm/min. The samples chosen are thick films in order to exaggerate surface topography. Similar surface structure occurs on a reduced scale at film thicknesses ($\sim$0.5 μm) suitable for optical waveguides. The slower growth rate was chosen for the latter purpose. The surface faceting observed in the micrographs runs parallel to the direction of the c axis of the (11$\bar{2}$0)-oriented ZnO film. In general, film surfaces are not sufficiently smooth for the light-processing functions of optical waveguides and must be polished to minimize attenuation of the guided light by surface scattering losses. Chemical polishing is usually not satisfactory because of preferential etching. The most successful method employed by the author has been the use of silica sol where the pH can be varied within limits allowing adjustment of the relative rates of chemical and mechanical polishing.

Postgrowth annealing.   Despite the care taken in the growth process and subsequent surface polishing of ZnO waveguides, the attentuation of coherent light propagating in the films has been found to vary over wide extremes, even in the case of lower-order propagation modes, when the film thickness is several times the optical wavelength and bulk scattering losses are dominant. Losses ranging from 5 dB/cm to 70 dB/cm have been observed. In all cases the preparative methods were apparently similar and the films were characterized as single crystal by the various analytical techniques. Evidently, one of the most stringent tests of crystalline quality is optical waveguiding loss measurement at nonabsorbing wavelengths. The influence of annealing on propagation loss depends upon temperature and ambient and can be dramatic in its effect. When polycrystalline films are annealed in H$_2$ at 800°C, a reordering process occurs. The films become transparent and single crystalline in character according to x-ray analysis. The time of annealing must be of short duration at this temperature because decomposition occurs in H$_2$. No appreciable improvement in propagation loss occurs in the case of single crystal films with this procedure. Likewise, annealing in N$_2$

has little effect. The most dramatic effects occur by heating in O$_2$ or air at temperatures from 800°C to 850°C. Single crystal samples which display losses as high as 70 dB/cm for the lowest-order propagation mode before annealing may display values as low as about 1 dB/cm after annealing. This step is essential to the fabrication of reproducible low-loss ZnO optical waveguides. The duration of the anneal can be up to 1 hr. The reason for the reduction in propagation loss with annealing is not well understood, but it is interesting to note that Hammer and coworkers (1972) have tentatively suggested the presence of excess Zn ions associated with the carrier concentration as a possible source of scattering centers. The influence of annealing in an oxidizing ambient, in contrast to a reducing or inert ambient, in modifying these losses suggests a correlation between stoichiometry and propagation loss. It is also interesting to note that Hammer and coworkers observed that the bulk film scattering losses increased with higher carrier concentration.

Postgrowth compensation.   High-resistivity films are essential for successful operation of active optical waveguides in deflectors and modulators. As already

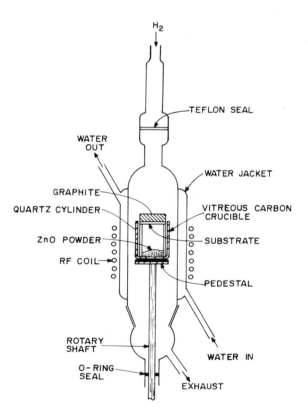

FIGURE 4.17   Diagram of ZnO epitaxial growth apparatus. After Channin et al. (1975)

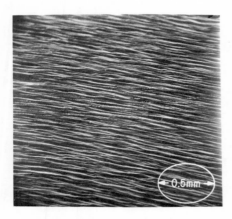

FIGURE 4.18   Scanning electron micrograph of 100-μm ZnO film showing dependence of surface structure on growth rate. Fast rate of growth (~10 μm/min)

indicated, ZnO films are semiconducting, n-type, with resistivity of <1.0 Ω-cm, a carrier concentration of about $10^{17}/cm^3$, and mobility in the range 30–100 cm²/V-s, depending upon growth conditions and film thickness. The compensation of epitaxial ZnO films by Li or Na has been reported by Galli and Coker (1970), who obtained resistivities greater than 10^5 Ω-cm. Their technique involved surface coating the films with a saturated solution of Li_2CO_3 or Na_2CO_3 followed by a drying and heating process in air at 650°C for 24 hr. The author has used LiOH and $LiCO_3$ in a similar manner and obtained comparable results. However, the lithium source proved very corrosive, attacking both the exposed sapphire surface and the ZnO film and made subsequent surface polishing essential and difficult. Attempts to compensate the films with Li during the growth process were unsuccessful since the resulting films were either of inferior quality or polycrystalline. In an alternative approach, the samples were prevented from making direct contact with the lithium source, and compensation occurred via vapor transport. The lithium source in this case is provided (Duffy et al., 1973) by heating LiOH in air on a silicon carbide–coated graphite susceptor until molten at about 450°C. The liquid spreads uniformly over the inductively heated susceptor and is then heated at higher temperatures (~600°C) until dehydration is complete and a white powder coating remains. The temperature is then raised to about 1100°C for approximately 15 min before cooling to room temperature. The samples are placed, film downward on a 5-mm-high quartz spacer on this susceptor, which is then heated in O_2 or air by RF induction (Figure 4.20). The samples are indirectly heated by the susceptor, the temperature of the susceptor being about 850°C and that of the samples from 650 to 700°C. This process results in minimal adverse effects on the film and substrate surfaces and permits the compensation of

films in about 1 hr which previously took about 24 hr. Apparently, the limiting factor in the compensation process is the availability of lithium; higher temperatures favor increased vapor pressure of source material. This cannot be accomplished when the lithium source is placed directly on the film surface because of chemical attack and because compensated high-resistivity films heated to this temperature (>850°C) are returned to the low resistivity values of as-prepared films. In addition, films heated to about 900°C or above react with the substrate surface, evidence of which can be observed after etching off the films. The vapor transport technique proves superior in all respects for the compensation of ZnO films. However, this source absorbs water and CO_2 from the air upon exposure and must be recycled through the above temperature treatment before each usage if exposed to air. Compensation in the presence of O_2 is preferable to compensation in air since much higher resistivities are obtained and in a shorter time. It is evident from annealing and compensation studies that O_2 is an important ambient in determining the optical scattering losses and electrical properties of both compensated and uncompensated ZnO films. Postgrowth annealing and compensation can be accomplished in a single step when O_2 is used as the ambient during Li diffusion, and low waveguide losses can be obtained in a reproducible manner.

Variation in resistivity of compensated ZnO films. It has been observed (above references) that the apparent resistivity of compensated heteroepitaxial ZnO films on sapphire decreased upon exposure to air. Routine qualitative resistivity measurements on ZnO films were performed by the 4-point-probe technique (Valdes, 1954). It was found that samples, which displayed resistivity values of about 10^6 Ω-cm immediately after compensation, usually reverted to lower val-

FIGURE 4.19   Scanning electron micrograph of 100-μm ZnO film showing dependence of surface structure on growth rate. Slow rate of growth (~1 μm/min)

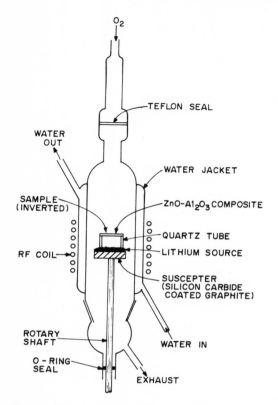

FIGURE 4.20   Diagram of apparatus for Li compensation of ZnO films

ues in the range 10^2 to 10^4 Ω-cm after exposure to room ambient for a few days. Heating the samples on a hot plate in air partially restored the initial high values, whereas at higher temperatures of about 650°C the resistivity again became ~10^6 Ω-cm. The cycle then repeated itself. From this behavior and the well known surface reactivity of ZnO, it is apparent that surface conduction mechanisms are involved, probably resulting from variations in surface stoichiometry and the effects of adsorbants. When compensated samples were annealed in N_2 at 850°C, the resistivity dropped to very low values, whereas annealing the samples subsequently in O_2 at the same temperature restored the initial high values. This behavior could then be repeated, indicating that the oxygen content, at least at the film surface, had a pronounced effect on conductivity.

Some qualitative measurements have been performed in order to determine the sensitivity of the films and a possible solution to the fabrication of high-resistivity waveguides. Figure 4.21 shows the behavior of a 2.0-μm-thick Li-compensated ZnO sample which had been exposed to room ambient for about 6 months. The resistivity after compensation had been in excess of 10^6 Ω-cm. It can be seen that the apparent resistivity had dropped as low as 300 Ω-cm in room

illumination. The apparent resistivity rises gradually with time after the sample is shielded from room lighting. The photoconductive properties of ZnO are well known. However, the response rate in this case is slower than expected, suggesting a light-sensitive surface adsorption or desorption process which is reversed in the dark.

In order to determine the depth within the film to which surface effects extend, the same film was polished slightly with silica sol and the apparent resistivity determined both in the presence and absence of room lighting as represented by curve A in Figure 4.22. The apparent resistivity has increased in both segments of the curve and the photoconductive response is more rapid, reaching a steady state in a relatively short period as compared to the response shown in Figure 4.21, thus indicating a different mechanism. The film thickness was approximately 1.9 μm after polishing, representing a removal of about 1000 Å of film surface. The surface material removed was more than sufficient to eliminate the surface mechanism associated with long exposure of ZnO films to local ambient as depicted in Figure 4.21.

Successive polishing and measurement were conducted in similar fashion. The data are also presented in Figure 4.22 as curves B, C, and D. The corresponding film thickness was approximately 1.75 μm (curve B), 1.7 μm (curve C), and 1.5 μm (curve D), representing further surface film removal of about 1500, 500, and 2000 Å, respectively. The general response indicated by each curve is similar to that of curve A, but the magnitude of the response increases dramati-

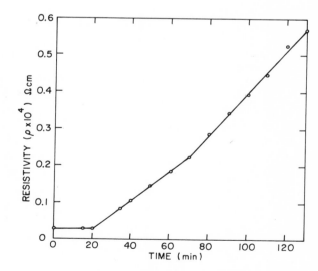

FIGURE 4.21   Apparent resistivity of compensated ZnO film after exposure to room ambient for about 6 months. Horizontal segment: resistivity vs. time in room illumination; sloping segment: resistivity vs. time after shielding from light

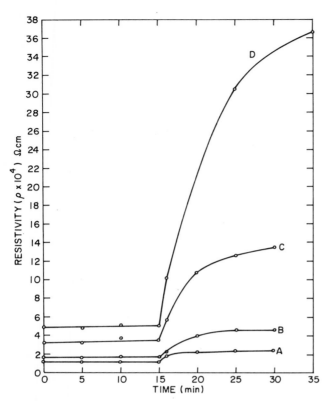

FIGURE 4.22
Apparent resistivity of a ZnO sample (referred to in
Figure 4.21) after repeated polishing and measurement

cally as surface material is removed. Curve D corresponds to a total surface material removal of about 5000 Å, and the apparent resistivity of the sample in the absence of light has not reached the $>10^6$ Ω-cm value measured in room lighting immediately after compensation. In the 4-point-probe method of determining the resistivity it was assumed that the conductivity was uniform throughout the depth of the film. Clearly, this is not so in the present case; the resistivity increases with increasing distance into the film. Thus, the apparent value measured represents only an average value for the film. The results indicate that various surface mechanisms may be involved when Li-compensated samples are exposed to the local ambient and that large resistivity changes can occur to appreciable depths within the film.

Mass spectrometry measurements. Simultaneous resistivity and mass spectrometry measurements have been performed under high-vacuum conditions (Zanzucchi and Duffy, 1973) at pressures near 1×10^{-9} torr in an ion-pumped stainless steel–glass vacuum system. Prior to ion pumping, samples were baked at ~200°C for 24 hr, in situ, using a doubly trapped mercury diffusion pumping system which can maintain pressure near 5×10^{-9} torr. The samples under study were ion pumped, after baking, for four days at 2×10^{-9} torr. A 4-probe feedthrough with spring-loaded nichrome wire was used to contact the ZnO surface. No contacts were evaporated onto the film, thus avoiding further treatment of the ZnO surface. Changes in resistivity were registered as a resistance change between two of the probes on a high-impedance ($>10^8$ Ω) output monitor under the high-vacuum conditions maintained in the system. Mass spectrometry data were obtained with a Varian partial pressure gauge capable of scanning the mass range 1 to 70. Mass spectra indicated that water vapor and nitrogen were the chief background constituents in the vacuum cell containing the sample.

The resistance of one of the samples chosen for this study was too high to measure ($>10^8$ Ω) under the prevailing vacuum conditions in the system prior to the bake-out step. After bake-out the resistance had dropped to ~3×10^7 Ω. Figure 4.23 shows the effect of incident light on the measured resistance of the film after bake-out. The sample was irradiated with a polychromatic mercury radiation source. The resistance dropped sharply by two orders of magnitude while a sharp rise in the partial pressure of CO_2 in the system occurred simultaneously, indicating the evolution of CO_2 from the film surface. The rate of resistivity change was a function of the intensity of the incident radiation (i.e., the higher the intensity of the incident radiation, the sharper and greater the change in resistivity). Similarly, the evolution of carbon dioxide occurs more readily with higher-intensity radiation. Only part of the change is shown in Figure 4.23

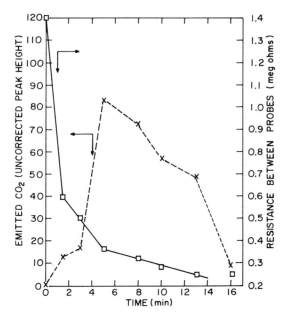

FIGURE 4.23   Simultaneous monitoring of sample resistance and CO_2 evolution as sample is irradiated with a polychromatic light source in high vacuum ($\sim 10^{-9}$ torr). After Zanzucchi and Duffy (1973)

the mass spectra. Spectrum A was recorded after 13 min of UV irradiation; spectrum B was recorded 1 min after the light source was turned off; spectrum C, recorded on a scale ~ 10 times more sensitive than for A or B, is representative of the vacuum background prior to UV exposure and shows small amounts of CO or N_2 and water vapor. Spectrum A shows the presence of CO_2 (mass 44), CO (mass 28), H_2O (mass 18), O^+ (mass 16) and C^+ (mass 12). Hydrogen was also present but is not shown on this scale. The principal constituents evolved were CO_2 and CO. The nature of the ZnO surface, the influence of adsorbates as extrinsic surface states on the semiconducting properties of ZnO crystals, and the effects of photodesorption have been discussed by several authors (Collins and Thomas, 1958; Kokes and Glemza, 1965; Heiland and Kunstmann, 1969; Williams and Willis, 1968; Morrison, 1969, 1970; Arijs et al., 1969, 1973; Levine et al., 1972; Iyengar and Codell, 1972); Arijs and Cardon, 1973; Leysen et al., 1973; Fiermans et al., 1973. Though most of the referenced work deals with the (0001) ZnO polar surfaces of bulk single crystals, there is sufficient evidence to indicate that the surface reactivity of thin heteroepitaxial (11$\bar{2}$0)-oriented films on sapphire is qualitatively similar.

Surface passivation.   The surface reactivity problems associated with ZnO films necessitate a surface passivation process in order to efficiently utilize the electrooptic and piezoelectric properties of the films when equipped with interdigital electrode structures. The

because of the difficulty in obtaining data points during the initial rapid change upon illumination of the sample. The signal corresponding to evolved CO_2 has not been corrected for the pumping removal rate. An interesting observation is that the resistivity change is irreversible in vacuum when the light source is removed. This is in contrast to the case in which the films are exposed to atmospheric pressure in air. Thus, the combined film resistance and mass spectrometry measurements show that a species desorbed from the ZnO surface can cause large changes in film resistivity. It has already been demonstrated in annealing and compensation studies that O_2 can also have a major effect on the electrical properties of ZnO. Consequently, in separate experiments O_2 and CO_2 were admitted to the high vacuum system to study film resistance changes (after irradiation) resulting from surface adsorption. Because of the nature of the vacuum experiment, however, the addition of O_2 was carefully controlled so that a pressure of 10^{-7} torr was not exceeded, and no effect on resistivity of the sample was observed with the small quantities admitted. On the other hand, CO_2 was admitted to the cell until atmospheric pressure was reached. The addition of CO_2 produced a sharp initial increase in resistivity followed by a steady gradual increase as shown in Figure 4.24. Again the initial increase was not recorded because of the short duration of the change.

Other species are also evolved from the ZnO surface during the irradiation process. Figure 4.25 shows

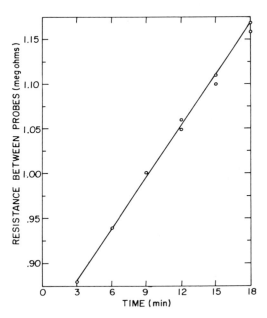

FIGURE 4.24   Effect of CO_2 adsorption on film resistivity after irradiation with a mercury light source in high vacuum. After Zanzucchi and Duffy (1973)

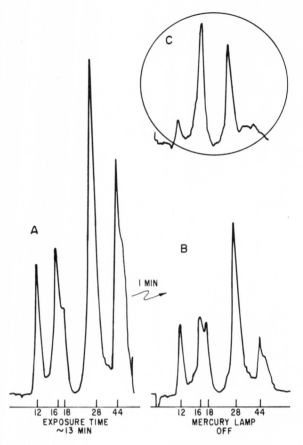

FIGURE 4.25   Mass spectra showing evolution of different species from ZnO surface after irradiation in high vacuum. After Zanzucchi and Duffy (1973)

author has tested various thin-film dielectrics associated with MIS technology, such as amorphous Al_2O_3, Si_3N_4 and SiO_2. Of these, Al_2O_3 is best suitable since the other dielectrics result in lower surface resistivity. In particular, the chemical vapor deposition of amorphous Al_2O_3 overlay films was performed at 500°C by the pyrolysis of Al-isoproproxide (Duffy et al., 1970, 1971; Carnes and Duffy, 1971). Film thicknesses ranged from 800 to 1500 Å. Deposition was performed immediately after the compensation step. The deposition process resulted in reduced sample resistivity when prior compensation was performed in air. However, this did not happen when the samples had been compensated in O_2. In addition, passivated samples with diminished resistivity, because of prior annealing in air, could be restored to the initial high-resistivity values by subsequent annealing of the composite structures in O_2 at 800°C (but not in air). Composite structures not so passivated were adversely affected by photolithographic processing. Passivated samples can be stored under local ambient conditions without noticeable change in surface resistivity and show

reversible photoconductive behavior under ultraviolet illumination.

Piezoelectric properties.   Most measurements on the piezoelectric properties of ZnO films have been performed on sputtered samples. Consequently, a brief discussion of sputtered films will be included here. These films are usually polycrystalline in character but with preferred (0001) orientation irrespective of the substrate material. The crystallites display azimuthal disorder, with the c axis of individual crystallites deviating by some angle from parallelism with the substrate surface normal. Chubachi (1976) has reported a Gaussian angular distribution. Certain precautions are necessary during the sputtering process to prevent c-axis polarity inversion. Individual crystallites can also be misaligned relative to each other due to rotation about the c-axis direction.

Sputtered films have performed well in acoustic wave applications with a coupling efficiency up to 90 percent that of single crystal ZnO (Hickernell, 1976). Optical propagation losses in sputtered films, however, tend to be high and for this application heteroepitaxial layers are favored. Rozgonyi and Polito (1969) have reported on the growth of heteroepitaxial layers by sputtering when the substrate temperature is raised to 300–500°C. For heteroepitaxial growth, substrate temperature and crystallographic orientation, in addition to the sputtering conditions, become important factors in determining film crystallinity and orientation, in similar fashion to heteroepitaxial growth by chemical vapor transport. In addition, ZnO films, grown by either process, become increasingly semiconducting as the substrate temperature is raised to obtain heteroepitaxy, whereas preferentially oriented films sputtered near room temperature are sufficiently insulating for the above applications. The resistivity of sputtered single crystal films can be much higher, however, than that of films prepared by chemical vapor deposition. Rozgonyi and Polito (1969) have reported resistivity values between 10^4 and 10^6 Ω-cm, depending upon oxygen content in the sputtering system, for (0001) oriented films on (0001) oriented substrates of sapphire and CdS, and Shuskus and coworkers (1973) have reported resistivities of 10^6 or greater for undoped (0001) and (11$\bar{2}$0) oriented sputtered films on (0001) and (01$\bar{1}$2) sapphire substrates. Films prepared by chemical transport on the other hand can have resistivity values as low as about 0.1 Ω-cm, and must be compensated, usually by lithium or sodium diffusion, to obtain suitably high resistivities ($\sim 10^6$ Ω-cm). This process may not be compatible with device fabrication. Thus, the preparative method depends on the intended application and the mode of operation. As noted above, sputtering favors preferentially oriented films with c-axis alignment approximately normal to the substrate surface. This orientation is required for longitudinal acoustic bulk wave transducers. Denburg (1971) has reported a coupling factor of $k_t = 0.25$ for films of this type which compares favorably with a value of $k = 0.28$ reported by Reeder and Winslow (1969) for single crystals of ZnO. Bulk shear wave transducers require that the c axis of the crystallites be aligned parallel to the plane of the substrate and parallel to each other. This is a much more difficult task to accomplish. By the addition of certain impurities to the sputtering system and tilting of the substrate with respect to the sputtering source a certain degree of success has been obtained (Foster, 1969A, 1969B; Hada et al., 1971; Vuilloud, 1972). Foster has reported a coupling coefficient k_{15} of between 0.12 and 0.18 for shear wave transducers which is approximately half the value reported for ZnO single crystals ($k_{15} = 0.32$). In general, much of the work on piezoelectric films has been performed in connection with the study of longitudinal bulk acoustic wave transduction, using films of about one-half acoustic wavelength in thickness (Olson, 1970). The discussion here is concerned primarily with thin dispersive heteroepitaxial $(11\overline{2}0)$-oriented ZnO films grown on $(01\overline{1}2)$-oriented sapphire substrates with the directional relationships described previously. Most preparative work on this configuration has been performed by chemical vapor transport.

Collins and coworkers (1970) have studied the surface acoustic wave properties of relatively thick heteroepitaxial ZnO films with the above configuration prepared by chemical vapor transport. They report a value of $k^2 = 0.31$ percent for the electromechanical coupling coefficient at a film thickness of about 50 μm. The operating frequency was about 175 MHz and the corresponding acoustic wavelength about 15 μm. The acoustic velocity was determined to be 2.65×10^5 cm/sec. The coupling coefficient was lower, however, than the value of $k^2 = 1.1$ calculated from Campbell and Jones data (1970) for bulk crystals. In these measurements the film thickness–to–acoustic wavelength ratio, t/λ, was between three and four and past the dispersion region ($t/\lambda < 2$).

Pizzarello (1972) has carried out an extensive study on thin dispersive films of ZnO also prepared by chemical vapor transport. His data show the dependence of phase velocity, group velocity, and coupling coefficient on the ratio t/λ, in addition to delay time versus frequency for dispersive delay lines. The coupling coefficient dependence is shown in Figure 4.26 together with a theoretical curve determined by Lakin and Penunuri (1971). The experimental data show a maximum corresponding to $k^2 \approx 2$ percent at $t/\lambda = 0.78$ and an auxiliary maximum corresponding to $k^2 \approx 0.9$ percent at $t/\lambda = 0.13$. The theoretical and experimental data are in good agreement except for the asymptotic value at $t/\lambda = 2$. The theoretical value appears somewhat high while the experimental value

appears low if a value of $k^2 = 1.1$ percent is expected based on data from Campbell and Jones. These results attest to the quality of the piezoelectric films prepared in this manner. Shuskus and coworkers have reported a value for k^2 of 0.4 percent obtained at $t/\lambda = 0.11$ on sputtered heteroepitaxial films of the same orientation on sapphire substrates which were heated to 700°C during the sputtering process. This data point also lies close to the experimental curve in Figure 4.26 for the corresponding value of the abscissa. The velocity dispersion data, according to Pizzarello, is shown in Figure 4.27.

4.3. Device applications

In the area of interest here, both active and passive waveguides for surface acoustic wave, electrooptics, and acoustooptics applications may be fabricated in heteroepitaxial films grown on sapphire substrates with the prospect of combining silicon semiconductor functions on the same substrate through materials integration. In practice, however, the number of options are limited because of materials incompatibility with respect to preparation and properties. In particular, epitaxial aluminum nitride, zinc oxide, and perhaps gallium nitride are potentially useful for microsonics applications. Zinc oxide has the added advantage that it can be used for active optical waveguides and acoustooptic interactions and has interesting photoconductive properties. Both aluminum nitride and gallium nitride have rather high optical propagation losses and are, thus, unsuitable as optical waveguides at the present time; the propagation loss in the visible region for epitaxial AlN films on sapphire

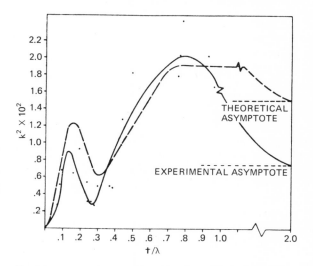

FIGURE 4.26   Experimental and theoretical curves for coupling constant k^2 of the Rayleigh mode as a function of t/λ for ZnO/Al$_2$O$_3$. After Pizzarello (1972)

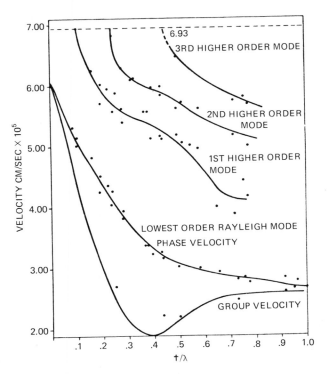

FIGURE 4.27   Plot of experimental values of phase velocity vs. t/λ for the Rayleigh mode and three higher-order modes on heteroepitaxial ZnO. Also included is the group velocity curve for the Rayleigh mode. After Pizzarello (1972)

is usually too high to measure and that for GaN films about 30 dB/cm or greater. The following discussion is a review of device applications, primarily for AlN and ZnO films, as reported in the literature.

Aluminum nitride

The interest in AlN as a piezoelectric material arises from the fact that (1) it has a surface acoustic wave velocity about double that of the better known piezoelectric materials such as lithium niobate and can be operated at double the frequency for given transducer pattern dimensions. This extends the frequency range for purposes of signal processing over that possible with the better known piezoelectrics using conventional photolithography for device fabrication; (2) aluminum nitride can be grown in side-by-side configuration with silicon on the same sapphire substrate, thus providing the basis for a large number of signal processing functions in a monolithic structure through materials integration. Fortuitously, the materials crystallographic orientations involved are those of greatest current interest to surface wave generation and propagation direction and to SOS technology; (3) aluminum

nitride is insulating when not intentionally doped and is reasonably compatible with semiconductor processing. Certain precautions have to be taken, however, in the integration of AlN and Si on the same sapphire substrate. If Si is grown on a substrate on which AlN was previously grown and removed by etching, the resulting films have very much reduced Hall mobilities and are considerably more p-type than would otherwise be the case on regular substrates. It has been indicated previously that there is evidence of chemical reaction between AlN and sapphire substrates depending upon growth temperature. Silicon films can also be grown heteroepitaxially on AlN films but also with degraded semiconducting properties. A mobility of 150 cm²/V-s has been reported by Lakin and coworkers (1974) for n-type films. Likewise, if silicon is grown as the initial layer and AlN subsequently grown in side-by-side configuration, the semiconducting properties of the silicon film are degraded. Compensation of n-type films occurs probably because of Al diffusion from the pyrolysis of Al (CH₃)₃ source material at the high temperature (~1250°C) of AlN growth. The best approach has been to grow AlN while masking that portion of the substrate on which good quality Si is subsequently required. The masking film may be silicon itself which is grown over the entire surface, then selectively removed from part of substrate by standard photolithographic procedure. Aluminum nitride is then grown on this structure and removed from the silicon region by preferentially etching the silicon, thus causing "lift off" of the superimposed AlN layer. Silicon can now be grown on the exposed substrate at ~1000°C with greater immunity than otherwise feasible. The author has obtained electrical properties characteristic of SOS films in this manner except for a Si strip, along the boundary between both films where compensation occurs, as indicated in Figure 4.28. This, however, is an important region from the point of view of interfacing between the piezoelectric properties of the AlN film and semiconducting properties of the silicon depending upon the application, and more complex processing may be required in certain cases in order to protect semiconducting properties to the boundary edge. The properties of AlN grown on sapphire from which silicon has been removed do not appear to be affected by prior Si epitaxy. The selective etchants used in processing are compatible with semiconductor technology. For example, AlN can be selectively etched in hot (~120°C) phosphoric acid, while silicon can be selectively etched by the well known HF + HNO₃ acid mixture.

The potential advantages of AlN in a planar integrated technology for signal processing has led to the study of various test structures incorporating these films both separately and in conjunction with silicon, with varying degrees of success. The thickness limitation of the films has confined the frequency range to the region between about 200 MHz and 1 GHz. Perhaps the most successful application to date has been in programmable analog matched filters reported by Hagon and coworkers (1973B) for spread spectrum applications. In previous work, Wrigley and coworkers (1972) and Hagon and coworkers (1973A) had studied hybrid structures of ST-cut quartz tapped delay lines (TDL) with wire-bonded interconnections to silicon-on-sapphire (SOS) integrated control circuits to provide programmable tapped delay lines (PTDL). By an extension of this work, Hagon and coworkers were able to generate SOS control circuits intraconnected with AlN multiply tapped acoustic delay lines by conventional photolithography. This was accomplished by materials integration in side-by-side configuration on the same substrate. They reported on the performance of a 63 tap PTDL operating with a center frequency of 195 MHz and a chip rate of 20 MHz. Apart from the ability of a PTDL to process different codes, SOS technology offers high-speed capability necessary for real-time programming. A tapped delay line, functioning as a matched filter for an incoming phase-coded waveform, may be switched rapidly enough to match another code without a required time interval between the two received codes. The low parasitic capacitances associated with SOS integrated circuits, by virtue of the insulating sapphire substrate, give a distinct advantage over other integrated circuits in this respect as well as a low-power capability. In addition, the switching of individual tap outputs (to change phase) poses certain restrictions on switch capacitance in the open state for proper isolation because of the high output impedance of the transducer taps. Again, silicon on sapphire is suited to this purpose because of the low capacitance of vertical junction diodes in the thin Si films ($\sim 1\ \mu$m) performing the switching function. The spread spectrum application is of particular military interest and the real-time programming capability offers security, immunity to multipath problems, and multiplexing capability.

Hagon and coworkers (1972) have also demonstrated the successful utilization of AlN on sapphire in the fabrication of wide-band UHF compression filters and noted certain advantages over other piezoelectric materials. For example, the high surface wave velocity of AlN allows about a two-to-one increase in center frequency and bandwidth over quartz or lithium niobate for a given interdigital transducer geometry. For high-resolution radar system applications the wide bandwidth is advantageous. Bulk mode generation by interdigital transducer patterns on AlN/sapphire is lower than for quartz, lithium niobate, or bismuth germanium oxide. This was also observed by Bristol and coworkers (1971). Bulk waves may introduce spurious signals which appear at the output transducer

and may seriously degrade surface wave device performance.

Adkins (1972) studied nonlinear interactions between oppositely directed surface waves on convolvers fabricated from LiNbO$_3$ and AlN/sapphire. The results of his investigation indicate that the convolved signal derived from the interaction of the input surface waves can be a significant source of spurious bulk wave mode generation in LiNbO$_3$, while evidence of bulk wave generation was not found in the case of the AlN/Al$_2$O$_3$ convolvers. In this respect epitaxial AlN holds a favorable position for nonlinear device application. Apart from the disadvantage of spurious signals, bulk wave generation constitutes an energy drain on the desired convolved signal. However, the efficiency of the convolution process must also be considered. Adkins estimated a parametric conversion loss of 46 dB for AlN as compared with a value of 31 dB for LiNbO$_3$. He indicated that the nonlinear interaction for AlN might be enhanced if thicker films were available. It is also unlikely that the films were of optimum quality at the time of testing.

Adkins (1973, 1974) extended his studies to AlN and Si on sapphire monolithic strip-coupled convolvers. In this configuration the fields produced by oppositely directed surface waves on AlN are coupled onto an adjacent silicon film by means of multistrip metallic couplers located midway between the two input transducers and spanning the two films. The component of the electric field normal to the silicon surface and direction of propagation interacts nonlinearly with the charge carriers in the silicon giving rise to an electrical output which is proportional to the convolution function of the two input signals. Devices with the semiconductor placed above the surface wave interaction

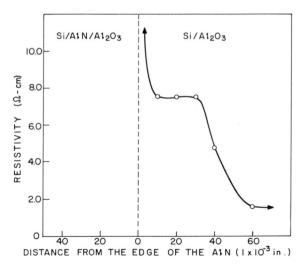

FIGURE 4.28   Resistivity, ρ, of Si as a function of distance from AlN-Al$_2$O$_3$ boundary

region and separated by an air gap (~1000 Å) had previously been shown to have higher efficiencies than surface wave materials alone (Kino et al., 1972; Smith et al., 1973). The utilization of a multistrip coupler (Marshall and Paige, 1971) to couple the strain-induced electric field to an adjacent region outside the beam path allows direct contact between the semiconductor and the strips without interfering with the acoustic wave and eliminates the difficult-to-control air gap. Both these latter approaches were examined by Adkins. The efficiencies of both structures were measured and found to be in reasonable agreement with theory. The evidence further indicated that, with optimum materials properties and design dimensions, efficiencies comparable to those reported (Smith et al., 1973) for LiNbO$_3$/Si air gap devices might be obtained. An efficient AlN/Al$_2$O$_3$ convolver incorporating heteroepitaxial silicon offers several possibilities for signal processing, such as pulse-compression filters and programmable correlators and imagers, and could be integrated with programmable tapped delay lines described above. Optimization of the device structure with respect to fabrication was complicated since it required considerable photolithographic processing. This resulted in low yields because of broken strip couplers.

Budreau and coworkers (1973, 1974) applied the AlN/Al$_2$O$_3$ system to the development of a novel UHF multiple-channel compact frequency synthesizer using surface wave filters. The goal was to generate a number of equally spaced frequencies of high precision suitable for secure communication and data-link systems. Such a system might substitute for the more conventionally used quartz oscillators with frequency multiplication to the microwave range. This technique (Walther et al., 1973) involves the use of just one stable clock to control the frequencies of a comb generator. The comb frequencies are separated into 21 continuously available output frequencies by a contiguous acoustic surface wave filter bank, contact printed on the piezoelectric surface and covering an area of 2 cm × 0.9 cm. No mixers or multipliers are required. The above authors were successful in fabricating an acoustic filter bank of this nature on lithium tantalate with frequencies in the 520 to 650 MHz band and with a channel separation of 5.36 MHz. The same patterns generated on AlN on sapphire should produce frequencies from 970 to 1210 MHz with center frequencies spaced 10 MHz, since the surface acoustic wave velocity of this material on sapphire (film thickness ≃ 1.5 μm) is approximately twice the surface acoustic wave velocity of lithium tantalate. Budreau and coworkers found that, although such filters performed well in selecting the required single frequency from a comb spectrum, thickness nonuniformity and velocity dispersion caused detuning and a shift in null rejection between input and output transducers. They indicate

that a velocity uniformity of better than 0.1 percent is necessary for their application, however, and that for less stringent applications AlN is a highly useful material.

In unpublished work, Hurlburt (1974) has studied the use of AlN/Al$_2$O$_3$ composites in SAW oscillators. In this case the AlN surface wave oscillator was connected with a microstrip matching network to an external amplifier mounted on the same base. The thickness to wavelength ratio for AlN was approximately 0.4. The same interdigital pattern was used on quartz and LiNbO$_3$ for comparison. The fundamental resonant frequencies for the different oscillators are shown in Figure 4.29, which also shows the variation in resonance frequency with temperature. The temperature dependence is in general agreement with the data reported by Liu and coworkers (1975).

Zinc oxide

Zinc oxide is considered one of the most versatile and useful overlay materials for acoustoelectric, acoustooptic, electrooptic, and photoconductive device applications. In a recent novel application [see Coldren (1976)], it has also been utilized as an integral part of MOS planar structures to serve as a charge-storage memory medium with optical and bias sensitivity. Despite the apparent usefulness of ZnO films,

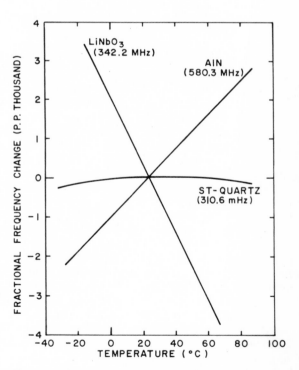

FIGURE 4.29   Frequency variation with temperature of SAW oscillators. After Hurlburt (1974)

TABLE 4.3   Results of Bulk and Surface Attenuation Measurements [After Channin et al. (1975)]

Sample	α_{b0} (dB/cm)	$\partial\alpha_s/\partial\eta_g$ (dB/cm)	Polish	Processing
3-2-72	23.0	300	Cr_2O_3	Substrate fired, film air annealed 900°C before polish
3-29-72B	2.6	166	Cr_2O_3	Slow deposition, film air annealed (900°C) after polish for 1 hr
10-5-72B	0.5	—	Cr_2O_3	Slow deposition, film air annealed (900°C) after polish for 1 hr
12-19-72C	2.2	153	Cr_2O_3	Slow deposition, film air annealed (900°C) after polish for 1 hr; light propagation perpendicular to c axis
12-19-72C	0.3	180	Cr_2O_3	Slow deposition, film air annealed (900°C) after polish for 1 hr; light propagation parallel to c axis
4-26-73A	3.8	41	Syton	Annealed at 850°C
4-26-73B	1.9	39	0.05-μm Linde	Annealed at 850°C

little information is available on the incorporation of these films into SAW programmable devices such as has been accomplished in the case of AlN films. This is probably due to film reactivity, low initial resistivity (except for sputtered films), and low surface wave velocity as compared with AlN. In contrast, epitaxial ZnO films on R-plane sapphire have received much greater attention as active optical waveguides where film crystallinity is more critical than for SAW applications. Hammer and coworkers (1972, 1973, 1975), Channin and coworkers (1975), and Duffy and coworkers (1975) have carried out a critical evaluation of the preparation and optical properties of ZnO films for this purpose. Channin and coworkers have performed a theoretical analysis of light scattering in these films. Table 4.3 summarized some of these results on different waveguides showing bulk and surface attenuation terms, α_{b0} and $\partial\alpha_s/\partial\eta_g$, respectively. Figure 4.30 shows attenuation of wave-guided light of wavelength 6328 Å in three samples from Table 4.3. Each data point corresponds to a particular TE mode and is plotted versus the corresponding waveguide index c/v. The experimental data are fitted by theoretical curves according to Channin. The term α_{b0} in Table 4.3 is the measured loss of the lowest order mode near $c/v = 2.0$ and the term $\partial\alpha_s/\partial\eta_g$ is the measured slope of the curve in the linear region. It is apparent from these results that surface scattering is the dominant factor in determining propagation loss at the higher-order modes and that bulk scattering losses

are strongly dependent on such factors as growth rate and annealing process. The attenuation losses were determined by measuring the intensity of light scattered out of the waveguide directly above the surface of the film over the guided light by means of an optical fiber probe and photomultiplier (see above reference). The intensity monitored by the probe is proportional to the intensity of the guided light in the ZnO film beneath the probe. Thus, a decrease in probe signal along the propagation path represents a corresponding decrease in waveguided light intensity. A curve depicting this is shown in Figure 4.31, for the case of a low-loss waveguide (lowest-order mode in this case) from Table 4.3. The slope of the curve gives the attenuation coefficient, which is the sum of the bulk and surface loss components for a given effective index. When waveguides are prepared as described previously, samples comparable to the best in Table 4.3, with respect to simultaneous minimization of bulk and surface scattering losses, can be obtained. It has also been demonstrated in these studies that bulk scattering losses show a λ^{-4} dependence on wavelength, indicating Rayleigh scattering, and that no appreciable difference in scattering loss occurs for propagation parallel and perpendicular to the film crystallographic c axis.

Hammer and coworkers (1973) have reported on a fast, efficient light deflector and modulator utilizing these films. The device is based on an electrooptically induced Bragg phase grating in the ZnO waveguide

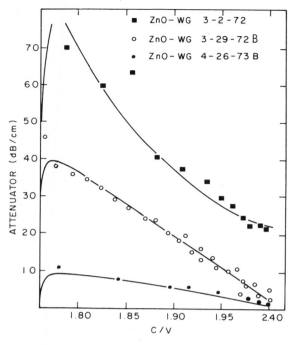

FIGURE 4.30   Attenuation vs. c/v for three samples showing theoretical curves fitted to the data and indicating the influence of processing changes on waveguide loss. After Channin et al. (1975)

structure, as depicted in Figure 4.32. Prisms are used to couple the laser light in and out of the waveguide. A film of amorphous Al_2O_3 about 1000 Å thick extends over the surface of the ZnO except under the prism couplers. When a voltage is applied between the terminals of the interdigital electrodes, a periodic variation in index of refraction is generated in the ZnO waveguide beneath the electrodes and acts as a dif-

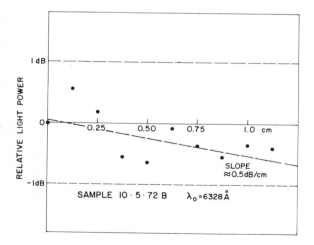

FIGURE 4.31   Measured scattered light intensity vs. propagation distance on a low-loss sample. After Channin et al. (1975)

fraction grating, causing some of the light to be diffracted through an angle in the waveguide plane. When the angle subtended by the guided light and the fingers of the interdigital electrodes, θ, is adjusted to the Bragg angle, θ_B, light is diffracted through an angle $2\theta_B$ from the direction of the main beam with applied electrode bias. The fractional diffraction as defined by the above reference is $I_1/I_0 = \sin^2 B'V$, where I_0 is the intensity of the output beam without applied voltage, and I_1 is the intensity of the diffracted beam when a voltage, V, is applied. The term B' is a function of the optical constants of ZnO and the aperture of the interdigital pattern. An example of the performance of a low-loss ZnO optical modulator (Hammer et al., 1974) is shown in Figure 4.33. The data points correspond to the TE$_0$ mode at the wavelength $\lambda_0 = 6328$ Å, and the solid curve is calculated from $\sin^2 B'V$ normalized to the TE$_0$ mode data at 60 V. The voltage required for 80 percent modulation is about 80 V. The theory predicts 50 percent modulation at $V = 41$ V (Hammer et al., 1973). This is an excellent result when compared with bulk modulators which, for the same interaction length of 0.3 cm, would require a voltage about ten times this value. When the angle, θ, as described above, deviates from the Bragg condition, the intensity of diffracted light decreases. An example is shown in Figure 4.34 for TE$_1$ and TE$_2$ modes as indicated. The halfwidth $\Delta\theta$, corresponding to 50 percent reduction in deflected intensity, as θ is varied about θ_B, is 0.6° for the TE$_1$ mode and close to the theoretical value of 0.58°.

Another potential application for ZnO/sapphire composites is in acoustooptic deflectors and modulators. In this case the guided light beam traverses across a surface acoustic wave which may be amplitude modulated by the signal to be impressed on the light beam. Diffraction of guided optical waves by acoustic surface waves has been studied in several film/substrate combinations, and acoustooptic device functions based on deflection, modulation, and mode conversion have been reported by several authors (Kuhn et al., 1970, 1971; Chubachi et al., 1973, 1974; Ohmachi, 1973A, 1973B; Schmidt et al., 1973; Hillman and Seguin, 1974; Douglas and Hamilton, 1974; Sasaki et al., 1974; Kushibiki et al., 1975; Chubachi, 1976; Lean et al., 1976). Acoustooptic interaction can occur in several ways. Kuhn and coworkers (1971) reported on optical mode conversion by colinear interaction between parallel propagating beams. In most experiments, however, the interacting beams intersect at near-normal incidence (i.e., with the path of the optical beam near parallel to the acoustic beam wavefront). The interaction of the incident beams has been treated quantitatively by Quate and coworkers (1965) and by Klein and Cook (1967). Bragg diffraction, consisting of zero and first order modes occurs when the path of the optical beam is incident at the Bragg angle,

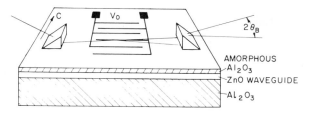

FIGURE 4.32   Schematic diagram of electrooptic waveguide grating modulator. After Hammer et al. (1975)

θ_B, with respect to the acoustic wavefront. The diffracted (or first-order) beam is deflected through an angle $2\theta_B$ relative to the undeflected (or zero-order) beam at the Bragg condition, as depicted in Figure 4.35. This is the condition most frequently employed in thin-film acoustooptic devices.

Some of the references cited above deal with Bragg diffraction in film/substrate composites where the optical waveguide is deposited on a piezoelectric substrate used to generate the surface acoustic waves. In other cases the waveguiding film is also piezoelectric and is deposited on a nonpiezoelectric substrate. The surface waves are generated on the film itself. Sputtered ZnO films have frequently been used in this latter case and efficient deflectors and mode converters have been fabricated. The optical propagation loss associated with sputtered ZnO films tends to be high, about 5 dB/cm or more, as compared with about 1 dB/cm for thin heteroepitaxial ZnO films on sapphire. This latter value represents a satisfactory level for practical device application. Despite the demonstrated useful acoustic and optical properties of heteroepitaxial ZnO, as discussed previously, relatively little seems to have been accomplished with regard to acoustooptic interaction in these films. Brady and coworkers (1973) have reported briefly on the optical and acoustic properties of heteroepitaxial ZnO/sapphire composites of different crystallographic orientation. They indicate that the materials system was not optimized. Difficulties were encountered with the preparation of low-loss $(11\bar{2}0)$ZnO/$(01\bar{1}2)$Al$_2$O$_3$ waveguides, and they obtained best results with $(11\bar{2}4)$ZnO/(0001)Al$_2$O$_3$ composites. In the latter case, they report optical attenuation coefficients of from 1 to 5 dB/cm for low-order TE modes and greater than 40 dB insertion loss for acoustic propagation along the c-axis projection on the plane of the film. No details of acoustooptic interaction were given.

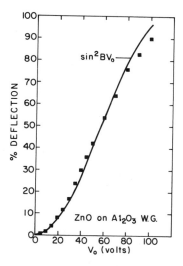

FIGURE 4.33   Percentage deflection vs. voltage for an epitaxial ZnO on sapphire waveguide modulator at $\lambda_0 = 6328$ Å. After Hammer et al. (1974)

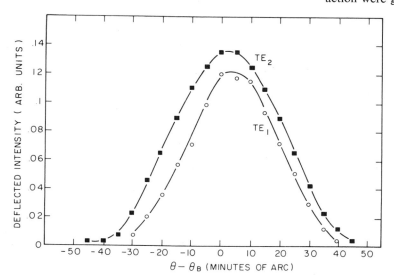

FIGURE 4.34   Deflected intensity (arbitrary units) as θ is varied around θ_B. Squares TE$_2$, $\Delta\theta = 0.73°$. Circles TE$_1$, $\Delta\theta = 0.60°$. After Hammer et al. (1974)

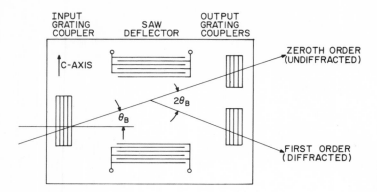

FIGURE 4.35
Schematic diagram of acoustooptic deflector

4.4. Discussion

The discussion here has been confined to AlN and ZnO heteroepitaxial layers on sapphire. This does not mean that other III–V and II–VI compounds in thin-film form are not of interest to microsonics and electrooptics. Relatively little work has been undertaken, however, in the study of heteroepitaxial films of these other materials for the above applications. It is of interest, at this point, to outline some of the factors influencing the usefulness of AlN and ZnO heteroepitaxial films on sapphire and those areas which require further development.

It is clear from the available data that these materials are potentially useful for the above applications. Aluminum nitride films have performed well in surface acoustic wave applications both separately and in combination with silicon on the same substrate. The major difficulties have been (1) lack of reproducibility in film piezoelectric properties not only in consecutive growth processes but also as a function of position on a given samle; (2) the occurrence of a porous faceted surface structure which impedes the growth of continuous films and requires a polishing step when film thickness exceeds about 1.5 to 2 μm; (3) residual strain in the film–substrate composition leading to film cracking at thicknesses greater than about 10 μm; and (4) the need for masking procedures to prevent undesirable doping of heteroepitaxial silicon on the same substrate during the growth process. The first three items are related in part to the growth parameters and purity of the preparative materials. Degradation in crystalline character and film properties can be expected in any heteroepitaxial system if optimum growth conditions are not established. Much remains to be accomplished in this respect in the case of AlN films. Consequently, variation in piezoelectric properties can be expected, and this does not appear to be a limiting factor. On the other hand, the surface structure of (11$\bar{2}$0)-oriented AlN films appears to be an inherent property of this orientation on sapphire and imposes limitations on the film thickness attainable. This also appears to be true irrespective of the choice of preparative materials, metal–organic or chloride, and appears to be related to the nucleation process as discussed previously. Some improvement in the thickness of coherent films can be obtained by consecutive growth and polishing steps. However, the films generally crack with thermal cycling and it is difficult to obtain greater than about 10 μm in coherent film thickness. Although it has been shown that residual strain is related to impurity content (such as silicon), differences in thermal expansion properties between film and substrate for this orientation also appear limiting in terms of film thickness attainable. By contrast, relatively smooth (11$\bar{2}$7)-oriented AlN films can be grown on the same orientation of sapphire to several tens of micrometers in thickness by the chloride process without cracking. In view of this dependence on orientation and the high surface acoustic wave velocity of AlN films, it appears at this time that (11$\bar{2}$0)-oriented films are best suited for high-frequency ($\approx 2 \times 10^8$–10^9 Hz, using conventional photolithography to define transducer patterns) SAW applications. To cover a wider frequency range, (11$\bar{2}$7)-oriented heteroepitaxial films should also be studied. These are relatively smooth and continuous but are subject to twinning, and the influence of the inclination of the c axis to the film surface has not been studied. The extent to which twinning occurs can possibly be controlled by controlling the misorientation of the substrate. In preliminary measurements these films appear to have similar piezoelectric properties to (11$\bar{2}$0)-oriented films but without the same thickness limitations. The masking procedures required to prevent doping of silicon films by the presence of AlN is inconvenient but not necessarily a limiting factor.

Zinc oxide heteroepitaxial layers are comparatively easy to prepare and exhibit a crystalline perfection much superior to that of the corresponding AlN structure. The major difficulties have been (1) low film resistivity requiring compensation by postgrowth diffusion for surface acoustic wave and optical waveguide applications; (2) instability of compensated films to the local ambient and the influence of extrinsic surface states, all of which necessitate a surface passivation process for long-term stability; (3) chemical reactivity of ZnO films in photolithographic processing; (4) the occurrence of a faceted surface which also requires a polishing step; and (5) incompatibility with silicon semiconductor device fabrication on the same substrate. As described previously, compensation and surface passivation are possible and, thus, not limiting factors. Chemical reactivity makes device processing inconvenient but patterns can be defined by photolithographic "lift off" methods. As in the case of AlN, surface faceting can be removed by polishing using the methods developed for polishing silicon wafers. Although both materials are easily polished by silica sol, thickness uniformity, in the case of dispersive films, can be important depending upon the specific application. In addition, ZnO is a relatively soft material and surface scratch marks can be introduced easily during the polishing process. However, neither factor is limiting in principle. One of the most appealing prospects for utilization of these materials is the possibility of materials integration with silicon on a common substrate. Since ZnO decomposes at silicon deposition temperatures in a H_2 ambient, the silicon film must be deposited first. Also, at the processing temperatures for silicon device fabrication, an outdiffusion of lithium occurs from Li-compensated ZnO. Consequently, the growth and compensation of ZnO must be performed subsequent to silicon device fabrication. The success of this operation then depends on the ability to sufficiently mask MOS devices against contamination with alkali species during the compensation of ZnO films. Although these steps are possible in principle, the difficulties involved may be considerable. Ion implantation of ZnO layers thin enough for integrated optics could possibly aid in this respect. Thick layers for surface wave and acoustooptic applications would probably have to be compensated by diffusion. In general, ZnO layers up to several tens of micrometers in thickness can be grown on sapphire. Films thicker than about 20 μm crack, however, when compensated and this forms an upper limit to the thickness attainable for device purposes.

In conclusion, heteroepitaxial AlN on sapphire has good potential for high-frequency ($\approx 2 \times 10^8 - 10^9$ Hz) surface acoustic wave application both separately and in combination with silicon-on-sapphire substrates. At present the crystalline perfection of the films is far from approaching the quality required for integrated optics. Zinc oxide heteroepitaxial films on sapphire have useful properties for lower surface wave applications than AlN and have excellent optical properties but have much less potential for integration with silicon semiconductor device functions on the same substrate. This is a serious limitation. Although SAW and acoustooptic applications are possible, these layers are outperformed by lithium niobate–lithium tantalate

waveguide structures for deflectors and modulators in integrated optics (Hammer and Phillips, 1974). Much more work is necessary to realize the full potential of heteroepitaxial III-V and II-VI compounds on sapphire.

References

Adkins, L. R., IEEE Ultrasonics Symposium Proceedings (1972) p. 292.

Adkins, L. R., IEEE Ultrasonics Symposium Proceedings (1973) p. 148.

Adkins, L. R., IEEE Ultrasonics Symposium Proceedings (1974) p. 228.

Arijs, E. and F. Cardon, *J. Solid State Chem. 6:* 310 (1973)

Arijs, E., F. Cardon, and W. Maenhout-Van der Vorst, *Surface Sci. 17:* 387 (1969).

Arijs, E., F. Cardon, and W. Maenhout-Van der Vorst, *J. Solid State Chem. 6:* 319 (1973).

Balkanski, M. and P. Lallemand (Eds.), *Photonics,* Paris: Gauthier-Villars (1973).

Ban, V. S., *J. Electrochem. Soc. 119:* 761 (1972).

Barnoski, (Ed.), *Fundamentals of Optical Fiber Communications,* Academic Press, New York (1976).

Brady, M. J., M. Levanoni, and A. Reisinger, *Optics Communications 7:* 391 (1973).

Bristol, T. W., L. Dyal, P. J. Hagon, and K. M. Lakin, IEEE Ultrasonics Symposium, Miami Beach, Florida (1971) Paper C-6.

Budreau, A. J., P. H. Carr, and K. R. Laker, IEEE Ultrasonics Symposium Proceedings (1973) p. 464.

Budreau, A. J., J. H. Silva, and K. R. Laker, IEEE Ultrasonics Symposium Proceedings, (1974) p. 299.

Callaghan, M. P., E. Patterson, B. P. Richards, and C. A. Wallace, *J. Crystal Growth 22:* 85 (1974).

W. J. Campbell and C. Grain, U.S. Bureau of Mines Report on Investigation No. 5757, 1961 (unpublished).

Campbell, J. J. and W. R. Jones, *J. Appl. Phys. 41:* 2976 (1970).

Carnes, J. E. and M. T. Duffy, *J. Appl. Phys. 42:* 4350 (1971).

Carr, P. H., *IEEE Trans. Microwave Theory Tech. MIT-17:* 845 (1969).

Channin, D. J., J. M. Hammer, and M. T. Duffy, *Appl. Optics 14:* 923 (1975).

Chu, T. L., *J. Electrochem. Soc. 118:* 1200 (1971).

Chu, T. L., D. W. Ing, and A. J. Noreika, *Solid-State Electron. 10:* 1023 (1967).

Chubachi, N., *Proc. IEEE 64:* 772 (1976).

Chubachi, N., J. Kushibiki, and Y. Kikuchi, *Electron. Letters 9:* 193 (1973).

Chubachi, N., J. Kushibiki, H. Sasaki, and Y. Kikuchi, *Japan Soc. Appl. Phys. Suppl. 43:* 199 (1974).

Coldren, L. A., *Proc. IEEE 64:* 769 (1976).

Collins, R. J. and D. G. Thomas, *Physical Review 112:* 388 (1958).

Collins, J. H., P. J. Hagon, and G. R. Pulliam, *Ultrasonics 8:* 218 (1970).

Cox, G. A., D. O. Cummins, K. Kawabe, and R. H. Tredgold, *J. Phys. Chem. Solids 28:* 543 (1967).

Cullen, G. W., M. T. Duffy, and C. C. Wang, Fourth International Conference on Chemical Vapor Deposition, Boston, MA, Oct. 8–11, Princeton, NJ: The Electrochemical Society, Inc. (1973) p. 247.

Denburg, D. L., *IEEE Trans. Sonics Ultrasonics SU-18:* 31 (1971).

Douglas, A. W. and M. C. Hamilton, *Appl. Phys. Letters, 24:* 159 (1974)

Duffy, M. T. and A. G. Revesz, *J. Electrochem. Soc. 117:* 372 (1970).

Duffy, M. T. and R. T. Smith, unpublished work (1974).

Duffy, M. T. and P. J. Zanzucchi, unpublished work (1973).

Duffy, M. T., J. E. Carnes, and D. Richman, *Metallurgical Trans. 2:* 667 (1971).

Duffy, M. T., C. C. Wang, G. D. O'Clock, Jr., S. H. McFarlane III, and P. J. Zanzucchi, *J. Elect. Mat. 2:* 359 (1973).

Duffy, M. T., D. J. Channin, and J. M. Hammer, U.S. Patent 3,924,020, 1975.

Dugger, C. O., *Mat. Res. Bull. 9:* 331 (1974).

Fiermans, L., E. Arijs, J. Vennik, and W. Maenhout-Van der Vorst, *Surface Sci. 39:* 357 (1973).

Foster, N. F., *J. Appl. Phys. 40:* 4202 (1969A).

Foster, N. F., *J. Vac. Sci. Technol. 6:* 111 (1969B).

Galli, G. and J. E. Coker. *Appl. Phys. Letters 16:* 439 (1970).

Hada, T., K. Wasa, and S. Hayakawa, *Thin Solid Films 7:* 135 (1971).

Hagon, P. J., L. Dyal, and K. M. Lakin, IEEE Ultrasonics Symposium Proceedings (1972) p. 274.

Hagon, P. J., F. B. Micheletti, R. N. Seymour, and C. Y. Wrigley, *IEEE MTT-21:* 303 (1973A).

Hagon, P. J., F. B. Micheletti, and R. N. Seymour, IEEE Ultrasonics Symposium Proceedings (1973B) p. 333.

Hall, R. O. A. Acta. *Crystal. 14,* 1004 (1961)

Hammer, J. M., D. J. Channin, M. T. Duffy, and J. P. Wittke, *Appl. Phys. Letters 21:* 358 (1972).

Hammer, J. M., D. J. Channin, and M. T. Duffy, *Appl. Phys. Letters 23:* 176 (1973).

Hammer, J. M., D. J. Channin, M. T. Duffy, and C. C. Neil, *IEEE J. Quantum Electron. QE-11:* 138 (1975).

Hammer, J. M. and W. Phillips, *Appl. Phys. Letters 24:* 545 (1974)

Heiland, G. and P. Kunstmann, *Surface Sci. 13:* 72 (1969).

Hickernell, F. S., *Proc. IEEE 64:* 631 (1976).

Hillman, G. D. and H. J. J. Seguin, *Can. J. Phys. 52:* 1096 (1974).

Holland, M. G. and L. T. Claiborne, *Proc. IEEE 62:* 582 (1974).

Hurlburt, D. H., unpublished work (1974).

Ilegems, M. and H. C. Montgomery, *J. Phys. Chem. Solids 34:* 885 (1973).

Iyengar, R. D. and M. Codell, *Advan. Colloid Interface Sci. 3:* 365 (1972).

Jacob, G., R. Madar, and J. Hallais, *Mat. Res. Bull. 11:* 445 (1976).

Kaminow, I. P. *An Introduction To Electroptic Devices,* Academic Press, New York (1974).

Kino, G. S., W. R. Shreve, and H. R. Gautier, Ultrasonics Symposium Proceedings, IEEE Cat. #72 CHO 708-8 SU (1972), p. 285.

Klein, W. R. and B. D. Cook, *IEEE Trans. Sonics and Ultrasonics SU-14:* 123 (1967).

Kokes, R. J. and R. Glemza, *J. Phys. Chem. 69:* 17 (1965).

Kosicki, B. B. and D. Kahng, *J. Vac. Sci. Technol. 6:* 593 (1969).

Kuhn, L., M. L. Dakss, P. F. Heidrich, and B. A. Scott, *Appl. Phys. Letters 17:* 265 (1970).

Kuhn, L., P. F. Heidrich, and E. G. Lean, *Appl. Phys. Letters 19:* 428 (1971).

Kushibiki, J., H. Sasaki, N. Chubachi, N. Mikoshiba, and K. Shibayama, *Appl. Phys. Letters 26:* 362 (1975).

Lakin, K. and D. Penunuri, IEEE Symposium, Miami Beach, (1971) Paper D-7.

Lakin, K. M., L. Liu, and K. Wang, Ultrasonics Symposium Proceedings, IEEE Cat. #74 CHO 896-ISU (1974) p. 302.

Lean, E. G. H., J. M. White, and C. D. W. Wilkinson, *Proc. IEEE 64:* 779 (1976).

Levine, J. D., A. Willis, W. R. Bottoms, and P. Mark, *Surface Sci. 29:* 144 (1972).

R. Leysen, G. van Orshaegen, H. van Hove, and A. Neyens, *Phys. Stat. Sol. 18:* 613 (1973).

Liu, J. K., R. B. Stokes, and K. M. Lakin, Ultrasonics Symposium Proceedings (1975) p. 234.

Maines, J. D., Ultrasonics Symposium Proceedings, IEEE Cat. No. 74 CHO 896-1SU (1974).

Manasevit, H. M. and W. I. Simpson, *J. Electrochem. Soc. 118:* 644 (1971).

Manasevit, H. M., F. M. Erdmann, and W. I. Simpson, *J. Electrochem. Soc. 118:* 1864 (1971).

Maradudin, A. A. and D. L. Mills, *Appl. Phys. Letters 28:* 573 (1976).

Marshall, F. G. and E. G. S. Paige, *Elect. Lett. 7:* 460 (1971).

Maruska, H. P. and J. J. Tietjen, *Appl. Phys. Letters 15:* 327 (1969).

Matsumoto, T. and M. Aoki, *Japan J. Appl. Phys. 13:* 1583 (1974).

McFarlane, S. H., III and C. C. Wang, *J. Appl. Phys. 43:* 1724 (1972).

Morrison, S. R., *Surface Sci. 13:* 85 (1969).

Morrison, S. R., *J. Vacuum Sci. Technol. 7:* 84 (1970).

Noreika, A. J. and D. W. Ing, *J. Appl. Phys. 39:* 5578 (1968).

Noreika, A. J., N. H. Francombe, and S. A. Zeitman, *J. Vacuum Sci. Technol. 6:* 194 (1969).

O'Clock, G. D. and M. T. Duffy, *Appl. Phys. Letters 23:* 55 (1973).

Olson, F. A., *Microwave Journal 13:* 67 (1970).

Omachi, Y., *Electron. Letters 9:* 539 (1973A).

Omachi, Y., *J. Appl. Phys. 44:* 3928 (1973B).

Onyshkevych, L. S., private communication (1975).

Onyshkevych, L. S. and M. T. Duffy, unpublished results (1975).

Pankove, J. I., *J. Luminescence 7:* 114 (1973).

Pankove, J. I., E. A. Miller, D. Richman, and J. E. Berkeyheiser, *J. Luminescence 4:* 63 (1971A).

Pankove, J. I., E. A. Miller, and J. E. Berkeyheiser, *RCA Review 32:* 383 (1971B).

Pankove, J. I., J. E. Berkeyheiser, and E. A. Miller, *J. Appl. Phys. 45:* 1280 (1974).

Pankove, J. I., S. Bloom, and G. Harbeke, *RCA Review 36:* 163 (1975).

Pastrnak, J. and L. Souckova, *Phys. Stat. Solidi 3:* K71 (1963).

Pastrnak, J. and L. Roskovocova, *Phys. Status Solidi 26,* 591 (1968).

Penunuri, D. and K. M. Lakin, IEEE Ultrasonics Symposium Proceedings (1972) p. 328.

Penunuri, D. and K. M. Lakin, Ultrasonics Symposium Proceedings (1975) p. 478.

Pizzarello, F., *J. Appl. Phys. 43:* 3627 (1972).

Pizzarello, F. A. and J. E. Coker, *J. Electron. Mat. 4:* 25 (1975).

Quate, C. F., C. D. W. Wilkinson, and D. K. Winslow, *Proc. IEEE 53:* 1604 (1965).

Reeder, T. M., and D. K. Winslow, *IEEE Trans. Microwave Theory and Techniques MTT-17:* 927 (1969).

Reisman, A. and J. E. Landstein, *J. Electrochem. Soc. 118:* 1479 (1971).

Reisman, A., M. Berkenblit, S. A. Chan, and J. Angilello, *J. Electron. Mat. 2:* 177 (1973)

Rozgonyi, G. A. and W. J. Polito, *J. Vacuum Sci. Technol. 6:* 115 (1969).

Sasaki, H., J. Kushibiki, and N. Chubachi, *Appl. Phys. Letters 25:* 476 (1974).

Schmidt, R. V., I. P. Kaminow, and J. R. Carruthers, *Appl. Phys. Letters 23:* 417 (1973).

Shiloh, M. and J. Gutman, *J. Crystal Growth 11:* 105 (1971).

Shuskus, A. J., D. J. Quinn, E. L. Paradis, J. M. Berak, and T. M. Reeder, NTIS Report M951337-6, (1973) p. 20.

Smith, W. R., *J. Appl. Phys. 42:* 3016 (1971).

Smith, J. M., E. Stern, and A. Bers, *Elect. Lett. 9:* 145 (1973).

Stern, E., IEEE Trans. Microwave Theory Tech. *MTT-17,* 835 (1969).

Tamir, T. (Ed.), *Integrated Optics,* Topics in Applied Physics, Vol. 7, New York: Springer-Verlag (1975).

Taylor, K. M. and C. Lenie, *J. Electrochem. Soc. 107:* 308 (1960).

Valdes, L. P., *Proc. IRE 42:* 420 (1954).

Vuilloud, J., *J. Vac. Sci. Technol. 9:* 87 (1972).

Wachtman, J. B. Jr., T. G. Scuderi, and G. W. Cleek, *J. Am. Ceram. Soc. 45,* 319 (1962).

Walther, F. G., A. J. Budreau, and P. H. Carr, *Proc. IEEE 61:* 1162 (1973).

White, R. M., *Proc. IEEE 58:* 1238 (1970).

Wickenden, D. K., K. R. Faulkner, R. W. Brander, and B. J. Isherwood, *J. Crystal Growth 9:* 158 (1971).

Wille, D. A. and M. C. Hamilton, *Appl. Phys. Letters 24:* 159 (1974).

Williams, R. and A. Willis, *J. Appl. Phys. 39:* 3732 (1968).

Wrigley, C. Y., P. J. Hagon, and R. N. Seymour, IEEE Ultrasonics Symposium Proceedings (1972) p. 226.

Yim, W. M. and R. J. Paff, *J. Appl. Phys. 45:* 1456 (1974).

Yim, W. M. and E. J. Stofko, *J. Electrochem. Soc. 119:* 381 (1972).

Yim, W. M., J. P. Dismukes, and H. Kressel, *RCA Review 31:* 662 (1970).

Yim, W. M., E. J. Stofko, P. J. Zanzucchi, J. I. Pankove, M. Ettenberg, and S. L. Gilbert, *J. Appl. Phys. 44:* 292 (1973).

Zanzucchi, P. J. and M. T. Duffy, unpublished results (1973).

Chapter 5

Characterization of Heteroepitaxial Thin-Film Semiconductors on Insulating Substrates

P. J. Zanzucchi

5.1. Introduction

The production of reliable solid state electronic devices from bulk (John, 1967), homoepitaxial, or heteroepitaxial (Koelmans, 1971) material is largely dependent on controlling the properties of these materials. For example, the impurity content and crystallinity of semiconductor materials will strongly influence device performance. Semiconductor technology could not have developed so rapidly without some understanding of the properties of materials involved in fabricating solid state devices. Over the past years many measurement techniques have been developed, and it is through the use of these techniques that the performance of semiconductor devices can be related to such properties as the dopant density, impurity content (uncontrolled doping), and lattice perfection.

Developments in solid state technology have been very rapid. The materials used, the design of semiconductor devices, and, consequently, the types of characterization methods required have changed rapidly. Since the 1950s, new processes have resulted in the steady reduction in the volume of semiconductor material required for devices. With the advent of heteroepitaxial films as starting materials for device fabrication, the amount of semiconductor used in typical device structures is of the order of microvolumes (μm^3). As a consequence of this, bulk characterization techniques are not suitable for many aspects of thin-film characterization; this requires analysis of surface and bulk composition, the latter as a function of film thickness. In the field of mass spectrometry, for example, impurity enrichment by thermal evaporation (Honig, 1953) and, based on the early work of Dempster (1936), spark source mass spectrometry (Gorman et al., 1951; Hannay, 1954; Hannay and Ahearn, 1954;

Ahearn, 1961) were developed to survey trace elemental impurities at the parts-per-million (ppm) level in bulk and thick-film semiconductor and related materials. Yet these methods were not completely suitable for many aspects of homoepitaxial or heteroepitaxial thin-film semiconductor characterization. In particular with these methods, the area or volume of material sampled is larger than the dimensions of discrete thin-film devices. The research of Honig (1958), Bradley (1959), Stanton (1960), and others concerning the sputtering of surfaces by ion beams predated the relatively recent development of secondary-ion mass spectrometry (Castaing and Slodzian, 1966; Benninghoven, 1970) and ion-scattering spectrometry (D. Smith, 1967). An especially important feature of these and other recently developed techniques is that the area or volume analyzed can be comparable to the micrometer dimensions of thin-film semiconductor device components. Surface analysis techniques have been critically reviewed by Honig (1976).

It is interesting to note that many bulk or surface techniques give information concerning the elemental composition of materials, yet only a limited number of techniques are available that give information concerning molecular composition. Spark source mass spectrometry, Auger electron spectrometry, and low- or high-energy Rutherford ion-scattering spectrometry all provide, to varying degrees, information about the bulk or surface-distributed elemental compositions of solids.* For the semiconductor industry, the interest

*For a summary description of bulk and surface analysis methods, see Kane and Larrabee (1972, 1974). The many physical processes measured by various instrumental techniques often show a strong dependence on the chemical state of the sample, and this is termed a "matrix" effect. Through matrix effects, information on the chemical states

in measuring the trace element content of materials reflects the need to control the unintentional doping of semiconductors by parts per million (ppm) to parts per billion (ppb) of impurities.

Elemental species are often present in semiconductor and related materials as molecular complexes (Newman, 1969). Furthermore, molecular species can be important in the preparation of thin-film semiconductors (Ban, 1971) as well as in device processing (Olberg, 1971). To identify molecular species, the techniques for bulk analysis, which include mass, infrared, and x-ray spectrometry as well as the more recently developed surface analysis techniques of electron spectroscopy for chemical analysis (ESCA) (Siegbahn et al., 1967), are well established. Because of the need for surface analysis, there have been many recent developments in the use of these and related techniques for thin-film analysis. One example is the development of inelastic electron-tunneling spectroscopy. From electron-tunneling data, the molecular vibrational spectra of surface material of less than 20–30 Å thickness can be obtained (Lambe and Jaklevic, 1968).

The identification of elemental or molecular species is only one aspect of semiconductor characterization. In this chapter, four selected aspects of semiconductor materials important in the preparation and production of semiconductor devices are reviewed. Measurement of film thickness, dopant density, lattice perfection, or impurity content contributes directly to understanding or improving the performance of solid state materials and devices. This chapter briefly reviews the importance of these factors in semiconductor technology, with respect to heteroepitaxial composites in particular, and gives a summary description of the various specific measuring techniques. Although many techniques are described for their general applications, it should be noted that the properties of heteroepitaxial semiconductors often are a function of film thickness. Consequently, understanding the properties of these films requires, in many cases, measurements that provide information as a function of depth. Because silicon films for silicon-on-sapphire (SOS) devices are nominally 0.5 to 1.5 μm thick, measurements of this type must provide information from film regions only a few hundred angstroms thick (e.g., by sputter etching combined with a surface analysis technique such as Auger spectroscopy). In characterizing heteroepitaxial films, one important consideration is the thickness of material required to provide information.

As all measuring techniques have inherent detection limits, some minimum depth (more precisely, a mini-

mum volume) of sample is required for a measurement. A general and somewhat approximate comparison of the average depth of material required for selected analytical methods is given in Figure 5.1. Closed rectangular boxes indicate surface methods (excluding the use of sputter etch techniques), as well as multilayer or bulk analysis methods, with fixed depth of sampling. The open rectangular boxes are used to indicate a variable depth of sampling for some of the measurement techniques. In these cases, depth of sampling is a parameter controlled by experimental variables. For example, the depth of ion beam penetration in Rutherford ion-scattering spectrometry is controlled by the energy of the incoming ion beam. In ellipsometry, the depth of light beam penetration is related to the absorption coefficient, which is often measured by this technique. For bulk analysis in which the sample must be in solution or other form prior to analysis (e.g., with atomic absorption spectrometry) the total weight of sample available for measurement is a more meaningful quantity than is depth of sampling.

5.2. Characterization of thin-film epitaxial semiconductors

Film thickness

Optical interference measurements. The film thickness of micrometer-thick semiconductors on insulating substrates must be measured for meaningful correlation of electrical data, for precise doping by high-temperature diffusion techniques, for precision etching of circuit elements, and for fabrication of protective layers. Although many techniques exist for measuring micrometer film thickness (see Table 5.1), the thickness of epitaxial semiconductors is most often measured by optical interference methods. These methods are nondestructive, accurate to a fraction of a wavelength of light, and relatively simple in procedure.

Interference optical methods date back to observations by Newton (Heavens, 1955) of color fringes produced by interference effects. Because the presence of interference fringes depends on film thickness, the color of thin films is a convenient way of estimating ~0.1–1.0 μm film thicknesses. Tables correlating color and film thickness have been published for SiO_2 films on silicon (Pliskin and Conrad, 1964; Corl and Wimpfheimer, 1964) and for Si_3N_4 films on silicon (Reizman and van Gelder, 1967). In recent years the optical properties of thin films, in particular measurement of film thickness, have been discussed in many publications. Reviews by Pliskin and Zanin (1970) and Gillespie (1967) are oriented toward epitaxial semiconductor thin-film measurements. More general discus-

---

of materials can be obtained indirectly from techniques primarily used for elemental analysis; see, for example, Castle and Epler (1974) with regard to matrix effects in Auger electron spectrometry.

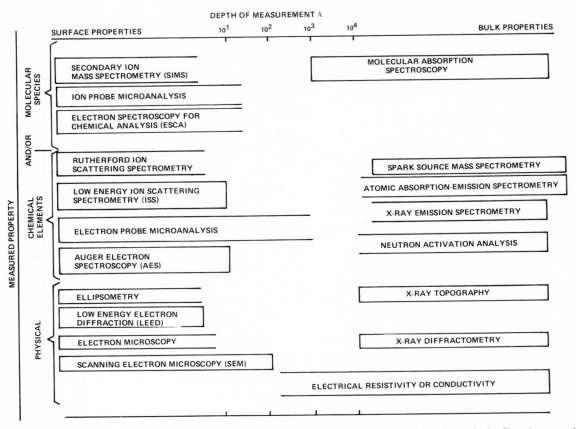

FIGURE 5.1   Comparison of measurement methods by the nominal depth of material required for analysis. Closed rectangular boxes indicate the surface measurement methods as well as the bulk methods that have fixed depth of sampling. See text for an explanation of open boxes

sions of the optical properties of thin films with comments on the measurement of film thickness can be found in the books by Tolansky (1948), Heavens (1955, 1970), and Vasicek (1960).

Optical interference involves the reflectance and transmittance of light, as shown in Figure 5.2. The film thickness, t, is related to the interference of reflected rays R_0 and R_1 by

$$t = \frac{N\lambda_0}{2(n_1^2 - \sin^2 \phi_0)^{1/2}}, \qquad (5.1)$$

where N is the fringe order. From Equation (5.1) it is clear that either the angle of incidence, ϕ_0, or the wavelength, λ_0, can be varied to determine film thickness. The characteristics of these two different optical interference procedures have been reported. Reizman and van Gelder (1967) as well as Corl and Wimpfheimer (1964) have discussed constant-angle reflection interference spectroscopy (CARIS); Pliskin and Conrad (1964) have discussed variable-angle monochromatic fringe observation (VAMFO). Popova and Jordanov (1970) and Jordanov and Popova (1971) have reported on features of the two procedures.

If the interference fringes are recorded over a wavelength range λ_1 to λ_2, which correspond to wavelengths for a maximum or minimum in fringe amplitude, Equation (5.1) can be rewritten as

$$t = \frac{X\lambda_1\lambda_2}{2(\lambda_1 - \lambda_2)(n_1^2 - \sin^2 \phi_0)^{1/2}}, \qquad (5.2)$$

where X is the number of complete fringes in the wavelength region λ_1 to λ_2. Use of this method for determining the thickness of homoepitaxial silicon on silicon was first reported by Keck (1961) as well as Spitzer and Tanenbaum (1961). A more detailed discussion of the method has been given by Albert and Combs (1962). A standard procedure for measuring the film thickness of silicon epitaxially grown on silicon is given by the American Society for Testing and Materials, ASTM Method F95-73 (1973).

Equations (5.1) and (5.2) imply that (1) in the interface region between the film and substrate there is a sharply defined junction between two materials with different refractive indices, (2) the substrate does not absorb at the wavelength of the radiation involved in the measurement (i.e., in the wavelength range λ_1 to

TABLE 5.1   Survey of Methods for Measuring Film Thickness

Minimum Measurable Thickness (nm)	Methods for Measuring Film Thickness		
	Optical	Mechanical	Electrical
10^0-10^1	Interferometry Ellipsometry X-ray absorption[a]	Stylus displacement	Quartz crystal Oscillator[b]
10^1-10^2			Conductivity[d]
10^2-10^3 [c]	Color comparison Variable-angle monochromatic fringe observation Constant-angle reflection interference spectroscopy Infrared reflection Optical absorption[g] Internal reflection[h]	Angle lap[e] Stacking fault dimensions[f]	
10^3-10^4 [c]	Light section microscopy	Cleaving across film[i]	

Source: Gillespie (1967).
[a]Requires calibration.
[b]Measures thickness of films evaporated onto oscillator crystal.
[c]Range of thickness for conventional semiconductor films.
[d]Requires calibration for each film material.
[e]Often by use of a calibrated microscope.

[f]Stacking fault dimensions are related to film thickness for silicon on silicon (Mendelson, 1964).
[g]Requires calibration for each film material.
[h]Requires use of a prism.
[i]Often by use of a calibrated microscope.

λ_2), and (3) the refractive index of the film and the film thickness are uniform over the measured film region. Reflection R_1 cannot occur if the film and substrate have the same refractive indices. Absorption of ray R_1 by the substrate causes the interface to be less sharply defined as the ray reflects partly from a depth in the substrate. Moreover, if the absorption coefficient of the substrate is significant, the integral phase change in R_1 implicit in Equation (5.1) does not occur; the phase change in R_1 becomes a function of the refractive and absorption indices of the substrate material. A corrective term for the phase shift must be included in Equation (5.1) when the substrate absorbs at the wavelength of the measurement. This can be done in two ways. Calibration curves can be used where film thickness measured by some independent means is correlated with the wavelength of interference maxima or minima. Alternately, the various n and k values can be determined from which the phase shift and correct film thickness are then calculated. Reizman (1965) has reported on the use of calibration curves for determining the thickness of silicon oxide on silicon whereas Schumann with various coworkers (Schumann et al., 1966; Schumann and Phillips, 1967; Schumann and Schneider, 1970; Schumann, 1970B), Pliskin

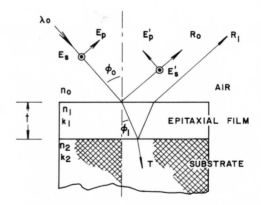

FIGURE 5.2   Diagram of the transmitted and reflected light rays, R_0 and R_1; optical interference of R_0 and R_1 is related to the film thickness, t. Note that ray vector, E_s, is directed into the plane of the paper indicated by the interior circle (i.e., E_s and E_p are perpendicular transverse components of the light ray). Legend: n, refractive index; k, absorption index; ϕ_0, angle of incidence; ϕ_1, angle of refraction; R, reflected ray; T, transmitted ray; E_p, parallel component of the electric vector; E_s, perpendicular component of the electric vector; λ_0, wavelength of the incident ray

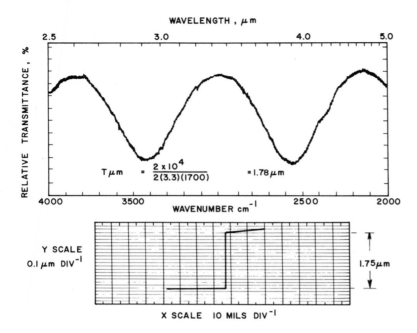

FIGURE 5.3
Comparison of optical interference and mechanical (stylus) data for determining the film thickness of a heteroepitaxial silicon film on sapphire

(1968), and Severin (1970) have discussed calculations of phase shift corrections for films on doped silicon substrates. In relation to this, Sato and coworkers (1966) and Schumann (1969) have discussed the significance of autodoping in homoepitaxial films by the substrate dopant. Autodoping causes the homoepitaxial film/substrate interference to be less sharply defined, because the refractive indices of the film and substrate are related to the doping density. Autodoping thus introduces errors in the measured film thicknesses. Although autodoping occurs with heteroepitaxial films such as silicon on sapphire (Dumin and Robinson, 1966) and silicon on spinel (Robinson and Dumin, 1968), the optical reflection from the film/substrate interface is largely unaffected because of the large discontinuity in optical properties between silicon and metal oxide substrates. The refractive index of silicon is approximately 3.3, while for sapphire or spinel the refractive index is approximately 1.7.

Comparison of mechanically and optically determined film thickness of silicon on sapphire (Figure 5.3) shows, as expected, that the measurements are in good agreement. Moreover, for wavelengths between approximately 0.2 and 5.0 μm, in which both sapphire and spinel are transparent, a substrate phase correction is not needed in order to calculate the film thickness. There is, however, the real possibility that mechanical and optically determined film thicknesses will not agree because of silicon's surface structure. As discussed in detail in previous chapters, nucleation and growth of the silicon film (on sapphire) proceed through various forms of island growth. Under certain conditions a poor, textured silicon surface will form and the effects of this cannot be ignored in measuring film thickness. For example, the stylus in displacement-measuring devices is typically a micrometer in diameter and is too large to resolve the surface texture of poor heteroepitaxial silicon films; i.e., it measures a film thickness relative to the "top" of the textured surface. In contrast, optical techniques measure, for example, interference fringes due to parallel boundaries, which will average to some region below the "top" of the textured surface. Thus, in general, the actual film thickness of a silicon film on an insulating substrate can be accurately measured by a variety of methods as described in this section only when the surface texture is negligible.

As indicated by Equation (5.1), the thickness of a film can be determined if the refractive index, n_1, as well as the angle of incidence and the wavelength are known. Refractive indices for many semiconductors are given in the publications by Moss (1959), Moses (1971), and Neuberger (1971). A review of the optical properties of III-V compounds has been published by Seraphin and Bennett (1967). The optical constants of silicon have been published by Salzberg and Villa (1957) and more recently by Schumann and coworkers (1971) for doped silicon. The optical properties of sapphire have been reported by Malitson and coworkers (1958) and the optical properties of magnesium aluminate spinel have been reported by Wickersheim and Lefever (1960) as well as Wang and Zanzucchi (1971). The optical properties of silicon on spinel have been published by Stein (1972).

The optical properties of crystalline, polycrystalline, and amorphous silicon on spinel and sapphire substrates have been studied by Kuhl and coworkers (1974). From the characteristics of interference fringes and Kramers–Kronig analysis of reflection data, the optical constants for the silicon films have been determined. These data show that the optical constants are related to the degree of crystalline disorder in the films. Note silicon films on sapphire which have good crystalline quality have optical properties similar to bulk silicon. Data reported by Hulthén (1975) are shown in Figure 5.4.

It is often desirable to grow epitaxial films to a specified thickness. Although the thickness of semiconductor films can be estimated from the duration of the deposition (with predetermined growth rates), this does not allow changes in other deposition system parameters. A novel optical interference method for the measurement of film thickness in situ has been reported by Dumin (1967) for the epitaxial deposition of silicon on sapphire. At temperatures near ~1200°C, the heated substrate emits predominantly infrared radiation. Therefore, Dumin used a detector with a

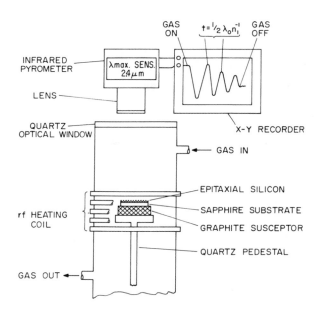

FIGURE 5.5   Schematic diagram of the apparatus used for measurement of heteroepitaxial silicon film thickness during deposition. By monitoring the radiation emitted from the heated substrate with an optical pyrometer, the film thickness of the growing epitaxial film can be monitored in situ as a function of time. From Dumin (1967)

peak sensitivity at 2.4 μm to monitor (as shown in Figure 5.5) the radiation from a heated sapphire substrate. As an epitaxial silicon film is grown on the sapphire, interference fringes are recorded. These fringes represent changes in the transmission of substrate radiation which are directly related to the film thickness of the epitaxial silicon. The film thickness of silicon per cycle of recorded interference, t (in μm/cycle), is $\frac{1}{2} \lambda_0 n_1^{-1}$, which can be easily derived from Equation (5.1) for normal incidence. Although interference measurements are generally made on a fixed film thickness by varying the wavelength of incident radiation, in this procedure the wavelength is fixed (by the substrate temperature) and the film thickness varies with time. Thus, at 2.4 μm for silicon, the film thickness per cycle of recorded interference fringes is found experimentally to be 0.36 μm/cycle. Note that the refractive index for silicon must be the value determined at 1200°C if theoretical calculations are to be made to compare to the experimentally determined t. The amplitude of the interference fringes is attenuated by the optical absorption, which follows an $e^{-n\alpha t}$ function (n is a whole number, α is the absorption coefficient, and t is the film thickness) and scattering which occurs in the film with increasing film thickness. Thus, it is found experimentally that the fringe amplitudes diminish significantly with films greater than several micrometers in thickness. Dumin (1967)

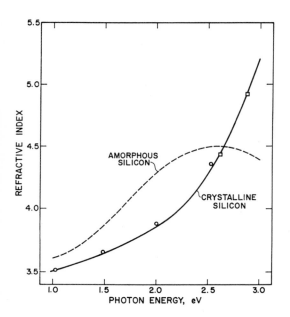

FIGURE 5.4   Refractive index of crystalline and amorphous silicon in the 1–3 eV region. [See D. T. Pierce and W. E. Spicer, *Phys. Rev. 5:* 3017 (1972) for optical data on amorphous silicon and H. R. Phillipp and E. Taft, *Phys. Rev. 120:* 37 (1960) for optical data on crystalline silicon.] Circles and squares are selected data points from Hulthén (1975) obtained by transmittance and photoconductivity measurements, respectively. These data show that the refractive index of heteroepitaxial silicon is identical to bulk crystalline silicon

reports that the film thickness of silicon films can be measured to $\pm$ 0.1 μm in the range of 0.1–15 μm. In general, silicon for SOS technology is 0.5 to 1.5 μm thick.

In general, interference optical measurements are simple to perform and accurate. The rate at which samples can be measured with dispersive spectrometers is limited by the time required to record the interference fringes, a minimum of 30–60 s. In connection with this problem, an interesting application of interferometry has been reported (Cox and Stalder, 1973) in which interference fringes are recorded using a Michelson interferometer. Fourier analysis of the data is required to obtain the film thickness. However, in comparison to measurements made with dispersive instruments, interferometry measurements can be made more quickly and with an improved signal-to-noise ratio. Reported average times for such thickness measurements are approximately 6 s with a long-term precision somewhat better than the interlaboratory precision of the ASTM Method F95-73 (1973). In addition, for heavily doped substrates the phase correction for free carrier absorption is easily included in the calculations.

Light section microscope. For films exceeding one micrometer in thickness, a "light section" microscope can also be used to determine film thickness. This instrument measures the interval between a slit image reflected from the film surface R_0 and film/substrate interface R_1. By mechanical adjustments, the distance, t_m, between the slit image rays R_0 and R_1 is measured and correlated to film thickness, t, by use of Equation (5.3):

$$t = t_m(2n^2 - 1)^{1/2}. \tag{5.3}$$

Using this technique, the thickness of transparent films can be measured directly. For the determination of the thickness of opaque films, a step must be formed in the film to profile the film thickness. For aluminum oxide and plastic films, Mansour (1963) has compared the accuracy of the light section microscope measurements to those obtained by other means. Good accuracy, better than 2–3 percent, is possible.

If a step is etched through a film to the substrate, and the surfaces are reflective (e.g., by evaporating a thin coat of metal), optical interference arising from the film/substrate reflection at fixed wavelength can be used to determine the step height (Tolansky, 1948). Although this procedure is destructive and depends on the ability to etch the film but not the substrate, it gives accurate results and is often used for opaque films.

Light ray polarization and film thickness, ellipsometry. The measurement of the phase angle and amplitude of the two components of a ray of incident and reflected light can be used to determine the film thickness of very thin films as well as thick films on specularly reflecting substrates. These ray components are designated p for the parallel ray component in the plane of incidence and s for the perpendicular component normal to the plane of incidence. If the phase angle is given as β and the amplitude as A, then Δ is defined as the change in phase while ψ is defined as the arctangent of the amplitude ratio change. The terms Δ and ψ are given by Archer (1962) as

$$\Delta = (\beta_p - \beta_s)_{\text{reflected}} - (\beta_p - \beta_s)_{\text{incident}} \tag{5.4}$$

$$\psi = \arctan\left[\left(\frac{A_p}{A_s}\right)_{\text{reflected}} - \left(\frac{A_p}{A_s}\right)_{\text{incident}}\right]. \tag{5.5}$$

Measurement of Δ and ψ by the ellipsometer involves angular adjustment of the incoming polarized ray to a reference position in the plane of incidence at the sample surface (see Figure 5.6). Changes in polarization on reflection of the polarized light by the sample surface are measured by use of a quarter-wave plate and the analyzer polarizer. Once the equipment is calibrated, the measurement involves adjusting the angular position of the quarter-wave plate and the analyzer prism until a setting is reached at which there is extinction of the reflected ray.

Both the thickness and optical constants of transparent, nonabsorbing films can be determined from Δ and ψ. The calculations, however, involve forms of the Drude equations and are tedious and complicated (Heavens, 1955; also Zaininger and Revesz, 1964) without some form of machine calculation. McCrackin (1969) published a useful Fortran program for calculating n, k, and film thickness from ellipsometry data. The program will evaluate the optical constants for a multifilmed surface. An interpretation of the program calculations is also included in the publication. When used in a time-sharing system ellipsometry measure-

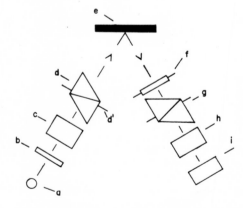

FIGURE 5.6   Schematic diagram of an ellipsometer [after Zaininger and Revesz (1964)]: a, mercury arc; b, filter; c, collimator; d, polarizer (Nicol prism); d', angular scale/positioning device; e, sample; f, quarter-wave plate; g, analyzer; h, telescope; and i, detector

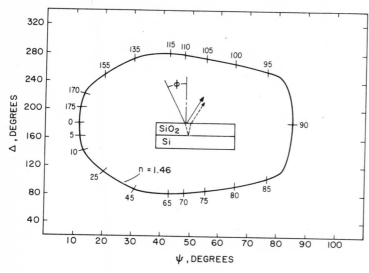

FIGURE 5.7
Ellipsometry data (incomplete) for an SiO_2 film on silicon. Ellipsometry curves are cyclic, repeating for certain multiples of film thickness. The film thickness is determined from the angular values of phase difference given on the curve by using Equations (5.6) or (5.7). The origin, 0, corresponds to a film-free surface

ments are greatly simplified and can be used to monitor surface properties of materials on a realistic time scale relative to on-going experiments. With absorbing films, a single ellipsometer measurement may not give a unique value for thickness or the refractive index (Archer, 1962).

The procedure of an ellipsometer measurement is outlined in detail by McCrackin and coworkers (1963). Calculations for various Δ and ψ values give a series of points corresponding to various film thickness of a given refractive index. Such curves are cyclic; the curves repeat for multiples of film thickness and therefore some independent measurement of film thickness is necessary for thicker films ($\gtrsim 200$ nm) depending on the refractive index (Archer, 1962). Ellipsometer data have been reported by Saxena (1965) for SiO_2 on silicon. Data for silicon oxide on silicon are given in part in Figure 5.7. The film thickness of the silicon oxide film is calculated from the angular values of phase difference given on the curve using Equation (5.6)

$$t = \left(\frac{\lambda}{2\pi}\right)(n_1^2 - n_0^2 \sin^2\phi)^{-1/2}(\delta), \qquad (5.6)$$

where t, the film thickness, is calculated from δ, the phase difference, which is experimentally determined by the ellipsometer measurement, as is n, the refractive index. The other terms, the angle of incidence, ϕ, and the wavelength of the incident light, λ, are known experimental parameters. Equation (5.6) is valid for thin films. This shows that the film thickness is such that δ represents a phase difference of 180° or less, the zero order. For thick films, the phase difference is a multiple of 180° (i.e., $m\pi$ where m is the mth order). In this case, the film thickness is given by

$$t = (m\pi + \delta)\left[\left(\frac{\lambda}{2\pi}\right)(n_1^2 - n_0^2 \sin^2\phi)^{-1/2}\right]. \qquad (5.7)$$

The origin 0° of Figure 5.7 corresponds to a film-free surface. Except for the extreme values of refractive index, measured ellipsometry values give unique solutions for the optical constants and thickness of a film. Zaininger and Revesz (1964) have reviewed the use of ellipsometry for surface studies, particularly for semiconductor surfaces. T. Smith (1968) has reported on ellipsometry measurements for surfaces with monolayer films. Bootsma and Meyer (1969), Meyer and coworkers (1971) and Meyer (1971) have reported on ellipsometer measurements of silicon or germanium surfaces covered with thin films of adsorbed gases. Ellipsometry tables for silicon oxide films on silicon using either mercury or He/Ne laser radiation have been published by Gergely (1971). Note that ellipsometry measurements are made on smooth surfaces—surface texture as can occur on SOS will influence the ellipsometry measurement.

Nonoptical techniques. Sapphire and spinel are relatively inert at room temperature and epitaxial silicon films can be readily etched from the metal oxide substrates without etching the substrate. This means that a step can easily be etched in the film allowing for mechanical step-height measurements, e.g., Talysurf measurements (Pliskin and Zanin, 1970).

A summary comparison of selected methods for measuring film thickness is given in Figure 5.8. These methods are compared by the range of film thickness associated with the use of each method. Inherently destructive methods appear on gray backgrounds. For SOS devices, where silicon of about 0.5 to 1.5 μm thick is used, both optical and stylus displacement methods can readily be used to measure film thickness. The light section microscope and similar techniques are generally not used for films less than a few micrometers thick. Finally, ellipsometry measurements are adversely affected by surface texture.

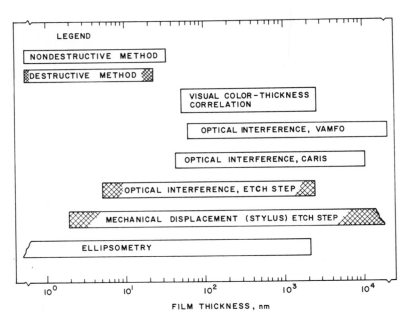

FIGURE 5.8
Comparison of film thickness measurement methods by thickness range for which the methods are conventionally used. Destructive measurement methods are indicated by the gray rectangular boxes. Nondestructive methods are indicated by the unmarked rectangular boxes

Dopant density

The electrical properties of semiconductors are established by the deliberate addition of n- or p-type dopants. The density of dopants in semiconductors for device purposes is often in the range 10^{15}–10^{20} cm^{-3}. Unintentionally added impurities are commonly less than 10^{16} cm^{-3} and can be as low as 10^{12} cm^{-3}. Thus, for elemental and compound semiconductors,* techniques are needed to detect one impurity atom in 10^2 to 10^{10} host atoms. For heteroepitaxial films, spatial resolution is an important consideration in determining dopant density and related properties such as carrier mobility. Many techniques (e.g., Hall and optical transmittance measurements) determine an average value that will not lead to an accurate understanding of the properties of heteroepitaxial films. In the previous chapters, the process and characteristics of autodoping by substrate materials have been discussed in detail. As a consequence of this and other effects, heteroepitaxial films will exhibit a variation of carrier properties as a function of film thickness. This variation is greater near the semiconductor film/substrate interface than in the bulk of the film. To measure the

*The electrical properties of compound semiconductors also depend on stoichiometry. For example, ZnO becomes an n-type semiconductor with an excess of zinc atoms above stoichiometry. Excess, ionizable zinc atoms can occupy interstitial sites in the ZnO lattice. In other types of oxide compound semiconductors, cation vacancies associated with excess oxygen lead to p-type properties. It should be noted that detecting deviations from stoichiometry can be a difficult problem because it requires detecting very small differences in the very large number of host atoms.

effects of autodoping, electrical techniques can be used whereby the depth of the measurement is determined by capacitance–voltage relations (Hilibrand and Gold, 1960). Electrical measurements are, however, nonspecific, and dopants cannot be identified. A much more serious problem is the need to chemically etch or otherwise destructively alter a semiconductor surface in order to provide for electrical connection tabs (or special sample geometry) required for electrical measurements. For economic reasons, nondestructive measurements are desirable. In certain cases, the optical properties of semiconductors such as free carrier absorption (Black et al., 1970) or the reflection minimum associated with the plasma frequency (Kudman, 1963) can be used to nondestructively measure bulk and surface dopant densities, respectively. In addition to these methods, MeV ion backscattering and channeling measurements have been used to determine the lattice site location and the in-depth distribution of ion-implanted or thermally distributed dopants (Mitchell et al., 1971B; Meyer and Mayer, 1970). This technique is specific and can be used to identify dopants. However, high-energy ion beams are employed in ion backscattering measurements and dopant atoms sometimes can be displaced from their normal lattice site. This effect must be taken into account with such measurements (Rimini et al., 1972). The general characteristics of C-V and other common measurements for measuring dopant density are discussed in the following section. Specific electrical measurements for silicon on sapphire have been discussed in Chapter 6.

Electrical measurements. The dopant concentration can be determined from the Hall effect (van der Pauw,

1958; R. Smith, 1959; Putley, 1960) or estimated from resistivity measurements assuming a value for carrier mobility. From an easily measured Hall voltage, V_H, generated in the presence of a current flow normal to a magnetic field (Figure 5.9) the dopant density can be calculated. The Hall constant, R_H, is given by

$$R_H = \frac{t V_H}{I} \boldsymbol{H}, \tag{5.8}$$

where t is the semiconductor film thickness, V_H is the average Hall voltage, I is the current, and $\boldsymbol{H}$ is the magnetic field. It can be readily shown [e.g., R. Smith (1959, pp. 100–107)], neglecting scattering effects, that

$$n, p = \frac{1}{R_H e}, \tag{5.9}$$

where n and p are the number of electrons or holes, respectively and e is the electronic charge.

The sample is etched to form the Hall bar, which usually destroys it for other purposes, and contacts placed at (1,2) and (3,4) (Figure 5.9b) for duplicate measurements to obtain an average Hall voltage V_H. The resistance is also determined from current and voltage measurements on the etched Hall bar. Because the free carrier density is a function of temperature, its value must be noted. From the measured electrical data and the known film thickness, average values of the Hall constant, carrier concentration, conductance of the sample, and the electron or hole mobility can be calculated using the following relationships, as they appear below,

where R_H is the Hall constant, t is the film thickness, V_H is the average measured Hall voltage, I is the current, $\boldsymbol{H}$ is the magnetic field strength, n is the carrier concentration, ρ is the conductance, R is the measured resistance of the Hall sample, w and l are the width and length of the Hall bar as shown in Figure 5.9b, and μ is the carrier mobility. Use of Equations (5.10)–(5.13) implicitly assumes that only one type of charge carrier is present in the test sample. The Hall and related measurements as well as sample preparation procedures are outlined in the American Society for Testing and Materials Method F76-73 (1973).

With regard to Hall measurements on heteroepitaxial films, it should be noted that Equation (5.10) assumes the current flow is uniform throughout the film thickness, which as discussed in Chapters 2 and 7 is not necessarily true for heteroepitaxial films. Evidence exists that, for silicon on sapphire, crystalline perfection (which affects carrier mobility) and carrier concentration are dependent on film thickness. Similarly, Equation (5.13) assumes the semiconductor is uniform in its crystalline quality throughout the film thickness, but as mentioned in the foregoing discussion, this is also not necessarily true for heteroepitaxial films. Finally, for Equation (5.10) the film thickness of the test sample must be measured and the calculated R_H will reflect the inaccuracies of this measurement. The calculated mobility is, however, independent of the film thickness [see Equations (5.10), (5.12), and (5.13)].

In addition to the Hall effect measurement, the

	Common* Usage	Units MKSA or SI	
Hall Constant			
$R_H = \dfrac{t(\text{cm}) \cdot V_H(\text{V})}{I\,(\text{A}) - \boldsymbol{H}\,(\text{gauss})\left[\dfrac{\text{V s} \cdot \text{cm}^{-2}}{10^8\ \text{gauss}}\right]}$	$\text{cm}^3\text{-C}^{-1}$	$\text{m}^3\text{-C}^{-1}$	(5.10)
Carrier Concentration			
$n = \dfrac{6.25 \times 10^{18}\ (\text{C}^{-1})}{R_H\ (\text{cm}^3\text{-C}^{-1})}$	cm^{-3}	m^{-3}	(5.11)
Resistivity			
$\rho = \dfrac{R(\Omega) \cdot t(\text{cm}) \cdot w(\text{cm})}{l(\text{cm})}$	Ω-cm	Ω-m	(5.12)
Hall Mobility			
$\mu_H = \dfrac{\lvert R_H \rvert}{\rho}$	$\text{cm}^2\text{-V}^{-1}\text{-s}^{-1}$	$\text{m}^2\text{-V}^{-1}\text{-s}^{-1}$	(5.13)

*When length is measured in centimeters and the magnetic flux is measured in gauss rather than tesla units, the set of units is not consistent and the conversion factor in Equation (5.10) must be used.

(a)

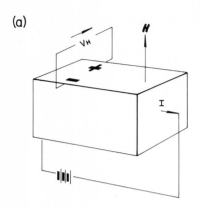

(b)

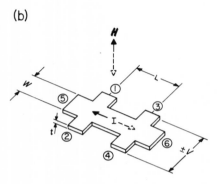

FIGURE 5.9   (a) Schematic representation of electric and magnetic vectors associated with the Hall effect. (b) Typical geometry of a Hall sample obtained by etching. The current across the bar is measured using tabs (5,6) while the induced voltage is measured at either tab (1,2) or (3,4). Values are taken at both voltage tabs with current or magnetic flux in two directions and these values are averaged to eliminate spurious potentials

resistivity of semiconductors such as silicon can be correlated to dopant concentration (Irvin, 1962). The basic theory and procedure for semiconductor resistivity measurements using a 4-point probe have been discussed by Valdes (1954). As shown in Figure 5.10, in the 4-point-probe resistivity measurement the two outer probes carry current while the two inner probes are used to measure the induced voltage drop across a distance s_2. The outer probes are spaced a distance s_1 and s_3 from the respective inner probes. For the simple case of a large-area semiconductor sample and a 4-point probe where $s_1 = s_2 = s_3 = s$, the resistivity, ρ, of the semiconductor is given by

$$\rho = \frac{2\pi V_f s}{I},\qquad(5.14)$$

where V_f is the measured floating potential and I is the measured current.

Resistivity can be measured using a 3-point-probe geometry as well. The 3-point probe, described by Brownson (1964), measures breakdown voltage of a surface probe diode. The breakdown voltage is related to resistivity.

The Hall measurement gives an average value of the dopant concentration which may be nonuniformly distributed through the thickness of the semiconductor. For bulk or thick semiconductor films, measuring the resistivity along a cleaved or etched edge which defines the sample thickness gives in effect a profile of the dopant distribution as a function of depth. For this type of measurement the probe must be small with respect to sample thickness in order to obtain good spatial resolution. For the conventional 4- or 3-point probe, the probe width is on the order of 0.1 cm, which does not allow for measurements along the edge of thin films. To reduce the probe size and thereby improve the spatial resolution of resistivity type measurements, the spreading-resistance technique was developed (Mazur and Dickey, 1966). By the nature of the spreading resistance probe, a volume of approximately 10^{-10} cm³ contributes to the measurement of silicon, insuring good spatial resolution on the order of micrometers.

In addition to the spreading resistance probe, the dopant's distribution in depth can be obtained with good spatial resolution by diode capacitance–voltage measurements. As discussed by Hilibrand and Gold (1960) as well as Greenberg and coworkers (1956) and McAfee and coworkers (1952), it can be shown that

$$A^2 N(x) = \frac{2}{e\epsilon}\left[\frac{d(1/C^2)}{dV_T}\right]^{-1},\qquad(5.15)$$

where A is the junction area, $N(x)$ is the net impurity concentration at the edge of the depletion region, e is the electronic charge, ϵ is the semiconductor permittivity, C is the capacitance, and V_T is the applied

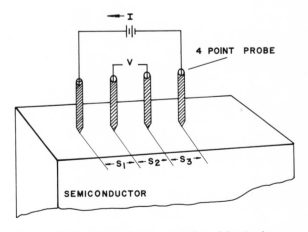

FIGURE 5.10   Schematic representation of the 4-point-probe geometry. The two outer probes carry current while the two inner probes measure the floating potential induced by the current flow. Probe spacings are typically on the order of 0.02 cm

potential. A depth-distributed profile of the dopant concentration as a function of distance x is obtained by measuring C and V_T. Distance x is calculated from $x = \epsilon A/C$. General considerations for C-V measurements have been discussed by Copeland (1969), Reutlinger and coworkers (1969), and Spiwak (1969). Note that a profile of electrically active impurity species is obtained and this is not necessarily the same as a profile of the total number of dopant atoms. C-V profiles of electrically active impurities may differ substantially from profiles obtained by such techniques as secondary-ion mass spectrometry which measure the total number of atoms.

Applications of the C-V technique have been reported by Thomas and coworkers (1962) for homoepitaxial silicon films. Redistribution of n- or p-type dopants during thermal oxidation of silicon has been studied in detail using C-V measurements by Grove, Deal, and coworkers (Grove et al., 1964; Deal et al., 1965; Deal and Grove, 1965). Impurity distribution in metal oxide–semiconductor (MOS) devices have been measured by C-V techniques as reported by van Gelder and Nicollian (1971) and Zohta (1973). A C-V measurement involving an MOS Hall bar for determining the depth distribution of carriers in heteroepitaxial silicon films grown on sapphire has been reported by Ipri (1972) and Ipri and Zemel (1973). A plot of the mobility as a function of distance from the outer surface of an n-type 1-μm-thick silicon film on sapphire is shown in Figure 5.11.

Both autodoping (Dumin, 1970; Dumin and Robinson, 1966) and lattice perfection have a significant influence on the depth-distributed electrical properties of silicon on sapphire or spinel. These effects have been discussed in some detail in Chapters 2 and 7. As will be discussed later in this chapter, data by other measurement techniques, particularly MeV ion backscattering and channeling experiments, also suggest that autodoping as well as lattice imperfection of silicon films on sapphire or spinel generally increase with depth. Therefore, interpretation of the data from electrical measurements discussed here must be done with some caution in the case of heteroepitaxial films.

Optical measurements. The reflection spectra of doped semiconductors show a minimum in reflectivity, generally in the low-energy infrared region, caused by free carrier–related dispersion effects (see Figure 5.12). The wavelength of the minimum is related to the dopant density. It can be shown from theory (Houghton and Smith, 1966) that with free carriers present the refractive index of a semiconductor is given by

$$n^2 = \epsilon_l - \frac{Ne^2}{m^*\epsilon_0\omega^2}, \tag{5.16}$$

where n is the real part of the refractive index, N is the carrier density, m^* is the carrier effective mass, ϵ_l

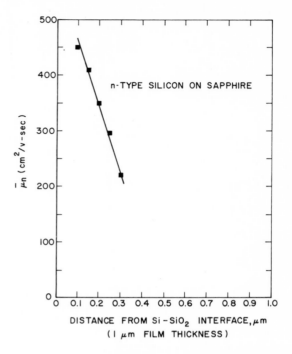

FIGURE 5.11   Carrier mobility as a function of distance from a metal oxide–semiconductor (MOS) Si-SiO₂ interface for heteroepitaxial silicon on sapphire. The presence of a large nonconducting region is inferred from these data. From Ipri (1972); reprinted courtesy of the American Institute of Physics

is the dielectric constant related to lattice vibrations, ϵ_0 is the dielectric constant for free space (vacuum), and ω is the angular frequency. When n is 1, the plasma edge is characterized by a reflection minimum:

$$\omega^2 = \frac{Ne^2}{m^*\epsilon_0(\epsilon_l - 1)}. \tag{5.17}$$

The plasma frequency occurs when the refractive index is zero,

$$\omega^2 = \frac{Ne^2}{m^*\epsilon_0\epsilon_l}, \tag{5.18}$$

and is characterized by a maximum in reflectivity. These changes occur over small changes in ω so that the minimum in reflectivity is well defined. The wavelength at which the minimum occurs can be correlated to N, the p- or n-type dopant density. Typical infrared reflection spectra for selenium-doped n-type epitaxial GaAs films are shown in Figure 5.12. As is evident from the spectra, dopant-related dispersion effects cause the shift of the minimum.

In general, the minimum in reflectivity exhibits a dependence on an n-type dopant density above approximately 10^{17} cm⁻³ and exhibits a dependence on a p-type dopant density above approximately 5×10^{18} cm⁻³. Curves that correlate n- or p-type dopant den-

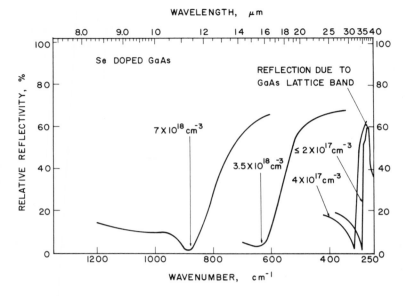

FIGURE 5.12   Infrared reflection spectra for selenium-doped n-type GaAs epitaxial films. The wavelength at which the minimum in reflectivity occurs is related to the dopant density

sity with the wavelength of the minimum in reflectivity have been reported for p-type GaAs by Murray and coworkers (1966), Riccius and Bertie (1966), and Rheinländer (1973); for n-type GaAs (see Figure 5.13) by Riccius and Ulbricht (1965), Jones and Hilton (1965), Okada and Oku (1967), and Schumann (1970A); and for p- and n-type silicon by Gardner and coworkers (1966) as well as Schumann (1970A).

By the plasma frequency technique, the dopant concentration measured is averaged over the depth of material sampled. This depth can be estimated from

$$t_p = \frac{1}{\alpha},\tag{5.19}$$

where t_p is the nominal penetration depth of a reflected light ray and α is the absorption coefficient. For shallow doped layers, the depth of penetration of the ray is a significant factor in the measurement. Procedures for measuring the dopant concentration in shallow layers have been discussed by Abe and Kato (1965) and Abe and Nishi (1968).

In general, the reflectivity method requires some independent means of calibrating the wavelength at which a minimum in reflectivity occurs in terms of dopant density. Once the calibration is established, the reflectivity method does not require other data in order to determine the dopant density.

By comparison, in order to correlate free carrier absorption to dopant density, the film thickness must be known. The film thickness is used to calculate the absorption coefficient,

$$\alpha = \frac{\ln (I_0/I)}{t},\tag{5.20}$$

where I_0 is the transmittance without the sample, I is the transmittance with the sample (corrected for reflection losses), and t is the film thickness of the doped semiconductor film (Black et al., 1970) Obviously, if the substrate is heavily doped, the absorption is very high and this method cannot be used. It is also assumed that the dopant density is uniform through the film.

As dopants can exhibit strong free carrier absorption in the infrared region, infrared imaging has also been used to measure the distribution of free carriers in semiconductors. This method is similar to an infrared technique used by Dash (1956) to observe copper precipitation at dislocations in silicon. With recently developed infrared methods, slightly magnified images of free carrier and defect absorption are obtained over large surface areas with high resolution (i.e., ~30-μm features can be resolved). Sherman and Black (1970) and Gupta and coworkers (1971) have employed a 3.39-μm or 1.1-μm infrared laser to raster scan a 1.2×1.2 cm² area with a 30-μm-diameter beam. Contrast changes on the oscilloscope-recorded image are due to differences in free carrier absorption or defects (e.g., inclusions which cause a change in transmittance). In this way a type of topograph is obtained characteristic of the uniformity of starting materials or arrays of completed semiconductor devices. Using this technique, Jungbluth (1970) has reported on the quality of arsenic-doped silicon; images he obtained clearly show arsenic precipitates, and these data correlate with x-ray diffraction topography. Sunshine and Goldsmith (1972) have shown that a silicon vidicon can be used with incandescent illumination of the sample to provide high-resolution visual monitoring of dopant distributions, precipitates, or decorated dislocations.

In summary, electrical methods can be used to measure the dopant density of epitaxial semiconductor films in discrete regions with varying in-depth resolution. Either surface or bulk dopant densities can be measured nondestructively by optical methods. Optical techniques can also be used to obtain unique infrared topographs of dopant as well as defect distributions.

Lattice perfection and stress

The crystalline perfection of either a substrate or an epitaxial film is influenced by factors such as the growth and processing conditions of the materials (Rai-Choudhury and Takei, 1969; also Dumin and Henry, 1971). In both homoepitaxial and heteroepitaxial film growth the quality of the substrate, in particular the substrate surface, is a critical factor in producing defect-free epitaxial films. It has been shown that in the homoepitaxial growth of silicon on silicon (Mendelson, 1964) stacking faults in the epitaxial layer form at defects such as microscratches or surface impurities on the silicon substrate surface. Using LEED–Auger techniques to study the sapphire surface, C. C. Chang (1971) has shown that annealing at temperatures near 1200°C is required to remove surface contaminants and to produce sharp LEED patterns of the surface structure. It is generally thought that the lattice perfection of the substrate limits the perfection obtainable in an epitaxial film.

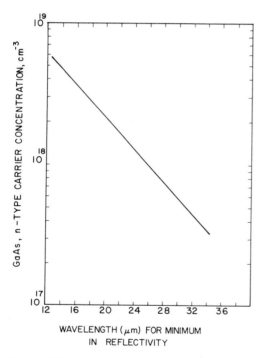

FIGURE 5.13   Comparison of n-type dopant density to the wavelength at which the minimum in reflectivity occurs. From Jones and Hilton (1965); reprinted courtesy of the Electrochemical Society

Lattice structure, crystalline perfection, and stress are related. Stress arises from lattice mismatch between the epitaxial film and substrate material and from differential contraction of the composite while cooling from the growth temperature. Lattice mismatch is often relieved by formation of dislocations and stacking faults at the growth temperature and thus limits the crystalline perfection of the epitaxial layer. Differential contraction usually leads to curvature of the composite structure because other mechanism to relieve stress, such as the formation of dislocations, cannot occur very readily at temperatures much below the growth temperature.

The electrical properties of epitaxial semiconductors are significantly affected by the degree of crystalline perfection and stress in the semiconductor composite. Dumin (1970, 1971) has shown that the carrier mobility in either heteroepitaxial germanium and silicon on sapphire or spinel substrates is dependent on the crystalline perfection of the epitaxial semiconductor. The crystalline defects introduce scattering centers which decrease carrier mobility. Jefkins (1970) and Schlötterer (1968) have reported theoretical and experimental data relating compressive stress of epitaxial silicon on insulating substrates (sapphire and spinel) to the properties of the silicon film. In particular, Schlötterer (1968) has discussed the effects of stress on the electrical properties of the semiconductor film.

Stress, in particular the stress-related curvature of heteroepitaxial composites, is readily measured by x-ray techniques and optical microscopy. Lattice perfection can be determined by x-ray measurements as well as by Rutherford MeV ion backscattering and channeling technique. From Rutherford scattering measurements, lattice perfection can be determined as a function of depth for epitaxial or bulk material. These techniques and measurements are discussed in the following section.

Displacement and x-ray measurements.   As previously mentioned, stress arising from thermal contraction can be partially relieved by curvature of the film/substrate composite.

The curvature is proportional to the stress and in general increases with film thickness (Dumin, 1965). The stress forces in silicon on sapphire are of the order of 10^9–10^{10} dyne/cm² and therefore the deformation of the composite is sufficiently large that the curvature can be measured by mechanical displacement techniques. Schlötterer (1968) reported using a commercial device, which by stylus tracing gives a recording representing the curvature of a surface. Dumin (1965) used a Zeiss light section microscope, described in principle by Glang and coworkers (1965), with which the height of the epitaxial silicon surface relative to a reference position was measured. An x-ray method

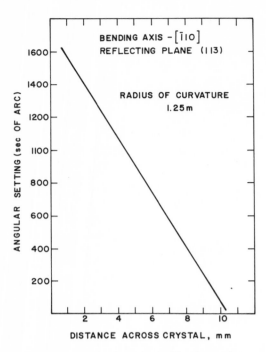

FIGURE 5.14   Plot of angular position for maximum x-ray intensity (reflection) as a function of distance across the film composite. The slope of the curve is inversely related to the radius of curvature of the film/substrate. From Wang et al. (1969); reprinted courtesy of the Metallurgical Society of AIME

belongs to the D_{3d}^6 rhombohedral space group and six of the 17 optical modes are infrared active. These allowed modes give rise to a reflectivity spectrum as shown in Figure 5.15a which compares the reflectivity spectrum obtained from single crystal sapphire to a transmittance spectrum obtained from powdered crystalline Al_2O_3. Approximately a 50–100 nm depth of Al_2O_3, as calculated from Equation (5.19), contributes to the reflection spectrum. The reflection spectrum as a function of beam polarization is shown in Figure 5.15b. Measurements with polarized infrared radiation give qualitative information about the orientation of a sapphire wafer.

Barker (1963) has shown that the forbidden Al_2O_3 lattice modes appear in the sapphire reflection spectrum when the Al_2O_3 surface is stressed or damaged (e.g. by a polishing procedure). Surface damage reduces the high symmetry of the Al-O modes and allow for normally forbidden modes to appear in the spectrum (see Figure 5.15c). By monitoring the Al_2O_3 reflection spectrum in the infrared region, the quality of the sapphire surface can be rapidly and nondestructively determined. In addition to this, most semiconductor materials are transparent to (or weakly absorbing of) infrared radiation in the Al_2O_3 lattice band region. Infrared reflection measurements can, therefore, be made to determine if the epitaxy process has introduced stress (or damage) at the sapphire surface.

Yim and coworkers (1973) have shown that in the case of epitaxial AlN on sapphire, Al_2O_3 forbidden modes are clearly observed after AlN epitaxy. As shown in Figure 5.15c, which compares the reflection spectrum obtained from the free Al_2O_3 surface and from the Al_2O_3 surface through the weakly absorbing AlN epitaxial film, forbidden modes appear as splittings of reflection bands. This indicates that the Al_2O_3 surface is physically affected by the AlN epitaxy such that the symmetry of the Al_2O_3 at the interface is reduced either by stress or lattice damage. Similar measurements have been made to determine the effects of silicon and gallium arsenide epitaxy on the sapphire surface as a function of deposition conditions.

In the context of the preceding discussion it should be noted that use of optical techniques to characterize surface damage, particularly defects which lead to surface texture, is well known, although not often used for semiconductors. In a series of papers, Bennett and Porteus have derived expressions which correlate the roughness of a plane surface to its specular reflectance (Bennett and Porteus, 1961; Bennett, 1963; Porteus, 1963). Donovan and coworkers (1963) have reported on the influence of surface damage and lattice disorder on the optical properties of bulk germanium. These techniques are being used at RCA Laboratories to characterize silicon on sapphire (Duffy et al., 1975/76).

using a Lang camera has also been reported for measuring curvature by McFarlane (Wang et al., 1969). In this technique, the angular position for maximum x-ray intensity is recorded over 0.5-m intervals across the epitaxial film/substrate composite. The slope of the curve is inversely proportional to the radius of curvature. (see Figure 5.14). This method relies on x-ray reflection from crystallographic planes which can be used to check if the measured deformation is isotropic with respect to the various crystallographic orientations. The deformation, δ_d, is calculated from

$$\delta_d = \frac{r}{2R}, \qquad (5.21)$$

where r is the *circular* substrate radius and R is the radius of curvature. The stress σ is given by

$$\sigma = \frac{E\delta_{ds}^2}{3(1-\nu)r^2 t_f}, \qquad (5.22)$$

where t_f is the semiconductor film thickness, t_s is the substrate thickness, E is the elastic modulus, and ν is the Poisson ratio of the substrate.

In a qualitative way, the presence of a stressed or damaged sapphire surface can be determined from infrared reflection measurements in the Al_2O_3 lattice band region, approximately 900–300 cm^{-1}. Sapphire

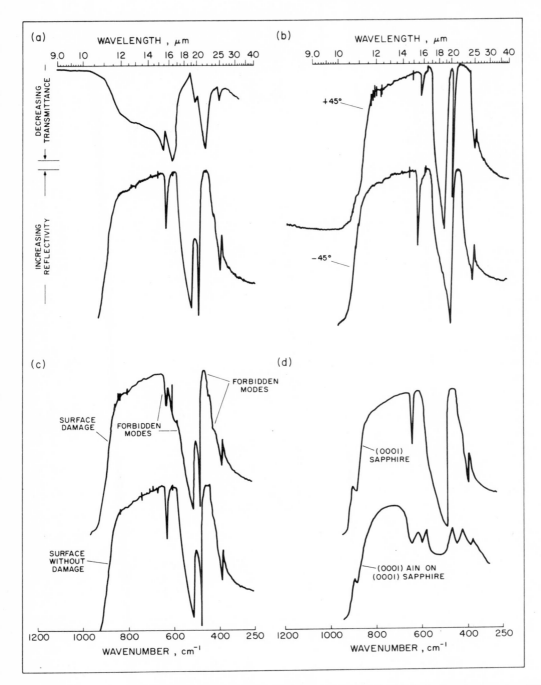

FIGURE 5.15   (a) Comparison of the transmittance and reflectance spectrum for Al_2O_3. (b) Polarized spectra of ($1\bar{1}02$)-oriented single crystal sapphire. (c) Comparison of the unpolarized reflectance spectrum of undamaged and polishing-damaged sapphire. The undamaged surface does not show the presence of forbidden modes; forbidden modes are clearly present in the spectrum of the poorly polished sapphire. (d) Comparison of the unpolarized reflection spectra of sapphire prior to and after AlN epitaxy. The free (0001) sapphire surface does not show the presence of forbidden modes. After epitaxy, forbidden modes are clearly present indicating surface stress or lattice damage occurred with epitaxy. From Yim et al. (1973)

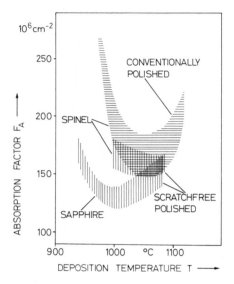

FIGURE 5.16   Optical absorption factor F_A compared to deposition temperature for epitaxial films on insulating substrates. The absorption factor decreases in value as the crystallinity of the silicon films improves. These data show that crystallinity is a function of the deposition temperature and substrate pretreatment, courtesy of Preuss (personal communication)

In related work, Druminski and coworkers (1974) have compared optical and nonoptical techniques for evaluating the quality of silicon films. From their optical measurements, the empirical absorption factor F_A was derived* and shown to be correlated to the deposition temperature of silicon films (see Figure 5.16). In this study, nearly 200 silicon films were prepared on conventionally polished and so-called scratch-free polished substrates. In each case, a minimum in the absorption factor is obtained corresponding to a deposition temperature for optimum film crystallinity. The spread in the absorption factor values is caused by the different pretreatments and characteristics of the substrates. In summary, the optical properties of heteroepitaxial silicon films can be correlated to crystalline quality.

The density of lattice defects is measured by techniques sensitive to either the deformation or discontinuity of lattice planes. X-ray transmission and reflection topography, electron diffraction, and more recently, keV to MeV ion backscattering and channeling spectroscopy have been used to characterize lattice defects of epitaxial semiconductors. The x-ray

*The absorption factor F_A is calculated from

$$F_A = \alpha(\nu, 1.7\mu\text{m}^{-1}) \cdot [\alpha(\nu, 2.2\mu\text{m}^{-1}) - \alpha(\nu, 1.6\mu\text{m}^{-1})],$$

where α is the absorption coefficient determined at various wavenumbers (Preuss, private communication, 1974).

techniques which are used to measure lattice defects depend on the attenuation of x-rays by two entirely different processes. Both the back-reflection and the transmission Berg–Barrett or Lang method depend on the absorption of x-rays. The Borrmann anomalous transmission method depends on nonabsorptive interference-like interactions which highlight defects of nearly perfect crystals. The experimental arrangements for each of these x-ray measurements, after Webb (1962), are shown in Figure 5.17. By these methods, a dislocation density of 10^6 cm^{-2}, or less, can be resolved. Newkirk (1958) and Schiller (1970) have discussed aspects of the back-reflection Berg–Barrett method. Lang (1959) has described the basic aspects of the transmission topograph method and Schwuttke (1962) has discussed aspects of the Borrmann anomalous transmission method. Schwuttke (1962) also has compared features of the extinction methods and the Borrmann anomalous transmission method. The characteristics of these x-ray topograph methods are summarized in Table 5.2.

McFarlane and Wang (1972) have reported Lang topographic studies of III-V heteroepitaxial films grown on both sapphire and spinel substrates. Paired topographs, one topograph related to the epitaxial film

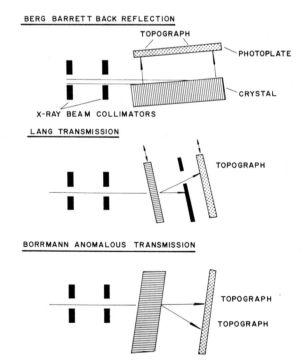

FIGURE 5.17   Experimental arrangements for various x-ray topograph methods. From Webb (1962); reprinted courtesy of Interscience Publications; published in *Direct Observation of Imperfections in Crystals,* (J. B. Newkirk and H. Wernick, Eds.) New York: Interscience Publications (1962)

TABLE 5.2   Characteristics of X-ray Topographic Methods

Method	Characteristics
Berg–Barrett reflection	Detects dislocation structure near or at a crystal surface with a nominal 1 μm depth of penetration.
Lang transmission	Good contrast requires $\mu t \leq 1$ where μ is the x-ray linear absorption coefficient for the film material and t is the sample thickness. A dislocation density of 10^6cm^{-2}, or less, can be measured with good resolution.
Borrmann anomalous transmission	Good contrast requires $\mu t > 20$. A dislocation density of $5 \times 10^3 \text{cm}^{-2}$ can be measured with good resolution.

and the companion topograph related to the substrate, show that the crystalline perfection of the semiconductor epitaxial film is significantly affected by such defects on the substrate surface as polishing scratches and growth striae. Figure 5.18 shows a pair of Lang topographs obtained for Czochralski (111) spinel with a 3.8-μm (111) GaAs epitaxial layer. Both the substrate and semiconductor material meet the criterion of $\mu t < 1$ for Lang topography. The topographs can be separated and resolved into a topograph representing the epitaxial layer and a topograph representing the substrate. This is possible because the lattice parameters for the epitaxial film and substrate are different and diffracting planes from each layer can be selected. Figure 5.18a shows regions misoriented from the nominal crystal orientation. In particular, lines caused by scratches are visible. As shown by Figure 5.18b, the GaAs grown over the scratches is misoriented from the rest of the layer as indicated by the white lines. Note that other line structure appear in the epitaxial layer, suggesting the presence of undetected scratches on the substrate surface. GaAs on these substrate defects is misoriented.

Reflection (high-energy) electron diffraction (RHEED) had been used extensively in the early phases of the development of silicon on insulating substrates. This method was particularly useful in examining the quality of the substrate surface finish (see Chapter 2). An advantage is that the beam samples only a few monolayers below the substrate surface and the diffraction patterns are not complicated

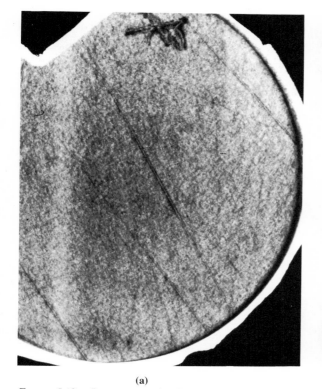

(a)

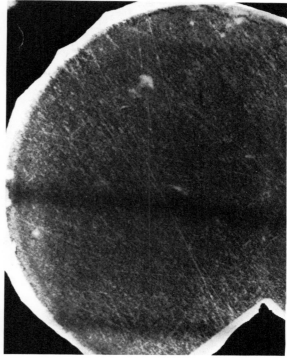

(b)

FIGURE 5.18   Lang topographs of Czochralski (111) spinel and a (111) GaAs layer grown on it, taken with silver $K_{\alpha 1}$ radiation: (a) spinel substrate, (440) diffracting planes; (b) GaAs layer, 3.8 μm thick, (220) diffracting planes. From McFarlane and Wang (1972); reprinted courtesy of the American Institute of Physics

by bulk crystallographic effects. The method is qualitative and can be used for an approximate estimation of the crystalline quality of the heteroepitaxial silicon. It has been recently used to examine the influence of deposition parameters on the crystalline nature of the silicon near (within ~0.2 μm) the substrate surface (Cullen et al., 1975). Here marked differences in the diffraction pattern are observable corresponding to variations in deposition conditions because the silicon near the substrate is more disordered than is material more remote from the interface.

MeV ion backscattering and channeling measurements.   Although x-ray topographic methods can be used to determine the density of such defect structures as dislocations, these measurements lack depth resolution. As indicated by electrical data (Ipri, 1972), the properties of epitaxial silicon on sapphire vary with film depth. For defect centers (defined as lattice displacements) as low as approximately 10^{15} cm^{-2}, keV to MeV ion backscattering and channeling measurements can be used to determine the depth distribution of the *defect sites*. As discussed in the following section, the same technique can be used to determine the depth distribution of *impurity sites*. Bøgh (1968) has reported on the theory and experimental conditions for these types of measurements. Rubin (1963) has reviewed the historical aspects of ion scattering measurements. Davies (1971) discusses Rutherford backscattering and channeling measurements as applied to the determination of surface and bulk impurities, film composition, and single crystal defect density. Morgan and Bøgh (1972) have reviewed the application of Rutherford scattering and channeling techniques with particular emphasis on the study of GaAs. W. K. Chu and coworkers (1973) have reviewed the principles and applications of ion beam techniques for structure and composition analysis of solids and thin films.

As shown in Figure 5.19, keV to MeV ions impinging on a crystal surface will either channel along rows of atoms or backscatter by collision with lattice atoms indicated by the black circles. The lattice atom can either occupy a normal lattice site, be absent, or be displaced as in the case of a defect lattice site.

For analytical purposes, the energy of collected backscattered ions is measured. This results in a spectrum as given in Figure 5.20. As indicated by the equations given with Figure 5.19, the backscattered ion energy will be determined by the mass of the scattering atom as well as the degree of orientation of the incoming ion beam to lattice planes. Thus, in Figure 5.20, which represents data for protons backscattered from a micron-thick epitaxial silicon film on spinel, the region between 400–600 keV represents backscatter of protons from silicon atoms. As shown by the inset, the deeper-penetrating ions have propor

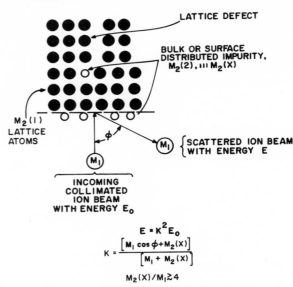

FIGURE 5.19   Schematic representation of Rutherford ion backscattering measurements. M_1 represents the incident ion, $M_2(1)$ represents the host lattice atoms. Note that the energy of the backscattered incident ions M_1 is measured; sputtering and backscattering of the host atom $M_2(1)$ or impurity atoms $M_2(2) \ldots M_2(x)$ are not processes measured in Rutherford scattering. The energy loss of the Rutherford backscattered ions is given by the equations which correlate energy loss of the incident M_1 ion with mass of the host lattice atom $M_2(1)$ and with the mass of impurity atoms $M_2(2) \ldots M_2(x)$. Note also that in certain crystal orientations the incident ion beam can channel along crystallographic planes. This process sharply diminishes the number of backscattered ions and is important in determining the density of lattice defects which disrupt the channeling process

tionately less energy. The oriented bulk silicon shows the lowest ion backscatter because of its relatively high crystalline perfection, and therefore, a high degree of channeling is possible. The silicon epitaxial film is less perfect in lattice structure and exhibits higher ion backscatter. The edge at approximately 400 keV corresponds to the silicon–spinel interface. Ion backscattering from the substrate produces another peak which is smaller than that observed for the epitaxial silicon. Ion backscattering from the substrate is lower because of its higher crystalline perfection.

The defect density of the silicon film is determined by comparing the measured ion backscattering yield to that expected from a perfect crystal lattice at a given depth. Depth is measured by the energy of the backscattered ions. Picraux (1972) reports that in bulk single crystals with a high degree of lattice perfection the fraction of the ion beam no longer channeled, $X_R(t)$, is related to the fraction dechanneled, $X_1(t)$, by

$$X_R(t) = X_1(t) + [1 - X_1(t)]P(t), \qquad (5.23)$$

where $P(t)$ is the probability that a channeled ion will scatter outside the critical angle for channeling. Picraux and coworkers (1969) have reported on the critical angle for ion channeling in diamond type lattices. At a depth t, the density of defects (scattering centers) $N^1(t)$, in, for example, epitaxial single crystal silicon films, is obtained from $X_2(t)$, the ratio of yield in a channeled and unchanneled orientation:

$$N^1(t) = \frac{NX_2(t) - X_R(t)}{1 - X_R(t)}, \qquad (5.24)$$

where N is the atomic density of the film. By this method, Picraux (1972) has shown that for epitaxial silicon on spinel or sapphire substrates the number of scattering sites increase with film thickness. More recent work by Picraux (1973) and Picraux and Thomas (1973) using both ion channeling–backscattering measurements and electron microscopy indicates that the imperfections in epitaxial silicon films on either sapphire or spinel substrates are mainly caused by stacking faults or microtwins. They show that, at equal distances from the interface, (111) silicon on

spinel and (100) silicon on sapphire exhibit *lower* densities of imperfections than does (100) silicon on spinel. It is reported that for the (100) silicon on sapphire the variable quality of the sapphire caused large variations in silicon defect density.

In summary, strain and lattice defects affect the electrical properties of epitaxial semiconductor films on insulating substrates. The magnitude of these physical defects can be estimated by measuring substrate curvature and film lattice perfection using such techniques as x-ray topography and keV to MeV ion-scattering and channeling spectrometry.

Impurity content (and composition)

It is well known that small amounts of electrically active impurities can significantly change the electrical properties of semiconductors. Unintentionally added elements are in general undesirable in semiconductor manufacture. This is a particular problem for heteroepitaxial semiconductors since the substrate is a potential uncontrollable dopant source for the semiconductor film. For example, aluminum atoms from a sapphire substrate dope the epitaxial silicon. Very much distinct from the problem of measuring the concentration of a deliberately added dopant, the question in this context is how to identify the impurities and determine their distribution and amount. (Note in the section on measuring dopant density such techniques generally are nonspecific and cannot be used to identify the electrically active species.)

There are many considerations in selecting an analytical technique to identify elemental impurities, such as sensitivity, selectivity, matrix effect, elemental coverage, time and ease of analysis, and overall cost. Because of the small volume of semiconductor material in thin-film heteroepitaxial devices, spatial resolution (both lateral and in-depth) is a very important consideration. As indicated in Table 5.3, the first monolayer of atoms constituting a surface can be analyzed by such techniques as ion-scattering spectrometry and static secondary-ion mass spectrometry. For measurements requiring analysis of a few atomic layers, a number of ion or electron beam techniques are available (e.g., Auger electron spectrometry). Finally, where in-depth resolution is not a critical factor [i.e., where the *average* film composition (bulk composition in this context) over micrometer depths is important], techniques such as electron probe microanalysis, solids (spark source) mass spectrometry, and infrared spectroscopy can be used. With respect to the latter technique, Newman (1969) has reported that many impurity species exist in semiconductor structures as molecular infrared-absorbing complexes. For example, oxygen in silicon is usually found in the form of SiO_x; similarly, carbon is often found in the form of CO_x or SiC depending on the overall oxygen content

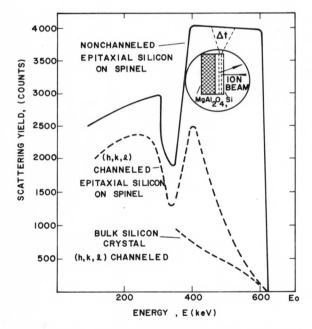

FIGURE 5.20   Plot of the number of backscattered ions as a function of backscattered ion energy. The scattering yield for a given energy range corresponds to ion scattering from some depth of silicon on spinel as indicated by the inset. The interface region is marked by the change in backscatter ion yield at 300–400 keV. The effects of crystalline perfection are evidenced by comparison of the scattering yield for oriented bulk as for oriented and nonoriented epitaxial silicon. From Picraux (1972); reprinted courtesy of the American Institute of Physics

TABLE 5.3   Spatial Resolution for Selected Methods of Thin-Film Analysis

Method[a]	Lateral Resolution[b] Diameter (μm)	In-depth Resolution[c] (Number of Atomic Layers)
Ion-scattering spectrometry	10^3	1^d
Rutherford backscattering and channeling spectrometry	10^3	$2-50^e$
Static secondary-ion mass spectrometry	10^3	1^d
Ion probe microanalysis	$1-300^f$	$10-20$
Auger electron spectrometry	$25-100$	$2-10^g$
Scanning Auger spectrometry	5	$2-10^g$
Electron probe microanalysis	1	$10^{3\,h}$
Solids (spark source) mass spectrometry	10	10^3

[a] Surface analysis techniques have been reviewed by Bauer (1972) and more recently by Benninghoven (1973).

[b] The smallest diameter conventionally sampled by the technique.

[c] The minimum number of atomic layers which provide analytical information. In-depth profiling requires ion sputtering by which successive atomic layers are removed for analysis.

[d] A fraction of a surface monolayer is detectable by this technique.

[e] In-depth resolution is a function of the energy resolution provided by the ion detector.

[f] Ion beam diameter is adjustable.

[g] In-depth resolution depends on the electron escape depth (see Powell, 1974).

[h] In-depth resolution is a function of the atomic number of the matrix and of the primary electron beam energy [i.e., penetration depth of the incident electrons, see Kyser and Murata (1974), also Birks (1971)].

of the silicon. Thus, infrared absorption of impurity complexes can be used to determine selected elemental impurities in semiconductors.

The following section describes these surface and bulk techniques and their applications in determining the impurity content of epitaxial layers.

Surface and in-depth measurements AUGER ELECTRON EMISSION AND SPECTROSCOPY. Auger electron emission can occur when inner shell electrons rearrange by nonradiative three-electron processes. For example, loss of a $L_{2,3}$ shell electron by ionization, schematically shown in Figure 5.21, is accommodated by a V shell electron which fills the $L_{2,3}$ shell by a nonradiative energy transfer process. The excess energy lost by the electron in the V to L shell transition allows a V shell Auger electron to be emitted (C. S. Chang, 1971). Alternately, the L shell vacancy could be filled by a V shell electron which emitted its excess energy as an x-ray photon. This later process is the basis of analytical x-ray fluorescence spectrometry (Bertin, 1975) and electron probe microanalysis (Beaman and Isasi, 1971).

By monitoring the energy of secondary electrons as shown in Figure 5.21, the discrete energy of the emitted Auger electrons can be determined. This energy is characteristic of the electrons involved in the energy transfer process and, therefore, of the element emit-

ting Auger electrons. The magnitude of the electron emission is related to the amount of an element present in the surface volume analyzed. Auger electron spectroscopy is especially useful for, but not limited to, detecting such low atomic number elements as carbon and oxygen, for which this method is particularly sensitive. Because the escape depth of Auger electrons is about two to ten atomic layers (see Table 5.3), the Auger electrons are characteristic of surface atoms.

Auger electron spectroscopy was first investigated by Lander (1953) for surface analysis applications. Lander observed the presence of a large, varying background superimposed on the very small Auger peak; he concluded the presence of the background limited practical application of the Auger electron emission to surface analysis. Harrower (1956) related Auger peaks to x-ray data in order to predict possible Auger transitions. Both Weber and Peria (1967) and Tharp and Scheibner (1967) showed that a LEED apparatus could be used for Auger spectroscopy. Harris (1968A,B) was the first to show that the derivative of the measured number of electrons with respect to electron energy, $dN(e)/dE$, gives sharply defined Auger peaks, allowing practical analytical applications. Palmberg (1972) reported on a device that allows for simultaneous sputter etching and Auger measurements by which depth profiles of chemical impurities can be obtained. Quantitative measurements of an element are possible if the derivative peak height can be related to the concentration of an element. Calibration techniques based on ellipsometry measurements have been reported by Vrakking and Meyer (1971) and Meyer and Vrakking (1972).

Surface images generated by recording the Auger electron emission over, typically, a scanned 200×200-μm area using a 5-μm-diameter primary electron beam have been reported (MacDonald and Waldrop, 1971). These images represent selected elemental distributions over the scanned surface area. As indicated previously, the escape depth for Auger electrons is a few atomic layers from the surface. Thus, the Auger images have a much improved depth resolution in comparison to x-ray images obtained from electron microprobe instruments (see Table 5.3). Features of the Auger scanning technique have been discussed by MacDonald and Waldrop (1971) and by Arthur (1973).

With the availability of commercial instruments for Auger spectroscopy, application of this technique to surface studies has greatly increased since the late 1960s and early 1970s. For example Grant and Haas (1970) have studied the Auger electron emission from clean silicon to establish the silicon Auger peaks. Uebbing and Taylor (1970) have reported on Auger electron emission of clean gallium arsenide to establish the true gallium arsenide Auger peaks. Contamination of silicon (111) and (100) faces by chemical etching and polishing methods has been studied by C. C. Chang (1970) using Auger spectroscopy. Auger surface analysis data show that carbon and oxygen are persistent surface contaminants of chemically cleaned silicon.

Recent studies of homoepitaxial silicon growth by Auger, low-energy electron diffraction (LEED), and high-energy electron diffraction (HEED) spectroscopy (Joyce et al., 1969; Henderson and Helm, 1972) show that there is a distinct correlation between the lattice perfection of silicon epitaxial films and the surface-distributed impurity content of the silicon substrates.*

Joyce and coworkers (1969) have studied silicon homoepitaxy by molecular beam decomposition of silane in ultrahigh vacuum. Using LEED and Auger measurements, they report that surface coverage of the silicon substrate by a specific impurity, such as carbon, changes the nucleation and growth of the homoepitaxial silicon layer. Henderson and Helm (1972) have studied silicon homoepitaxy using the pyrolysis of silane at approximately 800°C. From LEED and Auger measurements, their data show that it is possible to grow homoepitaxial silicon films free of stacking faults, growth pyramids, or crystallographic pits if the substrate surface impurity content is less than 10^{13} cm^{-2}. Surface carbon, arsenic, and transition metals such as nickel at concentrations greater than 10^{13} cm^{-2} invariably lead to the presence of lattice defects in the epitaxial film. The presence of carbon is correlated with an island-type growth, the incorporation of β-SiC, and facet formation in the epitaxial

---

*For a review of surface studies by electron diffraction, see Estrup and Macrae (1971).

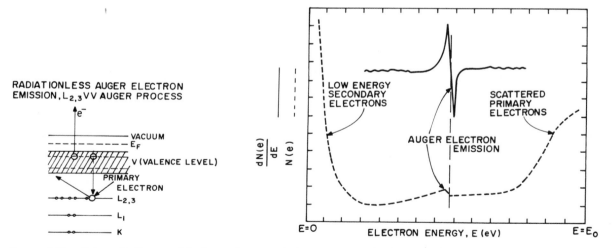

FIGURE 5.21   Schematic diagram of the Auger electron emission process and the characteristics of recorded Auger spectra. The number of Auger electrons is very small in comparison to the total number of electrons. For this reason, the peak in electron emission caused by the discrete Auger process is too small for accurate analytical measurements when the total number of electrons, $N(e)$, is measured as a function of the electron energy. By taking the first derivative of the measured number of electrons with respect to electron energy, $dN(e)/dE$, sharply defined peaks are obtained which are representative of Auger electron emission. Over the energy range E, the derivative is not affected by the large and gradually varying number of background electrons which hindered analytical application of early Auger measurements

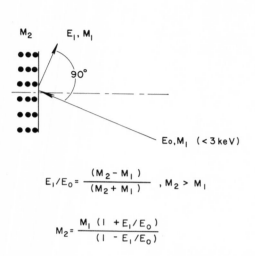

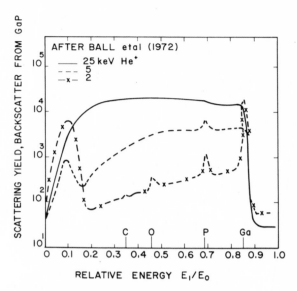

FIGURE 5.22   Composite figure showing the ion-scattering process and typical spectra for 25, 5, and 2 keV ions backscattered from a GaP layer. For ion energies less than 2 keV, ion scattering occurs at the surface of the solid (GaP) without significant penetration in depth. In the spectrum given, oxygen is detected as a surface impurity. From Ball et al. (1972); reprinted courtesy of the North Holland Publishing Company

films. When nickel is detected on the substrate surface the silicon epitaxial films show triangular defects. Stacking faults commonly associated with the presence of metallic impurities were not observed.

C. C. Chang (1971), using LEED–Auger techniques, has studied the heteroepitaxial deposition of silicon on sapphire. The LEED–Auger data show that in the preparation of sapphire for epitaxy, annealing at temperatures near 1200°C is required in order to remove surface contaminants, mainly carbon, and to produce the expected sapphire lattice structure at the substrate surface.

X-RAY APPEARANCE POTENTIAL SPECTROSCOPY. In addition to the Auger emission process, inner shell electrons can also fill a vacancy by giving up excess energy in the emission of an x-ray photon. Park and coworkers (1970), Houston and Park (1971), and Tracy (1972) have reported on a surface analysis technique that measures the total undispersed x-ray photons emitted from an irradiated surface. The measured ionization energies for inner shell electrons in a nondispersed total x-ray emission spectrum give information on the elemental composition of a solid surface. This technique is called x-ray appearance potential spectrometry and has been compared to Auger electron spectroscopy by Tracy (1971, 1972, 1973) for the study of sulfur, carbon, and oxygen contaminants on titanium films. Experimental data suggest that although the simplicity of appearance potential spectrometry is advantageous for measurement purposes,

the sensitivity of the technique is not as high or uniform as in Auger spectroscopy.

ION-SCATTERING SPECTROMETRY. The information obtained from ion-scattering or Rutherford scattering spectrometry depends on the energy of the incoming ions. With high-energy ion beams, keV to MeV energies, the ions penetrate into films or bulk material and backscatter from large depths, by which process the depth profiles of impurity distributions can be obtained. The measurement of impurity distributions is very similar in many respects to the previously discussed measurement of lattice perfection by ion scattering. Nicolet and coworkers (1972) have reviewed the analysis of materials by backscattering spectrometry. Ziegler and Baglin (1971) have reported on MeV ion backscattering measurements of the arsenic distributions in silicon. Morgan and Bøgh (1972) have discussed application of Rutherford scattering to the study of semiconductor surfaces, particularly GaAs.

Ball and coworkers (1972) have shown that a broad energy distribution for backscattered ions can substantially be reduced when the ion beam energy is below 5 keV (see Figure 5.22). In addition, below 5 keV the ion backscattering occurs mainly at the surface of a solid. This allows for elemental analysis of a single monolayer of surface atoms. Honig and Harrington (1973) have reported on the characteristics of ion-scattering spectrometry below 10 keV.

Historically, Tollestrup and coworkers (1949) were

among the first to report that ion scattering could be used for surface analysis. Energy analysis of data from protons backscattered by beryllium or lithium targets showed the presence of thin surface layers of oxygen and carbon. Based on this, Turkevich (1961) suggested using ion scattering of Cm^{244} alpha particles for analysis of lunar material in situ. For the lunar analysis, it was proposed that ion backscattering be detected by a solid state detector and an electronic pulse-height analyzer, both of which could be made small for use in a remotely controlled space vehicle. Following this, surface analysis using 0.1–3 keV noble gas ion beams has been demonstrated by D. P. Smith (1967, 1971). Strehlow and Smith (1968) confirmed that, with low-energy primary-ion beams, ion scattering occurs mainly from the first atomic layer at the surface of the sample. Following these developments, particularly the work of D. P. Smith, Goff (1972) reported on an instrument designed for surface analysis by low-energy ion beam scattering. This instrument allows for elemental analysis of the first monolayer of surface atoms, a measurement not possible with Auger spectroscopy because the Auger electrons are emitted from at least 2–10 atomic layers below the surface (see Table 5.3). Since the primary ion beam, an ionized noble gas, slowly sputters the sample surface, this technique has an inherent capability for providing information on the depth distribution of elements. With the development of commercial ion-scattering instrumentation, applications of this technique to the surface analysis of semiconductor and related materials have become practical [see Goff and Smith (1970); also D. P. Smith (1971)].

MASS SPECTROMETRY. While Auger, appearance potential, and low-energy ion-scattering spectroscopy give information about the elemental composition of surfaces, ion probe microanalysis (Castaing et al., 1960; Castaing and Slodzian, 1962; Long, 1965; Liebl, 1967, 1971) and secondary-ion mass spectrometry [Benninghoven, 1970, 1971; Fogel', 1972) can be used to obtain information about both the elemental and molecular composition of surface layers. These techniques sputter etch a surface from which depth profiles of composition are obtained [e.g., see the review by Honig (1973)]. Historically, the basic aspects of surface analysis using ion-sputtering techniques were reported by Honig (1958), followed by Bradley (1959) and Stanton (1960), as well as others.

Secondary-ion mass spectrometry refers to methods involving the mass measurement of ions which are generated by various forms of primary-ion bombardment. With low-energy primary-electron or ion beams, large-area surface analysis is possible. Alternatively with keV primary beams, small-area ($\geq$ 1-μm-diameter) microanalysis of surface and bulk mate-rial is possible (Liebl, 1974). Werner (1972) reviews the various types of secondary-ion mass spectrometry (SIMS) methods.

One form of secondary-ion mass spectrometry, static SIMS, involves the use of very low primary-ion currents densities (10^{-9} A/cm^2) on a large target area such that the time to remove the first monolayer of surface atoms by ion emission is very long compared with the measurement time. The species slowly removed from the surface by ion emission, the secondary ions, are detected by a mass spectrometer (Benninghoven, 1971). Measurements of the surface compounds averaged over the target area are possible by this technique.

Benninghoven and Storp (1973) have used secondary-ion mass spectrometry to investigate the reaction of molecular oxygen with atomically clean (111) silicon surfaces in ultrahigh vacuum. High vacuum is required to maintain an uncontaminated surface representative of the true sample. Mass measurements indicate that on exposure of a clean silicon surface to molecular oxygen, a large number of characteristic secondary ions are detected: SiO^+, Si_2O^+, O^-, SiO_2^-, SiO_3^-, and $Si_2O_5^-$. The presence of these species has been interpreted as relating to a two-step oxidation of the silicon surface. The initial step is reported to be the formation of a first monolayer of oxide which relates to the detected secondary ion species SiO_3^-. The second step is the slow growth of a second oxide layer which relates to the detection of secondary ion $Si_2O_5^-$. Data indicate that the number of secondary ions is proportional to the degree of surface coverage. This technique, therefore, is potentially useful for quantitative measurement of surface species.

Another important secondary-ion mass spectrometry technique, ion probe microanalysis, allows the composition of a microvolume, 2–300 μm in diameter, to be determined. Both elemental and molecular species can be identified. The sputter etch rate of a sample can be rapid in an ion microanalyzer and this allows for in-depth measurements on the order of micrometers. Simultaneously with the sputter etching, the secondary-ion beam is mass selected and a magnified image of elemental or molecular species distributed in the sputtered area is obtained. As shown by the schematic diagram (Figure 5.23), a primary-ion beam sputters a sample surface. Sputter rates vary from a fraction of a monolayer to etch rates as high as 500 Å/s. The sputtered ions are separated from the primary ions, mass analyzed, and imaged by selected mass.

Commercial ion microanalyzer instruments produce ion images of surface-distributed species in two completely different ways: (1) sequentially, by raster scanning of the primary-ion beam over the sample surface or (2) simultaneously, by preserving the original distribution of secondary ions as shown in Figure 5.24.

(AFTER CAMECA IMS 300)

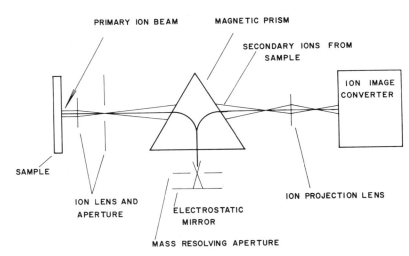

FIGURE 5.23
Schematic diagram showing the basic features of the Cameca IMS 300 ion microanalyzer

With both types of instruments, determination of a depth distribution requires special considerations. This is because of the geometry of the sputtered region. Ion etching of a surface creates a crater-like erosion of material, schematically shown in Figure 5.25. With erosion of material below the surface, the walls of the crater contribute secondary ions not meaningful to the measurement. Because these ions cannot be distinguished from ions originating at the crater center, the resolution of the depth measurement is limited. However, if a reasonably broad area is sputtered, then mechanical or electronic means can be used to determine the secondary ions sputtered from the center of the crater. The use of an aperture for direct imaging is shown schematically in Figure 5.25. In this way the impurity distribution is resolved, typically with an in-depth resolution of 20–100Å (Evans and Pemsler, 1970; Socha, 1971; Morabito and Lewis, 1973).

In general, for all analytical techniques involving production of secondary ions, the secondary ion yield is important in relation to reproducible and quantitative interpretation data. Andersen (1969, 1970) has discussed the correlation of ion probe microanalyzer data and secondary-ion yields. Andersen and Hinthorne (1973) have described a method of quantitative analysis using the secondary-ion microanalyzer and reactive gases.

Quantitative analysis with secondary-ion mass spectrometry techniques can, however, be difficult (or impossible) when matrix effects exist with respect to the production of the secondary ions. For example, oxygen is known to exert a strong influence on secondary-ion yields. As reported by Lewis and coworkers (1973) and Tsai and Morabito (1974), the presence of oxygen on the surface and bulk of single crystal silicon causes an *enhancement* of the secondary-ion

yield of silicon. In earlier work, Blanchard and coworkers (1972) reported on the use of an oxygen leak over samples to provide a high but uniform secondary-ion yield. However, it is not always possible to obtain quantitative secondary-ion yields when a strong matrix effect exists, as reported by Schubert (1974) for the analysis of iron and chromium in stainless steel. This aspect of secondary-ion mass spectrometry is particularly important with respect to the analysis of semiconductors on oxide substrates. A strong matrix effect is expected from the oxygen of the substrate and from oxygen present in the semiconductor.

With regard to matrix effects Mercier (1971) points out that it is difficult to interpret the ion microanalyzer data for silicon on sapphire unless such factors as crystalline perfection and crystalline orientation relative to sputtering yields are taken into account.

Using an ion probe microanalyzer, Mercier studied the qualitative in-depth characteristics of silicon grown on sapphire substrates where silicon was grown by the pyrolysis of silane at low temperatures ($\sim$900°C) in a helium atmosphere. Using a 250-μm-diameter ion beam with an estimated 100-Å depth resolution, Mercier reports that the films grown at low temperatures in helium show 10^4 less aluminum (as autodopant) and 10^2 less oxygen (also an autodopant from the sapphire) than in comparable films grown at a higher temperature ($\sim$1000°C) in hydrogen atmosphere. As found in earlier studies (Mercier (1970A,B)), reactions occur during deposition which apparently cause the formation of aluminosilicates at the silicon–sapphire substrate. With regard to these interface compounds, the ion probe microanalysis data are probably more conclusive than Mercier's infrared-attenuated total reflection (ATR) measurements. As reported by Harrick (1967), infrared-atten-

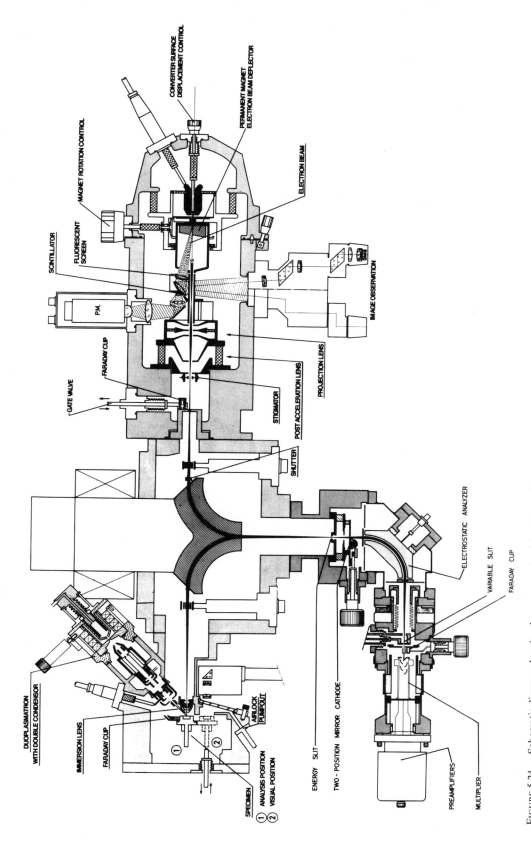

FIGURE 5.24 Schematic diagram showing the components of the Cameca IMS 300. The primary ion beam is generated by the duoplasmatron and focused onto the specimen. The beam is 1–300 μm in diameter. Both primary and sputtered secondary ions enter the electrostatic analyzer. Secondary ions are selected and enter the observation section. Images of the specimen can be obtained representing elemental or molecular distributions in a variety of ways. Courtesy of Cameca Instruments, Inc., Elmsford, NY

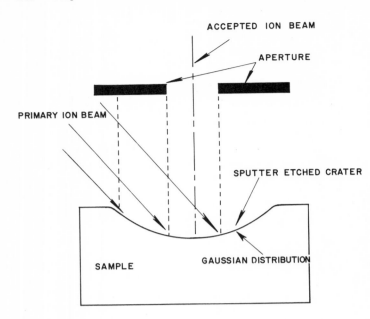

FIGURE 5.25
Diagram showing crater etched by an ion beam. Measurement of depth-distributed composition is affected by the contribution of wall material to the measurement. An aperture which selects ions sputtered from a localized central region of the crater can be used to eliminate detection of much of the sputtered wall material.

uated total reflection measurements are difficult to interpret when a material, such as sapphire, has a large and varying refractive index.

Bulk measurements. Often the average impurity content of an epitaxial film is important without regard to the distribution of the impurity. In this case well established measurement techniques can be used. Techniques such as spark source mass spectrometry, neutron activation analysis, or emission and absorption spectroscopy have been used to analyze the elemental content of thin films.

The general aspects of thin-film analysis by solids mass spectrometry have been discussed by Socha (1970). Both vacuum spark (Dempster, 1936) and laser (Honig, 1966) sources have been used to produce ions from solids for mass analysis. In this way, the elemental composition of a material is surveyed for all elements from major to parts-per-billion (ppb) constituents. With either source, the depth of solid sampled is relatively large in relation to the thickness of the film in present device structures. For example, the depth of penetration of the vacuum spark is about 1–5 μm, which is equal to the thickness of many heteroepitaxial films studied and therefore the average distribution of material in thin films is sampled.

Although the elemental impurity content of solids can be measured by mass spectrometry techniques to parts per billion, determining the molecular species, such as gases in thin-film solids, requires special techniques because of the inherently small amount of sample. Winters and Kay (1972) have reported on a mass spectrometric technique for determining residual gases in thin films. The gases trapped in films are released into a clean vacuum system by laser evaporation.

Noble gases and nitrogen are readily measured using a 0.5-mm^2 area of film.

Neutron activation methods depend on the measurement of γ or β emission from radioactive isotopes produced by neutron bombardment. This technique requires use of neutron irradiation facilities in addition to radioasay equipment. Neutron activation analysis is a means for trace, parts-per-billion analysis of many elements. In certain circumstances neutron activation methods are probably among the few analytical techniques that can measure impurity content at the low parts per billion level or less (Smales, 1967). However, it should be noted that for semiconductors, use of activation analysis for the simultaneous determination of several elements often requires laborious procedures. These procedures are necessary because the semiconductor elements are readily activated producing several radioactive isotopes yielding complicated γ-ray spectra. To sort out and identify the γ radiation it is necessary to use procedures involving chemical separations or standards with a matrix similar to the sample. In this context, the detection limit for a given element is largely determined by the type of host atoms present in the semiconductor sample and their activation products. Furthermore when semiconductors are activated and produced radioactive isotopes with long half-lives [e.g., GaAs (Wolfstirn, 1970)], elements with isotopes of short half-lives may be difficult to detect (e.g., Na, K, Al, and B).

A bibliography of activation analyses reported in literature through 1971 has been prepared by Lutz and coworkers (1972).

In summary, compositional analysis of semiconductors or processing films depends largely on the particular considerations which are often unique to each material. For heteroepitaxial semiconductors such as silicon on sapphire the presence of two materials with very different compositions is a matter for consideration in compositional analysis. For example, many investigations have been conducted to determine if silicon interacts with the sapphire during the early stages of deposition. Analysis of this interface region is very difficult. Oxygen in the sapphire strongly affects ion yields in secondary-ion mass spectrometry, and this can greatly distort the measured composition of the interface region. Furthermore, the interface region may not be uniform in composition. In any case, the need for a high degree of spatial resolution is apparent.

5.3. Characterization of films commonly used in device processing

Semiconductor device fabrication involves isolating electrically incompatible elements of a circuit. In practice, insulating films of silicon oxide, which are easily prepared by the oxidation of epitaxial silicon, have been used. With the development of chemical-vapor deposition (CVD) methods, silicon nitride, aluminum oxide, and glass films have also been used as insulators. To be effective insulators, these dielectric materials must contain few electrically active impurities and have well established and reproducible composition. It is also desirable that, insofar as possible, they be free of stress, impervious to diffusion of extraneous impurities, and resistant to damage by radiation fields. Often all these characteristics are not obtained because of deposition characteristics or the presence of impurities (among other problems).

Silicon oxide

In general, silicon oxide films are amorphous, may contain hydroxyl species, and often require thermal densification to improve structural quality and to relieve localized stress. Pliskin and Lehman (1965) and Pliskin and Castrucci (1968) have reported on the characteristics of silicon oxide films prepared by a wide variety of deposition methods. Using infrared, refractive index, and etch rate measurements, the properties of silicon oxide are found to depend significantly on the mode of preparation. Infrared measurements of the 9-μm silicon–oxygen vibrational mode show that shifts occur in the wavelength for the silicon–oxygen band center, as well as changes in the band half-width, depending on the mode of preparation. These effects are reported to be caused by the porosity of the films and the bond strain of the silicon oxide molecule. Allam and Pitt (1967/68) have shown that the optical properties of silicon oxide films can be correlated to electrical properties of the films. The

presence of hydrogen or free silicon in the silicon oxide is found to be significant.

Thermally grown silicon oxide films always have a positive charge which induces a negative charge in the underlying silicon (Revesz and Zaininger, 1968; Lamb, 1970). This positive charge has been associated with at least four types of charge centers involving fixed or surface potential sensitive states, mobile ions (such as sodium), or protons and ionized bonds (traps) within the silicon oxide structure. Because the residual positive charge affects device properties, other insulators, such as silicon nitride and aluminum oxide, have been studied for use with, or in place of, silicon oxide.

Silicon nitride

Silicon oxide films are not greatly resistant to sodium ion diffusion which seriously affects the electrical properties of the SiO_2-Si interface. Buck and coworkers (1967) have studied sodium diffusion in SiO_2 films using neutron activation and radiotracer techniques. They find that sodium will diffuse to the silicon oxide–silicon interface under certain temperature or bias conditions. Burgess and coworkers (1969) report a 1000-fold reduction in sodium diffusion with use of silicon nitride films. Dalton and Drobek (1968) find that silicon nitride films composed of small crystallites improve film resistance to sodium diffusion. Bibliographies of literature concerning the preparation and applications of silicon nitride have been recently published by Milek (1971, 1972).

A typical infrared spectrum of silicon nitride deposited on silicon is given in Figure 5.26. The silicon nitride spectrum is characterized by a strong band centered near 830 cm^{-1}, although the band center varies somewhat depending on the manner of preparation of the silicon nitride. Infrared spectra of silicon nitride films prepared by various techniques have also been reported by Levitt and Zwicker (1967), Swann and coworkers (1967), Bean and coworkers (1967), T. L. Chu and coworkers (1967), and Hu (1966).

The composition of silicon nitride pyrolytically deposited at 850°C has been determined by Gyulai and coworkers (1970, 1971A) using MeV ion backscattering measurements. These measurements show that for ammonia-to-silane ratios greater than 20, stoichiometric silicon nitride is formed. In-depth analysis of the films shows that the composition of the pyrolytically deposited silicon nitride is uniform over the entire film thickness. For ammonia-to-silane ratios less than 20, the deposited films become silicon rich. The chemical composition of the silicon nitride deposited with various ammonia to silane ratios is shown to be in close agreement with the electrical conductivity and etch rate properties of the films.

Similarly, Croset and coworkers (1971) and Meyer

and coworkers (1971) have used MeV ion backscattering experiments to determine the composition of silicon nitride formed by reactive sputtering in argon at 300°C and by glow discharge methods, respectively.

Croset and coworkers (1971) report films with uniform in-depth composition, although large amounts of argon are present in these films. Stoichiometry is a function of the nitrogen partial pressure. At small values of the nitrogen partial pressure an excess of silicon is detected in the films, while at large values an excess of nitrogen is found. Meyer and Scherber (1971) also found similar results for MeV ion backscattering analysis of silicon nitride films prepared by glow discharge decomposition of silane and nitrogen. The composition of silicon nitride films produced by this reaction is uniform in-depth while total pressure as well as silane or nitrogen partial pressure determine the composition of the silicon nitride.

The electrical properties of silicon nitride have been briefly reviewed by Lamb (1970); the characterization of silicon nitride films has been reported by Taft (1971).

Aluminum oxide

X-ray studies have shown that aluminum oxide films prepared by pyrolysis techniques are amorphous. The infrared spectrum of amorphous Al_2O_3 exhibits one broad absorption band from approximately 12 to 30 μm. Annealing these films at temperatures near 800°C can induce varying degrees of crystallinity. This is detected in the infrared spectrum by the presence of discrete bands in the 12–30 μ region (see Figure 5.27). Infrared measurements also show that annealing at high temperatures can cause the formation of silicon oxide at the silicon–aluminum oxide interface (see

Figure 5.27), as indicated by the presence of the absorption band near 1100 cm^{-1}.

The stoichiometry of hydrolytically deposited aluminum oxide has been measured by MeV ion backscattering techniques. Mitchell and coworkers (1971A) report that aluminum oxide films grown hydrolytically in the temperature range 600–830°C are stoichiometric and uniform in depth to 2500 Å, the nominal film thickness. Films deposited at the lower deposition temperatures tend to contain chlorine, which is present in the deposition reaction, up to 2 atomic percent.

Metal contacts

Contacts to semiconductor surfaces can interact with the semiconductor and thereby are a parameter in device performance. With ion-scattering techniques, the depth-distributed metal–semiconductor interactions can be measured. Gyulai and coworkers (1971B) have reported on the alloying of gold–germanium on GaAs. From 2-MeV helium ion backscattering measurements, the gold is found to penetrate the GaAs in depth to at least 3000 Å until consumed. The presence of germanium lowers the temperature for gold penetration into the GaAs. The authors note that for ohmic contacts on GaAs, gold films for alloying should be thin (i.e., less than 500 Å) and that the temperature and time for alloying the gold should be kept to a minimum.

Hiraki and Lugujjo (1972) and Hiraki and other coworkers (1971) have shown that at temperatures less than or equal to 400°C, silicon will migrate into silver, gold, or platinum contacts. MeV ion backscattering data show that the silicon migration depends on the silicon surface composition; for example, the pres-

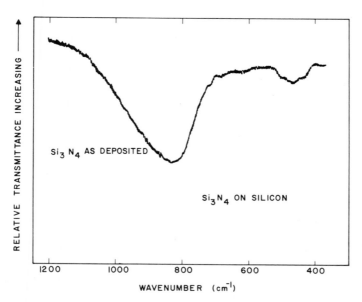

FIGURE 5.26
Infrared spectrum of amorphous, pyrolytic silicon nitride on silicon. The broad absorption bands occur at approximately 830 and 450 cm^{-1} In crystalline, β silicon nitride, the main absorption band occurs at approximately 930 cm^{-1} and below 600 cm^{-1} a series of sharp, discrete bands occur

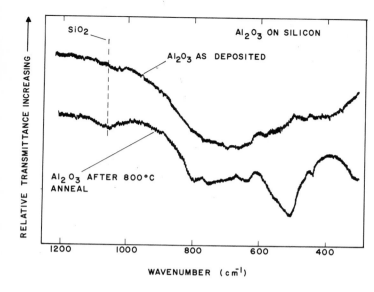

FIGURE 5.27
Comparison of aluminum oxide infrared spectra. The upper trace is an infrared spectrum of aluminum oxide "as-deposited," an amorphous film. The lower trace is an infrared spectrum of an aluminum oxide film after an 800°C bake. Band structure is evident in this trace and is indicative of the start of film crystallization

ence of a thin oxide layer inhibits silicon migration. The migration of silicon in platinum causes Pt_2Si or PiSi compounds to be formed, depending on temperature and annealing conditions. These effects probably relate to the mechanism of thermal compression bonding used to attach leads to semiconductor devices.

Both insulating films and metal contacts are often defective because of physical displacements: cracks, poor surface topology, etc. Scanning electron microscopy (SEM), which allows for highly magnified images of surfaces with good depth of field, is widely used to investigate physical surface defects (Kammlott, 1971). Using transmission electron microscopy (TEM) Ham and coworkers (1977) have studied the defect structure of silicon on sapphire composites. These data show the SOS structure can be prepared with good quality. Solid state devices are generally not flat, often having valley or plateau structures which are micrometers in dimension. With highly magnified images, the physical defects on semiconductor surfaces, insulating films, or contacts can be readily observed. This is not possible with an optical microscope largely because of the long wavelength of optical radiation. In the electron microscope, 1 to 10 kV, or greater, electrons have de Broglie wavelengths of Ångstroms or less. Because both resolution and depth of focus are related to the effective wavelength of the radiation, the electron microscope is able to resolve objects of Ångstrom dimension and provide a much larger depth of field than is possible with optical microscopes.

Because high-energy heavy ions have shorter de Broglie wavelengths than electrons, a scanning ion microscope would in principle improve resolution in comparison to an electron or optical microscope. Martin (1973) has reported on the feasibility of a scanning ion microscope.

The overall emphasis in this chapter has been to provide summary descriptions of widely used techniques for semiconductor analysis. Not all methods described here are commonly used in the characterization of heteroepitaxial semiconductors; some are included to describe techniques that provide the basis for the more specialized methods. Where possible the characterization of heteroepitaxial composites has been described using what is thought to be data characteristic of typical properties of these composites, particularly of silicon on sapphire.

ACKNOWLEDGMENTS. The author appreciates the assistance of associates at RCA Laboratories in preparing this chapter. The critical reading of and suggestions relating to this chapter by R. E. Honig are especially appreciated.

References

Abe, T. and T. Kato, *Jap. J. Appl. Phys. 4:* 742 (1965).

Abe, T., and Y. Nishi, *Jap. J. Appl. Phys. 7:* 397 (1968).

Ahearn, A. J., "Mass Spectrographic Analysis of Insulators Using a Vacuum Spark Positive Ion Source," *J. Appl. Phys. 32,* 1195 (1961).

Albert, M. P. and J. F. Combs, *J. Electrochem. Soc. 109:* 709 (1962).

Allam, D. S. and K. E. G. Pitt, *Thin Solid Films 1:* 245 (1967/1968).

Andersen, C. A., *Int. J. Mass Spectrom. Ion Phys. 2:* 61 (1969).

Andersen, C. A., *Int. J. Mass Spectrom. Ion Phys. 3:* 413 (1970).

Andersen, C. A. and J. R. Hinthorne, *Anal. Chem. 45:* 1421 (1973).

Archer, R. J., *J. Opt. Soc. Am. 52:* 970 (1962).

Arthur, J. R., *J. Vac. Sci. Technol. 10:* 136 (1973).

ASTM, *1973 Annual Book of ASTM Standards, Part 8,* Philadelphia, PA (1973): see F95-73, "Standard Method of Test for Thickness of Epitaxial Layers of Silicon on

Substrates of the Same Type by Infrared Reflectance,'' F-120-70T, ''Tentative Recommended Practices for Infrared Absorption Analysis of Impurities in Single Crystal Semiconductor Materials.''

Ball, D. J., T. M. Buck, D. MacNair, and G. H. Wheatley, *Surf. Sci. 30:* 69 (1972).

Ban, V. S., *J. Electrochem. Soc. 118:* 1473 (1971).

Barker, Jr., A. S., *Phys. Rev. 132:* 1474 (1963).

Bauer, E., *Vacuum 22:* 539 (1972).

Beaman, D. R. and J. A. Isasi, *Mater. Res. Stand. 11*(11): 8(1971).

Bean, K. E., P. S. Gleim, and R. L. Yeakley, *J. Electrochem. Soc. 114:* 733 (1967).

Bennett, H. E., *J. Opt. Soc. Am. 53:* 1389 (1963).

Bennett, H. E. and J. O. Porteus, *J. Opt. Soc. Am. 51:* 123 (1961).

Benninghoven, A., *Z. Physik 230:* 403 (1970).

Benninghoven, A., *Surf. Sci. 28:* 541 (1971).

Benninghoven, A., *Appl. Phys. 1:* 3 (1973).

Benninghoven, A. and S. Storp, *Appl. Phys. Lett. 22:* 170 (1973).

Bertin, E. P., *Principles and Practice of X-ray Spectrometric Analysis,* Second Edition, New York: Plenum Press (1975).

Birks, L. S., *Electron Probe Microanalysis,* Second Edition, New York: Wiley-Interscience (1971).

Black, J. F., E. Lanning, and S. Perkowitz, *Infrared Phys. 10:* 125 (1970).

Blanchard, B., N. Hilleret, and J. Monnier, *Mat. Res. Bull. 6:* 1283 (1971).

Bøgh, E., *Can. J. Phys. 46:* 653 (1968).

Bootsma, G. A. and F. Meyer, *Surf. Sci. 14:* 52 (1969).

Bradley, R. C., *J. Appl. Phys. 30:* 1 (1959).

Brownson, J., *J. Electrochem. Soc. 111:* 919 (1964).

Buck, T. M., F. G. Allen, J. V. Dalton, and J. D. Struthers, *J. Electrochem. Soc. 114:* 862 (1967).

Burgess, T. E., J. C. Baum, F. M. Fowkes, R. Holmstrom, and G. A. Shirn, *J. Electrochem. Soc. 116:* 1005 (1969).

Castaing, R. and G. Slodzian, *Advances in Mass Spectrometry,* Vol. 3 (W. L. Mead, Ed.) London: The Institute of Petroleum (1966); see also *J. Microscopie 1:* 395 (1962).

Castaing, R., B. Jouffrey, and G. Slodzian, *C. R. Hebd. Sean. Scad. Sci. 251:* 1010 (1960).

Castle, J. E. and D. Epler, *Proc. Royal Soc. Lond. A 339:* 49 (1974).

Chang, C. C., *Surf. Sci. 23:* 283 (1970).

Chang, C. C., *J. Vac. Sci. Technol. 8:* 500 (1971).

Chang, C. C., *Surf. Sci. 25:* 53 (1971).

Chu, T. L., C. H. Lee, and G. A. Gruber, *J. Electrochem. Soc. 114:* 717 (1967).

Chu, W. K., J. W. Mayer, M-A. Nicolet, T. M. Buck, G. Amsel, and F. Eisen, *Thin Solid Films 17:* 1 (1973).

Copeland, J. A., *IEEE Trans. Electron. Devices ED-16:* 445 (1969).

Corl, E. A. and H. Wimpfheimer, *Solid State Electron. 7:* 755 (1964).

Cox, P. F. and A. F. Stalder, *J. Electrochem. Soc. 120:* 287 (1973).

Croset, M., S. Rigo, and G. Amsel, *Appl. Phys. Lett. 19:* 33 (1971).

Cullen, G. W., J. F. Corboy, and R. T. Smith, *J. Crystal Growth 31:* 274 (1975).

Dalton, J. V. and J. Drobek, *J. Electrochem. Soc. 115:* 865 (1968).

Dash, W. C., *J. Appl. Phys. 27:* 1193 (1956).

Davies, J. A., *J. Vac. Sci. Technol. 8:* 487 (1971).

Deal, B. E. and A. S. Grove, *J. Appl. Phys. 36:* 3770 (1965).

Deal, B. E., A. S. Grove, S. H. Snow, and C. T. Sah, *J. Electrochem. Soc. 112:* 308 (1965).

Dempster, A. J., *Nature 135:* 542 (1935); see also *Rev. Sci. Instr. 7:* 46 (1936).

Donovan, T. M., E. J. Ashley, and H. E. Bennett, *J. Opt. Soc. Am. 53:* 1403 (1963).

Druminski, M., Ch. Kühl, E. Preuss, F. Schwidefsky, and J. Tihanyi, ''Quality Evaluation of Thin Silicon Films by Various Investigation Methods,'' Abstract No. 139, Electrochemical Society, Fall Meeting, Oct. 13–17, 1974.

Dumin, D. J., *J. Appl. Phys. 36:* 2700 (1965).

Dumin, D. J., *Rev. Sci. Instrum. 38:* 1107 (1967).

Dumin, D. J. *Solid State Electron. 13:* 415 (1970).

Dumin, D. J., *J. Vac. Sci. Technol. 8:* 235 (1971).

Dumin, D. J. and W. N. Henry, *Metall. Trans. 2:* 677 (1971).

Dumin, D. J. and P. H. Robinson, *J. Electrochem. Soc. 113:* 469 (1966).

Duffy, M. T., P. J. Zanzucchi, and G. W. Cullen, National Bureau of Standards Semiconductor Technology Program, Contract 5-35915 (1975/1976, 1976/1977).

Estrup, P. J. and E. G. McRae, *Surf. Sci. 25:* 1 (1971).

Evans, Jr., C. A. and J. P. Pemsler, *Anal. Chem. 42:* 1060 (1970).

Fogel', Ya M., *Int. J. Mass Spectrom. Ion Phys. 9:* 109 (1972).

Gardner, E. E., W. Kappallo, and C. R. Gordon, *Appl. Phys. Lett. 9:* 432 (1966).

van Gelder, W. and E. H. Nicollian, *J. Electrochem. Soc. 118:* 138 (1971).

Gergely, G. (Eds., with G. Forgacs, B. Szucs, and D. vanPhouc), Ellipsometric Tables of the Si-SiO$_2$ System for Mercury and He-Ne Laser Spectral Lines, Budapest: Alademiai Kaido (1971).

Gillespie, D. J., *Measurement Techniques for Thin Films* (B. Schwartz and N. Schwartz, Eds.) New York: Electrochemical Society Inc., Johnson Reprint Co. (1967) p. 102.

Glang, R., R. A. Holmwood, and R. L. Rosenfeld, *Rev. Sci. Inst. 36:* 7, (1965).

Goff, R. F., *J. Vac. Sci. Technol. 9:* 154 (1972).

Goff, R. F. and D. P. Smith, *J. Vac. Sci. Technol. 7:* 72 (1970).

Gorman, J. G., E. J. Jones, and J. A. Hipple, *Anal. Chem. 23,* 438 (1951).

Grant, J. T. and T. W. Haas, *Surf. Sci. 23:* 347 (1970).

Greenberg, L. S., Z. A. Martowska, and W. Happ, *IRE Trans. ED-3,* 97 (1956).

Grove, A. S., O. Leistiko, Jr., and C. T. Sah, *J. Appl. Phys. 35:* 2695 (1964).

Gupta, D. C., B. Sherman, E. D. Jungbluth, and J. F. Black, *Solid State Technol. 14*(3): 44 (1971).

Gyulai, J. O. Meyer, J. W. Meyer, and V. Rodriguez, *Appl. Phys. Lett. 16:* 232 (1970).

Gyulai, J., O. Meyer, and J. W. Mayer, *J. Appl. Phys. 42:* 451 (1971A).

Gyulai, J. O., J. W. Mayer, V. Rodriguez, A. Y. C. Yu, and H. J. Gopen, *J. Appl. Phys. 42:* 3578 (1971B).

Hannay, N. B., *Rev. Sci. Instr. 25,* 644 (1954).

Hannay, N. B. and A. J. Ahearn, *Anal. Chem. 26,* 1056 (1954).

Ham, W. E., M. S. Abrahams, J. Blanc and C. J. Buiocchi, RCA Review *38,* 351 (1977).

Harrick, N. J., *Internal Reflection Spectroscopy,* New York: Interscience (1967).

Harris, L. A., *J. Appl. Phys. 39:* 1419 (1968A).

Harris, L. A., *J. Appl. Phys. 39:* 1428 (1968B).

Harrower, G. A., *Phys. Rev.102:* 340 (1956).

Heavens, O. S., *Optical Properties of Thin Solid Films,* London: Butterworth Scientific Pub. (1955).

Heavens, O. S., *Thin Film Physics,* London; Methuen & Co. (1970).

Henderson, R. C. and R. F. Helm, *Surf. Sci. 30:* 310 (1972).

Hilibrand, J. and R. D. Gold, *RCA Review 21:* 245 (1960).

Hiraki, A. and E. Lugujjo, *J. Vac. Sci. Technol. 9:* 155 (1972).

Hiraki, A., M-A Nicolet, and J. W. Mayer, *Appl. Phys. Lett. 18:* 178 (1971).

Honig, R. E., *Anal. Chem. 25:* 1530 (1953).

Honig, R. E., *J. Appl. Phys. 29:* 549 (1958).

Honig, R. E., *Advances in Mass Spectrometry,* Vol. 3 (W. L. Mead, Ed.) London: The Institute of Petroleum (1966) p. 101.

Honig, R. E., "Analysis of Surfaces and Thin Films by Mass Spectrometry," 6th International Mass Spectrometry Conference, Edinburg (1973).

Honig, R. E., *Thin Solid Films 31:* 89 (1976).

Honig, R. E. and W. L. Harrington, *Thin Solid Films 19:* 43 (1973).

Houghton, J. T. and S. D. Smith, *Infrared Physics,* Oxford: Clarendon Press (1966) pp. 149–151.

Houston, J. E. and R. L. Park, *J. Vac. Sci. Tech. 8:* 91 (1971).

Hu, S. M., *J. Electrochem. Soc. 113:* 693 (1966).

Hulthén, R., *Physica 12:* 342 (1975).

Ipri, A. C., *Appl. Phys. Lett. 20:* 1 (1972).

Ipri, A. C. and J. N. Zemel, *J. Appl. Phys. 44:* 744 (1973).

Irvin, J. C., *Bell Syst. Tech. J. 41:* 387 (1962).

Jefkins, D. M., *J. Phys. D (Appl. Phys.) 3:* 770 (1970).

John, H. F., *Proc. IEEE 55:* 1249 (1967).

Jones, C. E. and A. R. Hilton, *J. Electrochem. Soc. 112:* 908 (1965).

Jordanov, B. and L. Popova, *Solid State Electron. 14:* 753 (1971).

Joyce, B. A., J. H. Neave, and B. E. Watts, *Surf. Sci. 15:* 1 (1969).

Jungbluth, E. D., *Appl. Phys. Lett. 17:* 437 (1970).

Kammlott, G. W., *Surf. Sci. 25:* 120 (1971).

Kane, P. F. and G. B. Larrabee, *Annual Review of Materials Science,* Vol. 2 (R. A. Huggins, Ed., R. H. Bube and R. W. Robert Asst. Eds.) California: Annual Reviews Inc. (1972), p. 33.

Kane, P. F. and G. B. Larrabee, *Characterization of Solid Surfaces,* New York: Plenum Press (1974).

Keck, P. H., *J. Electrochem. Soc. 108:* 262C (1961).

Koelmans, H., *Thin Solid Films 8,* 19 (1971).

Kudman, I., *J. Appl. Phys. 34:* 1826 (1963).

Kühl, Ch., H. Schlötterer, and F. Schwidefsky, *J. Electrochem. Soc. 121:* 1496 (1974).

Kyser, D. F. and K. Murata, *IBM J. Res. Develop. 18:* 352 (1974).

Lamb, D. R., *Thin Solid Films 5:* 247 (1970).

Lambe, J. and R. C. Jaklevic, *Phys. Rev. 165:* 821 (1968).

Lander, J. J., *Phys. Rev. 91:* 1382 (1953).

Lang, A. R., *J. Appl. Phys. 30:* 1748 (1959).

Larrabee, G. B. and J. A. Kennan, *J. Electrochem. Soc. 118,* 1351 (1971).

Levitt, R. S. and W. K. Zwicker, *J. Electrochem. Soc. 114:* 1192 (1967).

Lewis, R. K., J. M. Morabito, and J. C. C. Tsai, *Appl. Phys. Lett. 23:* 260 (1973).

Liebl, H., *J. Appl. Phys. 38:* 5277 (1967).

Liebl, H., *Int. J. Mass Spectrom. Ion Phys 6:* 401 (1971).

Liebl, H., *Anal. Chem. 46:* 22A (1974).

Long, J. V. P., *Brit. J. Appl. Phys. 16:* 1277 (1965).

Lutz, G. J., R. J. Bordeni, R. S. Maddock, and J. Wing (Eds.), *Activation Analysis: A Bibliography Through 1971,* NBS Technical Note 467, Washington, DC: U.S. Government Printing Office, (1972).

MacDonald, N. C. and J. R. Waldrop, *Appl. Phys. Lett. 19:* 315 (1971).

Malitson, I. H., F. V. Murphy, Jr., and W. S. Rodney, *J. Opt. Soc. Am. 48:* 72 (1958).

Mansour, T. M., *Mat. Res. Stands. 3:* 29 (1963).

Martin, F. W., *Science 179:* 173 (1973).

Mazur, R. G. and D. H. Dickey, *J. Electrochem. Soc. 113:* 255 (1966).

McAfee, F. B., W. Shockley, and M. Sparks, *Phys. Rev. 86:* 137 (1952).

McCracken, F. L., *A Fortran Program for Analysis of Ellipsometer Measurements,* NBS Technical Note 479, Washington, DC: U. S. Government Printing Office, (1969).

McCracken, F. L., E. Passaglia, R. R. Stromberg and H. L. Steinberg, *J. Res. NBS A67:* 363 (1963).

McFarlane III, S. H. and C. C. Wang, *J. Appl. Phys. 43:* 1724 (1972).

Mendelson, S., *J. Appl. Phys. 35:* 1570 (1964).

Mercier, J., *J. Electrochem. Soc. 117:* 666 (1970A).

Mercier, J., *J. Electrochem. Soc. 117:* 812 (1970B).

Mercier, J., *J. Electrochem. Soc. 118:* 962 (1971).

Meyer, F., *Surf. Sci. 27:* 107 (1971).

Meyer, O. and J. W. Mayer, *J. Appl. Phys. 41:* 4166 (1970).

Meyer, O. and W. Scherber, *J. Phys. Chem. Solids 32:* 1909 (1971).

Meyer, F. and J. J. Vrakking, *Surf. Sci. 33:* 271 (1972).

Meyer, F., E. E. DeKluizenaar, and G. A. Bootsma, *Surf. Sci. 27:* 88 (1971).

Milek, J. T., *Silicon Nitride for Microelectric Applications, Part I: Preparations and Properties,* Vol. 3 (1971) and *Part II: Applications and Devices,* Vol. 6 (1972) New York: IFI/Plenum.

Mitchell, I. V., M. Kamoshida, and J. W. Mayer, *J. Appl. Phys. 42:* 4378 (1971A).

Mitchell, I. V., J. W. Mayer, J. F. Kung, and W. G. Spitzer, *J. Appl. Phys. 42:* 3982 (1971B).

Morabito, J. M. and R. K. Lewis. *Anal. Chem. 45:* 869 (1973).

Morgan, D. V. and E. Bøgh, *Surf. Sci. 32:* 278 (1972).

Moses, A. J., *Handbook of Electronic Materials, Vol. I, Optical Material Properties*, New York: IFI/Plenum (1971).

Moss, T. S., *Optical Properties of Semi-Conductors*, London: Butterworths Scientific Publications (1959).

Murray, L. A., J. J. Rivera, and P. A. Hoss, *J. Appl. Phys* 37: 4743 (1966).

Neuberger, M., *Handbook of Electronic Materials, Vol. 2, III-V Semiconducting Compounds*, New York: IFI/Plenum (1971).

Newkirk, J. B., *Phys. Rev. 110:* 1465 (1958).

Newman, R. C., *Adv. in Phys. 18:* 545 (1969).

Nicolet, M-A, J. W. Mayer, and I. V. Mitchell, *Science 177:* 841 (1972).

Okada, K. T. and T. Oku, *Jap. J. Appl. Phys. 6:* 276 (1967).

Olberg, R. C., *J. Electrochem. Soc. 118:* 129 (1971).

Palmberg, P. W., *J. Vac. Sci. Technol. 9:* 160 (1972).

Park, R. L., J. E. Houston, and D. E. Schreiner, *Rev. Sci. Instr. 41:* 1810 (1970).

van der Pauw, L. J., *Philips Res. Rep. 13:* 1 (1958).

Picraux, S. T., *Appl. Phys. Lett. 20:* 91 (1972).

Picraux, S. T., *J. Appl. Phys. 44:* 587 (1973).

Picraux, S. T. and G. J. Thomas, *J. Appl. Phys. 44:* 594 (1973).

Picraux, S. T., J. A. Davies, L. Eriksson, N. G. E. Johansson, and J. W. Mayer, *Phys. Rev. 180:* 873 (1969).

Pliskin, W. A., *Solid State Electron. 11:* 957 (1968).

Pliskin, W. A. and P. P. Castrucci, *Electrochem. Technol. 6:* 85 (1968).

Pliskin, W. A. and E. E. Conrad, *IBM J. Res. Develop. 8:* 43 (1964).

Pliskin, W. A. and H. W. Lehman, *J. Electrochem. Soc. 112:* 1013 (1965).

Pliskin, W. A. and S. J. Zanin, Chapter 11, *Handbook of Thin Film Technology* (L. I. Maissel and R. Glang, Eds.) New York: McGraw Hill Book Co. (1970).

Popova, L. and B. Jordanov, *Solid. State Electron. 13:* 957 (1970).

Porteus, J. O., *J. Opt. Soc. Am. 53:* 1394 (1963).

Powell, C. J., *Surf. Sci. 44:* 29 (1974).

Putley, E. H., *The Hall Effect and Related Phenomena*, London: Butterworths & Co. (1960).

Rai-Choudhury, P. and W. J. Takei, *J. Appl. Phys. 40:* 4980 (1969).

Reizman, F., *J. Appl. Phys. 36:* 3804 (1965).

Reizman, F. and W. vanGelder, *Solid State Electron. 10:* 625 (1967).

Reutlinger, G. W., S. J. Regas, D. J. Sidor, and B. Schwartz, *Solid State Electron. 12:* 31 (1969).

Revesz, A. G. and K. H. Zaininger, *RCA Review 29:* 22 (1968).

Rheinländer, B., *Phys. Stat. Sol. 55:* K69 (1973).

Riccius, H. D. and J. E. Bertie, *Canadian J. Phys. 44:* 1665 (1966).

Riccius, H. D. and G. Ulbricht, *Z. f. angew. Physik 19:* 203 (1965).

Rimini, E., J. Haskell, and J. W. Mayer, *Appl. Phys. Lett. 20:* 237 (1972).

Robinson, P. H. and D. J. Dumin, *J. Electrochem. Soc. 115:* 75 (1968).

Rubin, S., *Treatise on Analytical Chemistry, Part 1, Theory and Practice*, Vol. 4 (I. M. Kolthoff and P. J. Elving, with E. B. Sandell, Eds.) New York: Interscience (1963) p. 2075.

Salzberg, C. D. and J. J. Villa, *J. Opt. Soc. Am. 47:* 244 (1957).

Sato, K., Y. Ishikawa, and K. Sugawara, *Solid State Electron. 9:* 771 (1966).

Saxena, A. N., *J. Opt. Soc. Am. 55:* 1061 (1965).

Schiller, C., *Solid State Electron. 13:* 1163 (1970).

Schlötterer, H., *Solid State Electron. 11:* 947 (1968).

Schubert, R., *J. Vac. Sci. Technol. 11:* 903 (1974).

Schumann, Jr., P. A., *J. Electrochem. Soc. 116:* 409 (1969).

Schumann, Jr., P. A., *Solid State Technol. 13:* 50 (1970A).

Schumann, Jr., P. A., *Silicon Device Processing*, NBS Spec. Pub. 337, Washington, DC: U. S. Government Printing Office (1970B) p. 234.

Schumann, Jr., P. A. and R. P. Phillips, *Solid State Electron. 10,* 943 (1967).

Schumann, Jr., P. A. and C. P. Schneider, *J. Appl. Phys. 41:* 3532 (1970).

Schumann, Jr., P. A., R. P. Phillips, and P. J. Olshefski, *J. Electrochem. Soc. 113,* 368 (1966).

Schumann, Jr., P. A., W. A. Keenan, A. H. Tong, H. H. Gegenwarth, and C. P. Schneider, *J. Electrochem. Soc. 118:* 145 (1971).

Schwuttke, G. H., *J. Appl. Phys. 33:* 2760 (1962).

Seraphin, B. O. and H. E. Bennett, *Semiconductors and Semimetals*, Vol. 3 (R. K. Willardson and A. C. Beer, Eds.) New York: Academic Press (1967).

Severin, P. J., *Appl. Optics 9:* 2381 (1970).

Sherman, B. and J. F. Black, *Appl. Optics 9:* 802 (1970).

Siegbahn, K. C. Nordling, A. Fahlman, R. Nordberg, K. Hamrin, J. Hedman, G. Johansson, T. Bergmark, S. E. Karlsson, I. Lindgren, and B. Lindberg, *ESCA, Atomic, Molecular and Solid State Structure Studied by Means of Electron Spectroscopy*, Uppsala: Almqvist and Wiksels Boktryckeri AB (1967).

Smales, A. J., *Trace Characterization* (W. W. Meinke and B. F. Scribner, Eds.) NBS Monograph 100, Washington, DC: U. S. Government Printing Office (1967).

Smith, D. P., *J. Appl. Phys. 38:* 340 (1967).

Smith, D. P., *Surf. Sci. 25:* 171 (1971).

Smith, R. A., *Semiconductors*, Cambridge: University Press (1959).

Smith, T., *J. Opt. Soc. Am. 58:* 1069 (1968).

Socha, A. J., *J. Vac. Sci. Technol. 7:* 310 (1970).

Socha, A. J., *Surf. Sci. 25:* 147 (1971).

Spitzer, W. G. and M. Tanenbaum, *J. Appl. Phys. 32:* 744 (1961).

Spiwak, R. R., *IEEE Trans. Instrum. Meas. IM-18:* 197 (1969).

Stanton, H. E., *J. Appl. Phys. 31:* 678 (1960).

Stein, H. J., *Solid State Electron. 15:* 1209 (1972).

Strehlow, W. H. and D. P. Smith, *Appl. Phys. Lett. 13:* 34 (1968).

Sunshine, R. A. and N. Goldsmith, *RCA Review 33:* 383 (1972).

Swann, R. C. G., R. R. Mehta, and T. P. Cauge, *J. Electrochem. Soc. 114:* 713 (1967).

Taft, E. A., *J. Electrochem. Soc. 118:* 1341 (1971).

Tharp, L. N. and E. J. Scheibner, *J. Appl. Phys. 38:* 3320 (1967).

Thomas, C. O., D. Kahng, and R. C. Manz, *J. Electrochem. Soc. 109:* 1055 (1962).

Tolansky, S., *Multiple Beam Interferometry of Surfaces and Films,* London: Oxford University Press (1948).

Tollestrup, A. V., W. A. Fowler, and C. C. Lauritsen, *Phys. Rev. 76:* 428 (1949).

Tracy, J. C., *Appl. Phys. Lett. 19:* 353 (1971).

Tracy, J. C., *J. Appl. Phys. 43:* 4164 (1972).

Tracy, J. C., *J. Appl. Phys. 44:* 4788 (1973).

Tsai, J. C. C. and J. M. Morabito, *Surf. Sci. 44:* 247 (1974).

Turkevich, A. L., *Science 134:* 672 (1961).

Uebbing, J. J. and N. J. Taylor, *J. Appl. Phys. 41:* 804 (1970).

Valdes, L. B., *Proc. IRE 42:* 420 (1954).

Vasicek, A., *Optics of Thin Films,* Amsterdam; North-Holland Pub. Co.; New York: Interscience Publishers Inc. (1960).

Vrakking, J. J. and F. Meyer, *Appl. Phys. Lett. 18:* 226 (1971).

Wang, C. C. and P. J. Zanzucchi, *J. Electrochem. Soc. 118,* 586 (1971).

Wang, C. C., G. E. Gottleib, G. W. Cullen, S. H. McFarlane III, and K. H. Zaininger, *Trans. Met. Soc. AIME 245:* 441 (1969).

Webb, W. W., *Direct Observation of Imperfections in Crystals* (J. B. Newkirk and J. H. Wernick, Eds.) New York: Interscience (1962) p. 29.

Weber, R. E. and W. T. Peria, *J. Appl. Phys. 38:* 4355 (1967).

Werner, H. W., *Vacuum 22:* 613 (1972).

Wickersheim, K. A. and R. A. Lefever, *J. Opt. Soc. Am. 50:* 831 (1960).

Winters, H. F. and E. Kay, *J. Appl. Phys. 43:* 789 (1972).

Wolfstirn, F. B., *J. Phys. Chem. Solids 31:* 601 (1970).

Yim, W. M., E. J. Stofko, P. J. Zanzucchi, J. I. Pankove, M. Ettenberg, and S. L. Gilbert, *J. Appl. Phys. 44:* 292 (1973).

Zaininger, K. H. and A. G. Revesz, *RCA Review 25:* 85 (1964).

Ziegler, J. F. and J. E. E. Baglin, *J. Appl. Phys. 42:* 2031 (1971).

Zohta, Y., *Soldi State Electron. 16:* 124 (1973).

Chapter 6

The Electrical Characterization of Heteroepitaxial Semiconducting Films

W. E. Ham

6.1. Introduction

Heteroepitaxial semiconducting films (HSF) have become of considerable practical interest in the last few years, primarily because of their use in integrated circuits. Several U.S. and foreign semiconductor companies are offering products based on silicon-on-sapphire (SOS) technology. This particular combination of semiconductor and insulator has emerged as dominant mainly because silicon has the most desirable overall properties compatible with integrated circuit processing and because of all the possibly suitable insulating substrates, sapphire offers the best resistance to shattering during processing and has the lowest cost. SOS technology offers small parasitic device and interconnection capacitance and provides nearly ideal electrical isolation between devices when islands of silicon are used. Because silicon islands are involved, it is necessary to cover the steps (island edges) with different dielectrics or metals to form the circuits. This step-coverage problem is minimized by using as thin a silicon film as is practical. On the other hand, thinner films can have inferior electrical performance and may be undesirable. Typically the film thicknesses range from 0.2 to 4.0 μm, with most circuits using 0.4- to 1.0-μm films.

It is nearly universally found that HSF are most suitable for applications that do not require high minority carrier lifetimes (e.g., MOS field effect transistors). Because the HSF may be very thin, it is possible to form MOS transistors which have no junctions (the so-called deep-depletion transistors) but which can be turned well off by removing mobil charge carriers from the silicon. Therefore, because the thickness of silicon from which these carriers can be removed by the surface field depends strongly on

the initial concentration of carriers, the doping level of the silicon is of primary interest. The doping level of the silicon is also important in determining the so-called threshold voltage of MOS transistors with junctions. For the presently used device geometries, the doping levels of greatest interest are those between $\approx 5 \times 10^{14}$ and $\approx 5 \times 10^{16}$/cm³. Other properties of HSF of interest are mobility and lifetime and in general any property that which may affect the operation of the devices.

Because HSF are to be used for subsequent processing, it is of considerable economic importance that some system to measure the significant electrical properties immediately after (i.e., before processing) film formation be available. These initial properties are of importance regardless of how the subsequent processing changes them since they are the starting point.

This chapter therefore is an attempt to consider the currently known methods for electrically characterizing HSF at various processing stages and to explore some of the less obvious aspects of these methods which may have important implications in interpreting the results.

6.2. General considerations

By the very nature of the thickness and growth techniques commonly used, the electrical properties of HSF are dominated by material properties that are of second-order importance in large single crystal (bulk) semiconductors. Even if the films were structurally perfect there are two abrupt and drastic changes in chemical composition on the top and bottom film "surfaces" or interfaces. This compositional distur-

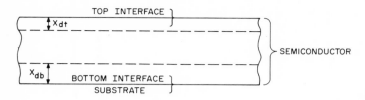

Xd = DISTURBED SURFACE REGION

FIGURE 6.1
Possibility of surface-dominated electrical
properties

bance necessarily creates a large perturbation in the band structure at the surface and for material within a Debye length of either surface, gross disruption of ideal electronic transport properties is possible. It is these transport properties that are usually used to characterize semiconducting materials in terms of resistivity, Hall coefficient, and mobility. In the case of HSF, however, the measured properties are frequently dominated by the properties of the interfaces, and unless special precautions are taken to ensure that the effects of the surface on the transport properties are well known, misleading information concerning the material properties can be derived from electrical measurements. Note that the effect of the surfaces is much stronger than simply the addition of a surface-scattering mobility term (although such a term is important in some cases) because the surface fields extend well into the semiconductor. A further complication arises when, as is usually the case, one interface is different from the other and therefore has a different surface band structure. These interfaces are illustrated in Figure 6.1.

Surface effects in bulk silicon have been known for some time and have been thoroughly discussed by Many and coworkers (1965) and by Frankl (1967), among many others. Recently, an attempt to include the effects of both surfaces on the conductivity of thin semiconductor films (Covington and Ray, 1974) was made. It was shown that if the interface potentials are quasisymmetrical (that is, both surface potentials above or both surface potentials below the "bulk" potential), then an extension of Greene and coworkers' (1960) treatment of the Boltzman transport equation for the case of symmetrical surface potentials is possible. These treatments require knowledge of the surface potentials by independent means. Another excellent treatment of the problem of electrical conduction in thin semiconducting films has been given by Anderson (1970) where only one surface was involved. Here, too, a knowledge of the surface potential is necessary.

It is of little comfort to the process control engineer to know that theoretical treatments that explain charge transport in ideal films exist when critical parameters are missing. Much of this chapter deals with some techniques used to control the surface potentials. Furthermore, as discussed in Chapter 2,

the attainment of structurally near-ideal films for practical purposes has not yet been achieved.

In addition to strong surface effects, it is usually found that the physical film structure and the electrical transport coefficients vary strongly with distance from the substrate, as illustrated schematically in Figure 6.2. Elliot and Anderson (1972) [using the theory of Petritz (1958)] have suggested a technique for dealing with the nonuniformities for the case in which the conducting portion of the film can be modified in a controlled way. The procedure is valid provided the usual concept of Hall coefficients and resistivities apply and the variations are not too extreme. The technique involves graphical differentiation and is tedious to apply. There is also the possibility of relatively large errors, as very precise measurements are needed.

In some cases the semiconductor defects are not representable as isolated point defects and impart some anisotropy to the transport coefficients. A typical edge dislocation space charge cylinder or a grain boundary are examples of commonly found HSF defects that are sometimes not considered when analyzing HSF electrical data. A specific example of this effect is when a grain boundary crosses between two Hall contacts. Because the physical structure of a HSF is usually much less perfect than bulk single crystal semiconductors, one expects the electrical transport properties to depend more strongly on treatments that alter physical structure. That is, HSFs are more sensitive to previous processing history than are more perfect semiconductors, and relationships

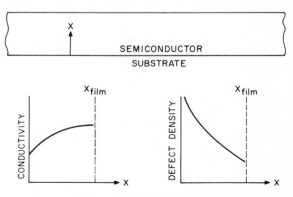

FIGURE 6.2   Variation of properties throughout the film

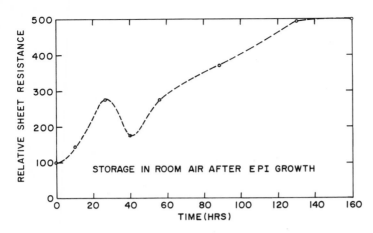

FIGURE 6.3   Unstable top surface. One-micron silicon on a sapphire n-type lightly doped ($\approx 10^{15}/cm^3$) film

between the structure and the transport coefficients are of paramount importance in the HSF case. Also, in nearly every case the HSF are strained because of the difference of thermal expansion coefficients between the substrate and the film and perhaps because of lattice mismatch.

Therefore, it is readily apparent that the task of relating a simple electrical transport coefficient measurement to a material property such as electrically active impurity concentration is much more complicated and likely to be misleading in HSF than in large, nearly perfect, single crystals.

The goal in the electrical characterization of HSF is to make a set of simple, quick, reliable, and nonde-structive measurements that produce a set of parametric values that not only characterize the HSF in its virgin state but also allow one to predict the changes in the significant properties after such processing steps as thermal treatments and oxidations.

There is no known set of measurements currently existing which effectively attains this goal. Part of the reason these measurements do not exist is that non-destructive techniques for reliably evaluating the physical characteristics of HSF are not available. If it is assumed that the structure of the HSF does not change appreciably for *nominally* identical film deposition procedures and that only the effects of electrically active impurities of differing concentrations are important, then one can limit the goal to determining the degree of change from a standard HSF which is known to give desirable results after processing. The validity of this assumption has yet to be proven.

On any unprocessed HSF the following problems may exist:
1. The exposed HSF surface may be electrically unstable.
2. There may be electrical contact problems caused by film thinness or light doping.
3. Only the average electrical parameter values (throughout the thickness of the film and over the area of the film sampled by the measurement) may be obtainable.
4. The unexposed HSF interface may not be reproducible.
5. Only very low signals may be available for measurement.
6. The interfaces may not be at flat band because of interface charges.

It is obvious that any electrical measurements should be made in an electrically shielded and optically dark ambient.

For the specific case of silicon on sapphire (for which most of this chapter is concerned), long-term instabilities in the exposed HSF surface have been noted (as illustrated in Figure 6.3). Here it is seen that

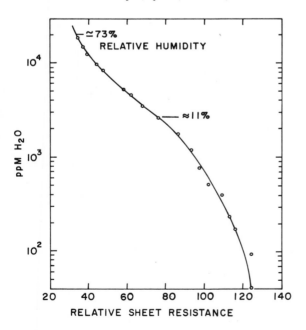

FIGURE 6.4   Effect of different water concentrations in N_2. Data were taken after a 5-min exposure to each water concentration beginning with the lower concentrations and progressing to the higher ones

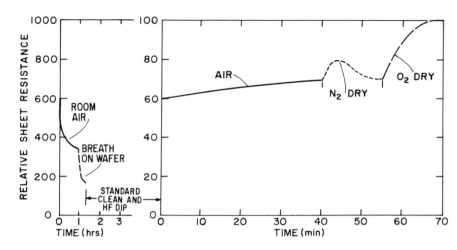

FIGURE 6.5   Some effects of chemical treatments on different gases and relative sheet resistance.

wide variations in the sheet resistivity occur upon simply allowing the film to remain in room air. The moisture content of the ambient is also seen to be an important factor (Figure 6.4). Certain chemical treatments can be used to stabilize the surface, such as standard cleaning (H_2O_2-NH_4OH solutions), HF dipping, or heating the film to remove moisture. Figure 6.5 illustrates the importance of chemical treatment history on the measured values. The actual state of the film surface is usually not known. However, if care is taken to exactly reproduce the surface treatment, a relative comparison of nearly identical films may be reasonable.

Except when optical or microwave techniques are employed, an electrical contact must be made to the HSF for use as a potential probe or a current source. In general, optical techniques are only useful for characterizing electronic properties of semiconductors with mobile impurity concentrations greater than 10^{17}/cm^3, which is higher than the impurity concentration of interest for most practical applications. Therefore, optical techniques will not be discussed further here. Several methods for making electrical contacts to unprocessed films are illustrated in Figure 6.6. By the very nature of a mechanical probe-type contact, it is not—strictly speaking—nondestructive. The destroyed area is assumed to be very small compared to the total film area. Regions of very high surface generation are created under the probe resulting, one hopes, in an ohmic contact. When high-resistivity (lightly doped) material is used, the space charge region under the probe may exceed the area of the damaged region and form an electrical barrier. Mechanical probes, therefore, are most satisfactory where a small amount of damage is tolerable and the material is not too lightly doped. The liquid contacts are usually completely nondestructive and have the additional property of being large compared to point probes (by several orders of magnitude in some cases). A disadvantage is that the size and exact position of these contacts is not usually accurately known. The requirements for specific contact resistance are lowered because of the large contact area. The contact spacing must therefore be large compared to the metal droplet size in order to maintain effective zero-dimension contacts. Removal of liquid metal contacts usually requires chemical treatment.

Perhaps the best controlled of all nondestructive contacts are the evaporated, sputtered, or electrodeposited metals. Here the interface between the semiconductor and the metal can be well controlled and the contact size is accurately known. The dimensions can be made relatively small using photolithographic techniques. If wires are to be attached to the film, a silver paste (metallic suspension in an organic binder) contact may be used.

For routine monitoring of HSF depositions either point probes or liquid contacts should be used, as

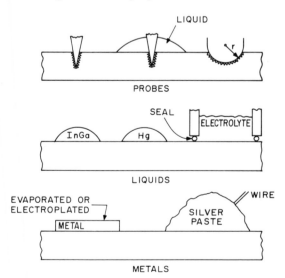

FIGURE 6.6   Some techniques for providing electrical connection to unprocessed HSF

considerable time and effort are required to form the metal contacts. Suitable metal contacts may require some surface preparation which can alter its properties. Overall, the best contacts employed by the author have been made with the combination of liquid droplets and point probes. Processed films usually have the option of using heavily doped regions, such as diffusions, for the contact.

In summary, the main general considerations for the electrical characterization of HSF are the influence of the surfaces, the nonuniform distribution of physical and electrical properties, and the technique used for electrical access to the film.

6.3. Electrical characterization of unprocessed films

Nondestructive techniques

Four-point probe. The most widely used and well known technique for the characterization of unprocessed wafers is the four-point probe technique. Average sheet resistance is sought, and the measurements have first-order dependence on the state of the surface. The method consists of forcing a current between two probes and measuring the resulting potential difference between the remaining probes. Usually the probes are equally spaced in a straight line. Correction factors for different probe spacing and sample shape have been computed (Valdes, 1954). In the absence of stray RF signals and light, the contact resistance of the potential probes is not important, as a very high impedance voltage sensing instrument is used; the contact resistance of the current contacts is not important because a current is forced. Both RF signals and light can produce undesirable DC voltages which interfere strongly with obtaining accurate measurements. The four-point probe method is very quick, but because of its strong dependence on the state of the surface, the measurements are of questionable value, especially for lightly doped and thin HSF. The particulars of the implementation of this method are discussed in Chapter 5.

Modified van der Pauw. It is possible to make Hall measurements on unprocessed wafers using the modified van der Pauw technique (MVDP) suggested by Buehler and Pearson (1966). This method, which is illustrated in Figure 6.7, consists of using the standard van der Pauw method (see Chapter 5), except that nonperipheral contacts are used. The probes must be placed *near* the edges of the film, however, because a Hall field can only be sustained if a nonconducting boundary exists. Therefore, the use of a square probe array which occupies only a small portion of the film area will not be suitable for Hall measurements, as seen in Figure 6.7c. Results using MVDP are averages not only throughout the thickness of the film but also across a wafer. The probes are not placed on the edges of the film to avoid strong effects from the nonuniform regions near the wafer edges. Comparison of MVDP measurements with actual parameter distributions will be discussed later. MVDP exhibits the same first-order dependence on the state of the surfaces as the four-point probe technique. Surface effects are much stronger in HSF than effects such as nonzero dimension contacts.

Microwave attenuation. A third nondestructive method that should be useful for characterizing unprocessed HSF is the microwave attenuation technique. Here the wafer is inserted in a microwave circuit or cavity and the resulting attenuation or absorption can be measured. The skin depth at usually obtainable microwave frequencies is much larger than the film thickness, so uniform absorption can be assumed over the conducting portion of the film. Microwave reflection techniques are not suitable because of the large skin depth. The microwave attenuation technique also suffers from possible surface problems, since the absorption occurs only in the conducting portion of

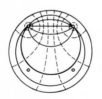  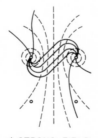

a) NO MAGNETIC FIELD   b) STRONG MAGNETIC   c) STRONG FIELD
                              FIELD   NO BOUNDARIES

FIELD LINES – – – – –
CURRENT LINES ————

FIGURE 6.7
The modified van der Pauw method

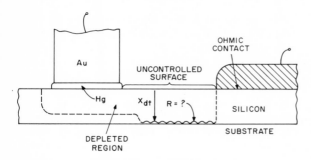

FIGURE 6.8   Mercury Schottky barrier method

the HSF. To the author's knowledge, this technique has not actually been used for the electrical characterization of HSF silicon but has been used for germanium on calcium fluoride (Covington and Ray, 1973).

There are at least two remaining methods for use on unprocessed films that are somewhat immune to the unstable surface properties. These are the mercury (or Al) Schottky barrier technique and the spreading resistance technique. In both of these methods the surface directly under the probes is an intimate part of the measurement and therefore may not be causing unknown errors. Let us consider first the mercury Schottky barrier.

Mercury Schottky barrier   The mercury Schottky barrier (MSB) is useful provided that suitable contact can be made to the silicon surrounding the mercury. If the HSF is thin or lightly doped, the electrical properties of the silicon interfaces can prevent suitable contact, as illustrated in Figure 6.8. Analysis of the reverse-biased capacitance–voltage characteristics yields reliable information concerning the doping profile near the top film surface. The area of the mercury probe can be controlled within reasonable accuracy by using the methods shown in Figure 6.8 or by using capillary methods. Both mercury and aluminum have been used on p-type SOS by the author. It is important that no heat treatment be necessary in order to form the barrier so that the method will be nondestructive. If good "back" contact can be made, the MSB is probably the most reliable and easy-to-use technique for unprocessed films. Note, however, that this technique does not measure transport properties directly.

Spreading resistance.   The spreading resistance technique involves measuring the resistance between two more or less spherical probes of the type illustrated in Figure 6.6. This resistance consists of the spreading resistance concentrated under each probe and the series resistance of the material between the probes. There is no provision for separating the dominant resistance component with this method. However, if the material between the probes is not dominant, the

measurements should be relatively insensitive to the state of the surface.

Commercially available probes ($r > 4 \ \mu m$) produce penetration of the order of 0.5 μm into SOS films and theoretically require large correction factors (Goldsmith et al., 1974) even for very thick films ($\approx 5 \ \mu m$). Nevertheless, the possibility exists that the spreading resistance could be used to monitor the reproducibility of HSF deposition and may provide a very valuable and practical, if not exact, method.

In order to test this idea, a set of n-type SOS films were deposited under as nearly identical conditions as possible except for the thickness which was determined by the deposition time. Also included in the reactor were bulk silicon wafers of the same type as the deposit.

Spreading resistance was measured at 50 or more points on each wafer; the results are shown in Figure 6.9. Above a thickness of approximately 2.5 μm there was a nearly constant difference between the SOS and the bulk. This probably reflects the thickness above which the resistance of the material between the probes on the SOS samples became insignificant compared to the spreading resistance. Below approximately 2.0 μm, large variations in the measured resistance were seen on the SOS samples. These data suggest that with currently available technology, spreading resistance measurements apparently cannot be reliably used to monitor the reproducibility of HSF depositions below approximately 2.0 μm which, as stated earlier, is the important thickness range.

The variations are probably caused either by exces-

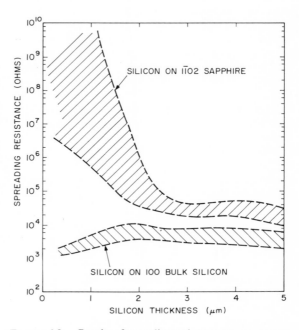

FIGURE 6.9   Results of spreading resistance measurements

sive probe penetration into the silicon or by variations in the resistance between the probes. Smaller-radius probes would undoubtedly extend the usable range of silicon thickness to smaller values both because higher spreading resistance is obtained and because less probe penetration could be achieved.

Sufficiently exhaustive experiments have not yet been performed to completely rule out the use of spreading resistance for practical HSF. The possibility of immunity to surface charges and the simplicity of implementation for both types of film make this method worthy of careful consideration.

Destructive techniques

The simplest destructive technique is to define an island of semiconductor on the substrate in a shape suitable for the test to be performed. This usually means defining a Hall bar for more localized measurements than the MVDP. A pattern can be defined without performing significant thermal treatment but some chemical surface treatment is necessary to form small patterns. Large patterns can be formed without chemical surface treatment using razor blade techniques. When patterns are defined, additional edge surfaces are exposed to the ambient and could possibly cause severe local disturbance at the edges of the pattern. These edge effects have not been reported on unprocessed HSF patterns, however, probably because the effects are small compared to the interface effects of the top and bottom surfaces. The primary advantage to using a defined pattern is that *localized* measurements can be performed.

Another destructive technique involves thinning the HSF by such techniques as sputtering or anodic oxidation. If surface effects can be suitably accounted for, this makes profiling the electrical properties of HSF possible.

6.4. Top semiconductor surface control techniques for electrical characterization

Low-temperature methods

Whole wafers.   The desired goal of simply characterizing HSF without processing seems somewhat elusive (with the possible exception of the mercury Schottky barrier). The main limitation in all of the methods in Section 6.3 is that the exposed top surface can cause unknown contributions to the measurements. Some methods for attempting to control the top surface potential without altering the film properties by heating above a reasonably low temperature (~350°C) are discussed in this section.

Using a specified chemical surface treatment history has already been mentioned and is probably the easi-

est to implement. It was found, however, that the treatments investigated to date (quite limited) still leave the surface somewhat unstable (see Figure 6.5).

Another technique involves carefully specifying the ambient in which the measurements are made. This ultimately produces a stable (i.e., non–time-varying) surface. Transient responses to varying ambients are seen in Figures 6.10 and 6.11 where only the moisture content of the ambient was varied. After a week of exposure to dry N_2 steady state has not been reached for the sample in Figure 6.11. Therefore, even though specifying the ambient may be useful for ultimately establishing reproducible surface potentials, the time responses can be prohibitively long in the dark at room temperature. The effects of elevated temperatures and optical excitation to increase the reaction rates have been considered but not yet attempted to the author's knowledge. Furthermore, when steady state is finally achieved, there is still no information concerning the *value* of the surface potential. Different ambients, of course, produce different steady state values and the essence of a control technique clearly exists. A great deal of further work is required on this method to find an appropriate ambient and a method to decrease the response time. Very preliminary work by the author indicates that wet O_2 may produce nearly flat band top surfaces on n-type SOS films.

Because both the chemical treatment and the ambient technique involve continuous transfer of charge from the HSF to the surroundings, the actual

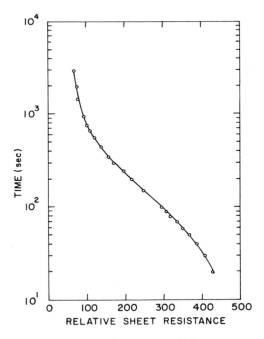

FIGURE 6.10   Transient response during exposure to wet N_2 (590 ppm H_2O). SOS film 1 μm thick $\approx 1 \times 10^{15}/cm^3$ n-type

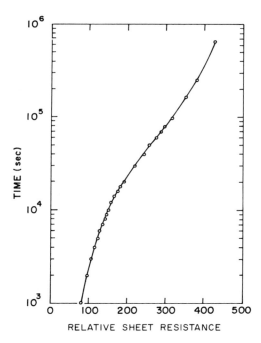

FIGURE 6.11   Recovery (exposure to dry N_2 after exposure to wet N_2) for the same film of Figure 6.10

state of the top interface is difficult to determine during an electrical measurement. In short, neither method presently allows reliable determination of the surface potential. Extensive studies of applying these methods to PbSe on NaCl substrates have been performed (McLane, 1971).

In order to overcome these difficulties, an insulating layer may be deposited or grown on the top HSF surface. The dielectric makes electronic transport between the ambient and the interface difficult and "fixes" the surface potential.

It is important that the formation of this dielectric not change the properties of the HSF in an unknown way. It is best if the film is not heated to anywhere near its growth temperature during the dielectric formation. In the case of SOS, this is at temperatures greater than 900°C. This deemphasizes the use of thermal oxidation for the dielectric since thermal oxides grow in reasonable times only above ≈800°C (at 1 atmosphere). Deposited silicon oxides, on the other hand, can be formed, for example, by the oxidation of silane at temperatures between 325°C and 450°C and have much less chance to alter the properties of the virgin HSF. This is not only due to the low temperatures involved but also to the fact that the silicon in the dielectric comes from the gaseous reactants rather than from the film. With appropriate low-temperature (<450°C) annealing treatments in atmospheres containing hydrogen, these deposited oxides can provide a reasonably stable top interface with a low value of fixed interface charge ($\approx 10^{11}$/cm²).

Deposited dielectrics are also useful for nonsilicon HSF where stable thermal dielectrics do not form. With the dielectric on the surface, one still does not know the top interface potential, since fixed interface charges exist and some charging of the exposed dielectric surface may occur. In order to investigate these effects, several SOS wafers were deposited under identical conditions except that the dopant flow (AsH_3) was varied. For each dopant flow, four or five wafers [(1$\bar{1}$02) single crystal sapphire] were used. The MVDP resistivity after deposition is plotted against the dopant flow during deposition in Figure 6.12. Approximately 1200 Å of SiO_2 was then deposited on these wafers, and small holes were cut in the oxide using photolithographic techniques for probing purposes. Opening these holes again exposes the semiconductor surface and can cause severe difficulty in obtaining good contact, as illustrated in Figure 6.13a and 6.13b. The method seen in Figure 6.13c is best, but it is difficult to achieve in practice. Such a contact can be approximated by applying an electric field such that dielectric rupture occurs under the probe.

The MVDP measurements were repeated with this deposited dielectric, with results as shown in Figure 6.14. The important points to note here are that the measurements with a deposited dielectric indicate a lower resistivity and show less scatter than those with a bare surface. The lower resistivity is probably caused by an increase in the fixed positive top interface charge which either causes accumulation or less depletion of the top surface of these n-type films. The decrease in scatter is small, but may indicate some "standardization" of the top interface. The deposited dielectric was then removed and small photolitho-

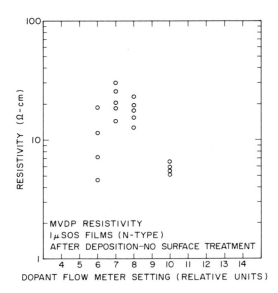

FIGURE 6.12   MVDP resistivity for a set of 1-μm n-type SOS wafers

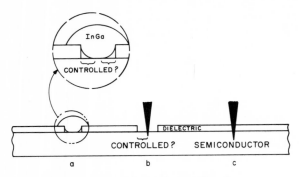

FIGURE 6.13   Some methods of contacting the semiconductor through a deposited dielectric

graphically defined van der Pauw structures were formed and subjected to normal measurements. The results of these measurements are shown in Figure 6.15. Here the resistivity has returned to approximately the same distribution it had originally. This is significant, because it implies that dielectrics can be deposited and removed without changing the films' properties. Unfortunately, the actual value of the semiconductor surface potential is still unknown, and it is necessary to know what the interface potential is to properly interpret transport coefficient measure-

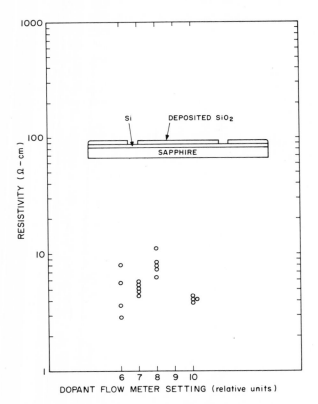

FIGURE 6.14   MVDP resistivity of the wafers of Figure 6.12 after dielectric deposition

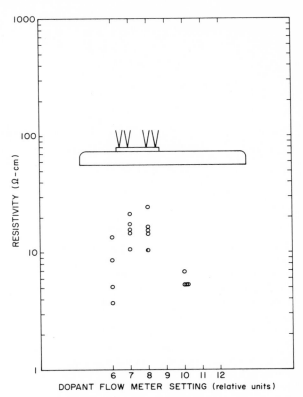

FIGURE 6.15   Van der Pauw resistivity near the center of the wafers of Figure 6.12 after stripping the deposited dielectric in HF

ments. In fact, a method of determining when the top interface potential is zero (flat band) is needed. Therefore, some means for *varying* the potential of the top of the dielectric must be implemented. The most obvious method is to deposit a metal on the dielectric. This, of course, forms the essence of the well known MOS structure which will be discussed in detail shortly. There are other techniques for modifying the potential of the top of the dielectric, however, which should be mentioned.

A metal plate can be placed in close proximity to (or directly on) the dielectric surface. This method is not good, however, because the contact between the plate and the dielectric cannot be sufficiently intimate. The plate surface finish would have to be flat to ≤ 1000 Å (0.1 μm) to avoid large inhomogeneities in the effective dielectric thickness. If one separates the metal plate from the dielectric such that surface inhomogeneities are small, the required voltages become quite large. Small wafer flatness variations can create very large dielectric thickness changes. In other words, it is very difficult to maintain a uniform flat band condition across a wafer with this method.

Another method, used by Williams and Woods (1973) for a different purpose, involves essentially

CORONA DISCHARGE

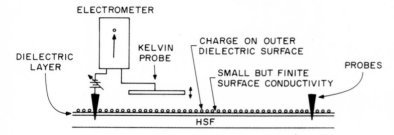

FIGURE 6.16
Modification of the surface charge on the dielectric with the use of a corona discharge

spraying the dielectric surface with charge from a corona. The dielectric surface potential is measured with a Kelvin probe which is a conductive coating on a small square of glass that oscillates vertically (≈0.5 mm) above the sample (see Figure 6.16). This probe is connected to an electrometer and the sample is connected to a reference voltage; when the reference voltage is the same as the surface potential, the electrometer reads null. This technique requires no metallization and is at least useful for determining the amount of surface effect possible. There is some problem with the dielectric surface potential changing with time (because of the dielectric's small surface conductance) and with determining the actual semiconductor surface potential. No experimental data are currently available with this technique, unfortunately.

The use of metal electrodes on whole wafers has not been reported and might be expected to be unsuccessful in any case since, unless the dielectrics were absolute free of pinholes, the structure would not be satisfactory. The corona technique does not suffer from this pinhole problem because of the dielectrics' low surface conductivity.

These are the nondestructive methods which have been considered to date. None of them are entirely satisfactory. The main problems are lack of knowledge of surface potential and that the large areas involved prohibit the application of intimate metallic contact to the dielectric surface. Therefore, a significant amount of research on the properties of smaller defined structures has been performed.

Defined structures. Several reports on the use of defined MOS structures for electrical characterization of HSF have appeared in the literature recently. Most of these investigations have used thermally oxidized dielectrics and therefore are characterizing only processed HSF. At least one report of the use of deposited dielectrics on HSF has appeared (Ham, 1972A). The basic idea in all of these investigations is to use surface capacitance v. surface potential data in conjunction with the usual Hall and resistivity data in order to (1) control (and know the value of) the top interface potential and (2) profile the electrical properties throughout the film. The basic structure used is illustrated in Figures 6.17 and 6.18, and the type of data obtained is shown in Figure 6.19. These structures are quite suitable for the reliable characterization of the top portion of HSF. Unfortunately, the amount of effort required to fabricate such a structure is comparable to fabricating an integrated circuit. Nevertheless, even though these structures do not answer the need for a simple characterization tool immediately after deposition, they are very useful in determining the properties of HSF after some processing has occurred.

By varying the outer dielectric potential (or gate voltage, as it is commonly called), the conduction in the film can be forced away from the top surface. This, therefore, allows the contribution of the top part of the films to be gradually and controllably removed. Only current carriers deeper in the film than the surface space charge layer can contribute to the current. The depth of film that can be depleted depends on the

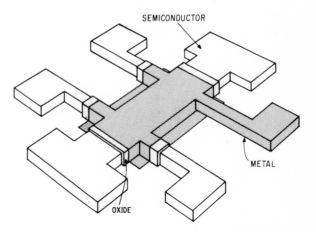

FIGURE 6.17   Schematic structure of one possible design for a gate-controlled Hall bar

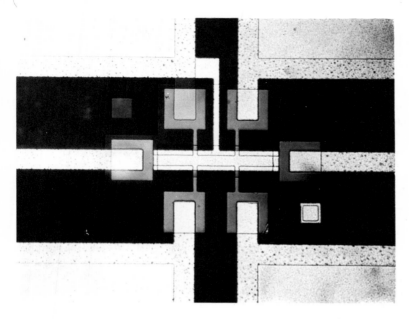

FIGURE 6.18
Photomicrograph of a gate-controlled
Hall bar

exact structure used. For the case of the simple gate-controlled Hall bar of Figure 6.17 the maximum depletion depth is determined by the doping of the film N, and the dielectric constant, $\epsilon_s\epsilon_0$, of the semiconductor

$$x_{d_{max}} = \left(\frac{2\epsilon_s\epsilon_0\phi_s(\text{inv})}{qN}\right)^{1/2}, \qquad (6.1)$$

where $\phi_s(\text{inv})$ is the surface potential for strong inversion (Grove, 1967).

Attempts to force the depletion depth further results only in obtaining charge at the top semiconductor surface in a so-called inversion layer. The inversion layer only appears under the gate metal and does not contribute to the observed current either because the gate does not extend to the ohmic contacts (if such are used) or because the contacts are of opposite type to the inversion layer (such as with diffused contacts). The depletion depth can be extended by making electrical contact to the inversion layer and applying additional voltage across the vertical dimension of the HSF. This technique has been suggested by Shannon (1971) for the case of bulk semiconductors and by Thianyi (1972) for the case of SOS. When contact is made to the inversion layer, the maximum depletion depth is limited by avalanche multiplication in the semiconductor. If conduction deep within the film is the only region of interest, a shallow surface junction may be formed to extend the surface depletion depth. This method does not allow examination of the conducting properties of the top part of the HSF. Both the shallow surface junction and the contacted inversion layer introduce possible additional sources of current which may interfere with accurate measurements. We shall not concern ourselves further with these techniques at this point since they are needed only when

deep penetration at the top surface space charge layer is required.

The link to the semiconductor surface potential from the applied gate voltage is provided by the analysis of the dependence of the surface capacitance on the applied potential. It is possible to determine from this data the gate potential at which there is no vertical field across the HSF or the so-called flat band conditions. These analyses are well documented in the literature [see Zaininger and Heiman (1970) and references therein] and will not be discussed in detail here. In addition, the depth of the surface space charge region can be determined from the surface capacitance values. Appropriate use of this C-V data depends critically on having a reliable value for the capaci-

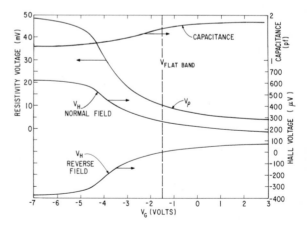

FIGURE 6.19   Data obtained from a gate controlled Hall bar. This film is approximately 1 μm thick. V_ρ is the voltage across the resistivity probes and V_H is the voltage across the Hall probes with applied magnetic field

tance, and a whole section of this chapter is devoted to this important subject.

Gate-controlled Hall structures provide, therefore, the possibility of accurately measuring the electrical properties of HSF under known top surface conditions.

Both the surface capacitance and Hall measurement can be used to determine the doping level of the HSF under certain conditions. The rate of change of the surface capacitance with surface potential is related to the number of ionized immobile species in the film under surface depletion conditions. In a simple, nearly perfect film in which no appreciable charges exist at the interface, each of these ionized species represents a donor or acceptor. Each donor or acceptor contributes one charge carrier under nondepleted conditions, provided sufficient thermal energy exists at the temperature of interest to ionize the species. This is the essence of the connection between surface capacitance and film doping. If the species is not ionized at the temperature of interest, it may still be ionized by varying the surface potential. In general, however, the surface capacitance is also affected by charge at the surface which is energetically distributed throughout the band gap (surface states). This affects the direct application of the C-V data to determine the ionized impurity density. The value of the surface capacitance can always be used to determine the *depth* of the surface space charge layer.

Hall measurements can provide information concerning the concentration and mobility of the charge carriers that are able to move at the temperature of the measurement. Only in the case in which all dopant species are ionized is there a direct relationship between a single Hall measurement and the film doping. The Hall measurements are averaged throughout the conducting part of the film whereas the C-V measurements are localized. Combinations of the two measurements allow calculation of the doping and mobility profile. Ham (1972B), Elliot and Anderson (1972), Ipri (1972), and Hynecek (1974) have all used this method to investigate the properties of SOS films.

These investigations all show the presence of a low-mobility region of the films near the substrate–semiconductor interface. When the current is forced to flow in this low-conductivity portion of the film, it becomes difficult to maintain sufficiently high currents to produce accurately measurable Hall voltages without impressing too much voltage across the current contacts. The voltage across the current contacts must be maintained at a low enough value to ensure a constant interface potential. The Hall Voltage is determined by the current (not the current density) and the film has low mobility near the bottom interface, both of which contribute to the extreme difficulty of making reliable Hall measurements in the lower part of HSF. No such limitation exists for resistivity measurements

in the lower part of the films. One can always combine the depletion depth from C-V measurements with resistivity measurements to obtain a resistivity profile.

A further practical problem exists when attempting to make Hall measurements below a certain thickness level. This problem manifests itself as an anomalous change in the potential between the Hall contacts (V_{AB}) as conduction is forced deeper into the film by the gate field. This problem only appears on films that can be depleted below a certain depth. Data on this subject from films of different thicknesses are shown in Figure 6.20. It is normal for V_{AB} to increase as the conduction is forced into the bottom part of the film because V_{AB} is caused primarily by small misalignments of the arms. It is not normal, however, in a structurally perfect film, for V_{AB} to first increase and then decrease.

The data in Figure 6.20 for the 2.5-, 1.5-, and 1.0-μm devices are approximately as expected. These devices were thicker than $x_{d_{max}}$. The thinner devices, however, show strongly anomalous behavior whose onset is observed progressively into accumulation as the film becomes thinner. We note in this figure and in Figure 6.19 that the voltage does not always approach zero as heavy accumulation is reached. This is due to zero offset drift in the voltage sensing system and is not due to properties of the film.

The cause of this phenomenon may be related to the physical structure of the films below a certain depth. In particular, the structure illustrated schematically in Figure 6.21 provides an intuitively satisfying explanation for the observations. While the conduction is allowed to remain in the normal scattering part of the film, normal V_{AB} behavior is observed. As the conduction is forced below this level, the potential developed across the barrier is added to the normal V_{AB} and

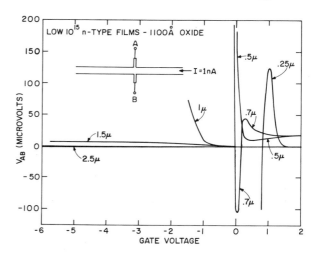

FIGURE 6.20 V_{AB} dependence on gate voltage with no applied magnetic field under constant current conditions for gate-controlled SOS Hall bars

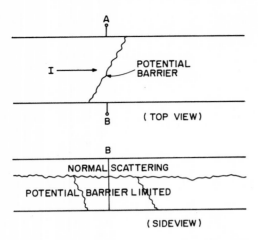

FIGURE 6.21   Conceptual cause for the dependence of V_{AB} in Figure 6.20 on gate voltage

appears as a larger or smaller value, depending on the way the barrier crosses between the arms. If barriers cross both ways at different depths into the film, some quasi oscillation in V_{AB} may be seen. Of course, it is not necessary that the barrier completely cross the Hall bar for this effect to be possible.

Electrical conductivity in the potential barrier region would be expected to be much less than in the normal scattering region. One could therefore effectively divide the film into a relatively highly conducting upper part and a less conducting lower part. This is discussed in detail in Section 6.8.

Some applications of low-temperature methods.   One of the most important applications of the low-temperature methods is to investigate the actual meaning of the measurement made with unprotected top surfaces. This section shows the applications in which the deposited-SiO₂ gate-controlled Hall bar was used to characterize virgin SOS films. A comparison between

the MVDP measurement and the flat band and depleted surface data was made on a nominally 1.0-μm p-type SOS film. The second application was investigating the applicability of the low-conductivity or "insulating" layer idea to the calculation of material properties in the normal scattering region. The word "insulating" is used occasionally in this chapter to refer to the relatively low-conductivity layer. It does not imply that the layer is totally nonconducting, only that the layer is nonconducting compared to the upper portion.

Figures 6.22 and 6.23 illustrate the distribution of resistivity, mobile concentration (Hall), and carrier concentration, and Hall mobility across a diameter of a 1.5-inch p-type SOS wafer. The silicon thickness across this wafer was not uniform and varied from 1.1 μm at the top of the wafer to ≈0.9 μm at the bottom. Also illustrated in these figures is the result of initial MVDP measurement. A relatively strong variation across the wafer was observed for both carrier concentration and resistivity. All parameters were calculated assuming a uniform 1.0-μm conduction. This involves no more than ±10 percent error caused by variation in the film thickness.

The MVDP data should ideally characterize the material near the center of the wafer. It appears that the MVDP characterizes the center material only if this material is depleted nearly to its maximum. Therefore, as found before (Ham, 1972A), the unprotected surface was apparently depleted when the MVDP measurements were made.

The second example of the use of the deposited-oxide gate-controlled Hall bar consists of attempting to apply the model of the insulating layer to an entire wafer. In this case the flat band Hall coefficient and resistivity were obtained from the gate-controlled Hall bar and the actual silicon thickness of each bar was measured. If the actual silicon thickness is t_m and the insulating layer thickness is t_{ins}, then define

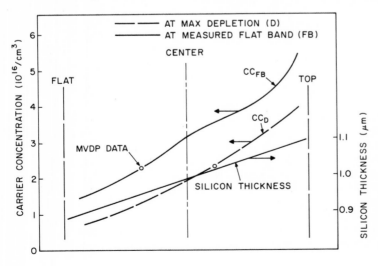

FIGURE 6.22
Flat band Hall carrier concentration (CC_{FB}), maximum top surface depletion Hall carrier concentration (CC_D), and the silicon thickness variation across a 1.5-inch-diameter SOS p-type wafer

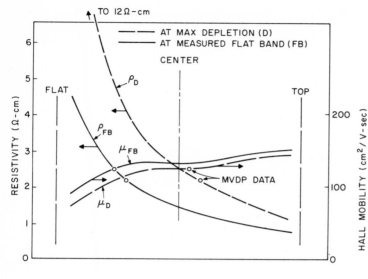

TO 12 Ω-cm

——— AT MAX DEPLETION (D)
———— AT MEASURED FLAT BAND (FB)

CENTER

FLAT   TOP

ρ_D

ρ_{FB}

μ_{FB}

MVDP DATA

μ_D

FIGURE 6.23
Flat band resistivity (ρ_{FB}) and Hall mobility (μ_{FB}) and maximum top surface depletion resistivity (ρ_D) and Hall mobility (μ_D) across the same wafer of Figure 6.22

$$R_{H_m} = \frac{V_H t_m}{IB}, \tag{6.2}$$

where V_H is the flat band Hall voltage, I is the current, and B is the magnetic field. The Hall coefficient in the normal scattering portion of the film is therefore

$$R_H = \frac{V_H(t_m - t_{ins})}{IB} = R_{H_m} \frac{(t_m - t_{ins})}{t_m}. \tag{6.3}$$

Similarly, the resistivity is

$$\rho = \rho_m \frac{(t_m - t_{ins})}{t_m}. \tag{6.4}$$

One can determine t_{ins}, R_H, and ρ from flat band measurements from two bars of different thickness, t_{m_1} and t_{m_2}:

$$\rho = \frac{\rho_{m_1}\rho_{m_2}t_{m_1} - \rho_{m_1}\rho_{m_2}t_{m_2}}{\rho_{m_2}t_{m_1} - \rho_{m_1}t_{m_2}} \tag{6.5}$$

$$t_{ins} = \frac{t_{m_1}t_{m_2}\rho_{m_2} - t_{m_1}t_{m_2}\rho_{m_1}}{\rho_{m_2}t_{m_1} - \rho_{m_1}t_{m_2}} \tag{6.6}$$

$$R_H = \frac{R_{H_{m_1}}R_{H_{m_2}}t_{m_1} - R_{H_{m_1}}R_{H_{m_2}}t_{m_2}}{R_{H_{m_2}}t_{m_1} - R_{H_{m_1}}t_{m_2}} \tag{6.7}$$

$$t_{ins} = \frac{t_{m_1}t_{m_2}R_{H_{m_2}} - t_{m_1}t_{m_2}R_{H_{m_1}}}{R_{H_{m_2}}t_{m_1} - R_{H_{m_1}}t_{m_2}}. \tag{6.8}$$

In this particular case, an n-type SOS wafer was used. The thickness ranged from 0.78 μm to 0.95 μm and approximately 40 different devices on the same wafer were measured. Application of the above formulas yielded $\rho = 0.0938$, $R_H = 34.4$, t_{ins} [eq. (6.6)] = 0.688 (μm), t_{ins} [eq. (6.8)] = 0.685, and $\mu = 366$ cm²/V-s; the carrier concentration near the top was 1.8×10^{17}. Initial MVDP measurements indicated $\rho = 0.45$, $R_H = 180$, $\mu = 400$ cm²/V-s; the carrier concentration was 3.4×10^{16}.

The values of ρ, R_H, and t_{ins} were used in Equations (6.3) and (6.4) to predict the observed values of ρ_m and R_{H_m}. These results are illustrated in Figures 6.24 and 6.25. A reasonably good fit with the observed data is seen. The difference between the MVDP data and the apparent actual HSF top surface properties is quite considerable. It appears that in this case the exposed surface was nearly at flat band conditions and that the discrepancy was caused almost entirely by the bottom portion of the film being insulating relative to the top portion. Some further discussion concerning the nature of this insulating region will be presented later in this chapter.

Capacitance–voltage methods

We have been discussing the use of *C-V* data on gate-controlled HSF structures as if it were a routine measurement. In many cases there is no need for special precautions. However, because contact with the semiconductor must be made essentially from the side of the gate metal on HSF MOS capacitors, the series resistance is *distributed* under the gate electrode. This requires a different analysis from that used for a bulk semiconductor MOS capacitor, in which the series resistance is essentially *lumped*. These problems have been considered by Ham (1974) and by Lehovec (1974) and will be discussed in detail in this section.

In the following analysis, both the lumped and distributed resistance are considered. It is of primary importance that one be aware of these effects when dealing with HSF MOS capacitors.

Nondistributed case. It may be readily appreciated that with HSF MOS capacitors a situation may arise in which the resistively conducting portion of the film may become very thin or eliminated entirely by bias-

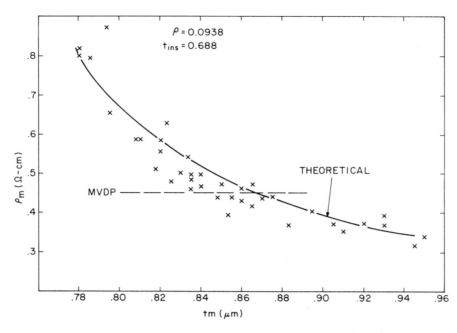

FIGURE 6.24
Theoretical [from Equation (6.4)] and experimental variations in the measured resistivity assuming a uniformly conducting film (ρ_m) as a function of actual film thickness. n-type SOS

ing into depletion (or out of heavy accumulation in some cases). When this happens, the simplest model is no longer applicable, because the series resistance of the structure may become a significant fraction of the total impedance. A small-signal steady state AC analysis is considered. Usually the conductance in parallel with the surface capacitance, C_m, is much smaller than ωC_m and the following analysis neglects it entirely.

As a first approximation, the circuit shown in Figure 6.26 will be analyzed. This is a lumped representation and describes the situation in bulk semiconductor MOS capacitors. V_S represents the voltage across a sense resistor and is assumed to be negligibly small. In practice, any current-detecting circuit can be used in place of R_S. Neglecting R_S and using standard notation:

$$I = \frac{V \angle 0°}{Z} =$$

$$\frac{V \angle 0°}{\frac{1}{j\omega C_m} + R_{ser}} = V \angle 0° \left(\frac{R_{ser}\omega^2 C_m^2 + j\omega C_m}{1 + R_{ser}^2 C_m^2 \omega^2} \right).$$

$$(6.9)$$

The 90° component of the current with respect to the driving voltage is given by

$$I \angle 90° = \frac{V_S \angle 90°}{R_S} = \frac{\omega C_m |V|}{1 + R_{ser}^2 C_m^2 \omega^2} \quad (6.10)$$

and the 0° component is

$$I \angle 0° = \frac{R_{ser}\omega^2 C_m^2 |V|}{1 + R_{ser}^2 C_m^2 \omega^2}. \quad (6.11)$$

Of course, the desired parameters should be available

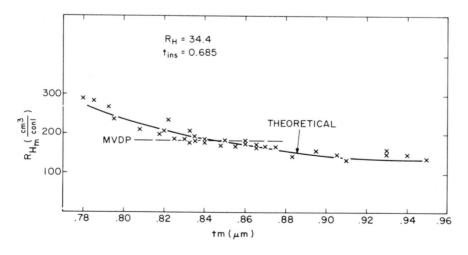

FIGURE 6.25
Theoretical [from Equation (6.3)] and experimental variations in the measured Hall coefficient assuming a uniformly conducting film (R_{H_m}) as a function of actual film thickness

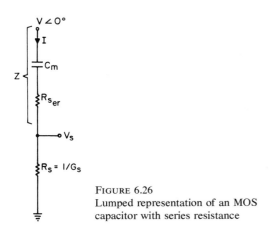

FIGURE 6.26
Lumped representation of an MOS
capacitor with series resistance

in terms of the experimentally observable quantities, $I \angle 0°$ and $I \angle 90°$. This is accomplished by using

$$\frac{I \angle 0°}{I \angle 90''} = R_{\text{ser}} \omega C_m. \qquad (6.12)$$

Since

$$C_m = \frac{I \angle 90°}{\omega |V|} (1 + R_{\text{ser}}^2 C_m^2 \omega^2) \qquad (6.13)$$

C_m will be accurate within 1 percent if $R_{\text{ser}}^2 C_m^2 \omega^2 < 0.01$. Thus, if $I \angle 0°/I \angle 90°$ is less than 0.1 (or equivalently, if the series resistance is less than 10 percent of the capacitive reactance), 1-percent accuracy of the simple expression

$$C_m = \frac{I \angle 90°}{\omega |V|} \qquad (6.14)$$

is maintained.
   In general,

$$C_m = \frac{I \angle 90°}{\omega |V|} \left(1 + \frac{I \angle 0°}{I \angle 90°} \right) \qquad (6.15)$$

or the capacitance given by Equation (6.14) is smaller than the actual value by the factor $1 + (I \angle 0°/I \angle 90°)$. The series resistance may be calculated to be

$$R_{\text{ser}} = \frac{|V| \dfrac{I \angle 0°}{I \angle 90°}}{I \angle 90° \left(1 + \dfrac{I \angle 0°}{I \angle 90°} \right)^2}. \qquad (6.16)$$

If, $I \angle 0°/I \angle 90° < 0.1$, then

$$R_{\text{ser}} \approx \frac{|V| \, I \angle 0°}{(I \angle 90°)^2}. \qquad (6.17)$$

   There are some general observations to be made concerning application of HSF MOS capacitors. Both C_m and R_{ser} are functions of the gate voltage V_G. C_m decreases as the gate voltage moves from accumulation to depletion and R_{ser} correspondingly increases. (For bulk semiconductor capacitors R_{ser} decreases.) Because $I \angle 0°/I \angle 90° = R_{\text{ser}} \omega C_m$, there is a tendency for $I \angle 0°/I \angle 90°$ to be more slowly varying with V_G than either C_m or R_{ser} alone. Therefore, lower values of capacitance than might at first sight be expected can be measured simply. R_{ser} increases continually and approximately linearly as V_G varies from heavy accumulation until near the flat band point is reached. When significant surface depletion appears, C_m begins to decrease rapidly and R_{ser} increases more rapidly. Consider the simple case shown in Figure 6.27. Electrical substrate contact is made from one side only and the distributed effect of the series resistance is neglected initially. It will be discussed in the next section.

$$\frac{1}{C_m} = \frac{t_{\text{ox}}}{\epsilon_{\text{ox}} A} + \frac{t_{\text{dep}1}}{\epsilon_{\text{Si}} A} \qquad A = WL \qquad (6.18)$$

where $t_{\text{dep}1}$ is the width of the surface space charge region.

$$C_m = \frac{1}{\dfrac{t_{\text{ox}}}{\epsilon_{\text{ox}} A} + \dfrac{t_{\text{dep}1}}{\epsilon_{\text{Si}} A}} = \frac{\epsilon_{\text{ox}} \epsilon_{\text{Si}} WL}{\epsilon_{\text{Si}} t_{\text{ox}} + \epsilon_{\text{ox}} t_{\text{dep}1}}. \qquad (6.19)$$

A reasonable first approximation of R_{ser} is

$$R_{\text{ser}} = \frac{\rho L}{W(t_{\text{film}} - t_{\text{dep}1})} \qquad (6.20)$$

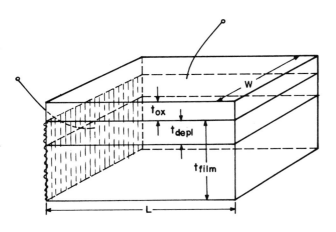

FIGURE 6.27
Model for SOS/MOS capacitor where electrical contact is made from one side only; t_{ox} is the thickness of the oxide layer

at flat band or depletion. ρ is the resistivity of the conducting silicon. Thus,

$$R_{\text{ser}} \omega C_m = \frac{\rho L^2 \epsilon_{\text{ox}} \epsilon_{\text{Si}} \omega}{(t_{\text{film}} - t_{\text{dep 1}})(\epsilon_{\text{Si}} t_{\text{ox}} + \epsilon_{\text{ox}} t_{\text{dep 1}})}. \quad (6.21)$$

Notice that *square* of the distance from the contact is involved. One can readily calculate the maximum surface depletion tolerable, $t_{\text{dep1 max}}$ for valid results (i.e., for $R_{\text{ser}} \omega C_m < 0.1$) given ρ, L, t_{ox}, and t_{film}.

$$t_{\text{dep1 max}} = -\left(\frac{t_{\text{ox}} \epsilon_{\text{Si}}}{2\epsilon_{\text{ox}}} - \frac{t_{\text{film}}}{2}\right) \quad (6.22)$$
$$+ \left[\left(\frac{t_{\text{ox}} \epsilon_{\text{Si}}}{2\epsilon_{\text{ox}}} - \frac{t_{\text{film}}}{2}\right)^2 + \frac{\epsilon_{\text{Si}}}{\epsilon_{\text{ox}}} t_{\text{ox}} t_{\text{film}} - 10\rho L^2 \epsilon_{\text{Si}} \omega\right]^{1/2}$$

Note that $t_{\text{dep1 max}} = 0$ if

$$10\rho L^2 \epsilon_{\text{Si}} \omega = \frac{\epsilon_{\text{Si}}}{\epsilon_{\text{ox}}} t_{\text{ox}} t_{\text{film}}. \quad (6.23)$$

That is, if

$$\rho\omega = \frac{t_{\text{ox}} t_{\text{film}}}{10L^2 \epsilon_{\text{ox}}}. \quad (6.24)$$

Thus, for $t_{\text{film}} = 1~\mu m$, $t_{\text{ox}} = 1000~\text{Å}$, and $L = 5$ mil (125 μm), $\rho\omega = 1.83 \times 10^6$. Then, at 1MHz, $\rho_{\text{max}} = 1.83 \times 10^6/6.28 \times 10^6 = 0.291~\Omega\text{-cm}$, and at 10 kHz, $\rho_{\text{max}} = 29~\Omega\text{-cm}$. In other words, there is no region of the $I \angle 90°$ v. V_G curve where the capacitance, in the range in which doping profiles are measured (i.e., in depletion), is not affected by series resistance if ρ is greater than ρ_{max} at the frequency of interest. These numbers for ρ_{max} can be increased if contact is made on all sides of the capacitor.

It is quite clear that the size and shape of frequently used bulk silicon capacitors (~1000 mil² circles or squares) is not practical for HSF. ρ_{max} as calculated above must be used with caution, however, because—as discussed earlier—in general ρ is not uniformly distributed throughout the film. If part of the film near the semiconductor–substrate interface is essentially insulating, the conducting portion of the film should be used as t_{film} to determine ρ_{max}. It is actually not the

resistivity that is important, however. The sheet resistance, whether in accumulation or depletion, is the important parameter.

These closed-form solutions are possible because a lumped analysis was used with R_{ser} intuitively obtained from Equation (6.20). The actual situation unfortunately is not quite as simple and requires an exact analysis with the resistance distributed under the gate.

Distributed case.   Referring to Figure 6.28,

$$dV(x) = V(x+dx) - V(x) = rdxI(x), \quad (6.25)$$

where r is the resistance per unit length x.
    Also

$$I(x) = \int_x^L dI(x) \quad (6.26)$$

$$V(x) = \int_0^x dV(x) = \int_0^x rI(x)dx, \quad (6.27)$$

$$V_i(x) = [V_{\text{app}} - V(x)]\frac{C_{\text{ox}}}{C_{\text{ox}} + C_{\text{Si}}} + V(x), \quad (6.28)$$

where C_{ox} and C_{Si} are the oxide and silicon capacitances per unit x respectively. Note that C_{Si} is determined only by the DC voltage across the silicon and therefore, unless the oxide leaks and allows DC current to flow, C_{Si} is *not* a function of x. The voltages are accordingly assumed to be small-signal AC voltages. Any charge arriving at the top surface of the conducting region will immediately (within a dielectric relaxation time) disperse throughout the vertical dimension. Therefore, because the measurement frequencies are much less than the dielectric relaxation time, vertical currents in the conducting region do not need to be considered.

$$dI(x) = [V_i(x) - V(x)] j\omega C_{\text{Si}} dx$$
$$= [V_{\text{app}} - V(x)]\frac{j\omega C_{\text{ox}} C_{\text{Si}}}{C_{\text{ox}} + C_{\text{Si}}} dx. \quad (6.29)$$

Let $C_{\text{ox}} C_{\text{Si}}/C_{\text{ox}} + C_{\text{Si}} = C_R$. Then,

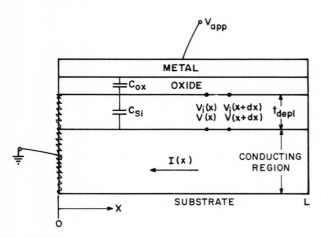

FIGURE 6.28
Model for the distributed analysis

$$I(x) = \int_x^L \left[V_{\text{app}} - \int_0^x rI(x)dx \right] j\omega C_R dx \qquad (6.30)$$

$$V(x) = \int_0^x r \int_x^L \left[V_{\text{app}} - \int_0^x rI(x)dx \right] j\omega C_R dx, \qquad (6.31)$$

$$\frac{dI(x)}{dx} = -\left[V_{\text{app}} - \int_0^x rI(x)dx \right] j\omega C_R dx, \qquad (6.32)$$

$$\frac{d^2I(x)}{dx^2} = rI(x)\, j\omega C_R, \qquad (6.33)$$

or

$$\frac{d^2I(x)}{dx^2} - A^2 I(x) = 0, \qquad (6.34)$$

where

$$A^2 = j\omega r C_R. \qquad (6.35)$$

The boundary conditions are at $x = L$, $I = 0$; and at $x = 0$ [from Equation (6.32)]

$$\frac{dI}{dx} = -V_{\text{app}} j\omega C_R = D = jK. \qquad (6.36)$$

This leads to solutions

$$I(x) = \frac{D}{A(1+e^{-2AL})} [e^{A(x-2L)} - e^{-Ax}], \qquad (6.37)$$

$$V(x) = \frac{Dr}{A^2(1+e^{-2AL})} [e^{A(x-2L)} + e^{-Ax} - (e^{-2AL+1}) - 1]. \qquad (6.38)$$

Realizing that

$$j^{1/2} = \frac{j(1-j)}{\sqrt{2}} = \frac{1+j}{\sqrt{2}}, \qquad (6.39)$$

A then becomes

$$A = G\frac{1+j}{\sqrt{2}} \quad \text{where } G = (\omega r C_R)^{1/2}. \qquad (6.40)$$

Although the distribution of current is given in Equation (6.37), it is the current at $x = 0$ that is measured and therefore is of the greatest interest. This current per unit width is

$$I(0) = \frac{D}{A(1+e^{-2AL})} (e^{-2AL} - 1) \qquad (6.41)$$

which reduces after a lengthy separation of real and imaginary parts to

$$I(0) = \frac{\sqrt{2}K}{2G} \times$$
$$\left\{ \frac{[e^{-2H}+2e^{-H}\sin(H-1)] + j[e^{-2H}-2e^{-H}\sin(H-1)]}{e^{-2H}+2e^{-H}\cos(H+1)} \right\}, \qquad (6.42)$$

where $H = \sqrt{2}\, GL$, or

$$I(0) = \frac{-\sqrt{2}}{2} \frac{\omega C_R V_{\text{app}}}{(\omega r C_R)^{1/2}} \left\{ \qquad \right\} \qquad (6.43)$$

$$I(0) = \frac{-\sqrt{2}}{2} V_{\text{app}} \left(\frac{\omega C_R}{r} \right)^{1/2} \left\{ \qquad \right\} \qquad (6.44)$$

As a check on the validity of the solution, it is clear that if $r = 0$,

$$I(0) = j\omega V_{\text{app}} C_R L. \qquad (6.45)$$

Let $M = (\omega C_R)^{1/2}$ then

$$H = \sqrt{2}\, r^{1/2}\, ML \qquad (6.46)$$

and

$$\frac{dH}{dr} = \sqrt{2}\, ML \frac{r}{2}^{-1/2} \qquad (6.47)$$

Applying l'Hospital's rule to Equation (6.44) one finds that the derivative of the bracketed numerator for $r = 0$ is $-j2\sqrt{2}\, ML$; and the derivative of $r^{1/2}$ times the bracketed denominator at $r = 0$ is 2. So,

$$I(0)_{r=0} = [-\frac{\sqrt{2}}{2} V_{\text{app}}(\omega C_R)^{1/2}] \times$$
$$\frac{-j2\sqrt{2}}{-2}(\omega C_R)^{1/2}L = jV_{\text{app}}\omega C_R L \qquad (6.48)$$

in agreement with Equation (6.45).

The desired capacitance is

$$C_m = C_R L. \qquad (6.49)$$

Then,

$$I(0) = \frac{-V_{\text{app}}}{2} \left(\frac{2\omega C_m}{Lr} \right)^{1/2} \left\{ \qquad \right\} \qquad (6.50)$$

and

$$H = \sqrt{2}\, GL = (2\omega r L C_m)^{1/2} \qquad (6.51)$$

Now,

$$I_{r=0}(0) \angle 90° = \omega V_{\text{app}} C_m. \qquad (6.52)$$

Therefore, define in the distributed case

$$\alpha_D = \frac{I(0) \angle 90°}{I_{r=0}(0) \angle 90°}$$
$$= -\frac{1}{H^{1/2}} \left[\frac{e^{-2H}-2e^{-H}\sin(H)-1}{e^{-2H}+2e^{-H}\cos(H)+1} \right] \qquad (6.53)$$

$$\beta_D = \frac{I(0) \angle 0°}{I_{r=0}(0) \angle 90°}$$
$$= -\frac{1}{H^{1/2}} \left[\frac{e^{-2H}+2e^{-H}\sin(H)-1}{e^{-2H}+2e^{-H}\cos(H)+1} \right] \qquad (6.54)$$

and in the nondistributed case

$$\alpha_{ND} = \frac{I \angle 90°}{I_{r=0} \angle 90°}$$
$$= \frac{\omega C_m V_{\text{app}}}{R_{\text{ser}}^2 C_m^2 \omega^2} \frac{1}{\omega C_m V_{\text{app}}} = \frac{1}{1 + R_{\text{ser}}^2 C_m^2 \omega^2}. \qquad (6.55)$$

$$\beta_{ND} = \frac{I \angle 0°}{I_{r=0} \angle 90°} \tag{6.56}$$

$$= \frac{R_{ser}\omega C_m}{1 + R_{ser}^2 C_m^2 \omega^2}.$$

If $R_{ser} = rL$,

$$R_{ser}\omega C_m = rL\omega C_m = \tfrac{1}{2}H^2 \tag{6.57}$$

and

$$\alpha_{ND} = \frac{1}{1 + \tfrac{1}{4}H^4} \qquad \beta_{ND} = \frac{\tfrac{1}{2}H^2}{1 + \tfrac{1}{4}H^4}. \tag{6.58}$$

Thus, both the distributed and nondistributed cases are controlled by the dimensionless parameter H. These functions are plotted in Figure 6.29. In Figure 6.29 the essential difference between the distributed and nondistributed case is clearly evident. These features are

Distributed Case
a. β_D peaks at ≈ 0.42.
b. The peak of β_D is not transversed by the α_D curve.
c. α_D and β_D are identical at high values of H.
   Nondistributed Case
a. α_{ND} peaks at 0.50.
b. The peak of β_{ND} is transversed exactly by α_{ND}.
c. α_{ND} and β_{ND} are not at all identical at high values of H.

Thus, by observing which features are present in a $I \angle 90° - V$ or $I \angle 0° - V$ plot it may be easy to determine whether distributed or lumped resistance is responsible. It may be necessary to adjust the frequency of the measurement in order to produce large enough H values to observe these features. Additional useful plots concerning α and β are shown in Figures 6.30, 6.31, and 6.32. These plots all refer to the capacitor configuration of Figure 6.28. H, of course, depends only on the sheet resistance (ρ_s) of the conducting layer (r is the same as ρ_s/W as mentioned earlier. The distribution of the resistivity throughout the film does not directly enter the calculations.

$$H = (2\omega rLC_m)^{1/2} = [2\omega L(\rho_s/W)C_m]^{1/2} = (2R_{ser}\omega C_m). \tag{6.59}$$

The change in the α's and β's as the gate voltage is varied (i.e., C_m and r are changing) adds another complexity. Unless a direct relationship between C_m and r is known [such as in Equations (6.18) and (6.21)], one cannot calculate either r or C_m unambiguously from only an $I \angle 0°$ and an $I \angle 90°$ measurement. If r is a known function of the conducting thickness, then a first-order simple relationship can be found. Thus, if appreciable series resistance is present, it is easy to detect but it is difficult to extract the desired parameters uniquely. Only the product rC_m can be determined. In both the distributed and the nondistribured cases the series resistance is apprecia-

ble if $\beta/\alpha > 0.1$ ($\gamma = \beta/\alpha = I \angle 0°/I \angle 90°$) (see Figure 6.29).

In the nondistribured case $\gamma_{ND} = H^2/2$ and in the distributed case

$$\gamma_D = \frac{e^{-2H} + 2e^{-H}\sin(H-1)}{e^{-2H} - 2e^{-H}\sin(H-1)}. \tag{6.60}$$

The values of H appropriate for the γ of interest are plotted in Figure 6.33. Note that higher values of H are useful for the distributed rather than for the nondistributed case.

If an additional electrical contact is placed on the side opposite to the one illustrated in Figure 6.28, the analysis for the single contact case can still be used but a different L is required. By symmetry $I(0) = -I(L)$ and $I(L/2) = 0$. Thus, the two-side problem is equivalent to the solution of a one-side capacitor of half the length and producing twice the current. (The equations and boundary conditions match.) The α and β variations for the capacitor with two contacts are identical to the α and β curves with $L = L/2$.

A distributed analysis yields results having a number of similarities to a lumped analysis. In both cases separation of the capacitance and the resistance is possible only if γ is less than 0.1 where the resistance does not affect the 90° current component appreciably (i.e., C_m is known) or if a relationship between the desired capacitance and the sheet resistance of the conducting region is known. The entire series resistance phenomena (for the cases analyzed specifically) can be described in terms of a dimensionless parameter $[= 2\omega\rho_s(L/W)C_m]^{1/2}$ whose value can be determined from the measurement of γ. If γ is less than 0.1, then a maximum of 1 percent errors in C_m is caused by series resistance.

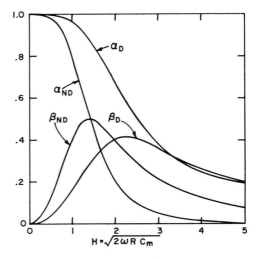

FIGURE 6.29   Graphical representation of Equations (6.53), (6.54), and (6.58)

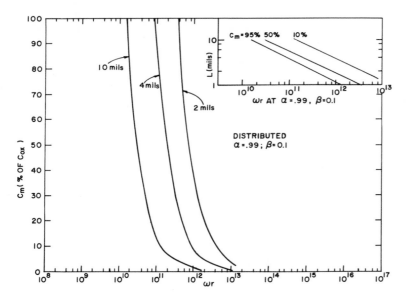

FIGURE 6.30
Values of ωr where $\alpha = 0.99$ and $\beta = 0.1$ for the distributed case (Figure 6.28) where $t_{ox} = 1000$ Å and $\epsilon_{ox} = 3.82$. Note that the length dependence follows nearly a square law as in the lumped case

Some experimental results. The application of the theoretical considerations to an actual MOS capacitor can be seen in Figure 6.34. These data were acquired from a 0.5-μm-thick n-type SOS film at a measurement frequency of 10 kHz. A significant part of the 90° current component was provided by a nonvoltage-varying capacitance ($C_{overlap}$) in parallel with the desired capacitance. This additional current biases the 90° current upward. The current discussed in the theoretical section includes only the current above this value. The capacitor illustrated exhibits both accumu-

lation and inversion because both n- and p-type contacts were used.

Proceeding from accumulation to inversion one can see the transition from a situation in which series resistance is unimportant to a situation in which it dominates the impedance. By comparing the 0° and 90° curves it can be seen that the onset of serious series resistance effects occur at $\gamma \approx 0.2$ and that the first indication occurs, as predicted, at $\gamma \approx 0.1$. The axis labeled "capacitance" does not really apply in the region where the series resistance is appreciable.

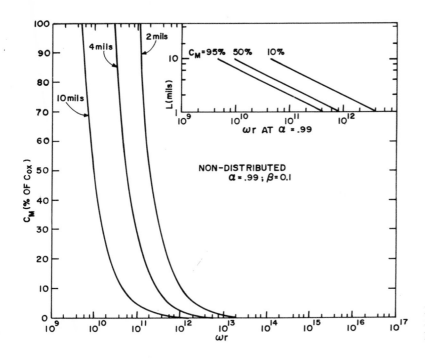

FIGURE 6.31
Same as Figure 6.30 except for the lumped case

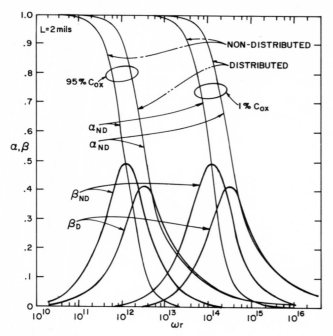

FIGURE 6.32

α and β variations for a 1000 Å thermal silicon oxide. Note that as C_m becomes a smaller percentage of C_{ox} (i.e., in depletion) larger values of r are still within the tolerable range. The configuration of Figure 6.28 is assumed with $L = 2$ mil

The data fit the theory for the distributed case since both the 0° and 90° currents are the same in the region in which H is very high. It is also of interest to note that the series resistance becomes important almost immediately after the flat band point is passed. This is consistent with the earlier discussion concerning the resistivity profile of SOS films.

Some further data on thinner films are shown in Figure 6.35, which shows the frequency dependence of the C-V curves. This capacitor is the same design as the one in Figure 6.34. Here again, the essential features of the theory are confirmed. This particular sample showed strong series resistance even in accumulation at the higher frequencies.

It is clear from the data in Figures 6.34 and 6.35 that a measurement of the 0° current is *essential* in order to properly interpret the data. Consider the interpretation one might give to the 100 kHz curve of Figure 6.35 without an awareness of series resistance effects and the importance of the 0° current.

The cause for the hysteresis in the curves is not currently known but is probably related to slow charge trapping at defects in the film. It is always observed when conduction is forced near the bottom of the film.

Design considerations. The series resistance effects indicate that the capacitor should be made as small as possible if it is to be used in a general application

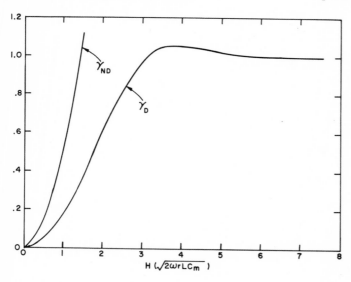

FIGURE 6.33

Plot for determining H from γ ($=I \angle 0 / I \angle 90$). γ is the observable

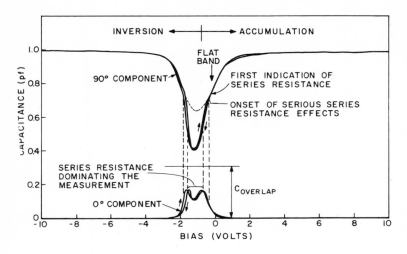

FIGURE 6.34
$I \angle 90°$ and $I \angle 0°$ measurements at 10 kHz on a 0.5-μm n-type SOS film

where many different resistivities and thicknesses of silicon are likely. Thus, for thermal silicon oxides (ϵ_{ox} = 3.82) of the order of 1000 Å thickness the capacitor area should be $\approx$5 mil². This gives maximum capacitances of the order of 1.0 pF. Larger capacitances may of course be obtained by using several parallel capacitors or an interdigitated structure. Lateral nonuniformities are then more likely to influence the measurements. With modern equipment the 1 pF maximum capacitance is not unduly small.

The optimum shape of the capacitor electrode is considered next. The area should be as insensitive to the fabrication process as possible while still minimizing size to maintain minimum series resistance. During the fabrication process a certain deviation from the design dimensions will occur. This is denoted by ΔL. ΔL will be removed (or occasionally added) approximately uniformly around all of the peripheries of the electrode by the processing.

Consider the general rectangular form in Figure 6.36. The area is $a = \alpha L^2$, and to first order $\Delta a = 2$

$(\alpha L \Delta L + L \Delta L) = 2L\Delta L(1+\alpha)$. The area sensitivity is

$$\frac{\Delta a}{a} = \frac{2\Delta L}{L} \frac{1 + \alpha}{\alpha}$$

$\Delta a/a$ is minimum for $\alpha \to \infty$ and has the value $2\Delta L/L$. For the square ($\alpha=1$), $\Delta a/a = 4\Delta L/L$ or twice that of the long bar. The circle on the other hand has an area sensitivity of $\Delta a/a = 4\Delta L/d$ which is essentially the same as the square so it appears that there is no particular advantage of using a circle. Obviously, the area sensitivity is reduced by making the size (L or d) larger, but this increases the series resistance effects by the square of the dimension.

The smallest geometry is dictated by the degree to which the dimensional aspects of the process can be controlled. This not only include ΔL but contacting techniques (usually diffusions). ΔL is typically less than 2.5 μm for polysilicon or aluminum, and lateral diffusions are frequently of the order of 1 μm. Thus, for 5 percent accuracy the minimum fundamental

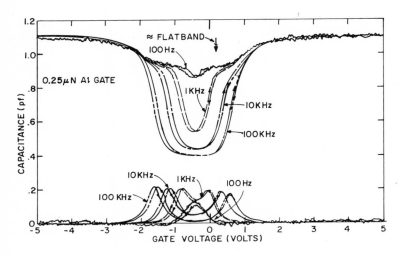

FIGURE 6.35
$I \angle 90°$ and $I \angle 0°$ measurements at various frequencies for a 0.25 μm n-type SOS film

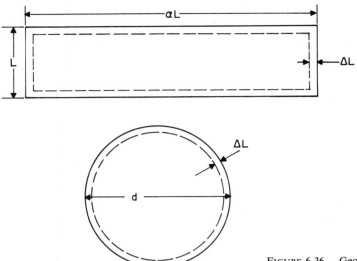

FIGURE 6.36   Geometries for area sensitivity calculations

dimensions are of the order of 50 μm. Dimensions as small as 6 μm have been used in some cases as discussed later. These small dimensions can cause difficulty. Attention is given to the square configuration for the remainder of this section, although by lengthening any of the designs such that $\alpha > \approx 4$ the long bar case may be formed.

ONE POSSIBLE DESIGN.   The 50-μm square can be implemented as shown in Figure 6.37. Note that there is 12 μm of undiffused silicon surrounding the oxide metal in an attempt to eliminate the effects of the lateral contact diffusion. The problems involved with this design are as follows:

1. The contacting probes can cause oxide failures by mechanically punching through the oxide. This problem is not severe if excessive probe pressures are not used.

2. The 50 × 50 μm pad is difficult to contact because it is quite small.
3. The probe can scrape the metal over the oxide causing shorts or errors in the metal size.
4. The dielectric surface between the oxide metal and the contact diffusion does not have a controlled surface potential. This problem will be discussed in more detail later.
5. The contact mask aligns exactly with the diffusion. Therefore, if the contact mask is slightly misaligned to the diffusion, the metal can contact the undiffused silicon resulting perhaps in an unsatisfactory contact.

One of the advantages of this capacitor structure is that there are no semiconductor steps that the oxide metal must cross. However, because the area of the step can be very small compared to the total metal area, no appreciable errors in the capacitance should

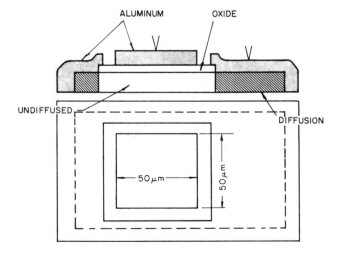

FIGURE 6.37
One possible HSF/MOS capacitor design

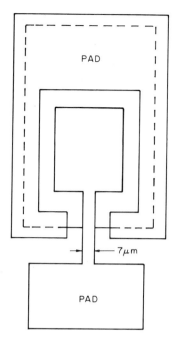

FIGURE 6.38   HSF/MOS capacitor where probing over the active region is not required

result by allowing the gate metal to cross a step unless an oxide dielectric failure at the step occurs. A capacitor with a narrow strip of metal connecting the oxide metal to a standard integrated circuit contact pad is shown in Figure 6.38. This design eliminates three of the problems above.

Problems 4 and 5 show up as either minority carrier sources or as series resistance. The minority carrier sources cause a rise in the inversion side of the C-V curve and are undesirable. This is particularly important if automatic testing is used, because the minimum point is a fundamental property of the C-V curve (as

discussed in a later section). The minority carrier sources can cause curves as shown in Figure 6.39. They may come from the following sources:

1. Stray illumination can create electron–hole pairs in the depletion region or under the unprotected oxide. As seen in Figure 6.39, the degree of illumination has a strong effect on the depletion part of the curve. This is a problem on MOS capacitors regardless of the capacitor design.

2. If the lifetime of the material is short enough and the measurement frequency is low enough, thermal generation can cause much the same effect.

3. Surface generation as is observed in bulk silicon MOS capacitors (Schroder and Nathanson, 1970) can supply minority carriers.

4. If the material is n-type and the aluminum metal contacts the undiffused silicon (i.e., a p-n junction exists), a generous supply of holes is available. In order for this to cause a rise in the depletion part of the curve, there must be either a very high-lifetime material (so the minority carriers can diffuse) or an inversion region under the unprotected oxide must exist. This can be caused by surface charges (ions) as discussed next.

In order to further investigate the nature of this inversion layer, a capacitor made from 0.5-μm n-type SOS (doped 2–4 $\times$ 10^{15}/cm^3) film using the design of Figure 6.37 with only one type of diffusion (same type as the silicon) was measured. These results are shown in Figure 6.40. This material should have a doping-thickness combination which allows the material to be completely depleted throughout. (With the "insulating" layer model for the film there is little doubt for this type of film.)

Capacitance (90° current component) measurements were taken first starting at point A. The accumulation capacitance based on area and oxide thickness should be ≈0.8 pF. The voltage was swept positively to point B, back to point A, then to C-D, D-

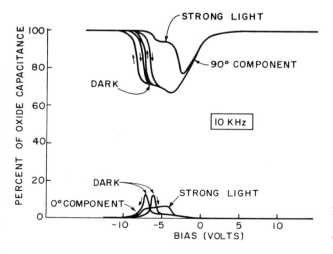

FIGURE 6.39
Some anomalous effects possible with the capacitor design of Figures 6.37 and 6.38

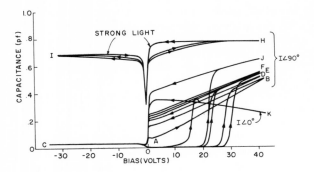

FIGURE 6.40   Surface ion effects on a 0.5 μm n-type SOS/MOS capacitor of the type shown in Figure 6.38

C-E, and *E-C-F.* These measurements were taken in the dark. Strong light was then shined on the capacitor and *H-I* was generated (two passes were made). Next *J-C-J* and the conductance (0° component) *K-C-K* were measured in the dark. The following can qualitatively explain the measurements.

Suppose that the region of uncontrolled surface potential was originally depleted and that charge could slowly leak from the gate to the surface of the oxide. This, of course, would change the surface potential under the oxide and is known to affect bulk silicon devices (Atalla et al., 1959; Shockley et al., 1964; Ho et al., 1967; Schlegel et al., 1968). As the voltage becomes more positive, a nondepleted region would form allowing better contact to the diffusion; thus, the curve rises (i.e., the series resistance is less). This situation is illustrated in Figure 6.41a. As the voltage is reduced (still positive), one might expect to see the capacitance measured to remain high, but it does not. It decreases with less slope than it rose with. This could be caused by partial capacitive (dV/dt) discharge of the oxide or by surface space charge. When biased in depletion (point *C,* Figure 6.40), both the region of uncontrolled surface potential and the region under the metal are depleted (inverted), resulting in a very low apparent capacitance (very high series resistance) (see Figure 6.41b). The slight rise of $C_{\min}$ is probably caused by thermal generation and capacitive coupling through the depleted regions. On sweeping from *C* to *D* the capacitance drops slightly near zero voltage. At this point minority carriers are no longer present under the oxide metal and the capacitive coupling switches to the position shown in Figure 6.41c. The curve remains low until sufficient charge can leak from the oxide metal to the oxide to remove part of the depleted region (Figure 6.41a), thus allowing resistive contact to the active region. When resistive coupling is available, the curve rises sharply and continues to point *D.*

The light apparently increased the conductivity of the region of uncontrolled surface potential and the curves look much more normal (curve *I-H*). There are

still some effects of the leaking charge as seen on the retrace from point *I* to *H.*

The magnitude of the $I\angle 0°$ curve shows the relative importance of series resistance caused by the region of undefined surface potential. At high positive bias the series resistance is less than it is when near zero bias, indicating that leaking positive charge is causing the series resistance to change. Another capacitor on the same wafer is illustrated in Figure 6.42. Here the curves have nearly the correct accumulation capacitance and the $I\angle 0°$ curve illustrates the series resistance quite nicely. Note that because the $I\angle 0°$ curve peaks at ≈ 0.5 of the maximum 90° current, the series resistance is mainly not distributed. This, of course, would be the case if the resistance were confined mainly to the outer ring where the surface ions (charges) are postulated.

One further test was made to ensure that the anomalous results were caused by a depleted region. A gate-controlled Hall bar on the same wafer as the capacitors of Figure 6.40 and 6.42 whose gate did not overlap the source and drain diffusions could not be made to strongly conduct in accumulation by varying the gate voltage. This also suggests that the region between the gate metal and the diffusions was initially insulating (depleted).

Apparently if the film can be depleted throughout, a

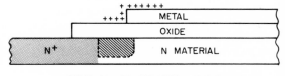

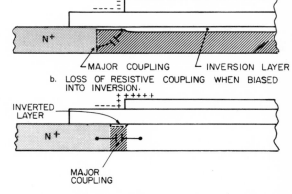

a.   REMOVAL OF DEPLETED REGION BY LEAKING CHARGE (BIAS IN ACCUMULATION) (RESISTIVE COUPLING)

b.   LOSS OF RESISTIVE COUPLING WHEN BIASED INTO INVERSION.

c.   COUPLING SWITCH WHEN BIASED INTO ACCUMULATION AFTER BIASING INTO INVERSION.

FIGURE 6.41   Conceptual model for the anomalous behavior shown in Figures 6.40 and 6.42.

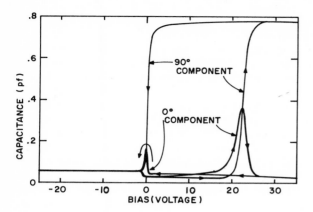

FIGURE 6.42   Experimental observation of surface ion effects

capacitor with no regions of uncontrolled surface potential should be used. (This applies to gate-controlled Hall bars as well.)

The method used to avoid the uncontrolled surface is simply to overlap the top electrode onto the diffused region. The metal overlapping the diffused regions contributes a voltage-independent capacitance to the measurement. This capacitance can simply be subtracted from the measurement to obtain the desired capacitance. The problem lies in determining the value of the voltage-independent capacitance. This capacitance depends on (1) the area of the overlap, (2) the thickness of the oxide over the diffused regions, and (3) the dielectric constant of the oxides.

As far as area is concerned, the overlap is at the worst possible point of the metal pattern, namely, on the perimeter. Thus, with a 6-μm overlap on a 50 × 50-μm active region, the area of the overlap is $\approx$200 × 6 μm^2 = 1200 μm^2 = one half of the active area. (Here then is a case where the 50-μm minimum fundamental dimension would cause a nearly unusable device. If a 50-μm overlap is used, the overlap capacitance is eight times the active area capacitance). For very accurate measurements the dimensions of each capacitor should be measured individually. This is inconvenient if automatic measurement (i.e., large numbers of capacitors) is used.

By making the diffusions symmetrical around the active region, small misalignments of the metal pattern (or of the diffusions) are compensated. Therefore, the only inaccuracy in the area of the overlap comes from the dimensional accuracy of the masks and the photoresist and etching processes. The amount of overlap is determined by at least two masks—the metal and the diffusion masks—instead of only one mask in the nonoverlapped capacitor. If two types of diffusion are used, such as illustrated in Figure 6.43, the overlap capacitance is determined by three masks.

The thickness of an oxide thermally grown on heav-

ily doped silicon depends on the degree of doping of the silicon, the temperature, and method of oxide growth (Deal and Sklar, 1965). Therefore, the oxide thickness under the overlap cannot be calculated directly from the accumulation capacitance and the area. There are several ways to handle this problem which involve making measurements on another test device.

The doping of the diffused silicon can be estimated by measuring the sheet resistance and referring to surface concentrations obtained after the particular diffusion schedule used. Using this surface concentration, the thickness of the oxide grown under well defined conditions can be estimated. The dielectric constants can be similarly estimated. This procedure is undesirable, however, because it requires independent and exact knowledge of the processing actually used, which may not be the same as the processing intended.

The second technique involves measuring the capacitance of a test capacitor with the oxide of interest as the dielectric. This method would work well (with appropriately compensation for errors in the metal area of course), because none of the processing need be known accurately. It requires the additional effort of making another capacitance measurement which has little intrinsic significance. The comparison of two capacitors with different and known amounts of overlap can also be used.

A comparison of the two capacitors discussed in this section is shown in Figure 6.44. The thickness of the oxide over the diffused regions was estimated and the dielectric constant of these oxides was assumed to be the same as ordinary thermal oxide. Under these conditions the accumulation capacitances measured on the two geometries agreed with each other to within $\approx$0.5 percent.

It is possible to minimize the problem of determining the overlap capacitance by using a self-aligned

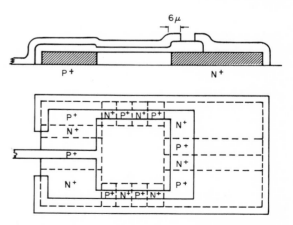

FIGURE 6.43   HSF/MOS capacitor design for observing both accumulation and inversion sides of the C-V curve

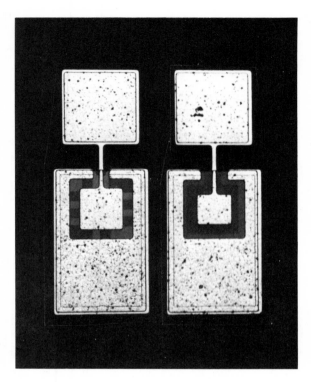

FIGURE 6.44   Photograph of actual HSF/MOS capacitors of the types seen in Figures 6.38 and 6.43

process. In such a process the gate metallization itself (which can be heavily doped polycrystalline silicon) provides a mask for the diffusions. With a self-aligned process the overlap capacitance is caused by diffusion under the gate, which will typically be of the order of 1.0 μm. It is not always possible to use a self-aligned process. It should be pointed out that the heavily doped regions may also be formed by ion implantation. For material evaluation a metal gate material may be used for an implantation barrier as no high temperatures need be involved. The implantation does not need to be fully activated. Therefore, self-aligned metal gate capacitors are possible.

Some of the discussion in this section is summarized in Table 6.1.

6.5. Electrical characterization of the film well below the top surface and adjacent to the substrate

We have discussed in detail the techniques used for control and electrical characterization of HSF near the top surface. The problems of examining the remainder of the film have been only briefly discussed, and it was assumed that the effects were at least constant with respect to the applied stimuli. For the measurements

previously considered, there were no applied stimuli that should alter the bottom interface potential, and therefore the assumptions are not unreasonable. There are at least four techniques that can be used to selectively examine the bottom portion of the film.

The first technique is to examine the characteristics of devices similar to those discussed in previous sections except that very thin films are used. Strictly speaking, this method is examining material near the outer surface, but this surface is in close proximity to the bottom interface with very thin films. If one wishes to measure the properties of the bottom portion of a thick film, this method may not be suitable because the characteristics of the material near the interface may be altered by the deposition of more material. It is known, for example, that for the case of silicon deposited on single crystal sapphire (by the pyrolysis of silane), the physical structure of layers as thin as 300 Å change markedly with subsequent deposition (Cullen and Corboy, 1974). Therefore, the very-thin-film device may not accurately represent the material near the bottom of thicker films.

In addition, there is very considerable difficulty in performing meaningful measurements on very-thin-film devices because some assumptions inherent in the analysis for thicker films do not apply. For example, the actual width of an inversion layer is typically 50–300 Å. The performance of these structures is largely controlled by the interaction of the charges in the semiconductor with the surface, and the relation of the measurements to properties of the semiconductor is somewhat indirect. Also, MOS capacitor measurements on SOS films as thin as 2500 Å are difficult to interpret (as seen previously in Figure 6.35). The observed current is related to the actual film properties through a very complicated mechanism involving the distribution of charge in the silicon, the surface fields, and the physical structure of the interfacial region, which in this case may include the entire silicon film. Even in the more nearly ideal case of bulk silicon with better known interface properties, the relationships between the observed currents and the material properties are not well understood in detail. Many effects may influence the observed current, among which are

1. Trapping of field-induced carriers into electronic states in which they satisfy the requirement for overall charge neutrality and Poisson's equations but nevertheless are not free to move under the influence of applied electric fields.
2. Scattering of carriers by fixed interface charges that are not free to move.
3. Scattering of carriers by localized potential barriers caused by localized space charge regions or ionized atoms or complexes.
4. Scattering of carriers by collision with phonons (lattice vibrations).

TABLE 6.1   Properties of Various HSF/MOS Capacitor Designs

Advantages	Disadvantages	Advantages	Disadvantages
A. Overlap capacitor with both types of diffusion available for contact		on only an oxidation step and material	slightly voltage dependent
1. No regions of undefined surface potential	1. Higher series resistance	3. Type conversion possible to detect and measure	3. Anomalous rises of C_{min} possible
2. Inversion point clearly shown	2. Measured capacitance depends on both the diffusion process and on oxidation schedule	4. Diffusions located away from active area	4. More light sensitive
3. Material type usually possible to detect	3. Overlap area determined by three photoresist steps		5. 4 of B, above
4. Thinner films may be measured more reliably	4. More difficult to analyze automatically	**D. Nonoverlap capacitor with only one diffusion**	
	5. Difficult to determine type if lightly doped	1. Same as C, above, except that type conversion here is indicated by a very low capacitance reading which is approximately independent of bias	1. Converted material impossible to characterize
	6. Diffusions located at active region edge may cause area distortion due to junction depletion regions		2. Same as C, above
	7. Lateral diffusion may affect the active area		
B. Overlap capacitor with only one type of diffusion		**E. Metal contained wholly on the top surface**	
1. 1 of A, above	1. Material type conversion very difficult to detect (impossible to measure)	1. Avoids edge problems	1. Small probe area
2. 4 of A, above			2. Probe may punch through dielectric
3. The overlap region is defined by only two photoresist steps	2. 2 of A, above		3. Metal area may be altered by probe
4. Automatic analysis is relatively easy	3. 7 of A, above	**F. Metal connected to a separate pad**	
	4. Inversion point not clear	1. Avoids disadvantages 1, 2, 3 of E, above	1. Additional stray capacitance
C. Nonoverlap capacitor with both diffusions			2. Metal crosses edge
1. Critical dimensions depend on only one photoresist step	1. Region of undefined surface potential exists (may cause loss of contact if depleted)	**G. Self-aligned metal-diffusions**	
2. Capacitance measurements depend	2. Stray capacitance is	1. Minimizes overlap problem	1. Does not allow either C or D type structures
		2. Area depends on only one photoresist step	2. May not be compatible with desired processing
		3. No regions of uncontrolled oxides	

5. Scattering of carriers by the collision with the surface.
6. Electric fields perpendicular to the surface which influence the distribution and amount of charge in the silicon.
7. Electric fields parallel to the surface which may influence the effective scattering or trapping cross sections by affecting the carrier velocity.
8. "Boxing in" of charge by enclosed potential barriers. Such charges are trapped in a sense but can move freely within the semiconductor between the barriers. Only those with sufficient energy to surmount the potential barrier can contribute to the observed currents.
9. Lateral nonuniformities.
10. Surface roughness.

The most easily accessible data from a very-thin-film device is the DC current flowing in a gate-controlled Hall bar, resistor, or transistor. Even though

the relation of this current to the material properties may be difficult, an empirical examination of these surface-dominated currents can provide a basis for comparison of devices of different thickness.

When very-thin-film devices are fabricated in SOS material, thermal oxidation can consume the entire silicon supply. Therefore, most, if not all, of the gate dielectric material must be deposited. An example of data from a narrow-channel MOS device fabricated on a 350-Å final silicon thickness film is shown in Figure 6.45. This device was fabricated by depositing 500 Å of nominally undoped silicon, thermally oxidizing until 300 Å of thermal oxide was produced and subsequently depositing $\approx$800 Å of SiO_2. This procedure allowed the film to coalesce before oxidation and consumed only a minimal amount of silicon. The thermal oxidation was used because thermal oxides produce better interfaces with silicon than do deposited oxides. Boron diffusions were used for the contacts. Because

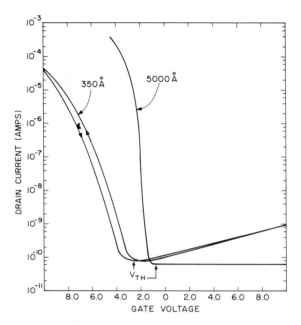

FIGURE 6.45 Conduction in very thin SOS gate-controlled resistors. See text for details

boron is a p-type dopant, only the hole flow contributes to the observed current. Note that the operation of a device of this type is similar to the case of the gate-controlled Hall bar except that no potential probes are used. The observed current increases strongly with negative gate field as holes are electrostatically induced into the interface region. The increase is much less, however, than in an identically fabricated device whose final thickness was ≈5000 Å, also shown in Figure 6.45. Other differences noted are the hysteresis in the thin-film device and the onset voltage for conduction, V_{th}.

The hysteresis is probably caused by a relatively rapid exchange of charge between the interface region and the heavily defected silicon. The cause of the higher onset voltage is not certain but a tentative explanation is given in Section 6.8. It strongly suggests that the silicon near the sapphire is not saturated with active aluminum. If it were, the onset voltage would be much more positive. Perhaps the most significant comparison to make here is the minimum current. The data suggest that when the electrostatically induced charges are of opposite type than the diffused contacts (i.e. are electrons in this case), the current-limiting mechanism may be the same for both film thicknesses. This may imply that the minimum currents flow primarily in the silicon less than 350 Å from the sapphire. These interpretations are only speculations at this point, but they illustrate the type of information that can be obtained from very-thin-film devices.

In the second method a gate-controlled Hall bar or gate-controlled resistor, with the gate biased such that the semiconductor is nearly depleted throughout, is used. This technique suffers considerably because the actual conducting thickness of the film is not accurately known, since the primary tool for determining this depth (the MOS capacitor) cannot be effectively used under these conditions. Also, as mentioned earlier, it is difficult to make Hall measurements in this region of the film.

A qualitative measurement of the electric profile near the bottom of the film can be achieved by essentially the same method used for thicker films described in Section 6.4 except that attention here is restricted to the resistivity. The measurement is qualitative because the actual conducting portion of the film is not accurately known. Also, these measurements are made with a low constant voltage instead of constant current for convenience.

Because large variations in current can be seen as the top surface potential varies from flat band to heavy depletion, a semilog plot of current vs. gate potential provides a convenient way to compare the *C-V* curve with the current. An example of such a plot is illustrated in Figure 6.46 for the case of a 0.6-μm n-type SOS film. For this particular film the current varies almost three orders of magnitude as the surface potential is varied from flat band to heavy depletion. We note that this is a much stronger variation than was seen in Figure 6.19, where the depletion region could not extend deeply into the film.

If one could be confident that all of the observed current were due to ohmic conduction in the nonde-

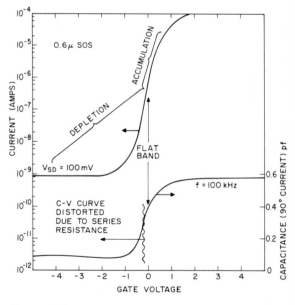

FIGURE 6.46
Comparison of conduction and capacitance on a 0.6-μm SOS gate-controlled resistor

pleted portion of the film, the analysis of Section 6.4. could be applied since a direct measurement of ρ_s (the sheet resistance of the conducting layer) is provided from the measured currents. However, especially near the minimum current (between $V_G = -5$ and $V_G = -2$ in Figure 6.46), there is the possibility that current thermally generated *in or near the depletion region* can also be appreciable. It is not immediately clear what the limiting mechanism for this minimum current is. The current immediately above the minimum current is at least partly ohmic since shrinking the depletion region provides both for a decrease in the depletion region volume (which obviously would decrease any thermally generated current) and for ohmic conduction through the previously depleted silicon. Therefore, in the range well above the minimum current, ρ_s is known and the analysis of Section 6.4 can be used.

It appears from the data of Figure 6.46 that a very strong resistivity gradient exists even near the surface of this relatively thin film. The current decreases a factor of 60 from flat band to $V_G = -0.5$ V and considerable depletion still occurs between -0.5 and -2.0 V. This suggests that the bottom interface is not imminently close at $V_G = -0.5$ V. Such a strong resistivity gradient deserves further discussion.

The current-limiting mechanism in such a gradient may be revealed by examination of the temperature dependence of the current in this region. A complete set of the reciprocal temperature dependence of the currents shown in Figure 6.46 is illustrated in Figure 6.47.

This type of data is valuable for the development of the profile of the conduction mechanisms throughout the film. The data in Figure 6.47 are described by a constant activation energy below flat band and by a progressively decreasing energy as heavy accumulation is reached. The behavior in accumulation is qualitatively what one would expect from the theory of surface transport (Garrett and Brattain, 1955). As the surface potential becomes closer to the conduction band edge energy, the activation energy decreases. On the other hand, all of the conduction below flat band is described by a single energy. This suggests that a single thermally activated process is the current-limiting mechanism and that the number of such limiting regions increase dramatically into the film. More will be said concerning this subject in a later section. The above discussion illustrates the type of information which may be obtained from this second method.

A third method for characterizing the silicon adjacent to the substrate surface, first used by Wrigley and Kroko (1969) and later by Cullen and coworkers (1973), involves applying an electric field perpendicular to the surface through the substrate. This technique is based on the same principles as the method used for top surface control. The application and inter-

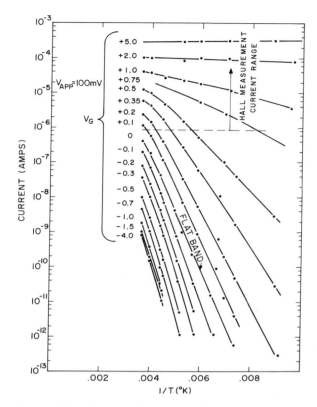

FIGURE 6.47   Reciprocal temperature dependence of the currents at different gate voltages for the same device used for the data of Figure 6.46

pretation is quite different however because silicon near the bottom interface is very defective and the substrate, which serves as the dielectric, is orders of magnitude thicker. Wrigley and Kroko found that the electrical conduction between diffused regions in SOS films can be significantly modified by the application of these fields. They attribute their data to inversion or accumulation layers at the silicon–insulator interface and note that the temperature dependence of these currents is characterized by an activation energy which varies with the degree of accumulation or inversion, similar to the variations in accumulation in Figure 6.47. By directly observing the currents, the actual transport properties near the interface can be inferred. Unfortunately, only very limited data are available on the bottom interface transport properties using this method. Because the substrate is always much thicker than the dielectrics discussed in earlier sections, very large voltages (several kilovolts) are required to induce significant charge into the semiconductor. Absolute dielectric integrity of the substrates is therefore required to avoid catastrophic destruction not only of the device under test but also of the sensitive input stages of sensing instrumentation. Also, great care must be exercised that the conductivity of the substrate is such that DC currents flowing not only through the substrate but also across its surface are small compared to the currents of interest in the device. These currents are in the picoamp range under some conditions (cf. Figure 6.47).

A fourth technique for the characterization of the interface silicon, recently reported by Goodman (1975), is closely related to Wrigley and Korko's method. Goodman measured the capacitance–voltage characteristics of the bottom interface by varying the field in the substrate. Because the substrate is very thick compared to the film depletion thickness, only very small variations in capacitance are measured and careful instrumentation is required. Information regarding the bottom interface potential can be obtained by analyzing the C-V data. As with C-V measurements from the top, surface transport properties are only indirectly inferred. Goodman's measurements on SOS films indicate that except for a marked increase in the so-called fast surface state density the bottom interface is qualitatively similar to the top interface. Measurement of C-V properties through the substrate provides conceptually a nondestructive method for evaluating HSF near the substrate after deposition. Even though the transport properties of this region may be radically different from those near the top of the film, a possibility exists that the ionizable doping density is reasonably uniformly distributed.

Of the methods described, only the method described by Wrigley and Kroko provides direct measurement of the electrical transport properties near the bottom interface on the actual film of interest with relative confidence in where the conduction is taking place in the film. The methods described in this section are merely modifications of the top surface control techniques and (except for different practical constraints) are not really distinctly different. The communication to the surface potential is still through the MOS capacitor and the sample current still provides the basic output concerning the transport properties of the conducting portion of the films. It must be borne in mind, however, when analyzing the electrical data, that the physical structure of the heteroepitaxial silicon film is much less perfect near the substrate than near the top (Chapter 2). Naive application of formulas and concepts applying only to more perfect structures may lead to erroneous conclusions. There is considerable evidence that the lower portion of the film is of importance in device operation and therefore is not to be overlooked. On the other hand, as illustrated in Figure 6.45, it takes stronger stimuli to obtain appreciable conduction in the bottom region than in the top region. Examination of Figure 6.46 shows rapidly decreasing conductivity as the bottom interface is approached. Therefore, in considering electrical data obtained by varying only the top surface potential, it is reasonable to assume that most of the

observed current is flowing in the "bulk" of the film or near the top surface. Studies on heavily defected semiconductors (such as polycrystalline silicon) or heavily dislocated single crystal silicon probably will find application in interpreting the electrical properties of the bottom portion of HSF. A brief discussion of this nature is presented in Section 6.8.

6.6. Automatic measurement techniques

The electrical characterization of HSF requires not only measuring average properties across the wafers and making detailed measurements on individual test devices but also examination of the distribution of electrical properties across the wafer surface. This is the primary motivation for using automatic testing since many repetitive tests are required. Another advantage of automatic testing is that a sufficient quantity of data can be generated to allow random effects to be averaged and rarely occurring phenomena to be found. Many different types of measurements can be made automatically. Two of the less common types which are of particular value for HSF are described here. These are the analysis of HSF/MOS C-V curves and the analysis of gate-controlled resistivity structures. The discussion of transport coefficient measurements is here limited to resistivity because it is technologically difficult to obtain any other type automatically across a wafer in a reasonable time. This is caused by the type of stimuli required (such as strong magnetic fields or controlled thermal gradients) and by the relatively weak response of the structures to these stimuli. It is, of course, possible to automate most measurements for individual structures. The main emphasis here, however, is on investigating the uniformity of properties across large areas of wafer surfaces. The basic instrument used in automatic testing is a computer-controlled integrated circuit tester with programmable voltage and current supplies and a mechanism for stepping the wafer under a set of fixed probes. This electrical interface requires that electrical access to the test structures be through standard integrated circuit contact pads. Use of these methods was recently described by (Crossley and Ham, 1973)

Capacitance voltage measurements

The capacitance measurement hardware is shown schematically in Figure 6.48. The capacitance meter is fed with bias voltage, to be applied to the capacitor, from the programmable power supplies of the DC tester with an accuracy ±1 mV. The capacitance measurement is made with a 15-mV AC signal added to the DC bias, and the meter output is an analog voltage which is measured by the programmable digital voltmeter (DVM) of the DC test set. For the work described here, the capacitance meter is set to 1 pF FSD range: the instrument is not range programmable by the computer. The measurement frequency can be between 10 kHz and 1 MHz, depending on the type of material being measured. The lower frequencies require care to ensure that the measurement is not unduly noisy.

Because tests may be made on capacitors with either conductivity-type semiconductor, the discussion which follows omits reference to voltage signs so that the statements made apply to both cases. For the purposes of explanation, reference is made to Figure 6.49, which is a possible C-V curve for n-Si. Testing proceeds as follows:

First, the accumulation and depletion capacitances, C_{acc} and C_{inv}, are measured with 10-V DC bias. This voltage is usually large enough to produce heavy accumulation or depletion without regard to the actual material doping. Other values may be appropriate in other cases. Note that capacitors with both inversion and accumulation contacts, such as in Figure 6.43, produce curves similar to those of Figure 6.34 and 6.35. These types of curves are much more difficult to analyze automatically than are the type in which only the accumulation side produces high capacitance (cf. Figure 6.42 and 6.49). The value of C_{acc} is noted and compared to the value expected for the geometry and oxide thickness used. If C_{acc} is not reasonable, the capacitor is assumed defective (C_{short}). To prevent spurious data from being retained, testing is halted if the C_{inv}/C_{acc} ratio is greater than 0.85 (a symptom of bad contacts or damaged gate metal). With C_{acc} and C_{inv} known, the following are calculated:

$$C_{1/2} = \frac{C_{acc} + C_{inv}}{2} \qquad (6.61)$$

and

$$C_{target} = 0.95 C_{acc}. \qquad (6.62)$$

To proceed further, it is necessary to find the voltage corresponding to C_{target} on the C-V curve. (This point is related to the flat band point and will be discussed later.) Under manual control of the test equipment, this is a trivial matter. With computer testing, however, an iterative procedure can be adopted. The applied bias voltage is altered (either positively or negatively depending on whether the measure-

ment is above or below the target value) in progressively smaller increments until the measured capacitance is within ±1 percent of the target value.

To calculate the doping density, N, from the C-V curve, the slope, dC/dV, must be measured. This is done by using the iterative technique to find the voltages corresponding to two capacitance values close to $C_{1/2}$, C', and C'' where

$$C' = C_{1/2} + 0.1(C_{acc} - C_{inv})$$
and corresponds to V'

and

$$C'' = C_{1/2} - 0.1(C_{acc} - C_{inv})$$
and corresponds to V''.

Then,

$$\left. \frac{dC}{dV} \right|_{C_{1/2}} = \frac{0.2(C_{acc} - C_{inv})}{V' - V''} \qquad (6.63)$$

Let $C_n = C_{1/2}/C_{acc}$. The following standard formula can then be used to calculate N:

$$\frac{dC_n}{dV} = \frac{(C_n)^3 \, \epsilon_{ox}^2}{\epsilon_{Si} q N t_{ox}^2} = \frac{1}{C_{ox}} \frac{dC}{dV}. \qquad (6.64)$$

Hence,

$$N = \frac{(C_n)^3 \, \epsilon_{ox}^2}{\epsilon_{Si} q t_{ox}^2 (dC_n/dV)}. \qquad (6.65)$$

Substituting,

$$N = \frac{(C_n)^3 C_{acc}^2}{a^2 \, \epsilon_{Si} q (dC_n/dV)} \qquad (6.66)$$

$$= \frac{(C_{1/2})^3 (V' - V'')}{(a^2 \, \epsilon_{Si} q) \, 0.2 (C_{acc} - C_{inv})}, \qquad (6.67)$$

where ϵ_{Si} is the dielectric constant of Si, a is the area of capacitor, and q is the electronic charge. Inserting the numerical values:

$$N = \frac{(C_{1/2})^3}{C_{acc} - C_{inv}} (V' - V'') (0.452 \times 10^{41}) \qquad (6.68)$$

for capacitances in pF and a 4-mil² active area.

The expression for N [Equation (6.68)] depends on the factor $(C_{acc} + C_{inv})^3/(C_{acc} - C_{inv})$. It might appear that a very accurate value of C_{inv} would be necessary in order to obtain an accurate value for the doping. C_{inv} is the most difficult capacitance to measure accu-

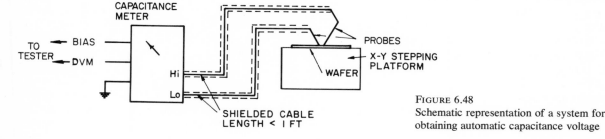

FIGURE 6.48
Schematic representation of a system for obtaining automatic capacitance voltage

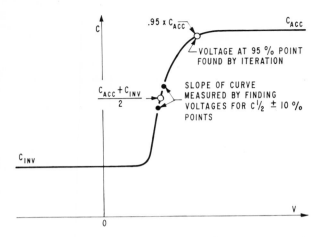

FIGURE 6.49   Significant points of the C-V curve for automatic analysis

rately since it is most sensitive to series resistance. In fact, however, the doping depends only very slightly on C_{inv}. The $\frac{1}{2}(C_{acc}+C_{inv})$ factor is the point at which the slope is measured. It does not matter exactly where this point is since the computer will find it accurately. The points V' and V'' are found where the capacitance values are as indicated in Figure 6.49 and the factor $1/(C_{acc}-C_{inv})$ accounts for any errors introduced by finding V' and V'' points that are not exactly as expected (caused by C_{inv} being in error). Thus, the only effect of an error in the C_{inv} value is that the impurity concentration may be measured at a different point on the C-V curve, and as long as there is no appreciable series resistance at this point, the measurement will be accurate. The C_{inv} value is used only to ensure that the slope is measured at a reasonable point on the C-V curve.

Figures 6.50 and 6.51 show the flow chart for these measurements. Values for the numerical constants are contained in the computer program controlling the test, and the measured values of C_{acc}, C_{inv}, V', and V'' are inserted into Equation (6.68) to calculate N.

In applying this measurement system, there are several points to be considered concerning its validity:
1. Series resistance effects.
2. The timing in this particular DC tester is such that the output from the capacitance meter is read, at the most, 36 ms after the bias voltage is applied to the capacitor: does the measurement then represent a true steady state value?
3. Is the control and calculation program valid (i.e., bug-free), and does it provide adequate guards against spurious data being recorded as valid?
4. Is the voltage-independent stray capacitance suitably accounted for?

These points are considered next.

The capacitance meter operates by measuring the quadrature component of current ($I \angle 90°$) flowing in

the device under test when the 15-mV AC signal is applied. This measurement reflects the desired capacitance only where $I \angle 0°/I \angle 90° < 0.1$ as discussed in detail in Section 6.4. The most reliable way to assure that series resistance is not affecting the measurement is to measure $I \angle 0°$ at each point. This, however, is very time consuming and requires measurement equipment capable of providing simultaneously $I \angle 0°$ and $I \angle 90°$ outputs. Such equipment exists but is very expensive compared to a single-capacitance meter. Therefore, it is frequently not practical to measure both $I \angle 0°$ and $I \angle 90°$. This requires that one establish a feeling for the series resistance problem in the particular HSF before attempting automatic measurements. It may not be practical to measure capacitors automatically on some types of HSF. In general, the series resistance problem may be diminished by (1) minimizing the resistive component by appropriate device design or (2) maximizing the reactive component by operating with the lowest possible frequency.

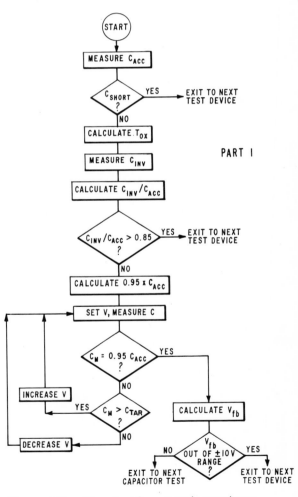

FIGURE 6.50   Flow chart for automatic capacitance measurements (Part 1)

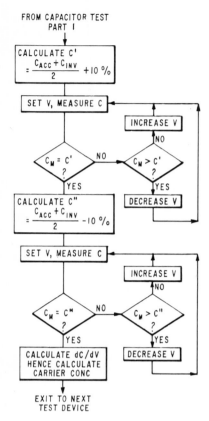

FROM CAPACITOR TEST
PART 1

CALCULATE C'
$= \dfrac{C_{ACC} + C_{INV}}{2} + 10\%$

SET V, MEASURE C

$C_M = C'$?

NO → $C_M > C'$?

NO → INCREASE V

YES → DECREASE V

YES ↓

CALCULATE C''
$= \dfrac{C_{ACC} + C_{INV}}{2} - 10\%$

SET V, MEASURE C

$C_M = C''$?

NO → $C_M > C''$?

NO → INCREASE V

YES → DECREASE V

YES ↓

CALCULATE dC/dV
HENCE CALCULATE
CARRIER CONC

EXIT TO NEXT
TEST DEVICE

FIGURE 6.51   Flow chart for automatic capacitance measurements (Part 2)

In practice, there is a doping level in a given thickness film at which the optimum device design cannot prevent measurement inaccuracies caused by series resistance. The region of validity of the measurement scheme has been investigated by comparing the carrier concentrations automatically measured at 1 MHz with those manually measured at 10 kHz. The agreement is good for surface carrier concentrations higher than $\approx 3 \times 10^{15}/cm^3$ in 1-μm n-type SOS layers and $\approx 10^{16}/cm^3$ in p-type material for the 4-mil² capacitor.

To investigate the possibility that transient effects at the application of the DC bias might interfere with the acquisition of valid data, the output of the capacitance

meter can be recorded on an oscilloscope. An example is shown in Figure 6.52. The trace shows that transient effects do indeed occur but that the steady state condition is well established prior to making the automatic measurement at 36 ms. The transient effects seen arise from two effects. The initial drop in capacitance (Figure 6.52a) is the response time of the capacitance meter—a few milliseconds. This response time is seen in the curve marked "+5 V." The rise from the minimum to the steady state value is caused by the finite rate of formation of the inversion layer, determined by the minority carrier generation rate and the fixed donor or acceptor concentration (Heiman, 1967)

$$T = 2T_g \frac{N}{n_i}, \qquad (6.69)$$

where T is the time constant for inversion layer formation, T_g is the time constant for minority carrier generation, and N is the donor/acceptor concentration. Inserting values for N and T,

$$N = 2 \times 10^{15}/cm^3 \qquad n_i = 1.5 \times 10^{10}/cm^3$$
$$T = 2 \text{ ms} \qquad T_g = 7.5 \text{ns}$$

a reasonable value for SOS films. The capacitance in Figure 6.52b does not show the recovery seen in Figure 6.52a because the measurement is series resistance dominated at the low capacitance values. In both cases, however, steady state is reached well before the measurement is taken. For capacitors with long recovery times, such as particularly with thick HSF films, it may not be practical to use automatic data acquisition. This effect should be carefully considered before attempting automatic measurements.

In order to ascertain that the program is functioning as intended, a C-V curve generated by the test set can be overlaid on a curve obtained from the same device taken manually with the C-V curve recorded on an x-y recorder. Figure 6.53 shows the results of this type of test. It is evident that the agreement is excellent. The disagreement resulted primarily from using different instruments to obtain the C-V plots.

The stray capacitance is accounted for by assuming that its value is known. This value is subtracted from any capacitance measurement to obtain the true capacitance. The value of the stray capacitance can be

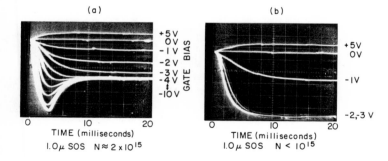

(a)

+5V
0V
-1 V
-2 V
-3 V
-4 V
-10V

GATE BIAS

0   10   20
TIME (milliseconds)
1.0μ SOS   N ≈ 2 x 10¹⁵

(b)

+5V
0V

-1V

-2,-3 V

0   10   20
TIME (milliseconds)
1.0μ SOS   N < 10¹⁵

FIGURE 6.52
Examples of SOS/MOS capacitor time response with the configuration of Figure 6.48.

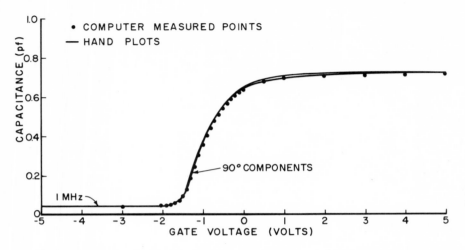

FIGURE 6.53
Comparison of automatically acquired data with manually acquired data

approximately determined by fabricating a wafer with only the metal pattern on the substrate and measuring the capacitance between the contact pads. This capacitance consists of probe-to-probe capacitance and intermetallic capacitance on the wafer. It is possible for the probe-to-probe capacitance to vary as the wafer is moved from device to device since the capacitive coupling through the wafer-holding mechanism changes. This can cause appreciable errors if the holding mechanism is not grounded. Other variations in the stray capacitance occur because of nonuniformities in the actual size of the capacitors and contact pads. These are usually small but are very difficult to account for. A related problem, as discussed in Section 6.4, is that the active area of the capacitor may vary across the wafer.

It was noted previously that an iterative procedure was used to find the voltages corresponding to the desired capacitance. This procedure was necessary because the capacitance measurement basically provides a "set V—measure C" function. If a procedure were available to provide a "set C—measure V" function, no iteration would be necessary and a great

deal of testing time could be saved. It is, in fact, possible and feasible to provide such a function with only minor additional effort. This scheme has actually been used by the author.

Basically an active control system is used wherein the capacitance meter itself provides the feedback. The actual output of the capacitance meter is compared with the desired output at the input to an operational amplifier. The output of this amplifier provides the bias for the capacitor. Because the amplification factor of an operational amplifier is very large, the input to the amplifier (the difference between the actual capacitance and the desired capacitance) must be very small or the output will be very large. Therefore, the operational amplifier will find whatever gate voltage is required to make the capacitance meter output equal to the desired value. This results in a "set C—measure V" function. The essential system is illustrated in Figure 6.54. An inverter is required for one or the other type of material depending on the actual capacitance meter used. This type of measurement is generally useful to "invert" measurement functions and can provide greater confidence in the results.

If the system settles to a reasonable value of bias, it *must* be the correct value. Any condition other than the meter output being equal to the set capacitance will result in a saturated operational amplifier. This type of measurement is also somewhat immune to noise effects. For example, an iterative measurement of the type described previously depends critically on *each* measurement correctly indicating the direction of change needed. Just one error from noise will cause the iterative procedure to fail. The operational amplifier method requires only that steady state be reached. Also, the iterative method will select only predetermined "quantized" numbers. The operational amplifier method provides a purely analog result.

It was previously mentioned that $C_{tar} = 0.95 C_{acc}$ was useful in determining the flat band voltage. This is

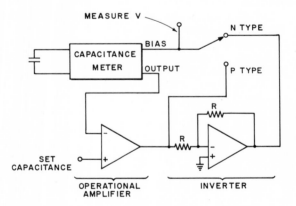

FIGURE 6.54   Use of capacitance meter as a feedback element to provide a "set C—measure V" function

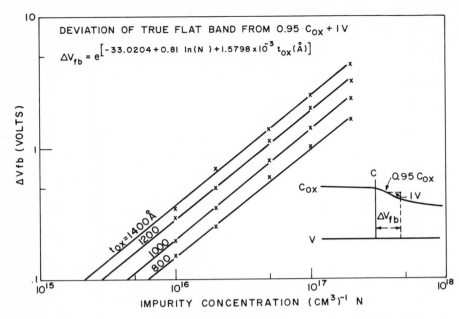

FIGURE 6.55
Correction factors for flat
band voltage determination

so because this value ±1.0 V provides a reasonable approximation to the flat band condition for a considerable number of oxide thickness–doping level combinations. For impurity concentrations greater than ≈5 × 10^15/cm^3 this empirical rule does not apply well and a correction factor, ΔV_{fb}, must be applied. This correction factor is illustrated in Figure 6.55 for a range of typically used oxide thicknesses. The empirical expression in Figure 6.55 for ΔV_{fb} can be calculated during testing from the measured doping density

(which does not depend on V_{fb}) and the oxide thickness which depends only on C_{acc}. An accurate value for V_{fb} can therefore be determined automatically. The correct value for V_{fb} is particularly useful for resistivity measurements.

Epitaxial material resistivity measurement

Automatic resistivity measurements can be made in the heteroepitaxial deposits using a gate-controlled van der Pauw device, although the general method would apply for any gate-controlled resistivity structure. The use of the gate to hold the surface potential at flat band is necessary, especially for lightly doped SOS layers, as discussed earlier. The use of the gate electrode places an upper limit on the potential which can be applied between the current drive electrodes, since a uniform flat band condition is assumed.

The flow chart for the resistivity measurement section of the test program is shown in Figure 6.56. The limit on the current-driving potential is taken to be 150 mV, and this is accomplished by applying 150 mV, measuring the resulting current, and then applying this current in the eight permutations possible with four electrodes. The gate current is first measured to ensure that the device is satisfactory for further measurement. A convenient circuit for supplying the low current necessary for this method is shown in Figure 6.57. With the same current applied cyclically to adjacent pairs of drive electrodes, the voltages appearing at the sense electrodes is measured. Averaging and calculation provides the average resistivity value, assuming the film thickness is known.

In addition, the carrier concentration value measured from the capacitor tests can be combined with

FIGURE 6.56   Flow chart for obtaining automatic van der Pauw flat band resistivity data

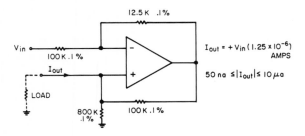

FIGURE 6.57   Current generator useful for supplying low currents for the measurements of Figure 6.56 from a programmable voltage source

the resistivity value to calculate a "relative mobility."

The data collected and reduced automatically may be displayed by a number of different techniques. Traditionally large volumes of data are condensed into a histogram where a numerical distribution can be seen at a glance. Such a plot is illustrated in Figure 6.58. Complementary to the histogram is the actual distribution of parameter values which can be repre-

sented by a quasi–three-dimensional presentation. A typical distribution of flat band resistivity and carrier concentration across a 1-inch-diameter SOS wafer is presented in Figure 6.59. These data are not necessarily typical of current technology; they merely represent an example of one set of measurements and the data display method.

6.7. Influence of the edge regions on the electrical characterization of HSF films

It was noted in Section 6.3 that a problem may exist when Hall bars (or other devices) are fabricated in the semiconductor because new interfaces are exposed by the definition procedure. These interfaces are chemically much the same as the top interface but differ at least in the orientation exposed and in their surface-processing history. These edges are approximately the same width as the film thickness and should be considered a fundamental part of the material since they are nearly always present in actual devices; that is, the

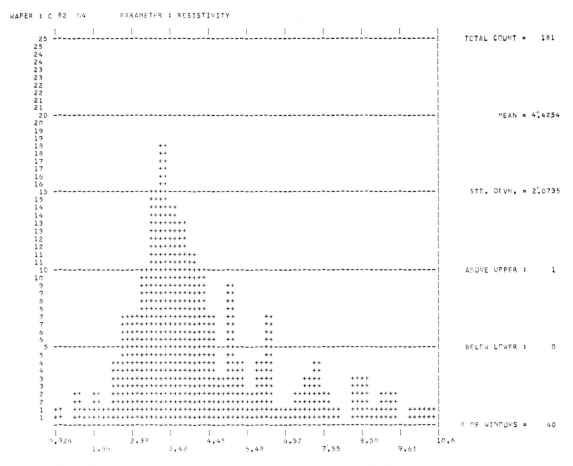

FIGURE 6.58a   Histograms of resistivity (left) and relative mobility (right) obtained from automatic van der Pauw measurements

ability to form a suitable edge region is an important requirement for the fabrication of devices in hetero-epitaxial silicon. These edge regions cause difficulties because they can conduct appreciably and with different properties from those of other interfaces.

Direct observations of preferential edge conduction have been made by Ham (1972B), Flatley and Ham (1974), and Tango and coworkers (1974) for the case similar to the gate-controlled Hall bar. Typically the edge interface has a different surface potential caused by fixed (i.e., unable to move under the influence of lateral electric fields) charges in the interface region different from those in the top interface. This difference arises from unknown causes, but the following are likely candidates:

1. The orientation of the edge interface is different from the top surface and therefore has a different surface band structure.
2. The edge region has been either mechanically or physically attacked during the definition process.
3. The semiconductor on the edge is exposed, resulting in a greater defect density than that of the top surface.

4. Mechanical stresses on the edge region are different from those near the top.
5. The edge surface is not usually as flat as the top surface and in fact is not a single crystallographic plane.

A typical edge region is illustrated in Figure 6.60 where approximately 1100 Å of thermal oxide was grown as a dielectric.

There is every reason to believe that unprotected edge surfaces would have different properties from those of unprotected top surfaces but there has been no report on an experimental investigation on this subject to date.

One of the necessary conditions for reliable Hall and resistivity measurements is that one know where the current is flowing. This is where the edge conduction can cause problems, as illustrated in Figure 6.61. If the edge impedance becomes comparable to the sample impedance, additional shunt paths for the current through the edges exist. The usual measurements will not reflect the properties of the bulk film in this case. In most cases the edge currents are small (< 1 μA); however, especially for measurements near the

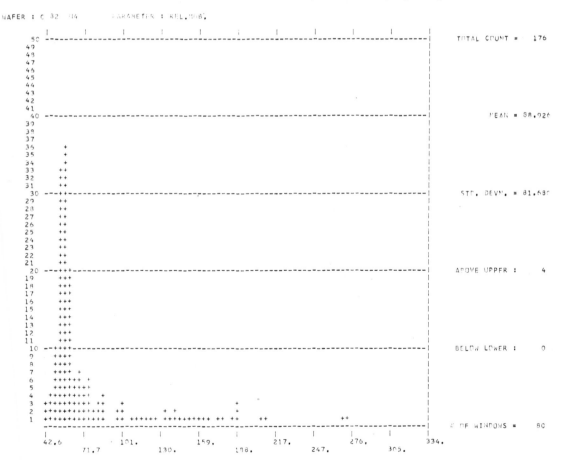

FIGURE 6.58b

RESISTIVITY (OHM-CM)   WAFER C-82

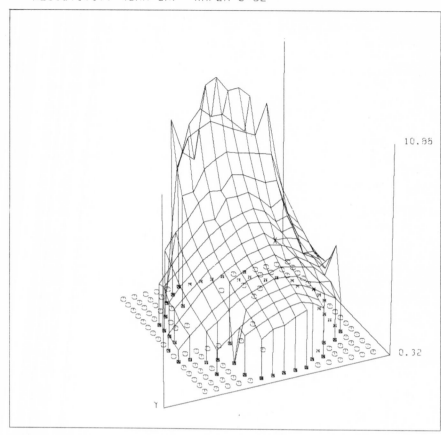

10.85

0.32

Y

FIGURE 6.59a
Three-dimensional representation of automatically acquired data. a) Flat band resistivity (left) varying from 0.3 to 10.8 Ω-cm (ρ_m of Figure 6.24 and b) carrier concentration (right) derived from automatic analysis of an SOS/MOS capacitor ranging from 0.17 to 5.0×10^{16}/cm^3 on the same n-type SOS 1-inch-diameter wafer are illustrated. From Ham and Crossley (1972)

bottom of HSF, these edge currents can be dominant, as illustrated in Figure 6.62. There is no simple way to detect this problem without a gate-controlled structure. In heteroepitaxial silicon in which a well defined type is present, the edge regions may be preferentially doped opposite to the contact type. This requires

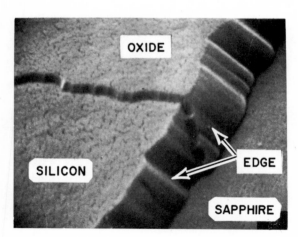

FIGURE 6.60   Typical edge region of a thermally oxidized 1-μm SOS film. The oxide thickness is $\approx$1100 Å

additional processing, but it effectively eliminates edge conduction if the junction between the doped edges and the contact is good.

The most reliable method of testing for edge current is to simultaneously fabricate both devices with and devices without edges; any difference in characteristics is thus caused by the edges. The scanning electron microscope has recently been reported as an alternate detection method (Gates and Griffith, 1975).

6.8. A model for carrier transport in heteroepitaxial silicon on sapphire derived from electrical data

The discussion of Sections 6.3 and 6.4 suggest that the properties of slightly processed SOS films can be crudely represented by a model in which the lower portion of the film is nonconducting compared to the upper portion. In addition, the top surfaces are unstable when exposed to atmospheric conditions and tend to be depleted of mobile carriers (especially for p-type films). Charges at the silicon–sapphire interface may exist but they are not important in the electrical characterization of the unprocessed films since they only

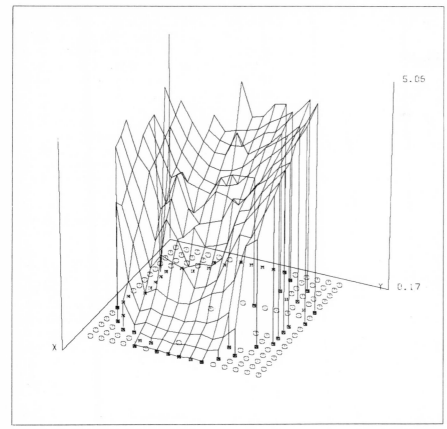

CARRIER CONCENTRATION (XE16)   WAFER C-82

FIGURE 6.59b

modify the properties of the film in the relatively nonconducting lower portion.

The film can therefore be divided into an upper section, in which the electrical properties are determined in much the same way as in bulk semiconductors, and a lower portion, in which the electrical properties are vastly different and are determined by mechanisms possibly peculiar to HSF. It will be shown that the channel mobility, which is determined solely by top surface charge transport, would be expected to vary much less than would the normal Hall or drift mobility with film thickness. The previous discussions suggest that a drastic reduction in Hall or drift mobility is seen as far away as 7000 Å from the bottom interface, whereas devices with nearly the same channel mobility as bulk silicon devices can be made in films as thin as 2500 Å. The difference is a result of the concentration of charge involved in the current transport. The channel concentration is much higher than the usual film concentration, and these concentrations are similar near the surface in films of differing thicknesses. The remaining discussion focus-

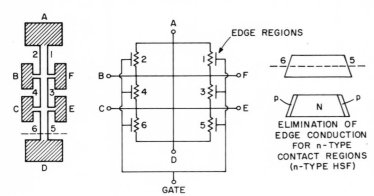

FIGURE 6.61
Equivalent circuits for the edge region on a defined HSF Hall bar

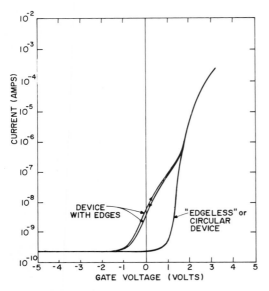

FIGURE 6.62   Illustration of preferential edge current in a gate-controlled resistor

es on the mechanisms governing the Hall or drift mobility.

The upper portion of processed films has been examined by Pödor (1970), Elliot and Anderson (1972), Ham (1972A), Ipri (1972), and Hynecek (1974). The general conclusion is that the Hall mobility decreases into the film and extrapolates to zero at a distance well above the silicon–sapphire interface. The scattering mechanisms suggested by these workers include those found in bulk single crystal semiconductors, such as phonon, neutral impurities, and localized space charges (e.g., ionized impurities, defects, complexes). None of these mechanisms explains the temperature dependence of currents in the lower portion or the anomalously high voltage across the Hall contacts when conduction is forced into the lower portion. It must also be explained why conduction in very thin films can be quite appreciable under the application of high surface fields (as seen in Figure 6.45), given that the lower portion has such a small effective mobility.

The key to the model proposed here for conduction in the lower portion of the film lies in the electrical properties of defects in semiconductors. An excellent treatise on the properties of defects in bulk semiconductors has been given by Matere (1971). These properties will not be reviewed here except to point out that in the vicinity of defects, particularly edge dislocations, a space charge region may exist. This region derives partially from the "dangling bond" which exists in an edge dislocation and partially from the stress field surrounding the dislocation.

The scattering properties of such space charge regions have been previously discussed by Weisberg

(1962), as applied to III-V HSF, and Schlötterer (1968), as applied to heteroepitaxial silicon. The principle result is that the scattering cross section increases with decreasing doping and that at high doping the space charge region essentially collapses. An isolated defect such as an edge dislocation does not drastically inhibit charge flow as is observed in the lower portion of the SOS films.

It is also relevant to the performance of p-n junctions in HSF that as the doping of the film is increased the thermal generation lifetime increases (due to the reduced space charge region volume). This lifetime can be the dominant source of junction leakage. Films grown with a heavily doped layer near the bottom can exhibit lifetimes of the order of 10^{-6} s, whereas more lightly doped films typically have lifetimes of the order of 10^{-9} s (McGrievy and Viswanathan, 1974); Schroder and Rai-Choudhury, 1973).

It has been suggested by Ernisse and Norris (1973), for the case of SOS, and by others for the cases of CdS (Neugebauer, 1968; Waxman et al., 1965) and InSb (Anderson, 1973; Juhasz and Anderson, 1967) that a situation may exist in HSF silicon in which defects coalesce to form a space charge surface. If the space charge surface is not parallel to the applied field, it is necessary for charge to cross *through* the space

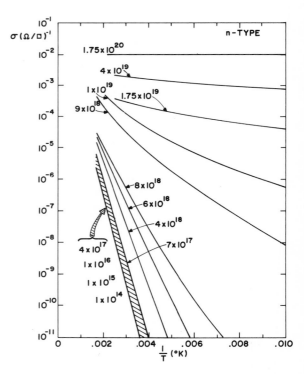

FIGURE 6.63   Conductivity of ion implanted polysilicon (5000 Å) as a function of n-type implant dose (given in cm^{-3} as calculated after redistribution at 1000°C) and reciprocal temperature

charge region in order to contribute to the observed current. This, of course, can drastically affect the observed current transport. This type of limitation may be considered more as a reduction of available current carriers than of mobility since only the carriers with sufficient energy are able to cross the barriers. Those carriers that can move have reasonably long mean free paths and their effective mass has not been significantly changed; thus, the usual sense of mobility applies to the carriers that are able to surmount the barriers.

A relatively well controlled example of this type of conduction can be found in the case of polycrystalline silicon. In this case, well defined closed-defect surfaces exist at the grain boundaries. By using ion-implantation techniques one can control the doping level of the silicon without severely disrupting the physical structure of the films. Therefore, it may be valuable to examine the properties of the electrical conduction in polycrystalline silicon in order to apply the results to the interpretation of the behavior in SOS films.

The conductivity of a thin film (5000 Å) of polycrystalline silicon with a grain size of approximately 1000 Å as a function of doping level and reciprocal temperature is illustrated in Figures 6.63 and 6.64. In many ways these plots look very similar to those of Figure 6.47, in which the "doping" was field induced. Below a certain level the conductivity is independent of the dopant. At high dopings the conductivity is essentially temperature independent. In the intermediate region the activation energy is continually decreasing as the doping increases. The same qualitative behavior is seen on both n- and p-type films. Figures 6.63 and 6.64 indicate that at low doping levels conduction is being controlled by a thermally activated process, and as the dopant concentration increases, the conduction mechanism changes into an almost temperature-independent process. A model for conduction in polysilicon recently proposed by Rai-Choudhury and Hower (1973) can be used to explain this behavior.

The model is based on the band diagram shown in Figure 6.65. The diagram is based on the inhomogeneous film model which was mentioned by Kamins (1971) and Salama and coworkers (1967) in discussion of polycrystalline silicon and other polycrystalline materials. The model assumes that polycrystalline material is composed of regions of low resistivity, the crystallites, separated by regions of high resistivity, the grain boundaries. It is assumed that because of the defects in the structure of the material the Fermi level (μ_F) of the grain boundaries is moved far away from either band edge. This is produced by a substantial number of electronic traps at the boundaries. When carriers are trapped at the boundary, the boundary becomes charged and the situation depicted in Figure 6.65 exists. A charged grain boundary surrounded by a

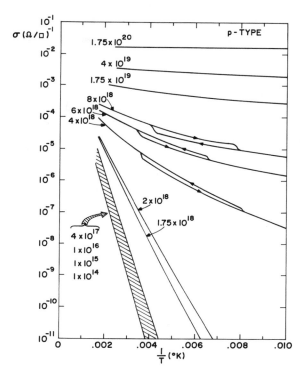

FIGURE 6.64   Conductivity of ion implanted poly-silicon (5000 Å) as a function of p-type implant dose (given in cm^{m}s^3 as calculated after redistribution at 1000°C) and reciprocal temperature

space charge region of opposite polarity is present, and the grain boundary acts as a potential barrier to conduction.

Rai-Choudhury and Hower (1973) assume that conduction across the boundaries is similar to Schottky barrier–type conduction exhibited in metal–silicon junctions. They adapt the analysis used by Yu (1970) for the contact resistance of n-type metal–silicon junctions. When a voltage is applied to the sample, it will divide among the grains and grain boundaries. Within each grain there will be a component V_g across the bulk portion and V_b across the boundary depletion region. Resistivity of the film can be written as:*

$$\rho = \frac{1}{l_g} \frac{V_g + 2V_b}{J}, \tag{6.70}$$

where J is the current density in the film and l_g is the grain size. This equation can be separated into two parts:

$$\rho = \frac{V_g}{l_g J} + \frac{2V_b}{l_g J}$$
$$= \rho_{\text{bulk}} + \rho_{\text{barrier}}. \tag{6.71}$$

---

*Equations (6.70) and (6.72) are derived in the Appendix.

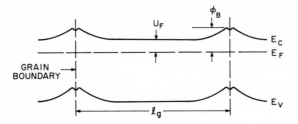

FIGURE 6.65   Band diagram for the potential barrier discussion

The effective resistivity of the barriers is then

$$\rho_{\text{barrier}} = 2\,\frac{R_C}{l_g}, \tag{6.72}$$

where R_C is the barrier resistance (in Ω-cm^2). At higher doping levels the depletion region surrounding the grain boundaries decreases. It eventually becomes small enough so that the carriers can tunnel their way through the barrier. Yu (1970) describes this region of conduction as the field emission region. In this region the value of R_C is almost independent of T and behaves as $\exp[\phi_b/(N_D)^{1/2}]$, where N_D is the dopant concentration.

As the doping level becomes lower, the depletion region becomes wider and carriers need additional energy to tunnel through the barrier. This energy can be provided by thermal means. This region of conduction is called the thermionic field effect region where carriers with some thermal energy tunnel through the barrier. In this region conduction becomes much more temperature dependent.

Eventually the depletion region becomes so wide that almost no carriers can tunnel their way through. The only way for conduction to continue is by carriers hopping over the barrier. This is called the thermionic emission region and in this region conduction is very much dependent on temperature and independent of doping level.

In the field emission region Yu derives an equation for R_C:

$$R_C = \left[\frac{A\pi q}{kT\,\sin(\pi C_1 kT)} \times \right.$$
$$\left. \exp\left(\frac{-\phi_b}{E_{00}}\right) \frac{-AC_1 q}{(C_1 kT)^2} \exp\left(\frac{-\phi_b}{e_{00}} - C_1\mu_F\right) \right]^{-1}, \tag{6.73}$$

where

$$A = \frac{4\pi m^* q (kT)^2}{h^3} \tag{6.74}$$

is the Richardson constant times T^2. E_{00} is a characteristic energy defined by

$$E_{00} = \frac{q\hbar}{2}\left(\frac{N_D}{m^*\epsilon}\right)^{1/2}, \tag{6.75}$$

where q is the electronic charge, ϵ is the dielectric constant of the material, m^* is the effective mass, and $\hbar$ is Planck's constant divided by 2π. C_1 is defined by

$$C_1 = \frac{1}{2E_{00}} \ln \frac{4\phi_B}{\mu_F}. \tag{6.76}$$

The equations above are valid if

$$1 - C_1 kT > \frac{kT}{(2E_{00}\mu_F)}^{1/2}. \tag{6.77}$$

In the thermionic field effect region R_C is given by:

$$R_C = \frac{kT}{qA}\,\frac{kT}{[\pi(\phi_b + \mu_F)E_{00})]^{1/2}}\cosh\left(\frac{E_{00}}{kT}\right)$$
$$\times \left(\coth\frac{E_{00}}{kT}\right)^{1/2} \times \exp\left(\frac{\phi_b + \mu_F}{E_0} - \frac{\mu_F}{kT}\right), \tag{6.78}$$

where

$$E_0 = E_{00} \coth\left(\frac{E_{00}}{kT}\right). \tag{6.79}$$

Equation (6.78) is valid if

$$\frac{\cosh^2\left(\frac{E_{00}}{kT}\right)}{\sinh^3\left(\frac{E_{00}}{kT}\right)} < \frac{2(\phi_B + \mu_F)}{3E_{00}}. \tag{6.80}$$

E_0 is a measure of the tunneling probability in the thermionic field effect region.

In the thermionic emission range the contact resistance is given by

$$R_C = \frac{kt}{qA}\exp\left(\frac{\phi_b}{kT}\right). \tag{6.81}$$

In Figure 6.66 theoretical plots of ρ vs. dopant and σ vs. $1/T$ obtained from Equations (6.73), (6.78), and (6.81) are shown in comparison with the actual data obtained experimentally. Qualitatively, the model and the data are similar; quantitatively, they differ considerably. This difference could be primarily a result of a failure in the model to account for changes in the barrier height, ϕ_b, as doping level changes. Kamins (1971) points out that as the doping level increases, the traps at the grain boundaries begin to become saturated. At this point the barrier height begins to decrease. The model, because it is based on metal–silicon junctions in which the barrier height is independent of doping level, contains no provision for a change in barrier height.

In an attempt to check Kamins' assumption, his equation for the relationship of ϕ_b and N_D was used to calculate different barrier heights at changing doping levels. These values were used in the model to generate the theoretical results displayed in Figure 6.67. It was assumed that ϕ_b does not begin to change until the levels at the grain boundary are filled by the

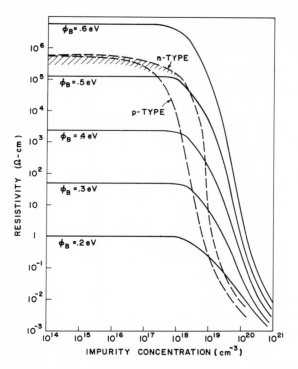

FIGURE 6.66   Comparison of theoretical and experimental results on ion-implanted polysilicon. The potential barrier height was not allowed to change with doping level in the theoretical curves

addition of charge carriers by increased doping. The exact doping level at which this occurs was estimated.

Figure 6.67 shows that if ϕ_b is allowed to decrease, the theoretical plot closely resembles part of the experimental curve for n-type material. The largest errors are obtained at higher doping levels. In the intermediate doping levels both the boundaries and the grains are important. In p-type films the barriers begin to lower at lower doping levels than in n-type films. The activation energy and barrier height are slightly lower in the p-type films. Lack of exact quantitative agreement may be caused in part by nonuniform distribution of dopant within the grains. Because the data and the model do show the same trends, however, it supports the essence of the model that conduction in polycrystalline silicon is controlled by the potential barriers at the grain boundaries and that tunneling is dominant at high doping levels and thermionic emission is dominant at low doping levels.

Of course, the properties of polycrystalline material do not necessarily directly apply to the case of single crystal heteroepitaxial silicon; however, the type of behavior exhibited by planar defects may be expected to be similar. Many authors have reported the existence of twin boundaries in HSF (Schlotterer and Zaminer, 1966; Naber and O'Neal, 1968; Yasuda and Ohmura, 1969; Yasuda, 1971; Stein, 1972) and the

large densities of edge dislocations may give rise to close-packed lineage, all of which are essentially planar in nature. Cross-sectional transmission electron micrographs such as shown in Chapter 2 strongly suggest planar defects. Ion-channeling experiments also indicate the presence of planar defects (Picraux and Thomas, 1973). Therefore, a strong case is made for the possibility of the existence of planar space charge regions in the lower portion of SOS films. A schematic representation of a possible distribution of space charge caused by planar defects is illustrated in Figure 6.68. When a distribution of this type exists, very different electrical properties are measured depending on whether the charge flow is confined to the bottom portion of the film or not. This type of distribution is a good illustration of a case in which C-V measurements do not predict the actual transport properties. The C-V measurement, whether of the top or bottom surface, should exhibit normal-looking curves and are useful for measuring surface potential and ionized impurity densities, but these parameters are of secondary importance compared to the barrier height in determining the current flow.

The data from deeply depleted gate-controlled SOS resistors of Figure 6.49 are not quite the same as the data from the polysilicon. In the SOS case the activation energy did not change until after the flat band

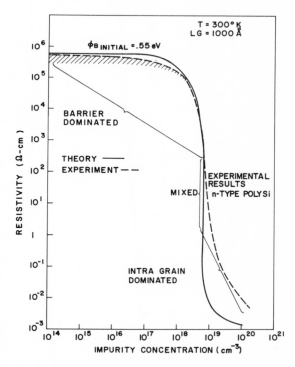

FIGURE 6.67   Comparison of theoretical and experimental results on ion-implanted polysilicon. The potential barrier was allowed to decrease as the doping increases in the theoretical curves; l_g is the grain site.

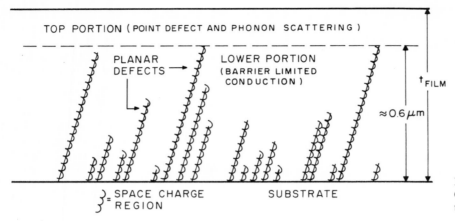

FIGURE 6.68
Conceptual model for electrical transport in SOS films

point was reached (proceeding toward accumulation), whereas in the polysilicon case the activation energy began to decrease immediately after departing from the dopant-independent regime (proceeding toward heavier doping). The model for the SOS case is essentially as follows:

When the conduction is confined to the lower portion of the film and the film is not accumulated, the dominant current limitation is by planar (or at least surface-like) space charge regions surrounding the planar defects. Therefore, the current exhibits an exponential temperature dependence. The temperature dependence is similar to the case in which thermally generated currents are important (e.g., silicon p-n junctions) but is the result of an entirely different mechanism. This activation energy is nearly the same as that seen in polycrystalline silicon. The observed current is strongly dependent on the gate voltage because the density of planar defects increases dramatically as the bottom interface region is approached. Therefore, as noted in Equation (6.72), when the effective "grain" size decreases, strong increases are seen in the resistivity. Proceeding to a region of the film where more planar defects are present effectively lowers the "grain" size and hence increases the resistivity. Because the density of defects has been observed to be strongly dependent on the distance from the interface (see Chapter 2), there is a strong dependence of the resistivity on the gate voltage, as seen in Figure 6.46. Below a certain point the space charge regions may merge as the defect density increases, and no further effect of varying the top surface potential would be seen. This model does not directly address the situation in which the entire film is depleted (by the gate field or by the space charges surrounding the defects). Conduction in this case would be dominated by thermal-generation currents.

There is some reason to believe that the space charge regions have not merged by the time the gate field-independent current is reached. This minimum current is higher in narrower devices and is not a strong function of film thickness (Ham, 1973). One is therefore led to the conclusion that the same dominant conduction mechanism persists all the way to the minimum currents. It is not clear at this time where the space charge edge is in the film when the minimum current is measured. Perhaps the inclined defects (Figure 6.68 and Chapter 2) screen the lowest portion from the top field and cause an effective maximum depletion depth which is less than the entire film thickness. Relatively heavily doped regions near the interface, caused by autodoping from the substrate, could also—by the application of top surface fields—cause the depletion region to not reach the interface. The possibility of a horizontal junction near the bottom interface in n-type material also exists (Hynecek, 1974).

Further support for this essential model is found in the data of Wrigley and Kroko (1969) who found a field effect mobility of 20 cm²/V-s in the conducting region near the bottom interface. This compares with typical mobilities in heavily doped polysilicon of approximately 20 cm²/V-s. The temperature dependence of the bottom interface current measured by Wrigley and Kroko suggests that the field-induced charge affects the barrier height in much the same manner as impurity-induced charge affects the barrier in polysilicon. The increase of threshold voltage in the very-thin-film devices (Figure 6.45) and the fact that appreciable conduction can be induced by the creation of large densities of field-induced charge (therefore lowering the planar defect barrier) are in fact predicted by the model.

Therefore, the SOS films in this model can be represented by three modes: (1) the upper portion where bulk-like conduction is found, (2) the intermediate portion where polycrystalline-like (barrier-limited) conduction is found, and (3) the ill-defined lowest portion where the minimum currents are determined. This lowest region may be a simple extension of the intermediate region. Careful specification of the region in which the data are acquired is necessary.

6.9. Summary and conclusions

This chapter has attempted to outline the problems and some of the methods involved with the reliable and significant electrical characterization of HSF. The entire situation with HSF is different from that of bulk single crystal semiconductors because of two main facts: (1) the interface between the HSF and its surroundings are in good communication with the body of the films and (2) the physical structure of the films may be widely variable and inhomogeneous, particularly in the vertical dimension. Therefore, because interpretating electrical data of any type can be done much more intelligently if an accurate physical model is available, the first order of business should be obtaining all the information possible on the physical structure of the particular HSF of interest. Without this information gross misinterpretations of "standard" electrical measurements are quite likely. The second order of business should be the implementation of some technique which at least considers that surface effects may be significant. The more lightly doped and the thinner the films of interest, the more difficult the attainment of significant electrical data.

The type of measurements that are most important depends entirely on the intended use for the HSF and the type of HSF being considered. The MOS transistor parameters of greatest interest are the drive capability, which is related to the surface mobility; the threshold voltage, which is related to the doping level; the top interface properties, and possibly the structure in thin HSF; and the minimum current, which is related either to junction properties or to the conduction mechanisms in the lower part of the HSF, depending on the type of device. Because this chapter deals primarily with the materials aspects of HSF, properties of heavily charged surfaces (inversion layer transport and accumulation layer transport) and p-n junction–related phenoma (generation rates, multiplication rates, lateral depletion widths, drift lengths etc.) have been largely ignored. Standard methods, such as the transient response of MOS structures to pulses (cf. Figure 6.52) (MGrievy, 1973; Kokkas, 1973) or the use of gate-controlled diodes [Meyer (1972); Kranzer (1974) and references therein], are the primary tools for the evaluation of junction effects. The application of these methods to HSF is not significantly different from their application to bulk semiconductors. Heavily charged surface transport is difficult to relate directly to material properties (cf. Section 6.5) and was therefore not pursued here.

The most important methods for the reliable electrical characterization of HSF have been from the gate-controlled structures in which the conducting portion of the HSF can be inferred from surface capacitance measurements. Unfortunately, because these structures require considerable effort to fabricate and are destructive, they are quite unsatisfactory for practical use in determining the electrical properties of unprocessed HSF.

The characterization of unprocessed HSF has been almost exclusively by the use of the 4-point probe and MVDP techniques. These methods provide at best a monitoring technique for reproducibility. They are unable to account for nonuniform electrical properties and are subject to first-order error from the uncontrolled top surface. Some workers have abandoned initial measurements altogether and have relied on the properties of fabricated devices to infer the characteristics of the initial deposits. This technique, apart from being practically prohibitive, does not account for the possibility of changes of the film characteristics during device fabrication. Others assume that the *addition* of known amounts of dopant *after* the initial deposition, such as by ion implantation, obviate the need for initial measurements. Such a method can be successful only if the initial deposition is very lightly doped or the doping is accurately known. *This* must be ascertained by measurement. When the film is reasonably thick, and N is $> 10^{16}/cm^3$, the 4-point probe and the MVDP provide valuable information concerning the reproducibility of initial depositions. The actual transport properties are measured by these methods.

Further investigation of the capacitance voltage techniques on the initial deposits may prove fruitful for characterizing the ionizable impurity density. Both the top surface Schottky barrier method and the bottom surface method through the substrate can be used nondestructively with relative insensitivity to the uncontrolled top surface. These methods are somewhat insensitive to the physical structure of the films and relation of results of these methods to actual current transport requires additional information concerning the conduction mechanisms.

The corona discharge method using a deposited dielectric is one of the most promising techniques for producing a controlled top surface. This method could be combined with a liquid metal MOS capacitor on the same dielectric. The electrical contact to the capacitor could be provided by accumulating the HSF with the corona charge. Under these conditions the surface potential, the surface capacitance, *and* the transport properties could be simultaneously measured on unprocessed films. Future work will reveal the usefulness of this method.

When the thickness of the HSF of interest is reduced to the point where the model presented in Section 6.8 applies, the significant electrical characterization of the HSF depends equally on the physical structure and on reliable electrical data. The relationship between electrical measurement and device properties are somewhat obscure in this thickness range. Either some nondestructive method for determining the physical structure must be implemented or the

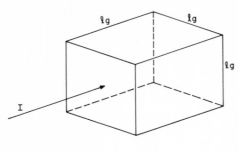

GRAIN

FIGURE 6.69   Simple analysis of current transport through a cubic grain

structure must be known from previous experiments. This area of study is perhaps the most important for further development of methods for determining early in the process what the characteristics of finished devices will be.

APPENDIX. SIMPLE DERIVATION OF EQUATIONS (6.70) AND (6.72).   Referring to Figure 6.69,

$$R^{\text{of one grain}} = \frac{V_g + 2V_b}{I},$$

$$\rho = \frac{Ra}{l} = \frac{V_g + 2V_b}{I}\,(a/l),$$

$$a = l_g^2,$$

$$l = l_g,$$

$$J = \frac{I}{l_g^2},$$

$$\rho = \frac{V_g + 2V_b}{J l_g} = \text{Equation (6.70).}$$

Barrier resistance $= R_b = R_C/a$, (i.e., the actual barrier resistance is inversely proportional to the barrier area). Because $\rho = Ra/l$,

$$\rho^{\text{barrier}} = \frac{2R_b l_g^2}{l_g} = \frac{2R_C}{l_g} = \text{Equation (6.72).}$$

Note only half of the actual boundary region applies to a given grain at each grain edge.

References

Anderson, J. C., *Advances in Physics 19:* 311 (1970).

Anderson, J. C., *Thin Solid Films 18:* 239 (1973).

Atalla, M. M., A. R. Bray, and R. Lindner, *Suppl. Proc. Inst. Elec. Engrs. (London) Pt B106:* 1130 (1959).

Buehler, M. G. and G. L. Pearson, *Solid State Electronics 9:* 395 (1966).

Covington, D. W. and D. C. Ray, *Solid State Electronics 16:* 301 (1973).

Covington, D. W. and D. C. Ray, *J. Appl. Phys. 45:* 2616 (1974).

Crossley, P. A. and W. E. Ham, *J. Electronic Materials 2:* 465 (1973).

Cullen, G. W. and J. F. Corboy, *J. Electrochem. Soc. 121:* 1345 (1974).

Cullen, G. W., J. F. Corboy, and·A. G. Kokkas, Technical Report AFAL-TR-73-200, June, 1973.

Deal, B. E. and M. Sklar, *J. Electrochem. Soc. 112:* 430 (1965).

Elliot, A. B. M. and J. C. Anderson, *Solid State Electronics 15:* 531 (1972).

Ernisse, E. P. and C. B. Norris, *Solid State Electronics 16:* 315 (1973).

Flatley, D. W. and W. E. Ham, Paper presented at Electrochemical Society Meeting, New York (1974); Abstract: *J. Electrochem. Soc. 121:* 290C (1974); Extended abstract: *74-2,* No. 198 (1974).

Frankl, D. R., *Electrical Properties of Semiconductor Surfaces,* New York: Pergamon Press (1967).

Garrett, C. G. B. and W. H. Brattain, *Phys. Rev. 99:* 376 (1955).

Gates, J. L. and O. K. Griffith, *Appl. Phys. Letters 27:* 43 (1975).

Goldsmith, N., R. V. D'Aiello, and R. A. Sunshine, NBS Special Publication 400-10, Spreading Resistance Symposium, Gaithersburg, MD, June, 1974.

Goodman, A. M., *IEEE Trans. Electron Devices ED-22:* 63 (1975).

Greene, R. E., P. R. Frankl, and J. N. Zemel, *Phys. Rev. 118:* 967 (1960).

Grove, A. S., *Physics and Technology of Semiconductor Devices,* New York: John Wiley and Sons (1967).

Ham, W. E., *Appl. Phys. Letters 21:* 440 (1972A).

Ham, W. E., unpublished RCA Laboratories, Princeton, NJ (1972B).

Ham, W. E., Paper presented at Electron Device Conference, Washington, DC (1973) Abstract No. 8.7 (Late News).

Ham, W. E., Paper presented at Electrochemical Society Meeting, New York (1974); Abstract *J. Electrochem. Soc. 121:* 285C (1974); Extended abstract: *74-2,* No. 144 (1974).

Ham, W. E. and P. A. Crossley, Paper presented at AIME Conference on Electronic Materials, Boston, 1972.

Heiman, F. P., *IEEE Trans. Electron Devices ED-14:* 781 (1967).

Ho, P., K. Lehovec, and L. Fedotowsky, *Surface Science 6:* 440 (1962).

Hynecek, J., *J. Appl. Phys. 45:* 2806 (1974).

Ipri, A. C., *Appl. Phys. Letters 20:* 1 (1972).

Juhasz, C. and J. C. Anderson, *Radio and Electronic Engineer,* April, 1967, p. 223.

Kamins, T. L., *J. Appl. Phys. 42:* 4357 (1971).

Kokkas, A. G., personal communication, RCA Laboratories, Princeton, NJ (1973).

Kranzer, D., *Appl. Phys. Letters 25:* 103 (1974).

Lehovec, K., *Appl. Phys. Letters 25:* 279 (1974).

Many, A., Y. Goldstein, and N. B. Grover, *Semiconductor Surfaces,* Amsterdam: North Holland Publishing Company (1965).

Matere, H. F., *Defect Electronics in Semiconductors*, New York: Wiley-Interscience (1971).

McGreivy, D. J., Ph.D. dissertation, University of California, Los Angeles, 1973.

McGreivy, D. J. and C. R. Viswanathan, *Appl. Phys. Letters 25:* 505 (1974).

McLane, G. T., Report No. 72-02, University of Pennsylvania, Moore School of Electrical Engineering (1971).

Meyer, J. E., Jr., Ph.D. dissertation, Rutgers University, New Brunswick, NJ (1972).

Naber, C. T. and J. E. O'Neill, *Trans. AIME 242:* 470 (1968).

Neugebauer, C. A., *J. Appl. Phys. 39:* 3177 (1968).

Petritz, R. L., *Phys. Rev. 110:* 1254 (1958).

Picraux, S. T. and G. J. Thomas, *J. Appl. Phys. 44:* 594 (1973).

Podor, B., *Phys. Stat. Sol. 2:* K193 (1970).

Rai-Choudhury, P. and P. L. Hower, *J. Electrochem. Soc. 120:* 1761 (1973).

Salama, C. A. T., T. W. Tucker, and L. Young, *Solid State Electronics 10:* 339 (1967).

Schlegel, E. S., G. L. Schnable, R. F. Schwarz, and J. P. Spratt, *IEEE Trans. Electron Devices ED-15:* 973 (1968).

Schlötterer, H. and Ch. Zaminer, *Phys. Stat. Sol. 15:* 399 (1966).

Schlötterer, H., *Solid State Electronics 11:* 947 (1968).

Schroder, D. K. and H. C. Nathanson, *Solid State Electronics 13:* 577 (1970).

Schroder, D. K. and P. Rai-Chaudhury, *Appl. Phys. Letters 22:* 455 (1973).

Shannon, J. M., *Solid State Electronics 14:* 1099 (1971).

Shockley, W., W. W. Hooper, H. J. Queisser, and W. Schroen, *Surface Science 2:* 277 (1964).

Stein, H. J., *Solid State Electronics 15:* 1209 (1972).

Tango, H., Y. Nishi, K. Maesuchi, and J. Iwamura, Paper presented at fifth IEEE Semiconductor Interface Specialist Conference, Puerto Rico, December, 1974.

Tihanyi, J., *Siemens Forsch. Entwickl 1:* 263 (1972).

Valdes, L. B., *Proc. IRE __:* 420 (1954).

Waxman, A., V. E. Henrich, F. V. Shallcross, H. Borkan, and P. K. Weimer, *J. Appl. Phys. 36:* 168 (1965).

Weisberg, L. R., *J. Appl. Phys. 33:* 1817 (1962).

Williams, R. and M. H. Woods, *J. Appl. Phys. 44:* 1026 (1973).

Wrigley, C. Y. and L. J. Kroko, *Semiconductor Silicon,* New York: The Electrochemical Society (1969) p. 329.

Yasuda, Y. and Y. Ohmura, *Japanese J. Appl. Phys. 8:* 1098 (1969).

Yu, A. Y. C., *Solid State Electronics 13:* 239 (1970).

Zaininger, K. H. and F. P. Heiman, *Solid State Tech., 13* (5): 49 (1970).

Chapter 7

An Analysis of the Gas-Flow Dynamics in a Horizontal CVD Reactor

S. Berkman, V. S. Ban, and N. Goldsmith

7.1. Introduction

The epitaxial growth of silicon by chemical-vapor deposition (CVD) techniques has become an important process for the production of a variety of solid state device structures. Because of the commercial importance of silicon, the fundamental aspects of the deposition process have been studied in some detail. There is an abundant amount of literature available to the reader concerning many aspects of the CVD process, such as chemical kinetics, reactor design, and hydrodynamic conditions and their effects. The reactor design most amenable to CVD research is the horizontal epitaxial system. The horizontal reactor is simple to construct and modify for experimental purposes, it is clean from a semiconductor purity standpoint, and the operation is reasonably efficient. Horizontal epitaxial reactors are a popular choice for the reasons given; but this is not to say that the horizontal reactor is necessarily the best choice for large-volume production of a specific epitaxial product. A schematic representation of a horizontal epitaxial system is shown in Figure 7.1.

The basic role of the theory of epitaxial growth is to predict conditions necessary for the achievement of a satisfactory deposit in terms of thickness and doping uniformity, chemical efficiency, etc.

The theory of epitaxial growth, as understood in horizontal reactors, can be applied to other epitaxial systems. For example, the gas-flow conditions of the barrel-type reactor are very similar to those of the horizontal reactor. The barrel reactor has been used for both the hetero- and homoepitaxial growth of sili-

con. One of the most perplexing problems associated with epitaxial growth theory is the precise theoretical interpretation and reconciliation of the existing theories with experimental results. This is the topic of this chapter. It will be demonstrated that suitable mathematical formulas can be derived from hydrodynamic flow theories to explain the epitaxial growth process in a horizontal CVD reactor. Our purpose is to provide a practitioner of epitaxy with a set of readily usable engineering formulas, rather than a rigorous mathematical analysis of the transport phenomena in the reactor. The mathematical and physical interpretation of the formulas provide a basis for the reconciliation of existing disagreements in current theories and also reinforces the most acceptable of the existing theories.

There are many approaches that can be taken in attempting to formulate a set of guidelines for the fabrication of an epitaxial reactor. These approaches have as their two extremes a fully developed mathematical treatment involving the solution of a set of nonlinear differential equations drawn from hydrodynamic theories and a purely descriptive set of "rules" based on experience. For this chapter we have tried to find the middle ground between these two extremes. We begin by first outlining the characteristics of an ideal reactor and then presenting the practical reasons as to why this reactor cannot exist. Nevertheless, having defined how a real reactor must operate in order to obtain the epitaxial layers that are desired, we then proceed to a detailed development of a simple mathematical model which can be readily used to provide guidelines for selecting the operating conditions for a reactor. Finally, we attempt to assess the

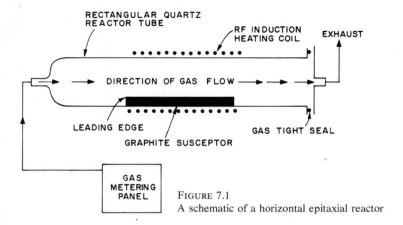

RECTANGULAR QUARTZ
REACTOR TUBE

RF INDUCTION
HEATING COIL

EXHAUST

DIRECTION OF GAS FLOW

LEADING EDGE

GRAPHITE SUSCEPTOR

GAS TIGHT SEAL

GAS
METERING
PANEL

FIGURE 7.1
A schematic of a horizontal epitaxial reactor

limitations of the model and its applicability to the goals of obtaining epitaxial layers of uniform thickness and doping.

7.2. The ideal reactor

For our purposes, an epitaxial reactor is simply a chamber which holds the substrates on which we wish to grow our layers. By one means or another these substrates can be heated uniformly to a desired temperature. Reactants are introduced into one, or more, ports of the chamber and the spent gases pass out through another port. Almost without exception, the major constituent introduced is hydrogen. Although it is true that hydrogen is an active ingredient in all of the reactions of interest, it is in such large excess that it is a convenient fiction to treat it as only a carrier gas. The function of the carrier gas is to distribute the reactants in such a way that all of the growing surfaces are effectively exposed to the same reactant mixture. The simplest reactor is a tube with reactants entering at one end and the spent gases leaving at the other end. For reasons which will be discussed in the following treatment, the tube cross section in such a reactor should be rectangular.

The ideal reactor would allow one to grow, on as many substrates as desired, layers that are identical in chemical, mechanical, and electronic properties with a (thermodynamic) efficiency of 100 percent. Of course, no such reactor exists. In order to see why, it is instructive to examine the extremes of operation of a reactor.

As one extreme, consider what must take place in a reactor as the carrier gas velocity approaches zero. Under such conditions, if chemical kinetics does not limit the deposition rate, all of the reactants will combine to form the solid at the leading edge of the reactor. No deposit will be formed on substrates located more than one or two reactant diffusion lengths from the leading edge of the heated zone since the reactants will have been stripped from the gas phase.

As the carrier gas velocity is increased, an increasing amount of deposit will be observed downstream. At the same time we can anticipate that the rate of growth at the leading edge of the susceptor will begin to decrease. This decrease will be noticed when the residence time becomes shorter than the reaction time constant. In the limit, as the velocity approaches infinity, there is no deposit at all and the yield goes to zero. This sort of argument assumes, of course, that all of the reactant in the space over substrates is available for deposition. As we shall see, this is not always true. Nevertheless, from the extremes of the problem, the solution to obtaining the uniform layers we desire appears to lie in arranging the conditions of flow in such a way that the growth rate on the downstream wafers is unaffected by the reactants having been removed from the gas stream. The practical effect of such a requirement is to impose a limitation on the overall efficiency at which a reactor can be operated.

Thus far in our discussion we have not provided any considerations of how the reactants reach the heated substrates. In our simplified discussion we have assumed total and complete mixing of the gas in the space above the wafer. In hydrodynamic terms, this is equivalent to requiring that the reactor operate under conditions of turbulent flow. Turbulent flow, however, is not obtained under the flow conditions normally encountered in an epitaxial reactor. Further, the quantities of carrier gas needed to produce turbulent flow and the necessary short residence times make it impractical to even consider designing a reactor about such constraints. Instead, most reactors are operated in a manner that produces essentially laminar flow. Under conditions of laminar flow, mixing occurs by diffusion rather than by forced convection.

Having arrived at the point where we are considering descriptions of reactors where we must account for the diffusion of the reactants to the growing surface, it proves to be more useful to abandon the narrative approach and proceed to the development of a mathematical model. In the sections that follow, the discussion is primarily oriented toward the problem of the epitaxial growth of silicon. In the epitaxial growth of silicon, a wealth of experimental information is available for comparison between theory and practice. The equations that are developed in the following sections are not restricted in any way, however, to the problem of silicon epitaxy but apply generally to epitaxial growth problems that meet the boundary conditions under which the equations were developed.

7.3. Previous models for the design of CVD reactors

It is appropriate to briefly review some of the models applied to gas flow in horizontal CVD reactors. Bradshaw (1967) proposed that a laminar boundary layer of relatively static gas exists adjacent to the growing silicon surface. He further assumed that the space between the inner reactor tube wall and the outer extremity of the boundary layer was filled with completely mixed gas that had essentially the same chemical composition as the gas introduced into the reactor. He then concluded that the reactive silicon-bearing chemical species diffused through the static laminar layer to the growing silicon surface and that the chemical reaction products diffused back through the boundary layer to the mixed gas region. This led to a growth rate equation that incorporated an expression, from classical laminar-flow theory, for the velocity boundary layer. By the use of an expression for the velocity boundary layer, Bradshaw's growth rate equation exhibited a square root of velocity dependence, and he was able to calculate the growth rate at one position near the leading edge of the susceptor.

Andrews and coworkers (1969) appealed to conventional laminar boundary layer theory to calculate an average rate mass-transport coefficient for the growth rate equation used by Grove (1967). Andrews' growth rate equation exhibited, as did Bradshaw's, a square root dependence on velocity. He applied his average growth rate equation to the extreme downstream end of the susceptor, where he considered the laminar flow to be fully developed. Thus, he was able to calculate an average growth rate at a specific location. Although this procedure gave a result for the one position, the same laminar-flow theory predicts a local growth rate that is half the average rate (Schlichting, 1955).

Rundle (1971) developed a growth rate equation by assuming laminar flow. He assumed that the velocity profile is constant across the reactor tube and that the depletion of reactive chemical species in the direction of flow is a result of the diffusion flux perpendicular to the direction of flow onto the substrate surface.

Because he does not assume any velocity profile from laminar boundary layer theory, his boundary layer thickness is the entire distance between the susceptor and the inside wall of the reactor tube. Therefore, his growth rate equation does not exhibit a square root of velocity dependence. Rundle's solution of the appropriate form of the diffusion equation does show the proper form of the exponential loss of chemical reactants in the direction of flow.

Eversteyn and coworkers (1970) devised a novel growth model which is supported, to a great extent, by elegant experimental work. He calls it the stagnant layer model. Both Eversteyn and Bradshaw assume a relatively stagnant boundary layer of gas adjacent to the susceptor surface and a completely mixed region of gas between the upper edge of the boundary layer and the inner surface of the reactor tube wall. From his experiments Eversteyn developed an empirical equation which has a square root of velocity dependence for the thickness of the stagnant boundary layer. Eversteyn included his boundary layer equation in the solution of the appropriate differential equation for diffusion and bulk carrier gas flow in a horizontal reactor. With this, he was able to describe, in general terms, the growth rate as a function of distance from the leading edge of the susceptor and specifically demonstrate the velocity dependence of the stagnant boundary layer thickness. As a result, equations are derived in a manner so as to describe the growth rate variations on a susceptor when it is placed in a horizontal plane or when the susceptor is tilted at a small angle to the horizontal plane. By the use of the equations it was demonstrated that the uniformity of the silicon growth on the susceptor was substantially improved by tilting the susceptor a small amount.

7.4. Theory

It was already mentioned that the nature of gas flow in the reactor significantly affects the transport phenomena. It is difficult to predict the nature of flow a priori. Although the low Reynolds numbers (typically below 200) and the proportionality of the growth rate to the square root of gas velocity suggest a laminar flow, the existence of the sharp temperature and density gradients implies possible disturbances of this laminar nature. These gradients are caused by the hot susceptor. Because the heating is from below, an unstable situation develops, with heated, more buoyant gas rising and possibly upsetting the laminar flow. In addition, one should consider possible entry effects. Because susceptors are relatively short (40 to 50 cm), a significant portion could be in the so-called entry region, where thermal and velocity profiles are not fully developed. In such cases one could expect the transport phenomena which were significantly different from the transport in fully developed flow situations.

In order to establish the nature of the flow in horizontal reactors we did a literature search, as well as conducted experimental studies. In this section we present the results of the work. Among the relevant articles are those of Kamotani and Ostrach (1976) and Hwang and Cheng (1973), who studied gas flow in a horizontal channel with heating from below. In their experiments, special precautions were taken to insure a fully developed laminar velocity profile (e.g., Ostrach and Hwang made their entry region as long as 183 cm). In such cases, one therefore deals with only thermally developing flows, where gas temperature increases with distance x, but the laminar nature of flow remains essentially unperturbed. Such a situation presumably holds for Re (Reynolds) and Ra (Rayleigh) numbers similar to those encountered in CVD reactors. On the other hand, Sparrow and coworkers (1959), who investigated simultaneous effects of buoyancy and viscous forces upon gas flow, predict that the flow will be perturbed in some manner. They devised the Gr/Re^2 (Gr = Grashof number) ratio as criteria for assessing the relative effect of buoyancy and viscous forces. The criteria are as follows:

$$0 < Gr/Re^2 < 0.3 \quad \text{Forced convection}$$
(i.e., viscous forces predominant)
$$0.3 < Gr/Re^2 < 16 \quad \text{Mixed flow}$$
$$16 < Gr/Re^2 \quad \text{Free convection}$$
(i.e., buoyancy forces predominant)

These criteria apply for the case of aiding flow (i.e., where buoyancy and viscous forces act in the same direction). When this is not the case, and in horizontal CVD reactors it is not, mixed flow characteristics can be observed at Gr/Re^2 ratios as low as 0.06 (Sparrow et al., 1959). In this mixed flow some motion perpendicular to the flow direction can be observed. Dismukes and Curtis (1973) studied flow characteristics in CVD reactors by means of temperature measurements and TiO_2 smoke-flow visualization. Their main conclusion is that buoyancy forces have an effect at much lower values of Ra numbers than in stationary systems. They base this conclusion on the observation of pronounced temperature fluctuations and turbulent smoke patterns, which occur in the Ra number range of 10^2 to 10^3.

Our experimental setup for studying transport phenomena was described in some detail by Ban and Gilbert (1975). Briefly, it consisted of a somewhat modified horizontal CVD reactor, equipped with a port for introduction of various instruments. The momentum transport was studied by means of flow visualization (TiO_2 smoke), the heat transport by means of temperature measurements at various locations in the reactor, and the mass transport by means of a specially designed mass spectrometric probe,

capable of determining the gas-phase composition at various locations. The flow-visualization experiments covered the Gr/Re^2 range from 0.16 to 12.5. The main observations are summarized in Figure 7.2. When Gr/Re^2 was approximately greater than 0.5 a fully developed spiral was observed, clearly demonstrating the presence of buoyancy forces. Similar behavior was observed in experiments by Takahashi and coworkers (1972). (Flow patterns shown in Figure 7.2 were obtained by tracing photographs of TiO_2 smoke trails; the photographs themselves were, unfortunately, too pale for direct reproduction.)

Results of temperature measurements are given in Figures 7.8, 7.9, and 7.10. The obtained results will be discussed in more detail later; for now, we point out the relatively sharp temperature drop in the first 10 to 15 mm above the susceptor and then the relatively flat temperature gradient above this area. Figure 7.3 shows the direct temperature profile recording and the relatively large temperature oscillations in the area 10 mm or more above the susceptor.

Figure 7.4 gives a typical concentration profile above the susceptor. In this case, one is dealing with $SiCl_4$ in H_2, but the qualitative picture should hold for any Si-containing gas. Again, note the relatively flat concentration gradient in the upper half (i.e., > 15 mm above the susceptor) of the channel and the relatively sharp concentration gradients close to the susceptor. Of course, the concentration gradient for the HCl species which are the product of the Si deposition reaction has a different sign (i.e., the concentration falls with the distance from the susceptor). Figure 7.5 shows the direct concentration profile recording and, again, quite noticeable concentration oscillations in the area ten or more millimeters above the susceptor.

The purpose of the above described experiments was to determine the nature of flow in the horizontal CVD reactor. Our observations can be summarized as follows:

1. There are two distinct areas in the space above the susceptor. Closer to the susceptor, relatively sharp temperature and concentration gradients are

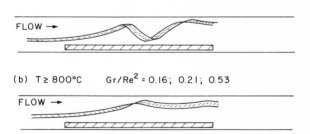

(a) $T \leq 800°C$ $Gr/Re^2 = 0.47; 0.56; 4.14; 12.5$

FLOW →

(b) $T \geq 800°C$ $Gr/Re^2 = 0.16; 0.21; 0.53$

FLOW →

FIGURE 7.2 TiO_2 smoke streamlines observed at various Gr/Re^2 ratios and susceptor temperatures; these drawings were obtained by tracing the actual photograph of the flow

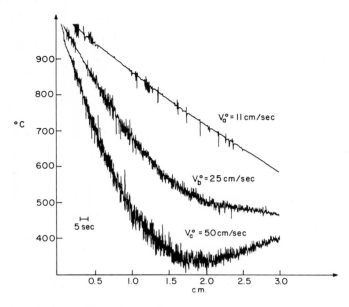

FIGURE 7.3
Oscillations in recording of temperature profiles in He. T_s = 1000°C; distance from the heating edge, x = 15 cm. Flow velocities for curves (a), (b), and (c) are marked

observed, while further from the susceptor, these gradients are essentially flat.

2. Significant temperature and concentration oscillations were observed in the upper area. These oscillations increase with the value of Re number.

3. The effect of buoyancy forces is observable at all practical Ge/Re² ratios.

From these observations we conclude that we are dealing with the mixed-flow situation. Close to the susceptor, the gas is hotter and more viscous and assumes characteristics similar to laminar flow. In the upper region the mixed character is more pronounced, with significant amount of gas motion perpendicular to the direction of flow; this leads to the observed flat temperature and concentration gradients. The cause of this cross-flow motion is probably twofold: the buoyancy forces and the entry effects. Concerning entry effects, we should point out that in the range of Re numbers used in our experiments, the entry length l varies from 10 to 35 cm (l = 0.029 ReD_h). This means that in all cases the entry length covered most of our susceptor. (D_h = hydraulic diameter.)

Figure 7.6 gives a schematic representation of the flow model suggested by our experiments. In the following text we shall discuss the transport phenomena in the horizontal CVD reactor based on this model. Equations will be developed from a treatment of laminar-flow theory that could not be found in the literature. A comparison of established laminar-flow theory and experiment will show that this treatment provides a physical model with which proper mass transport equations for a composite of laminar and turbulent flow can be developed.

Consider the molecular diffusion process from a volume of gas with a fully developed laminar parabolic velocity profile as shown in Figure 7.7. An arbitrary unit volume element contains a host carrier gas and a concentration, C, of reactive chemical species. The diffusion flux of C in the y direction is then given by the classical equation (Perry, 1963)

$$N_y = -D \frac{\partial C}{\partial y}. \tag{7.1}$$

The solution of Equation (7.1), in its most simplified

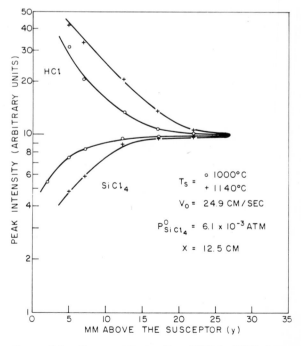

FIGURE 7.4   Concentration profiles of SiCl₄ and HCl; the experimental conditions are given in the figure

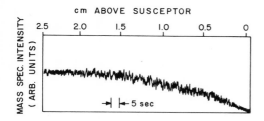

FIGURE 7.5   Direct mass-spectrometric recording of the concentration profile of $SiCl_4$. $T_s = 1140°C$; $P_0(SiCl_4) = 6 \times 10^{-3}$ Atmosphere; $V_0 = 27$ cm/s.

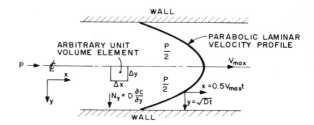

FIGURE 7.7   The laminar-flow diffusion model from a fully developed parabolic velocity profile

$$\delta_E = \left(\frac{DX}{0.5\,V_{max}}\right)^{1/2} = 1.414\left(\frac{DX}{V}\right)^{1/2}, \quad (7.3)$$

where D is the average diffusion coefficient for the system. For epitaxial systems, > 97 percent of the molecules would be those of the host carrier gas. An alternate equation for δ_E can be written using the coefficient of kinematic viscosity, ν, from the literature (Foust et al., 1960), $D = (6/5)\nu$ for gases. Therefore

$$\delta_E = 1.549\left(\frac{\nu X}{V}\right)^{1/2}. \quad (7.4)$$

Equation (7.4) can be used for comparison with established laminar flow theory. From Schlichting (1955),

$$\frac{U}{U_\infty} = 0.51676 \quad \text{when } n = y\left(\frac{U_\infty}{\nu X}\right)^{1/2} = 1.6.$$

Schlichting uses notation in which U is the point velocity on the parabolic laminar-flow velocity profile; U_∞ is the maximum velocity on the parabolic laminar-flow velocity profile; and n is the coefficient of $(\nu X/U_\infty)^{1/2}$. When $n = 1.549$ (from Schlichting), $U/U_\infty = 0.5003$. From Equation (7.4), $n = 1.549$ when $V/V_{max} = 0.500$.

It is obvious from the agreement between the constants from classical laminar-flow theory and the constants for δ_E derived from this diffusion model that a valid relationship does exist. It is still necessary, however, to show that δ_E corresponds to the diffusion boundary layer from classical laminar flow theory. The most relevant solution from established laminar-flow theory is the Pohlhausen solution that was used by Andrews. There are two equations for the rate mass transport coefficients (Andrews et al., 1969; Schlichting, 1955).

The average rate equation is given as a function of the total susceptor length (L), and the local rate is given as a function of the local position (x).

$$\text{Nu}_{(L)} = 0.664(S_c)^{1/3}[\text{Re}_{(L)}]^{1/2}, \quad (7.5)$$
$$\text{Nu}_{(x)} = 0.332(S_c)^{1/3}[\text{Re}_{(x)}]^{1/2}. \quad (7.6)$$

Nu is the Nusselt number and is also equal to (hL/D_1) or (hx/D_1). S_c is the Schmidt number ($u/\rho D_2$), and the

terms, for the transport of the reactive chemical species to the wall where its pressure is zero is

$$N_y = \frac{DP}{RT\delta}, \quad (7.2)$$

where P is the partial pressure of reactive species (in atmospheres); D is the diffusion coefficient of the reactive species in the host carrier gas; R is the universal gas constant ($= 82.06$ cm^3-atmosphere/g-mole); T is the absolute temperature (in Kelvins); δ is a specific diffusion distance, which needs to be defined, in the y direction; and N_y is the arrival rate of reactive chemical species to the wall in g-moles/s-cm^2. The assumption of zero pressure at the wall is justified for the growth from silane.

In order to define the diffusion distance, δ, we must consider that the average arrival rate of molecules to the lower wall must be from a unit volume element that possesses the average characteristics of the parabolic velocity distribution. The theoretical average velocity of the parabolic laminar-flow profile is given in the literature as $0.5\,V_{max}$, where V_{max} is the maximum free gas stream velocity. If the arbitrary unit volume element in Figure 7.7 is placed on the parabolic velocity profile and made to coincide with the point where $V = 0.5\,V_{max}$, then δ would be the distance a molecule could diffuse in the y direction (normal to the bulk gas flow) while it is traveling in the x direction (the direction of the bulk gas flow) with a velocity equal to the average velocity $0.5\,V_{max}$. Thus, the effective diffusion boundary layer thickness (δ_E) is by definition $y = (Dt)^{1/2}$; where $X = 0.5\,V_{max}t$.

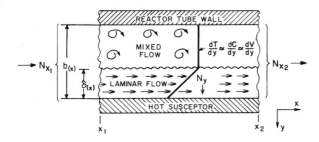

FIGURE 7.6   A schematic of the diffusion boundary layer model for a horizontal epitaxial reactor

Reynolds number is equal to $(\rho VL/u)$ or $(\rho Vx/u)$. h is the mass transport coefficient; L is the total length of the susceptor; D_1 is the diffusion coefficient of chemical species in the host carrier gas; x is the local position on the susceptor as measured from the leading edge; V is the total gas flow velocity in the direction of increasing $x;$ u is the average coefficient of viscosity for the total gas flow; ρ is the average density for the total gas flow; and D_2 is the average diffusion coefficient for total gas flow. The ratio u/ρ is the relationshiship for the kinematic viscosity, ν, of the gas. Thus, Equation (7.6) becomes

$$\frac{hx}{D_1} = 0.322 \left(\frac{\nu}{D_2}\right)^{1/3} \left(\frac{V}{x\nu}\right)^{1/2} \quad (7.6a)$$

Using the approximate relationship $D_2 = (6/5)\nu$, and solving Equation (7.6a) for h, we have

$$h = 0.332(0.94)D_1 \left(\frac{V}{x\nu}\right)^{1/2}. \quad (7.7)$$

The silicon epitaxial growth rate equation used by Grove and Andrews is

$$G = \frac{C}{N} hY, \quad (7.8)$$

where C is total number of molecules per cubic centimeter in the gas; N is the number of silicon atoms in a cubic centimeter of solid silicon (5×10^{22}); Y is mole fraction of silicon-bearing chemical species in the total flow of gas; h is the mass transport coefficient from the flowing gas to the susceptor; and G is the growth rate of silicon (in cm/s. From the combination of Equation (7.7) and (7.8), the growth rate can be written as

$$G = \frac{C}{N} Y 0.332(0.94)D_1 \left(\frac{V}{x\nu}\right)^{1/2}. \quad (7.9)$$

From $Pv = nRT$, $C = 6.02\times10^{23}P/RT$; $R = 82.06$ $(cm^3)(atmosphere)/(g\text{-}mole)(K)$. The mole fraction Y can be written directly in terms of partial pressure P. Thus,

$G_{(x)}$ (in cm/s of deposited silicon)
$$= \left(\frac{(6.02\times10^{23})(0.332)(0.94)D_1P_{(x)}}{5\times10^{22}\, RT}\right)\left(\frac{V}{x\nu}\right)^{1/2}. \quad (7.10)$$

It should be noted that the term (0.94) is approximate. For many applications to systems involving gases, $(u/\rho D_2)^{1/3} \approx 1$.

Equation (7.10) can be compared directly with a similar growth rate equation that can be developed from the previous diffusion theory.

From Figure 7.3, the arrival rate of silicon-bearing chemical species to the lower wall from the flowing gas was given by Equation (7.2). However, in Equation (7.2) P should be replaced by $\frac{1}{2}P$ because in laminar flow the transport to the lower wall is by diffusion from stream to stream. That is, half of the concentration of chemical-bearing species belongs to the upper wall and half belongs to the lower wall. This statement can be further justified by considering practical applications of laminar-flow theory. Air can be forced through nominal-sized ducts at moderate or room temperatures with laminar-flow velocities of many hundreds of feet per minute. For example, if V_x = 1000 ft/min ($\approx0.5\times10^3$ cm/s), assuming that most molecular species diffuse in air with room-temperature diffusion coefficients of $D \approx0.1$ cm^2/s, then the ratio of the distance a molecule travels in the x direction to the distance it diffuses in the y direction (in unit time) is

$$\frac{x}{y} = \frac{Vt}{(Dt)^{1/2}} = \frac{0.5\times10^3}{(0.1)^{1/2}} \approx 10^3.$$

Hence,

$$N_y = \frac{DP}{2RT\delta_E} \left[\frac{g\text{-mole}}{s\text{-cm}^2}\right]. \quad (7.2a)$$

Obtaining δ_E from Equation (7.4),

$$N_y = \frac{(0.3228)DP}{RT} \left(\frac{V}{x\nu}\right)^{1/2} \left[\frac{g\text{-mole}}{s\text{-cm}^2}\right].$$

Converting from g-mole/s-cm^2 to cm/s of deposited silicon, one can write

$G_{(x)}$ (in cm/s of deposited silicon)
$$= \left(\frac{(6.02\times10^{23})(0.323)DP_{(x)}}{(5\times10^{22})RT}\right)\left(\frac{V}{x\nu}\right)^{1/2}. \quad (7.11)$$

For all practical purposes, Equations (7.10) and (7.11) are identical, and it can be concluded that δ_E is the diffusion boundary layer thickness for laminar flow. It has been shown that the theoretical coefficients are derived from the physical distribution of the chemical species in the host carrier gas and their relationship to the laminar parabolic velocity profile.

The diffusion theory can now be applied to the physical model, shown in Figure 7.6, necessary to describe the mass transport for a system containing gases flowing with characteristics of laminar and mixed flow concurrently. Only two obvious assumptions are needed. First, it must be assumed that the concentration of chemical reactants within the space $(b\text{-}\delta)$ is uniform because of the mixing of the bulk gas flow in that region; second, it is assumed that the molecular diffusion transport to the hot susceptor is through a laminar boundary layer whose velocity profile has the average velocity characteristics of both laminar and turbulent flow. The average velocity for laminar flow has already been given as 0.5 V_{max}. The average velocity of the mixed flow is difficult to predict, but it is probably similar to the turbulent-flow velocity; the average velocity for turbulent flow is 0.83

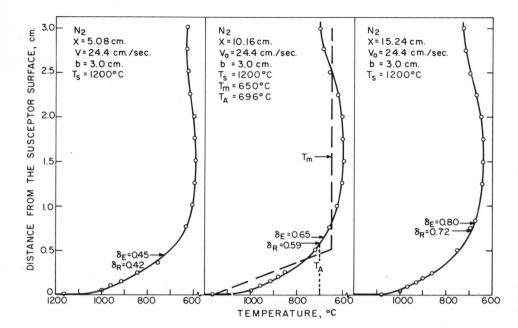

FIGURE 7.8 Experimental temperature profiles in N_2 for a resistance-heated cold-wall reactor. The profiles were taken at three points on the susceptor; $x = 5.08$, 10.16, and 15.24 cm. The broken line represents a linear approximation of the experimental temperature profile.

V_{max} (Perry, 1963). Hence, the average velocity for laminar and turbulent flow is

$$\tfrac{1}{2}(0.5\ V_{max} + 0.83\ V_{max}) = 0.665\ V_{max}.$$

The new equation for the desired diffusion boundary layer, δ_E would be

$$\delta_E = \left(\frac{Dx}{0.666\ V_{max}}\right)^{1/2} = 1.226\left(\frac{Dx}{V_{max}}\right)^{1/2}. \quad (7.12)$$

Ban and Gilbert (1974) have been able to confirm the general validity of Equation (7.12) in a resistance-heated horizontal reactor. Their relevant experiments are briefly discussed.

The experimental reactor used by Ban and Gilbert consisted of a rectangular quartz tube fitted with a precision translation mechanism for positioning a small-diameter thermocouple. The thickness of the resistance-heated slab was varied to effect a suitably flat temperature profile over its 15 cm length. A water-cooled copper coil was wrapped around the quartz tube to simulate the cooling and reflective effects of an RF induction heating coil. Gas temperature profiles were measured from the hot slab surface to the inner surface of the reactor tube wall. The distance between the hot slab surface and the inner reactor tube wall was 3.0 cm. Some of the measured temperature profiles are shown in Figures 7.8, 7.9, and 7.10 for nitrogen and helium. The experimental conditions are noted on the figures. The meanings of x, b, and V have been defined; the meanings of T_s, T_m, and T_A are given in Equation (7.27) and in Figure 7.14. The arrows on the figures point to the values of δ_E calculated by the use of Equation (7.12) and the mathematical relation-

ship $\alpha = (5/4)D$ (Foust, 1960), where α is the thermal diffusity and D is the mass diffusion coefficient.

Because this section deals with the theory, detailed examples of numerical calculations of the temperature curves will be deferred to the next section of this work. In order to proceed with the development of the theory, it is necessary to make a few qualitative obser-

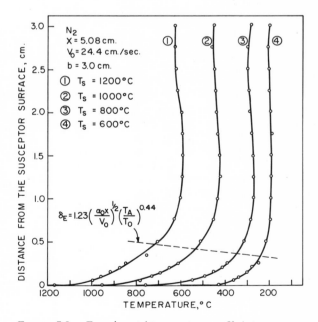

FIGURE 7.9   Experimental temperature profiles at susceptor temperatures of 600, 800, 1000, and 1200°C. The broken line is the theoretical temperature dependence of the diffusion boundary layer thickness from Equation (7.12).

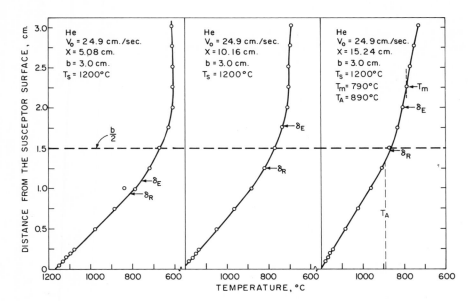

FIGURE 7.10
Experimental temperature
profiles in He for a
resistance-heated cold-wall
reactor. The profiles were
taken at three points on the
susceptor; $x = 5.08$, 10.16,
and 15.24 cm.

vations from Figures 7.8, 7.9, and 7.10. Aside from the obvious disagreement between theory and experiment for the one helium temperature curve at $x = 5$ cm, reasonable agreement is observed between values calculated [using equation (7.12)] and the experimental data points (Figures 7.8, 7.9, and 7.10). Although Equation (7.12) seems adequate to describe the average diffusion distance for situations where $\delta_E << b$, where b is the reactor height between the top of the susceptor and tube wall, it is not sufficient to describe the diffusion boundary layer for practical silicon epitaxial systems. Practical silicon epitaxial growth processes are characterized by the use of hydrogen carrier gas, high average reactor temperatures, and relatively narrow channel heights, b. With the use of Equation (7.12) for the conventional epitaxial process, δ_E would become larger than the reactor channel height for modest susceptor lengths. This does not pose a dilemma, but it does make the theory a bit more complicated.

From laminar-flow theory, the displacement boundary layer, δ^*, is defined (Schlichting, 1955) as the distance the free streams in the bulk gas flow are displaced because of the formation of the velocity boundary layer. The thickness of δ^* from the literature is given as $\delta^* \simeq 1/3\ \delta$; δ is the laminar velocity boundary layer. $\delta = 5.0\ (\nu x/V_\infty)^{1/2}$ (Foust et al., 1960) and $\delta^* \simeq 1.667\ (\nu x/V_\infty)^{1/2}$. It has been shown from the previous discussion of the laminar flow diffusion model (figure 7.7) that $\delta_E = 1.549\ (\nu x/V_{max})^{1/2}$. Therefore, $\delta_E \simeq \delta^*$ for laminar flow.

The bulk gas flow is displaced by a distance at least equal to δ_E. As δ_E increases, the channel height, b, for the bulk gas flow decreases to a dimension equal to $(b-\delta_E)$. This has the effect of increasing the bulk gas flow velocity which tends to decrease δ_E, and so on until some equilibrium value of δ_E is reached at each position of x. The physical model for this effect is shown in Figure 7.11. An equation for the real value of $\delta_{R(x)}$ can be derived in the following manner. From the continuity equation, the total flow through planes x_1, x_2, and x_3, is $Q = a_1 V_1 = a_2 V_2 = a_3 V_3$. The change in velocity through plane x_2 caused by the displacement diffusion layer δ_1 is dV. The change in δ_1 through plane x_2 is then $d\delta$.

$$\frac{dv}{dx} = V_2 - V_1 \qquad V_2 = \left(\frac{b}{b - \delta_1}\right) V_1.$$

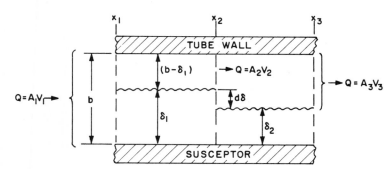

FIGURE 7.11
A schematic representation for the
mathematical derivation of the diffusion
boundary layer thickness with the reactor
wall as a physical constraint

By substitution and rearrangement,

$$\frac{dv}{dx} = V_1 \left(\frac{\delta_1}{b - \delta_1} \right).$$

From Figure 7.11, $\delta_2 < \delta_1$. Therefore, $d\delta/dx$ is negative, and

$$-\frac{d\delta}{dx} = (\delta_2 - \delta_1)$$

$$\delta_1 = 1.23 \left(\frac{Dx}{V} \right)^{1/2} \qquad \delta_2 = 1.23 \left[\frac{Dx}{(b/b - \delta_1) \, v_1} \right]^{1/2}.$$

By substitution and rearrangement, then

$$-\frac{d\delta}{dx} = \delta_1 \left[\left(\frac{b - \delta_1}{b} \right)^{1/2} - 1 \right].$$

By combining $(d\delta/dx) \cdot (dx/dv) \cdot (b/b)$,

$$-\frac{d\delta}{d} = \frac{b}{V} \left[\left(1 - \frac{\delta}{b} \right)^{3/2} - \left(1 - \frac{\delta}{b} \right) \right].$$

The term $[1 - (\delta/b)]^{3/2}$ can be expanded in the series

$$(1-Z)^n = 1 - nZ + \frac{n(n-1)Z^2}{2!}$$
$$- \frac{n(n-1)(n-2)Z^3}{3!} + \frac{n(n-1)(n-2)(n-3)Z^4}{4!}$$
$$- \frac{n(n-1)(n-2)(n-3)(n-4)Z^5}{5!} + \cdots .$$

Then by similar expansion, it can be shown that

$$d\delta = -\frac{b}{V} \left[-\frac{1}{2} \left(\frac{\delta}{b} \right) + \frac{3}{8} \left(\frac{\delta}{b} \right)^2 + \frac{3}{48} \left(\frac{\delta}{b} \right)^3 \right.$$
$$\left. + \frac{3}{128} \left(\frac{\delta}{b} \right)^4 + \frac{3}{256} \left(\frac{\delta}{b} \right)^5 \right] dV.$$

With the substitution, $\delta = 1.23 (Dx/V)^{1/2}$, each term can be integrated separately with respect to V. The limits of integration can be written with the aid of Figure 7.11. At x_1, when $V = V_1$, $\delta = \delta_1$. As δ_1 approaches the dimension b, V_2 approaches ∞. The integrated result is Equation (7.13).

$$\delta_{R(x)} = 1.226 \left(\frac{Dx}{V} \right)^{1/2} - \frac{0.564}{b} \left(\frac{Dx}{V} \right)$$
$$- \frac{0.0768}{b^2} \left(\frac{Dx}{V} \right)^{3/2} - \frac{0.0265}{b^3} \left(\frac{Dx}{V} \right)^2$$
$$- \frac{0.01298}{b^4} \left(\frac{Dx}{V} \right)^{5/2}, \quad (7.13)$$

where $\delta_{R(x)}$ is the real diffusion boundary layer thickness.

The characteristics of Equation (7.13) can be seen with the aid of Figure 7.12. Inspection of Equation (7.13) reveals that the first term is positive and is equal to δ_E. The remaining terms of the infinite series are negative. For practical epitaxial systems, five terms are sufficient to demonstrate the convergence of Equation (7.13). Because the boundary conditions for

the integration, from Figure 7.11, were from $\delta = \delta_1$ to $\delta = b$, δ_R should converge to approximately the value of $b/2$ when δ_E is equal to b.

$$\delta_E = 1.226 \left(\frac{Dx}{V} \right)^{1/2} = b \qquad \frac{Dx}{V} = \frac{b^2}{1.503}$$

with $b = 2$; $Dx/V = 2.6612$. Hence, from Equation (7.13), $\delta_R = 2.0 - 0.7505 - 0.0834 - 0.0234 - 0.0116 = 1.13$.

It also should be noted from Figure 7.12 that $\delta_R \simeq b/2$ when $(b/2) \leq (Dx/V) \leq 3b$. Hence, for the practical epitaxial systems that will be considered for this work, the condition can be written that $0 \leq \delta_R \leq b/2$. The experimental temperature profiles shown in Figure 7.10 tend to support this result. In reference to the apparent disagreement in Figure 7.10 between the calculated values of δ_R and the experimental temperature curves near the leading edge of the hot slab, it should be considered that the He gas is receiving heat from the hot quartz reactor tube upstream from the hot slab. Therefore, $x = 0$ for the formation of the boundary layer on the hot slab does not physically coincide exactly with the leading edge of the hot slab. It is also probable that the partially heated gas upstream from the hot slab is approaching the leading edge with a nearly laminar parabolic velocity profile. This would produce a thicker diffusion layer, which is predicted from Equation (7.3), near the leading edge of the hot slab. This is true for the nitrogen curves in Figures 7.8 and 7.9 as well, but the heavy N_2 molecules, which diffuse relatively slowly, are not affected as much as are the light He molecules, which diffuse more rapidly. Typically, in the epitaxial processes that are being considered here, H_2 is used as a carrier gas and the reactor dimensions are less than $b = 3.0$ cm. Under these conditions, it is evident from the previous discussion that δ_R reaches an equilibrium value of approximately $b/2$ very near the leading edge of a hot

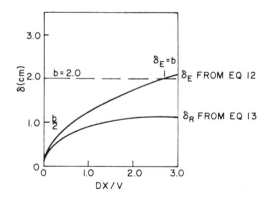

FIGURE 7.12   The effective [δ_E, from Equation (7.12)] and real [δ_R from Equation (7.13)] boundary layer thickness as a function of the diffusion parameter Dx/V for a reactor channel height, b, of 2 cm.

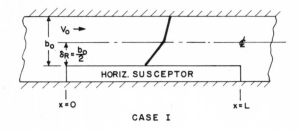

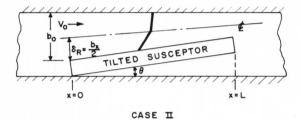

FIGURE 7.13   A schematic representation of a horizontal (Case 1) and tilted (Case 2) susceptor

susceptor. Thus, the flow can be considered fully developed with $\delta_R \simeq b/2$.

Suitable mass transport equations can now be developed to predict the epitaxial growth rate of silicon on a hot susceptor in a horizontal reactor. Growth rate equations for the two cases shown in Figure 7.13 will be developed. Case 1 applies to the situation, in Figure 7.13, in which the susceptor is placed in a flat, horizontal position in the reactor tube and the gas flow channel height, b_0, is constant with respect to position x. Case 2 applies to the situation in which the susceptor is tilted slightly from the horizontal plane so that $b_{(x)}$ decreases as position x increases from $x = 0$ to $x = L$. For simplicity, it will be assumed that $\delta_{R(x)} = b_{(x)}/2$ and that T_A is constant for the entire length of the susceptor. This is not exactly true near the leading edge of the susceptor, but the final calculations will show that it is justified for the purpose of demonstrating this model.

Case 1: Horizontal susceptor. Referring to Figures 7.6 and 7.13, and restating Equation (7.2), the following equations can be written for the untilted susceptor:

$$N_y = \frac{DP}{RT\delta} \qquad (7.14)$$

is the arrival rate (in gm-mole/s-cm²) of reactive chemical species to the susceptor by diffusion through δ.

$$N_{x_1} = \frac{QP_{x_1}}{RT} \qquad (7.15)$$

is the input of chemical reactive species through plane at x_1.

$$N_{x_2} = \frac{QP_{x_2}}{RT} \qquad (7.16)$$

is the output of chemical reactive species through plane at x_2.

$$Q = aV = (b)(\text{unit width})(V) = bV. \qquad (7.17)$$

For epitaxial conditions, the carrier gas usually contains less than a few percent of the reactive chemical species. Therefore, the total volume flow, Q, does not change through planes at x_1 and x_2. Because all the change in N_x, as the gas moves from x_1 to x_2, is caused by the diffusion in the y direction for deposition on the hot susceptor, the equation can be written

$$\frac{dN_x}{dx} = N_y \qquad (7.18)$$

$dN_x = -(N_{x_2} - N_{x_1})$, and since $N_{x_2} < N_{x_1}$, dN_x is negative. Combining Equations (7.15), (7.16), and (7.17),

$$dN_x = -\left(\frac{bVP_{x_2}}{RT} - \frac{bVP_{x_1}}{RT}\right) = -\frac{bV}{RT}(P_{x_2} - P_{x_1}), \qquad (7.19)$$

$$dP = (P_{x_2} - P_{x_1}), \qquad (7.20)$$

$$dN_x = -\frac{bV}{RT}dP.$$

Combining Equations (7.14), (7.18), and (7.20)

$$-\frac{bV}{RT}\frac{dP}{dx} = \frac{DP}{\delta RT}, \qquad (7.20a)$$

$$\int_{P_0}^{P_{(x)}} \frac{dP}{P} = -\int_0^x \frac{D}{bV\delta}dx,$$

$$\ln\frac{P_{(x)}}{P_0} = -\frac{Dx}{bV\delta},$$

$$P_{(x)} = P_0 e^{-(Dx/bV\delta)}. \qquad (7.21)$$

With the relationship $\delta = b_0/2$, Equation (7.2) can be combined with Equation (7.21) and the resulting equation can be converted from (gm-mole)/s to microns of solid silicon per minute. Then the growth rate of silicon on a flat susceptor with $0 \leq x \leq L$ can be expressed by Equation (7.22):

$$G_{(x)} = \left(\frac{(6.0 \times 10^5)(6.02 \times 10^{23})D_T P_0}{(5.0 \times 10^{22})(82.06)T_A b_0/2}\right)$$

$$\times \exp\left[\frac{-D_T x}{b_0 V_T (b_0/2)}\right]$$

$$G_{(x)} = \left(\frac{(1.76 \times 10^5)D_T P_0}{T_A b_0}\right)\exp\left(\frac{-2D_T x}{V_T b_0^2}\right) \qquad (7.22)$$

$$D_T = D_0\left(\frac{T_A}{T_0}\right)^{1.88} \qquad V_T = V_0\left(\frac{T_A}{T_0}\right).$$

Case 2: Tilted susceptor. The same basic equations, theory, and figures that were used to derive $G_{(x)}$ for Case 1 can be used for Case 2. For a tilted susceptor, b, V, and δ are a function of position, x. Therefore, the development of the equations for Case 2 can start with Equation (7.20a) from Case 1. Equation (7.20a) can be written

$$-\frac{dP}{dx} = \frac{D_T P}{b_{(x)} V_{T(x)} \delta_{(x)}},$$

where

$$b_{(x)} = b_0 - x \sin \theta, \qquad V_{T(x)} = V_T \frac{b_0}{b_{(x)}}$$

$$\delta_{(x)} = \frac{b_{(x)}}{2}.$$

Then,

$$\int_{P_0}^{P_{(x)}} \frac{dP}{P} = \frac{2D_T}{(\sin \theta) b_0 V_T} \int_0^x \frac{d(-x)}{(b_0/\sin \theta - x)}.$$

From integral tables,

$$\int \frac{dx}{(a + bx)} = \frac{1}{b} \ln (a + bx).$$

Finally,

$$P_{(x)} =$$

$$P_0 \exp\left\{\frac{2D_T}{(\sin \theta) b_0 V_T} \left[\ln\left(\frac{b_0}{\sin \theta} - x\right) - \ln\left(\frac{b_0}{\sin \theta}\right)\right]\right\}$$

$$(7.23)$$

and

$$G_{(x)} = \frac{(1.76 \times 10^5) D_T P_0}{T_A (b_0 - x \sin \theta)}$$

$$\times \exp\left\{\frac{2D_T}{b_0 V_T \sin \theta} \left[\ln\left(\frac{b_0}{\sin \theta} - x\right) - \ln \frac{b_0}{\sin \theta}\right]\right\},$$

$$(7.24)$$

where

$$D_T = D_0 \left(\frac{T_A}{T_0}\right)^{1.88} \qquad V_T = V_0 \left(\frac{T_A}{T_0}\right).$$

7.5. Calculations and results

Calculation of δ_E for Figures 7.8, 7.9, and 7.10

The temperature dependence of diffusion coefficients at elevated temperatures is known to obey the relationship

$$D_T = D_0 \left(\frac{T_A}{T_0}\right)^m \ldots \qquad (7.25)$$

where the value of m is $1.75 \leqslant m \leqslant 2$. Because of the high temperatures and temperature gradients present in epitaxial systems, and for the sake of consistency, the average value of $m = 1.88$ will be used. First, the diffusion coefficients will be calculated at room temperature because the mathematical formulas that exist are known to give excellent agreement with room-temperature experimental measurements; then the high-temperature values will be calculated from Equation (7.25) with $m = 1.88$. The self-diffusion coeffi-

cient for nitrogen for Figure 7.8 can be calculated from the following equation (Foust, 1960):

$$D_0 = \frac{2.628 \times 10^{-19} (T_0^3/M)^{1/2}}{P\sigma^2}. \qquad (7.26)$$

From tables (Foust, 1960), $\sigma = 3.749 \times 10^{-8}$ for N_2; T_0 is the room temperature (300 K); M is the molecular weight of N_2 (28); and P is the pressure (1.0 atmosphere).

$$D_0 = \frac{2.628 \times 10^{-19} [(300)^3/28]^{1/2}}{(1)(3.749 \times 10^{-8})^2} = 0.184 \text{ cm}^2/\text{s}.$$

The average reactor temperature, T_A, can be calculated with the aid of Figure 7.8. A practical reason for using T_A in the mass-transport calculations is discussed in the summary of this article. An approximation of the reactor temperature curve at $x = 10.16$ cm can be represented by the broken lines. Then T_A can be calculated from the relationship

$$T_A = \frac{\delta [T_s - (T_s - T_m/2)] + (b - \delta) T_m}{b}, \qquad (7.27)$$

where T_s is the susceptor temperature; T_m is the mean gas temperature in the region $(b - \delta)$; δ is the arbitrary boundary layer thickness at discontinuity in approximate temperature curve; and b is the reactor height between susceptor and upper reactor tube wall. At $T_s = 1200°C$, $T_m = 650°C$, and $x = 10.16$ cm (Figure 7.8), $T_A = (0.5 \text{ cm})(925°C) + (2.5 \text{ cm})(650°C)/3.0 \text{ cm}.$

$$T_A = 696°C \simeq 1000K$$

From Equation (7.25)

$$D_T = 0.184 \left(\frac{1000}{300}\right)^{1.88} = 1.769 \text{ cm}^2/\text{s}$$

The bulk gas velocity (V_0) over the cold susceptor was 24.4 cm/s. The velocity (V_T) over the hot susceptor caused by the volume expansion of the hot gas is $V_T = 24.4 (1000/300) = 81.33$ cm/s.

From Equation (7.12),

$$\delta_E = 1.226 \sqrt{\frac{D_T x}{V_T}}$$

for mass transport. Thus,

$$\delta_E = 1.226 \left(\frac{\alpha_T x}{V_T}\right)^{1/2}$$

for heat transport. For gases $D = (6/5) \nu$, $\alpha = {}^3\!/_2 \nu$.

$$\alpha_T = \frac{5}{4} D_T = \frac{5}{4} (1.769) \text{ cm}^2/\text{s} = 2.21 \text{ cm}^2/\text{s},$$

$$\delta_E = 1.226 \left(\frac{(2.211)(10.16)}{(81.33)}\right)^{1/2} = 0.645 \text{ cm}.$$

δ_E has been calculated in a similar fashion for the curves in Figures 7.8 and 7.10. The temperature dependence of Equation (7.12) is shown on Figure 7.9. If

$$\delta_E = 1.226 \sqrt{\left(\frac{\alpha_T x}{V_T}\right)^{1/2}} \qquad \alpha_T = \alpha_0 \left(\frac{T_A}{T_0}\right)^{1.88}$$

$$V_T = V_0 \left(\frac{T_a}{T_0}\right)$$

the combination of these equations yields

$$\delta_E = 1.226 \left(\frac{\alpha_0 x}{V_0}\right)^{1/2} \left(\frac{T_A}{T_0}\right)^{0.44}.$$

Calculation of δ_R for Figures 7.8 and 7.10

For mass transport,

$$\delta_R = 1.226 \left(\frac{D_T x}{V_T}\right)^{1/2} - \frac{0.564}{b} \left(\frac{D_T x}{V_T}\right)$$
$$- \frac{0.0768}{b^2} \left(\frac{D_T x}{V_T}\right)^{3/2}.$$

For heat transport,

$$\delta_R = 1.226 \left(\frac{\alpha_T x}{V_T}\right)^{1/2} - \frac{0.564}{b} \left(\frac{\alpha_T x}{V_T}\right)$$
$$- \frac{0.0768}{b^2} \left(\frac{\alpha_T x}{V_T}\right)^{3/2},$$

where $\alpha_T = (5/4) D_T$. If $x = 10.16$ cm, $b = 3.0$ cm, $V_T = 81.33$ cm/s, $D_T = 1.769$ cm²/s, and $\alpha_T = 2.211$ cm²/s,

$$\left(\frac{\alpha_T x}{V_T}\right) = \frac{(2.211)(10.16)}{(81.33)} = 0.2762$$

$$\delta_R = 1.226(0.2762)^{1/2} - \frac{0.564}{3}(0.2762)$$
$$- \frac{0.0768}{9}(0.2762)^{3/2}$$

$$\delta_R = 0.6444 - 0.0519 - 0.0012 = 0.59 \text{ cm}.$$

Example calculations for epitaxial growth rates for Figures 7.15 and 7.16

The equations that have been developed can now be applied to calculate epitaxial growth rates from the diffusion model that has been described in the theory. The equations will be applied to Eversteyn's experimental silicon epitaxial growth conditions and the results will be compared to his experimental growth curves (Eversteyn et al., 1970). The reader is referred to Eversteyn's original work for a detailed description of his theoretical and experimental work as it pertains to his horizontal system. For the purpose of this work, only a brief description of his epitaxial system and the necessary process variables will be described as they are needed.

Eversteyn's reactor was a conventional RF-heated horizontal system with a water-cooled quartz reactor tube and a 30-cm-long susceptor. The reactor was used to epitaxially deposit silicon on suitably prepared single crystal silicon substrates from a mixture of silane (SiH_4) and a carrier gas of hydrogen. Deposition experiments were carried out on flat and tilted susceptors with various carrier gas velocities and input concentrations of silane. The reactor height, b_0, above the flat and tilted susceptor was 2.0 cm.

Eversteyn measured the approximate temperature in the convective mixed-flow region between the extremity of the diffusion layer and the reactor tube wall (Eversteyn and Peek, 1970). $T_m = 700K$ is the measured temperature in the mixed region. $T_s = 1350K$ is the susceptor temperature measured with an optical pyrometer and corrected for the emissivity of silicon. Then the average reactor temperature can be calculated with the use of Figure 7.14 and Equation (7.27).

If $\delta_R = b/2$, and $b = 2.0$ cm, then $\delta_R = 1.0$ cm. From Equation (7.27),

$$T_A = \frac{(1.0 \text{ cm})(1025K) + (1.0 \text{ cm})(700K)}{2.0 \text{ cm}} = 863K.$$

Before the mass transport of SiH_4 to the susceptor can be calculated, the diffusion coefficient of silane in hydrogen must be known. As stated by Eversteyn, on the basis of molecular weight and size, silane should diffuse in hydrogen at room temperature at a rate (D_0) of 0.6 cm²/s. A search of the literature indicates that nearly all comparable molecular species diffuse in hydrogen at comparable rates. The diffusion coefficient of silane in hydrogen can be calculated from a well known equation for the simultaneous equal molar diffusion of two gases (Foust et al., 1960).

$$D_{ab} = \frac{2.628 \times 10^{-19}[T_0^3(1/2)(1/M_a + 1/M_b)]^{1/2}}{P \sigma_{ab}^2 \Omega_2},$$

where P is the reactor pressure (1 atmosphere); M_a is the molecular weight of H_2 (2); M_b is the molecular weight of SiH_4 (32); T_c is the critical temperature of SiH_4 ($-4.0°C$ or 269K); P_c is the critical pressure of

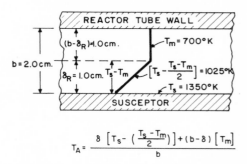

$$T_A = \frac{\delta \left[T_s - \left(\frac{T_s - T_m}{2} \right) \right] + (b - \delta)\left[T_m \right]}{b}$$

FIGURE 7.14   A schematic temperature diagram for data. After Eversteyn et al. (1970)

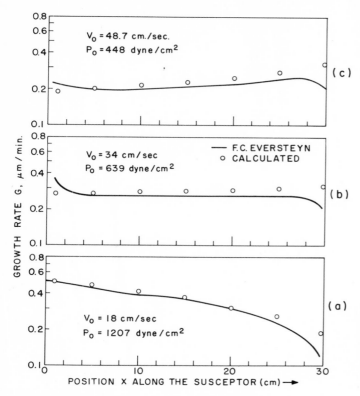

(c)

(b)

(a)

FIGURE 7.15
Growth rate as a function of the position on a tilted (2.9°) susceptor calculated from Equation (7.24) for various gas-flow velocities and partial pressures of silane. The calculated results are compared to the experimental data of Eversteyn et al. (1970)

SiH₄ (47.8 atmosphere); σ_a is the collision diameter of H₂; σ_b is the collision diameter of SiH₄; σ_{ab}, Ω_2, T^*_{ab} are Lennard–Jones constants; and ϵ_a/K, ϵ_b/K are constants for the Lennard–Jones potential.

$\sigma_a = 2.915 \times 10^{-8}$, from tables (Foust et al., 1960),

$$\sigma_b = 2.44 \left(\frac{T_c}{P_c}\right)^{1/3} \times 10^{-8} = 2.44 \left(\frac{269}{47.8}\right)^{1/3} \times 10^{-8}$$
$$= 4.34 \times 10^{-8},$$

$$\sigma_{ab} = \frac{1}{2}(\sigma_a + \sigma_b) = \frac{10^{-8}}{2}(2.915 + 4.34)$$
$$= 3.628 \times 10^{-8},$$

$$\frac{\epsilon_{ab}}{K} = \left[\left(\frac{\epsilon_a}{K}\right)\left(\frac{\epsilon_b}{K}\right)\right]^{1/2},$$

$\frac{\epsilon_a}{K} = 38$, from tables (Foust et al., 1960),

$\frac{\epsilon_b}{K} = 0.77\, T_c = (0.77)(269) = 207.13$,

$\frac{\epsilon_{ab}}{K} = (38 \times 207.13)^{1/2} = 88.72$,

$T^* = \frac{K}{\epsilon} T_0 = \frac{300}{88.72} = 3.38$,

$\Omega_2 = 0.92$, from tables (Foust et al., 1960),

$$D_{ab} = \frac{2.628 \times 10^{-19}\,[(300)^3(1/2)(1/2 + 1/32)]^{1/2}}{(1)(3.628 \times 10^{-8})^2(0.92)}$$
$$= 0.5812\ \text{cm}^2/\text{s},$$

$D_0 = 0.6$ cm²/s will be used for the room-temperature diffusion coefficient of SiH₄ in H₂. From Equation (7.25)

$$D_T = 0.6\ \text{cm}^2/\text{s} \left(\frac{863}{300}\right)^{1.88} = 4.4\ \text{cm}^2/\text{s}.$$

The theoretical values for the epitaxial growth rate curves shown in Figure 7.15 for a tilted susceptor have been calculated with the use of Equation (7.24) and the conditions in Table 7.1

$G_{(x)}$ (in μm/min)

$$= \frac{1.76 \times 10^5 D_T P_0}{T_A(b_0 - x\sin\theta)} \exp\left\{\frac{2D_T}{b_0 V_T \sin\theta}\right.$$
$$\left. \times \left[\ln\left(\frac{b_0}{\sin\theta} - x\right) - \ln\frac{b_0}{\sin\theta}\right]\right\}$$

TABLE 7.1   Physical Parameters Used to Calculate Growth Rate from Equation (7.24)

Figure	θ	T_A (K)	b_0 (cm)	$D_T(SiH_4)$ (cm²/s)	V_0 (cm/s)	V_T (cm/s)	$P_0(SiH_4)$ (dyne/cm²)	$P_0(SiH_4)$ (atmosphere)
15a	2.9°	863	2.0	4.4	18.0	51.8	1207	1.14×10^{-3}
15b	2.9°	863	2.0	4.4	34.0	97.8	639	6.04×10^{-4}
15c	2.9°	863	2.0	4.4	48.7	140.1	448	4.23×10^{-4}

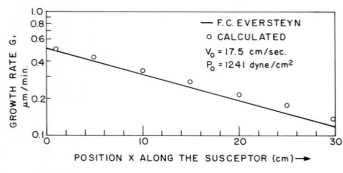

FIGURE 7.16
Growth rate as a function of the position on a horizontal susceptor. The calculated results are compared to the experimental data of Eversteyn et al. (1970)

The theoretical values for the epitaxial growth curve shown in Figure 7.16 for a flat susceptor have been calculated with the use of Equation (7.22) and the conditions that follow it:

$$G_{(x)} \text{ (in } \mu\text{m/min)}$$
$$= \frac{(1.76 \times 10^5)D_T P_0}{T_A b_0} \exp\left(\frac{-2D_T x}{V_T b_0^2}\right),$$

$T_A = 863\text{K}; \ b_0 = 2.0 \text{ cm}; \ D_T = 4.4 \text{ cm}^2/\text{s}; \ V_0 = 17.5$ cm/s; $V_T = 50.5$ cm/s; $P_0 = 1.17 \times 10^{-3}$ atmosphere.

Applications to reactor designs

For the purposes of designing an epitaxial reactor, it is useful to extract from the previous equations two quantities: the critical angle for uniform deposition and a figure of merit for the efficiency of the reactor. To derive the first of these quantities we start by rewriting Equation (7.24) in terms of D_0, T_0, and V_0, to obtain, after simplification,

$$G_{(x)} = \frac{(1.76 \times 10^5)D_0 P_0 T_A^{0.88}}{(b_0 - x \sin \theta)T_0^{0.88}}\left(\frac{b_0 - x \sin \theta}{b_0}\right)^M$$

where

$$M = \frac{2D_0}{b_0 V_0 \sin \theta}\left(\frac{T_A}{T_0}\right)^{0.88}.$$

If M is set equal to unity, then the two terms containing $(b_0 - x \sin \theta)$ will cancel. This means that $G_{(x)}$ is independent of x and that uniform deposits can be obtained along the entire length of the susceptor.

The value for θ_c, the critical angle condition for obtaining uniform growth along the full susceptor length, can then be found as

$$\sin \theta_c = \frac{2D_0}{b_0 V_0}\left(\frac{T_A}{T_0}\right)^{0.88}. \tag{7.28}$$

This equation will apply as long as the approximation that $\delta_R \simeq b_s/2$. Note that all of the constants from

Equations (7.22) and (7.24) that were specific to the growth of silicon have cancelled out. A plot of θ_c vs. T_A with V_0 as a parameter and $D_0/b_0 = 0.3$ is given in Figure 7.17.

One way to define a figure of merit for a reactor is to take the ratio of the total amount of material deposited on the susceptor to the total amount of reactant input to the reactor, both quantities being taken over the same time period. The total amount of material deposited requires the integration of $G_{(x)}$. If $G_{(x)}$ is a constant, as is true when $\theta = \theta_c$, then the expression immediately simplifies and one can write a figure of merit, M_c, for operation of a reactor at the critical angle as shown in the inset below.

The product, $V_0 \, b_0 W$, in the denominator is the flow rate through the system. When multiplied by P_0 it is the flow rate of the reactant species. The numerator is the total volume of solid deposited. Defined in this way M_c is not an efficiency, as it does not account for molar ratios in the reactant and deposit.

The length of a susceptor, L_c, which will physically fit in a given reactor when the tilt angle is θ_c is

$$L_c = \frac{b_0}{\sin \theta_c}.$$

Substitution into Equation (7.28) yields

$$L_c = \frac{b_0^2 V_0}{2D_0}\frac{T_0^{0.88}}{T_A}. \tag{7.29}$$

(The real susceptor must be somewhat shorter than this length; otherwise, it would touch the upper wall of the reactor.)

If we assume that the susceptor length is maximized, while maintaining uniform growth, we can solve for M_c as

$$M_c = \frac{G_{(0)}L_c}{V_0 b_0} = \frac{(1.76 \times 10^5)D_0 P_0 T_A^{0.88} b_0^2 V_0 T_0^{0.88}}{(1.67 \times 10^2)b_0 T_0^{1.88}2D_0 T_A^{0.88}V_0 b_0 P_0}$$

$$M_c = \frac{1.06 \times 10^3}{2 T_0} = \text{Constant}.$$

$$M_c = \frac{\text{growth rate } (G) \times \text{susceptor length } (L) \times \text{susceptor width } (W)}{\text{gas velocity } (V_0) \times \text{reactor height } (b_0) \times \text{reactor width } (W) \times P_0}.$$

The factor 1.67×10^2 has been introduced to keep the units of the numerator and denominator the same.

The unexpected result of this analysis is that all reactors designed to produce uniform layers will have the same figure of merit if the susceptor length is maximized. Further, this result is also independent of the deposition temperature or the diffusion coefficient of the reactant. The interested reader can verify this by referring to Figure 7.18, which is a plot of growth rate vs. temperature with D_0/b_0 as a parameter, in conjunction with Figure 7.17. Figure 7.18 has been computed for an initial input pressure of 1×10^{-3} atmosphere.

Because these conclusions seem to contradict intuition, some discussion is in order. As a first point, we remind the reader that the assumption was made that the concentration of reactant at the growth interface is zero in order to derive Equation (7.2). This assumption is certainly justified for silane. Nevertheless, it is equivalent to assuming 100 percent reaction efficiency for *any* species which diffuses to the interface. This restrictive assumption can be relaxed somewhat by assuming that the partial pressure of reactant is not zero but is instead the equilibrium partial pressure obtained from the thermodynamic equilibrium constant for the particular reaction. The resulting growth rate equations are somewhat more complex and will not be treated here.

For the case of complete reaction, consider the consequences of doubling the value of b_0. From Equation (7.29) uniform deposition can now be obtained on a susceptor which is four times longer. However, the growth rate, which is inversely proportional to b_0 [Equation (7.22)] will decrease by a factor of two. The net effect is to double the total amount of silicon deposited in a unit time. Doubling of b_0 also requires that the total amount of gas introduced into the reactor be doubled if V_0 is to be kept constant. Because both the amount of silicon deposited and the amount of reactant introduced have been increased by the same factor, the efficiency is unchanged. Similar arguments can be made for the effect of changing D_0 or T_A while maximizing the susceptor length.

The conventional experimental approach is to operate a reactor with a fixed susceptor length. If the susceptor length is kept fixed but the value of b_0 doubled, then uniform growth will be obtained at the same angle as before only if the velocity is halved. This follows from Equation (7.28). The growth rate, as given by Equation (7.22), also decreases by a factor of two. The total volume of gas per unit time supplied to this reactor is unchanged because the product of $V_0 b_0$ has been kept constant. With half the growth rate for the same chemical input, the efficiency will fall by a factor of two. On the other hand, if T_A or (equivalently) D_0 is increased, uniform growth on a fixed-length, fixed-angle susceptor can be obtained by a proportion-

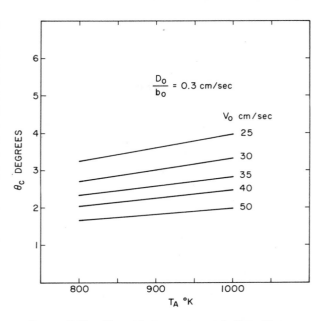

FIGURE 7.17   The critical susceptor angle for uniform silicon growth (θ_c) as a function of the average temperature in the boundary layer (T_A). The gas stream velocity (V_0) is taken as a parameter

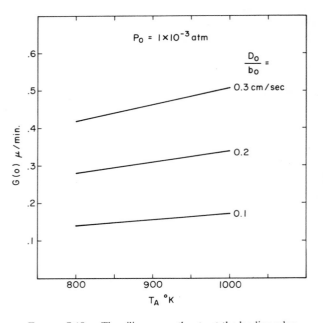

FIGURE 7.18   The silicon growth rate at the leading edge of the susceptor [$G_{(0)}$] as a function of the average boundary layer temperature (T_A). The ratio of the room temperature diffusion coefficient to the free height above the susceptor D_0/b_0 is taken as the parameter

al increase in V_0. The growth rate will increase, but because of the proportional increase in V_0 the efficiency will remain unchanged.

The reader should note that although the efficiency remains the same for all designs operating at θ_c, the total consumption of carrier gas increases directly with b_0. If, in the design of a reactor, an attempt is made to maximize the number of wafers that a given system will hold by making it very long (low values of θ_c) the cost and difficulty of supplying the large volumes of high-purity carrier gas that are needed can very quickly make the most efficient design impractical.

One additional point to be made in relation to reactor design involves the problem of doping uniformity. In order to obtain layers with the desired semiconducting properties, it is necessary to provide a suitable dopant. The dopant must reach the growth surface in a manner similar to that of the reactant. If the diffusion constant for the dopant is significantly different from that of the main reactant, it will be impossible to obtain uniform doping down the length of the reactor. If the dopant diffusion constant is smaller than that of the reactant the net concentration in the growing layer should increase down the length of the reactor and vice versa.

7.6. Summary and limitations of the model

A number of assumptions and simplifications have been used for the development of the model presented here. The most important assumptions are those that lead to the formulation of Equation (7.2) and the introduction of the concept of average temperature and velocity to replace the explicit values. Equation (7.2) is obtained when the concentration of reactive species at the growing interface is taken to be zero and the rate of change of concentration with distance is taken to be constant. The first of the conditions is equivalent to requiring that the reaction-rate constant, associated with the process of transferring a molecule from the gas phase to its final position and chemical state in the solid, is extremely large at the temperature of interest. Under these conditions, the limiting step in the rate of growth is the transfer of material from the gas stream to the growing interface. This transfer occurs by diffusion through a region of laminar flow which has a parabolic velocity profile. As with the previous models for diffusion-controlled transport, the present model predicts an approximate linear variation of the growth rate with temperature. This model is thus not applicable directly to those conditions in which the growth rate is best represented by an exponential dependence upon reciprocal temperature. Under these conditions the reaction is not governed

by mass transport but rather by the rate of reaction at the growing surface.

In practice, heteroepitaxial reactors are purposely operated in a temperature regime in which the limiting step is the supply of material to the growing surface. This is done in order to obtain control of the thickness with the gas input and deposition duration. The reduced temperature sensitivity also provides for greater reproducibility.

One of the most useful characteristics of the theory developed here is that there are no freely adjustable parameters. The required diffusion coefficients can be obtained from the existing literature, calculated from statistical mechanics, or measured at room temperature. The only experimental value that is needed is that for T_A, the average temperature in the region of laminar flow. From Equation (7.27), with the added condition that $\delta_x \simeq \frac{1}{2} b_x$, T_A can be written as

$$T_A = \frac{1}{4}(T_s + 3T_m)$$

so that one need only determine T_m for a given T_s and flow conditions.

In simplifying the final form of the equations the assumption has been made that $\delta_x \simeq \frac{1}{2} b_x$. This assumption is not restrictive. The development of the final equations has been presented in detail so that it is obvious where integrations have been performed with δ held constant. The reader is cautioned to distinguish carefully between the values of D to be used in the various equations. The values of D to be used in Equation (7.13) are those of the carrier gas, whereas those in Equations (7.22) and (7.24) are those of the diffusing reactant.

ACKNOWLEDGMENTS.   The authors are appreciative of the technical help of S. L. Gilbert. The authors would also like to recognize the helpful discussions with G. W. Cullen.

References

Andrews, R. W., D. M. Rynne, and E. G. Wright, *Solid State Technology,* October (1969) pp. 61–66.

Ban, V. S. and S. L. Gilbert, *J. Crystal Growth 31:* 284 (1975).

Bradshaw, S. E., *Int. J. Electron. 23:* 381 (1967).

Dismukes, J. P. and B. J. Curtis, *Semiconductor Silicon 1973* (H. R. Huff and R. R. Burgess, Eds.) Princeton, NJ: The Electrochemical Society (1973) p. 250.

Eversteyn, F. C. and H. L. Peek, *Philips Res. Rpt. 25:* 472 (1970).

Eversteyn, F. C., P. J. W. Severin, C. H. J. v.d. Brekel, and H. L. Peek, *J. Electrochem. Soc. 117:* 925 (1970).

Foust, A. S., L. A. Wenzel, C. W. Clump, L. Maus, and L.

B. Andersen, *Principles of Unit Operations,* New York: Wiley (1960) p. 105 and 106; 186–187.

Grove, A. S., *Physics and Technology of Semiconducting Devices,* New York: John Wiley and Sons, Inc. (1967).

Hwang, G. J. and K. C. Cheng, *J. Heat Transfer, Trans. ASME 95:* 72 (1973).

Kamotani, Y. and S. Ostrach, *J. Heat Transfer, Trans. ASME 98:* 62 (1976).

Perry, J., *Chemical Engineers' Handbook,* New York: McGraw Hill Book Co. (1963) pp. 5–8, Fig. 5-11; p. 14–12.

Rundle, P. C., *J. Crystal Growth 11:* 6 (1971).

Schlichting, H., *Boundary Layer Theory,* New York: Pergamon Press (1955) p. 26 and 109; 107; 256.

Sparrow, E. M., R. Eichorn, and J. L. Gregg, *Physics of Fluids 2:* 319 (1959).

Takahashi, R., Y. Koga, and K. Sugawara, *J. Electrochem. Soc. 119:* 1406 (1972).

Chapter 8

Misfit, Strain, and Dislocations in Epitaxial Structures: Si/Si, Ge/Si, Si/Al$_2$O$_3$

J. Blanc

8.1. Introduction

The growth of an epitaxial semiconductor layer on an insulating substrate (or vice versa) will in general lead to the development of strain in the composite structure because of lattice mismatch between the two substances or because of differential thermal contraction between the two layers. The strain may or may not be relieved, in whole or in part, by introduction of strain- or misfit-relieving dislocations, or cracks.

This chapter does not present a general theory or model of these phenomena. Rather, it points out phenomenological factors whose importance may well vary from case to case, sometimes in subtle ways. The position adopted here is that there is no generally satisfactory conceptual approach to these problems; thus, the intelligent acquisition of appropriate data is still a necessary prerequisite before either a predictive or descriptive theory can be formulated.

An evaluation of the current status and prognosis for first-principle theories of epitaxial strain is given in Section 8.2; the essentially phenomenological theories of van der Merwe for introduction of misfit dislocations are briefly described in section 8.3; I point out the limitations with regard to the cases of interest here. Section 8.4, discusses (in some detail) observations on strain and dislocations in layers of Si on sapphire, of epitaxial Ge on Si, and of undoped silicon on heavily doped silicon substrates (and vice versa); although the latter two hardly constitute insulating–semiconducting composites, they are probably the systems most closely allied for which enough data are available to illustrate the complexities and subtleties of the introduction of strain and dislocations in semiconductor epitaxy. It should be pointed out at the outset that the case of silicon on silicon is the simplest

and most thoroughly studied of these three cases but shows a bewildering array of phenomena. The previous discussion is summarized, and types of experiments which might reasonably be done in the future to elucidate some of the problems presented are outlined, in Section 8.5. I have reluctantly chosen not to discuss factors influencing electrical properties at and near the heteroepitaxial interface; this is because, in spite of their great practical significance, there are far too few data to evaluate what the important factors are in determining these properties.

8.2. "First-principle" theories

A systematic long range effort to construct a first-principle theory of heteroepitaxial growth of silicon on insulators has been reported (Ruth et al., 1973). It is fair to say, I believe, that this effort has not provided viable solutions to the problems that were attacked. Without underestimating the value of these efforts, I am pessimistic about such an approach leading to an understanding of heteroepitaxy in the foreseeable future, and I think it important to point out the factors limiting such theoretical work.

A first-principle attack presumably starts with only quantum mechanics, statistical mechanics, and thermodynamics (reversible and irreversible) as assumptions. Even without dealing with questions of nucleation and growth (which are beyond the scope of this chapter), a theory of epitaxial misfit must deal, as is demonstrated in the next section, with at least the following questions: (1) What is the strength of the bonding between substrate and overgrowth? (2) What is the equilibrium structure of the overgrowth and what is the energy of a "strained" as compared to an "unstrained" overgrowth? (3) What stress level is needed to nucleate dislocations? (4) What is the rate at which a dislocation will move via diffusion of atoms or defects?

Even if the above stringent requirements are relaxed and it is assumed that bulk parameters are available from experiment for both substrate and overgrowth, the problems cannot be directly attacked. This can be seen by rewording the problems: (1) What is the surface tension between two phases of arbitrary, but known, composition? (2) Given two phases of the same composition but different known structures, what are the thermodynamics of the phase transition? (3) Given a material of known structure and thermodynamics, how much stress must be applied for what period of time to obtain a given deformation? (4) What is the relationship between gross material properties and the coefficient of self-diffusion?

Each of these isolated problems has been seriously investigated by theoretical physicists, chemists, and metallurgists. Although appreciable progress has been made toward solving of some of these questions, it is also true that there are no general methods for obtaining numerical answers for any of these generalized problems. There is simply no computer large and fast enough to solve the appropriate equations, even in those cases in which it is known exactly what equations to put down. Under these circumstances, it seems not feasible to try and construct a genuine first-order theory of a system as complex as silicon on sapphire.

A reasonable theoretical program might evolve by study of these questions on "simple" substances (e.g., argon on neon or KCl on KBr), where the binding forces are relatively well understood. Such theoretical studies would have to be accompanied by parallel experimental studies to show which theoretical approximations are justifiable and which are not. Because there is no guarantee that approximations suitable for noble gases or alkali halides would prove adaptable to semiconductors, insulators, or metals, and because the former systems are themselves of no practical interest, the impetus for seriously studying these simple systems appears to be at a minimum.

It should be mentioned here that the extensive research on the deposition of metals on substrates such as the alkali halides has been successfully analyzed in terms of phenomenological models, but problems dealing with interfacial bonding in such cases have been discussed in an ad hoc fashion, if they have been dealt with at all.

In view of all this, the prognosis for having a workable first-order theory of silicon heteroepitaxy seems very dim in the foreseeable future. I thus turn out of necessity, and with relief, to phenomenological models and empirical relations with which one may hope to order the experimental data.

8.3. Models of coherent growth and misfit dislocations

Consider two crystals (e.g., Ge and Si) of the same cubic structure but differing lattice parameters. If the two crystals are placed in an epitaxial relationship with one another, an unequal number of atom planes will be found on the two sides of the bounding interface; in other words, the linear density of atom planes in a crystal is inversely proportional to its lattice parameter, so that in our *gedanken* experiment there can be no one-to-one correspondence between atom planes in the two halves of the bicrystal. Such an interface is said to be "incoherent." The presence of missing or extra planes is recognized in dislocation theory as a particular type of dislocation: the pure edge dislocation. (Dislocations are generally characterized by their directions in the crystals and by their Burgers vectors, **b**, but these points are of little direct consequence in what follows.) In our application, they are often referred to as "misfit dislocations." A simple geometrical construction will show that the density of dislocations, n_l, per unit length of interface is given by

$$n_l = \frac{|f|}{\lambda},\qquad(8.1)$$

where f, the misfit, is given by

$$f = \frac{\lambda_1 - \lambda_2}{\bar{\lambda}} = \frac{a_1 - a_2}{\bar{a}},\qquad(8.2)$$

λ_1 and λ_2 are the interplanar spacings of interest in the two crystals, a_1 and a_2 the lattice parameters in the two crystals; and $\bar{\lambda}$ and $\bar{a}$ are $(\lambda_1\lambda_2)^{1/2}$ and $(a_1a_2)^{1/2}$. For small f, which is of concern here, $\bar{\lambda}$ can be numerically taken as either λ_1 or λ_2 without significant error, and a_1 or a_2 for $\bar{a}$. Both n_l and f are recurrent quantities in this field although with a confusing variety of symbolism; also often used in the literature is the "dislocation spacing," which is simply the reciprocal of the linear dislocation density: $1/n_l$. [Much of the conceptual matter discussed in this section is treated more precisely and elaborately in a recent book by Matthews (1975). The interested reader is urged to look at this material, particularly Chapter 6 by van der Merwe and Bell and Chapter 8 by Matthews.]

Any dislocation has a strain field associated with it, and this causes an elastic relaxation of the crystal around it. Frank and van der Merwe (1949) realized that in thin epitaxial layers the energy associated with misfit dislocations could possibly be higher (less favorable) than that associated with a coherent overgrowth (sometimes also called "pseudomorphic" overgrowth) with "strained" lattice parameters (i.e., where the epitaxial layer rather than having its own "natural" lattice parameter adopts that of the substrate). van der Merwe and others have elaborated greatly on his initial work. I sketch here the results of Jesser and Kuhlmann-Wilsdorf (1967), as this chapter presents its results in readily interpretable form and has comprehensive references to earlier work. The main results of this work are (1) that for overgrowth layers less than a certain thickness, h_c, no misfit dislocations should be formed, the energy of the substrate–overgrowth composite being minimized by formation of a "coherent" overgrowth (i.e., compression or expansion of the overgrowth lattice parameters to conform to that of the substrate); (2) as the thickness of the overgrowth increases beyond h_c, the calculations indicate that there is a rapid, but not discontinuous, increase of misfit dislocation density with layer thickness followed by a much slower increase, such that the n_l calculated from Equation (8.1) is reached only for "infinitely" thick overgrowths. Although I have phrased the above discussion, implicitly, for the case of a continuous overgrowth on an infinitely thick substrate, similar conclusions apply for island growth or substrates of finite thickness.

For cubic crystals and an infinitely thick substrate, in the limit of small strains, Jesser and Kuhlmann-Wilsdorf give the following expression for h_c, the epitaxial layer thickness above which misfit dislocations are expected to occur:

$$h_c = \frac{G_2 b^2}{2\pi a_2 |f|(1 + 2\nu)(G_1 + G_2)}$$
$$\times \ln\left[\frac{(G_1 + G_2)G(1 - \nu)}{2\pi G_1 G_2 |f|}\right]. \quad (8.3)$$

Here G, G_1, G_2 are the shear moduli of the "interface," of the overgrowth and of the substrate respectively; b is the magnitude of the Burgers vector; a_2 is the lattice parameter of the substrate; and ν is Poisson's ratio. In principle, all quantities in this expression are measurable on bulk material except G, the interfacial bulk modulus, a quantity which is supposed to be "a measure of the bonding across the interface." In practice, G must be viewed as an empirical fitting parameter. Similar expressions for critical radii, r_c, for island overgrowths can be derived from Jesser and Kuhlmann-Wilsdorf.

For semiquantitative estimates, the rather formidable Equation (8.3) can be replaced (Blanc, 1973) by

$$h_c \cong \frac{b}{2|f|}. \quad (8.4)$$

This expression can be empirically approximated from Equation (8.3) by noting that the logarithmic dependence on f is much weaker than $1/f$ and by assuming that $G = G_1 = G_2$, which for the case of strong bonding or similar crystals is reasonable. In addition, Equation (8.4) has a relatively simple and attractive physical interpretation. Because b is always of the order of the unit cell dimension, Equation (8.4) implies that coherent overgrowth can occur until the accumulated misfit ($h \cdot |f|$) exceeds about half of the unstrained unit cell; this appears plausible. It is interesting that Equation (8.4) is analogous to the "Brooks criterion" developed for the "critical radius" beyond which second-phase precipitates in crystals will lose coherence (Brooks, 1952).

An important experimental consequence of the notion of coherent overgrowth is that the bicrystal composite will bend to accommodate the elastic strain in the coherent overgrowth (Sugita et al., 1969A,B). For equal elastic moduli in substrate and overgrowth, the radius of curvature, R, is predicted to be

$$\frac{1}{R} = \frac{(6t_f t_s f)}{(t_s + t_f)^3}, \quad (8.5)$$

where t_f is the thickness of the overgrowth and t_s that of the substrate. For $t_s \gg t_f$, the usual experimental situation, Equation (8.5) reduces to

$$\frac{1}{R} \cong \frac{6ft_f}{t_s^2}. \quad (8.6)$$

I would like to stress that these general notions, in various guises and elaborations, have been qualitatively verified for many metallic epitaxial systems. A striking example [see Olsen and Jesser (1971) and references quoted there] of the utility of this conceptual framework is the epitaxy of Fe on either Cu or Ni. Both copper and nickel crystallize in the face centered cubic (fcc) form, whereas bulk iron is stable in the fcc form only in the temperature range of 906–1401°C; below and above this range, its stable form is body centered cubic (bcc). Deposition of Fe thin films (less than ≈ 25 Å) on either Cu or Ni at low temperatures yields coherent (bulk unstable) films of fcc Fe; the films are under 1.2 percent tension on copper and 1.5 percent compression on nickel. In these two cases, the coherent overgrowth not only strains the "natural" lattice of Fe, but the overgrowth goes over to a crystal structure which is thermodynamically unstable in the bulk. By studying the fcc $\rightarrow$ bcc transformations which occur on both substrates after the film thickness exceeds h_c, and by comparing results of films under tension to those under compression, Olsen and Jesser were able to elucidate in interesting detail the mechanisms of the fcc $\rightarrow$ bcc transition in iron.

Models of the van der Merwe type give conditions for equilibrium (the equilibria are "mechanical" rather than "thermodynamic"); one presumes that the thermodynamic equilibrium is not significantly different from the mechanical (zero temperature) one, but I am not aware of detailed theoretical justification of this. It is generally true in statistical physics and chemistry that it is far easier to specify conditions for equilibrium than to calculate realistic kinetic paths for reaching equilibrium from arbitrary initial conditions. Unfortunately, this is also true in the present case.

Conditions for dislocation equilibrium are much

more easily met in metals than in covalent crystals, so that a model which works admirably for metals cannot, without hazard, be extrapolated to nonmetallic crystals. The reasons for this difference are very much related to crystal bonding forces. Dislocations are produced in a crystal in response to forces tending to deform the crystal; the actual mechanisms for nucleation of dislocations ("sources") are complex, and I shall return to them briefly in Section 8.4. Subsequent motion towards equilibrium occurs in two distinctly different ways: one is "slip" or "glide," the other is "climb." Slip involves the motion of entire atomic planes with respect one to another, as in a spread-out deck of cards. Climb, on the other hand, involves local mass transport to make or rearrange a dislocation, and is diffusive in nature. Slip is impeded by the so-called Peirls–Nabarro force which measures the force required to move one plane against another; although few quantitative estimates of this force have been made, it is intuitively attractive to believe that materials with short-range nondirected bonding forces (i.e., metals) have much smaller forces than materials with highly directional forces (covalent crystals) or with strong long-range interactions (ionic crystals). Climb, on the other hand, is intimately tied to self-diffusion processes, which vary enormously from crystal to crystal. Without undue generalization, I note that self-diffusion in many "soft" metals has an activation energy less than 3 eV, whereas self-diffusion Si is reported to have an activation energy of about 5 eV. It may thus be generally expected that dislocation motion will be easier, either by slip or climb, in metals than in covalent crystals, and that for covalent crystals, both slip under external stress and climb are expected to be the easier the higher the temperature. (Detailed discussion of these topics can be found in any text on dislocations.) Thus, in "soft" metals which are readily deformable at low temperatures, either the substrate will contain enough "native" dislocations so that when needed by the breakup of the coherency of the overgrowth, these can rearrange into misfit-relieving orientations, or dislocations can be introduced, at appropriate sources, under the influence of the stress caused by the overgrowth. The only serious modification (Jesser, 1970) to van der Merwe models which needs to be made, if the temperature is low enough that climb does not occur rapidly, is that "slip" dislocations are not usually of pure-edge type although they can relieve misfit. This effect requires essentially minor numerical modifications in calculations of both n_l and h_c.

In the limit that neither slip nor climb can occur, *no* dislocations can be introduced either by climb or slip, leading to overgrowths which are coherent to thicknesses much larger than calculated h_c. This gives rise to curvature of the grown composite and eventually to crack formation.

Unfortunately, semiconductor epitaxy falls in neither limit of very easy nor very hard dislocation introduction, and the conceptual situation is therefore muddy. Matthews and coworkers (1970) have made several interesting attempts to describe this intermediate regime, but as is demonstrated in Section 8.4, careful studies of Si heteroepitaxy have not so far yielded to this type of interpretation.

8.4. Some experimental results

Si/Al$_2$O$_3$

The silicon-on-sapphire system is the prototype, and currently most important, example of semiconductor heteroepitaxy on insulating substrates. The system presents serious experimental and conceptual difficulties. Sapphire is rhombohedral, but silicon is cubic, and therefore for growth on any one particular sapphire orientation the misfit strain is anisotropic, with mean strains of ≈ 0.1. In addition, the large difference in thermal expansion coefficient between the substrate and silicon ($\approx 4 \times 10^{-6}/°C$) leads to a strain of $\approx 3 \times 10^{-3}$ in cooling from a growth temperature of $\approx 1000°C$ to room temperature. These large strains make the preparation of specimens suitable for defect study by transmission electron microscopy difficult. Even if samples are successfully prepared, the observed contrast may be exceedingly complex because of large strains and high defect densities. In fact, the pioneering transmission electron microscope work of Cullis (1972) and Linnington (1974) failed to reveal *any* misfit dislocations near the Si/Al$_2$O$_3$ interface. Although these negative results can now be understood (see below), they appeared to pose a serious problem in understanding heteroepitaxy.

Before addressing the question of misfit dislocations, it is necessary to describe some results on the growth mechanism. Abrahams and coworkers (1976b) and Blanc and Abrahams (1976) have reported detailed experimental results and have modeled the early growth of 001 Si on 01$\bar{1}$2 Al$_2$O$_3$. The salient results are as follows.

a. The growth proceeds via island formation, as was known previously. The individual islands are predominantly of (100) orientation, with a rotational spread of ≈ 3 deg. The deposit contains a minor proportion of {110} material.

b. Individual islands appear to grow independently of one another, and the coalescence of two (100) islands is occasionally accompanied by the formation of a stacking fault at the juncture of the islands, presumably to accommodate the rotational misalignment.

c. The minor {110} constituent is eventually covered over as the growth attains full coverage, with presumed formation of grain boundaries.

The end result of this complex set of events is that a large number of stacking faults and microtwins is formed, and the density gradually decreases with distance from the interface [Linnington (1974), Abrahams and Buiocchi (1975), Ham and coworkers (1977)]. Near the silicon/sapphire interface, the mean distance between faults in the silicon is ≈100 Å. Besides their intrinsic interest, these results indicate why misfit dislocations were not found, even if present, by Cullis and Linnington. Contrast in foils of 2000 Å or more of Si would, except fortuitously, have been overwhelmed by features other than misfit dislocations.

In spite of these difficulties, misfit dislocations have been observed, but only in deposits with incomplete coverage (i.e., islands) [Abrahams and coworkers (1976a)], obviating much of the observational difficulty discussed above. The misfit dislocations occur with the density expected for the lattice misfit calculated; the dislocations are of pure edge character. It thus appears that in spite of the complications of growth mechanisms and introduction of high densities of faults, the case of Si/Al$_2$O$_3$ conforms very closely to van der Merwe's model for the accommodation of lattice misfit.

We have seen that the large lattice misfit is completely relieved at normal growth temperatures. There remains the relatively much smaller strain due to differential thermal contraction after growth. Although dislocation counts such as discussed above cannot experimentally distinguish the small increment in dislocation density to be expected from relief of this strain, it has long been known [Ang and Manasevit (1965)] that substantially all of this strain is retained elastically (i.e., without additional dislocation introduction). The principal reason for this is that at the low temperatures where most of the differential contraction occurs the silicon is so resistant to plastic deformation that new dislocations cannot be generated even though equilibrium theories call for them. Very recently, Abrahams and coworkers (1978) have shown that this normal strain (compression of Si interplanar spacings) is transformed to shear strains (distortion of angles between Si planes) at edges of Si/Al$_2$O$_3$ composites, such as one will almost inevitably encounter in device manufacturing. At the threefold intersection of Si, Al$_2$O$_3$, and vacuum (i.e., a corner), the shear strain reaches a maximum of ≈0.02, a magnification of a factor of 8 over the "normal" strain. The shear strain decays in an approximately exponential manner with distance away from the corner. The maximum shear strain observed is enormous from a number of points of view. It corresponds closely to the theoretical shear strength, so that corners must be close to incipient cracking or plastic flow, although this has not been observed. Such corners may well be

natural sources for dislocations throughout the Si overgrowth after subsequent processing. Furthermore, one would expect bandgap charges of the order of 0.1 eV under these high stresses, with possible consequences on the electrical properties of such regions. Abrahams and coworkers (1978) have given an approximate theoretical discussion of the strain distribution in the appropriate geometry. They conclude that the transformation of normal strain to shear strain, and resulting magnification of strain, should always be expected in the presence of a corner subjected to differential thermal contraction.

A detailed and comprehensible picture of defects of Si deposited on Al$_2$O$_3$ has thus emerged in the last several years. This should not obscure the fact that a great deal still remains to be discovered or understood. I point out only two such areas. It is still not agreed whether the predominant defects observed (stacking faults and microtwins) are inherent to the growth of silicon on sapphire or whether alterations in techniques of growth can affect these crystal imperfections; nor is it even clear whether such defects are beneficial or harmful to majority carrier devices which are the major use for these composites. More academically, it is far from obvious that we understand except in the vaguest way what physical or chemical factors are determinant in the orientation which silicon adopts on an arbitrary orientation of sapphire.

Ge/Si

The epitaxial deposition of germanium on silicon substrates is perhaps conceptually the "simplest" example of semiconductor heteroepitaxy. The mismatch in lattice parameter, however, is relatively large ($f \approx 4 \times 10^{-2}$), leading to a calculated coherent thickness of only ≈25 Å. It may thus be expected that growth with epitaxial layer thicknesses exceeding ≈100 Å should exhibit misfit dislocation densities ≈ 7×10^5/cm in the region of the Ge/Si interface.

Cullis and Booker (1970) and Briante and coworkers (1972) have grown epitaxial thin films by vacuum evaporation of Ge onto (111) Si substrates in the range of ≈700 to ≈900°C and examined the resulting layers by transmission electron microscopy. Insofar as the results overlap, the two investigations are in qualitative agreement; a critical comparison of the interpretation given by these two sets of careful workers immediately shows the difficulties in theoretical interpretation of this "simple" semiconducting system.

These workers report that the growth of Ge on Si proceeds by nucleation and growth of islands. Briante and coworkers report that a large number of small islands form at lower temperature compared to a smaller number of larger islands at higher temperatures. It is not clear from their data whether the substrate area covered by growth for constant average thickness changes appreciably for the different morphologies at the different growth temperatures. (This point will be of importance in the ensuing discussion.)

Cullis and Booker measured the dislocation and Moire spacings by transmission electron microscopy for specimens grown at various temperatures; the dislocations had the correct directions and Burgers vectors to be misfit relieving. These workers found that the density of dislocations *decreased* with *increasing* temperatures, the density expected for relief of misfit being approached only at the lowest temperatures. Briante and coworkers, on the other hand, rather than counting dislocations, performed "selected area" diffraction experiments and were thus able to find split diffraction spots from the Ge overgrowth and Si substrate. They found that at the lowest growth temperature the Ge islands had substantially the normal Ge lattice parameter, in agreement with the large dislocation density found by Cullis and Booker. At their highest temperature, the lattice parameter of the Ge islands had gone about halfway to that of Si; this is also qualitatively consistent with the small dislocation density found. Thus, the two groups of workers have both found, by two essentially different methods, that there is more misfit accommodated by dislocations for Ge grown on Si at lower temperatures than at higher temperatures.

The interpretations of these results by the two groups are, however, significantly different. Cullis and Booker believe that their results may indicate that at high temperatures interdiffusion occurs at the heterojunction and that "this will give an alloy zone which would reduce the mismatch and therefore increase the dislocation spacing." On the other hand, in their later work, Briante and coworkers argue that if an appreciable amount of diffusion had taken place, the lattice parameter of the overgrowth should continuously vary from that of pure Ge to that of pure Si and "it would not be possible to observe distinct Ge and Si patterns."

Although this argument is certainly very credible, adopting it leads to a paradox. If indeed the evidence of split diffraction spots indicates the absence of significant interdiffusion between Ge overgrowth and Si substrates, it then follows immediately that Ge tends to grow more coherently as the growth temperature increases. This is a paradoxical result for two reasons: (1) the coefficients of thermal expansion of Ge (6 × 10^{-6}/°C) and of Si (4 × 10^{-6}/°C) are such that the lattice misfit *increases* with increasing temperature, so that the driving force for introduction of misfit dislocations in the Ge overgrowth should be larger at the higher growth temperature; (2) both mechanisms for introduction of misfit dislocations (slip and climb) will

result, at a constant stress level, in an increasing rate of introduction as the temperature increases. Thus, the conundrum: on quite general grounds, one should expect more of the misfit to be accommodated as the temperature of the epitaxial growth increases, but that the reported experiments clearly show the reverse; unless the overgrowths are in fact alloyed, there appears to be no easy way out. [Attempts at measuring the actual composition of the overgrowth islands have been made, but have been unsuccessful (Booker, 1973).]

Cullis (1972) has suggested that for the thin deposits involved here, *surface* diffusion rather than bulk diffusion might be dominant. This could result in an overgrowth of effectively constant alloy composition. The actual composition would depend on the temperature through the temperature dependence of the diffusion coefficient. Although this plausible suggestion explains the experimental results, it is not as yet experimentally verified, and it may be quite difficult to do so.

Assuming that the experimental results (i.e., absence of interdiffusion) have been correctly interpreted, I now discuss possible sources of the difficulty. I have already noted that the island morphology changes as a function of temperature; it is conceivable that the smaller (in area) low-temperature islands are "tall" and have exceeded the critical thickness whereas the larger high-temperature islands are "short" and have not or have barely exceeded the critical thickness. This appears an unlikely explanation. Briante and coworkers (1972) have reported the "average" thickness of their deposits calculated from the rate of deposition; for island growth, the actual thickness of islands must be greater than this average thickness; for growths at 900°C, these workers report results at average thicknesses ranging from 25 to 600 Å, without any appreciable systematic change in the lattice parameter of the Ge overgrowth. It seems implausible in this case that a change by a factor of 24 in average thickness should not result in appreciable changes in actual island thickness, exceeding the critical thickness for the thicker deposits. Another possible explanation might be that an experimental boundary condition for the proper analysis of the experimental results has been implicitly violated (I am at a loss to propose a concrete possibility for these experiments, but see below for such an example in another growth system). Finally, if the coherent overgrowth does not have the atomic positions postulated by the van der Merwe model, it then seems possible that the coherence one might expect, and the associated strain, might well be a function of temperature. I discuss this possibility briefly in Section 8.5.

For the interesting case of the epitaxial growth of Ge–Si alloys on Si and an elaboration of van der

Merwe's model to semiconductors, see Kasper and Herzog (1977).

Si/Si

Prelude. The case of Si/Si homoepitaxy requires a brief discussion of the modes of dislocation introduction in crystals, although this discussion will be neither complete nor precise. The reasons for this are associated with the relatively low misfit strains ($< 10^{-3}$) associated with homoepitaxial growth of silicon and the low native dislocation densities in silicon substrates used conventionally.

Estimates for the stress necessary to nucleate dislocations spontaneously in an otherwise perfect crystal yield $\approx \mu/30$, where μ is the shear modulus of the crystal. Once nucleated, dislocations will typically move under applied stress of $\geqslant \mu/1000$. Numerical calculations show that the stress due to epitaxial misfit will exceed $\mu/30$ for misfits approximately greater than 0.02, although the actual values depend on the active shear system. Thus, in the example of the previous sections, there is no difficulty in visualizing sources of dislocations for possible misfit relief. However, for Si/Si, ($f < 10^{-3}$), this mechanism is not available. In such situations, one often finds stress raisers or stress concentrators (which may be precipitates, surface discontinuities, etc.) invoked in the metallurgical literature. The local stress may thus be considerably higher than the average applied stress. Dislocations may be nucleated at these sites and subsequently glide to relieve misfit. [Matthews et al. (1970) give a more precise discussion of this mechanism.] Except for the presence of intrinsic surface ledges in the growing crystal, sufficiently "clean" systems will have no such stress raisers. Whether expitaxial Si falls in this regime is not known, but it seems possible that the stress induced by misfit could be insufficient to nucleate new dislocations.

Another mechanism available for formation of misfit dislocations involves preexistent dislocations in the substrate. Those dislocations originally terminating at the free surface of the substrate will propagate into the overgrowth. After an appropriate critical thickness is reached, these will tend to bend into the interface to produce misfit relief. Matthews and coworkers (1970) have considered this process in detail. For Si, at temperatures of $\approx 1150°C$, they obtain a rate constant for this process of $\approx 10^{-3}/s$. For any growth exceeding 5 min, as in most conventional Si epitaxial methods, this bending of the grown-in dislocation should be essentially complete. However, for substrate dislocation densities of 100/cm², which is reasonably typical, even if all substrate dislocations were to bend for misfit relief, this is sufficient for relief of only $\approx 4 \times 10^{-6}$ or smaller. Thus, with substrates that are usually used in Si epitaxy, the substrate dislocations cannot provide enough dislocations for dislocation equilibrium even for very small values of misfit, unless some multiplication mechanism such as Frank–Read sources is available.

The only other evident source for dislocation generation and motion is via climb. As this is a diffusive process, it may be expected to be quite slow except in the presence of large supersaturations of point defects (interstitials or vacancies).

It is thus by no means clear by what process or on what time scales misfit will be relieved in silicon homoepitaxy.

Si:B/Si and Si/Si:B. Because of the availability of high-precision lattice parameters [from Horn (1955)] of B-doped silicon up to the saturation limit ($\approx 10^{20}$ B/cm³), the growth of undoped silicon on B-doped silicon and vice versa should be a splendid opportunity for the study of misfit relief in semiconductors as a function of various parameters: doping level, growth temperature, and method of growth. The lattice parameter of B-doped Si is smaller than that of pure Si. Sugita and his coworkers (1969A,B; also Tamura and Sugita, 1973), and we in our laboratories, have availed ourselves of this opportunity. Unfortunately, the two sets of experimental results are in serious discordance one with the other. I therefore take up each set separately and then try to point out what the significance of this disagreement is. Both in Sugita's work and ours, [111] and [100] Czochralski crystals were used as substrates, with diameters of 2.5 to 3 cm. Substrate dislocation densities were $<10^2/$cm². Except as otherwise noted, the surfaces to be grown on were polished by standard methods to a mirror finish. In our work, we used substrate wafers both with polished and lapped "back" surfaces.

THE WORK OF SUGITA AND COWORKERS. Sugita and his associates have studied growths of essentially undoped epitaxial layers on substrates with B concentrations varying from $\approx 2 \times 10^{18}/$cm³ to $\approx 6 \times 10^{19}/$cm³ ($5 \times 10^{-2} \geqslant \rho \geqslant 2 \times 10^{-3}$ Ω-cm). These correspond to f approximately between 10^{-5} and 3×10^{-4}. They grew their layers by reduction of silicon tetrachloride at 1140°C at a rate ≈ 1 μm/min.

In one series of experiments, Sugita chose a series of B-doped substrates of 200 μm thickness with resistivities from more than 2×10^{-2} to 2×10^{-3} Ω-cm (f approximately between 10^{-5} and 3×10^{-4}) and grew layers of undoped Si to an essentially constant thickness of 20 μm. Lang x-ray topography was utilized to detect the presence of misfit dislocations. They determined that for this thickness of overgrowths *no* misfit dislocations were introduced for substrate resistivities larger than 2×10^{-2} ($f \approx 3 \times 10^{-5}$). Equation (8.3) or (8.4) can be used to define, for constant overgrowth thickness, a misfit below which the overgrowth is expected to be coherent (with no misfit dislocations); for a thickness of 20 μm, Equation (8.4) yields a value for this "critical" misfit of $\approx 1 \times 10^{-5}$, which I take to be in substantial agreement with the experimental value. However, as Sugita and coworkers noted, the dislocation densities appeared to be much smaller than those expected for equilibrium relief of misfit; thus, for a sample with $f \approx 1.3 \times 10^{-4}$, n_l was observed to be 250/cm instead of the calculated $\approx 3 \times 10^3$/cm and for a sample with $f \approx 2 \times 10^{-4}$, n_l was measured to be 800/cm rather than the expected value of $\approx 5 \times 10^3$/ cm, a discrepancy of about an order of magnitude. With the sample of lowest resistivity ($f \approx 3 \times 10^{-4}$), with an expected n_l of $\approx 8 \times 10^3$/cm, individual dislocations could not be resolved, probably indicating an actual density in excess of 5×10^3/cm.

In another set of experiments to determine h_c, these workers prepared substrates from 7×10^{-3} Ω-cm B-doped Si, with thicknesses of either 10^3 or 2×10^2 μm, and grew undoped Si of various thicknesses. By observing in which overgrowths misfit dislocations were observed, they were able to determine h_c. For the thicker substrates, they measured $h_c \approx 5$ μm; for the appropriate misfit ($f \approx 9 \times 10^{-5}$) and the assumption of infinitely thick substrate, h_c should be ≈ 3 μm. For substrates of thickness 200 μm, they obtained $h_c \approx 7.5$ μm. This increase of h_c with decreasing substrate thickness is at least qualitatively expected as part of the misfit is expected to be taken up by elastic bending of the substrate [Equation (8.5)]. The behavior of the dislocation density with overgrowth thickness after h_c has been passed is extremely interesting. The data seem to indicate that for both thicknesses of substrate, the dislocation density rises quite abruptly after h_c to the (low) value of 5 misfit dislocations/cm; this is followed by an extremely gradual (apparently logarithmic) increase with further overgrowth thickness not approaching the theoretical limit until the overgrowth is ≈ 50 μm. This behavior is very strikingly different than that expected on theoretical grounds [see Figure 5 of Jesser and Kuhlmann-Wilsdorf (1967)] or what is actually observed in metallic epitaxy. The equilibrium theory suggests when the total overgrowth thickness reaches $2h_c$, about 90 percent of the n_l calculated by Equation (8.1) should have been made. This is not at all what the results of Sugita show for Si/Si.

I conclude this description of the experiments of Sugita and coworkers with another of their elegant experiments. Taking a series of substrates of constant thickness with $\rho = 3 \times 10^{-3}$ Ω-cm ($f = 1.9 \times 10^{-4}$), they measured the radius of curvature of epitaxial composites for epitaxial thicknesses less than the critical thickness (observed to be ≈ 2.5 μm, calculated ≈ 1 μm). By taking the derivative of $1/R$ with respect to the film thickness [see Equation (8.6)], they obtained a value of $f = 1.8 \times 10^{-4}$, in good agreement with that

expected. The values for f obtained directly from Equations (8.5) or (8.6), without differentiation, seem to show systematic differences from the values above.

I briefly summarize now the salient points of this work as it bears on misfit dislocations. (1) Critical thicknesses and misfits are in excellent agreement (to within factors of better than 2 or 3) with calculations of van der Merwe type models. (2) The dependence of the radius of curvature on epitaxial layer thickness in the coherent regime is accurately given by Equation (8.6), but absolute values are in some disagreement. (3) The initial abrupt rise in dislocation density after passing h_c can probably be accounted for by the Matthews mechanism of substrate dislocations bending over into the strained interface; after this point is passed, however, the supply of dislocations appears limited and equilibrium slow to be reached, until f exceeds 3×10^{-4}.

WORK AT RCA LABORATORIES.   The initial impetus for our work on Si/Si:B composites was twofold. We have had a long-standing interest in the study of misfit relief in epitaxial composites by means of transmission electron microscopy (TEM), which can provide more detailed information than x-ray topography; the work of Sugita and coworkers strongly suggested that the study of Si/Si epitaxial crystals by this method would provide a rich lode to mine, leading to a detailed understanding of the modes of introduction of dislocations. However, because of the much higher magnifications and consequent small fields of view used in TEM, the dislocation spacing ought not to be less than ≈ 1 μm in order to obtain statistically meaningful data. This implies an f approximately greater than 4×10^{-4}, a value higher than any studied by Sugita. The work reported in this section, therefore, deals exclusively with B concentration at or near the saturation limit (resistivity of 1×10^{-3} Ω-cm, B $\approx 10^{20}$/cm^3) leading to an expected f of 6×10^{-4} and $n_l \approx 2 \times 10^4$/cm.

Our growths were performed both by decomposition of dichlorosilane and silane at temperatures ranging from 1050 to 1150°C with growth rates between 1 and 2 μm/min. Substrates of 200 μm thickness, both in (100) and (111) orientation were used, with an epitaxial layer thickness between 25 and 100 μm.

I now summarize briefly the results of a long series of experiments. Observed by x-ray topography, epitaxial composites of undoped Si on substrates doped with 10^{20}/cm^3 B *never* showed a misfit dislocation density exceeding 30/cm, a discrepancy of almost three orders of magnitude from the expected result. (The dislocation density was far too low for examination by TEM.) This result was independent of growth temperature, method of growth, growth rate, postgrowth annealing, thickness of the epitaxial layer, or whether the back surfaces of the substrates were etched or lapped. In order to check whether this discrepancy was a result of the different method of growth used by Sugita and coworkers (SiH$_4$, SiH$_2$Cl$_2$ vs. SiCl$_4$), we examined a similarly doped composite produced by reduction of SiCl$_4$. This wafer had an n_l of 60/cm. Although this is a factor of two higher than the highest density observed in the other wafers, it is still orders of magnitude away from the expected density.

The *total* number of misfit dislocations is expected to be the same whether the misfit occurs at an abrupt interface or through a graded region as long as the free surfaces have the same misfit as an abrupt deposition [see Abrahams et al. (1969)]. We therefore verified that the resistivities of the surfaces of the epitaxial composite were characteristic of the dopant concentrations, with no appreciable contribution from out-diffusion from the back surface of the substrate nor long-range diffusion of B from substrate into epitaxial layer.

A possible reason for the paucity of misfit dislocations in our composites is the absence of suitable sources. We did a number of experiments to intentionally introduce local stresses. I describe here only the simplest and most convincing. We grew a layer on a lapped surface. The results are shown in the transmission topograph of Figure 8.1. Although cosmetically very poor, the results are very interesting. Long misfit dislocations (the straight black lines) running in the three $\langle 110 \rangle$ directions of the (111) surface can be seen; in addition, the noncrystallographic lap marks, reflecting the surface irregularities of the grown layer, can be observed as well as some precipitation in the substrate. The average density of *misfit* dislocations is still only about 30/cm. Thus, in spite of heavy surface

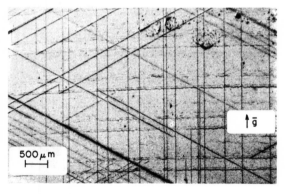

FIGURE 8.1   Transmission x-ray topograph of growth of undoped Si on heavily B-doped silicon substrate which had been lapped. Misfit dislocation lines are the long straight features running in the three $\langle 110 \rangle$ directions of the (111) growth plane. Lap marks can be seen as relatively fuzzy, roughly horizontal noncrystallographic features. Precipitates in the substrate can also be observed. The direction of the diffraction vector used ($\overline{g} = [2\overline{2}0]$) is indicated. Photograph courtesy of S. H. McFarlane

FIGURE 8.2   Transmission x-ray topograph of growth of heavily B-doped silicon on undoped Si substrate. Barely resolvable misfit dislocation lines can be seen running in the two $\langle 011 \rangle$ directions of the (100) growth plane. The white bands separate distinct topographs; this is because of the curvature of the composite. The direction of the diffraction vector ($\overline{g} = [022]$) is indicated. Photograph courtesy of S. H. McFarlane

damage and irregularities at the growth interface, we obtain the same dislocation density as in undamaged layers. We must therefore conclude that, in some sense, 30 dislocations/cm is all that is required to accommodate the net strain at the growth temperature in our growth systems.

I turn briefly now to the question of strain distribution in these layers. The observed radius of curvature was never smaller than 5 m; for epitaxial layer thicknesses of 50 μm (the midrange of our experiments), Equation (8.6) with a misfit of 6×10^{-4} leads one to expect a radius of ≈ 0.2 m. Both the large radius of curvature and the small density of misfit dislocations could be rationalized if in fact the misfit were much smaller than that expected from the impurity concentration in the substrate. Nonetheless, an attempt to etch off the substrate, to isolate the epitaxial layer for further examination, led to a complete shattering of the epitaxial layer after most of the substrate had been removed. This demonstrates, albeit qualitatively, that a considerable strain *had* been stored in the grown layer, presumably not in the form leading to the development of Equation (8.6).

All the experiments described above, as well as those of Sugita and coworkers, were performed by growing essentially undoped Si on B-doped Si substrates. In view of the perplexing results obtained, it is of obvious interest to see whether, in fact, dislocation introduction is similar in the reciprocal case of growing B-doped epitaxial layers on essentially undoped Si substrates. At the time of these experiments, the required B density of 10^{20}/cm^3 in the epitaxial layer was obtainable only with the SiH$_4$ system. The results can be summarized very briefly (see Figure 8.2): transmission x-ray topography showed an *unresolvable* density of dislocations (i.e., $n_l \geq 5 \times 10^3/\mathrm{cm}$) and a radius of curvature of ≈ 0.3 m in the sense expected for growth of a smaller lattice parameter film on a larger lattice parameter substrate.

The important points of our work are: (1) for substrate dopings of 10^{20} B/cm^3, the growth of undoped Si leads to much *lower* n_l's than one would have expected; (2) this result is *not* caused by an absence of dislocation sources; (3) there are indications that strain is stored in such composites, but not in the way predicted by van der Merwe type models of coherent epitaxy; and (4) there is an obvious asymmetry, not predicted by any model, between Si:B/Si and Si/Si:B composites.

Discussion. These experimental results must now be confronted, although I must admit that the ensuing discussion will involve more hand waving than I find comfortable.

I must point out an experimental variable whose existence neither we nor Sugita and coworkers were aware of at the time the experiments were carried out and whose consequences are extremely difficult to pinpoint. All discussions of misfit relief in epitaxy assume, in an essential way, that the substrate is unstrained. For conventional chemical-vapor deposition of Si on Si, this is simply untrue. In connection with another problem, Bloem and Goemans (1972) have pointed out that a wafer in a typical reactor will show appreciable thermal curvature in the absence of any growth. Although their argument is not precise, there can be no doubt of its essential physical validity. The argument briefly is as follows: Typically a Si wafer is placed on a heated substrate with the front surface facing a cold exterior wall; even without gas flow, radiation losses from this front surface to the cold wall will make the front surface colder than the back surface and the wafer will bend because of thermal contraction. Now, it is known that in silicon reactors using flowing gases, there is a stagnant layer of gas immediately adjacent to the front surface and that this layer supports a very large temperature gradient (Eversteyn et al., 1970). Bloem and Goemans, therefore, propose that the edges of the wafer, which have been lifted from the substrate by radiation cooling, are placed in regions of large temperature gradient which leads to further cooling and therefore further bending of the wafer. Using Bloem and Goemans's equations and their estimates for the parameters needed, I estimate that substrate wafers used by us and by Sugita would have had a radius of curvature of ≈ 0.1 meters, bending toward the cold exterior.

The importance of this result is twofold. Although this estimate of ''thermal'' curvature is surely not quantitatively correct, it *is* of the same order of magnitude as the calculated radii of curvature for coherent overgrowths of Si/Si, so that at the growth tempera-

ture both the epitaxial strain and the thermal strain somehow have to be taken into account. Also, it is now known (Eversteyn et al., 1970) that the hydrodynamics leading to the stagnant layer are very sensitive functions of the flow rate, gas pressure, and geometry. It seems plausible that a substantial part of the difference between our results and those of Sugita are attributable to differences in flow dynamics, although it will take a great deal of additional careful experimental work to verify such a conjecture.

The hypothesis of thermal curvature can qualitatively account in a simple way for some, but not all, of our observed asymmetric results for Si/Si:B and Si:B/Si. The growth of B-doped Si on pure Si should result, at the temperature of growth, in greater curvature than the growth of pure Si on B-doped substrates, as in the former case the overgrowth has a smaller lattice parameter than does the substrate; one may imagine that as the thickness of the deposit increases, the built-up stresses finally exceed the critical shear stress, introducing dislocations via slip both to relieve the lattice mismatch and to accommodate the initial thermal stress. The composite, having now undergone plastic deformation, will when cooled to room temperature show both the misfit dislocations and a residual curvature caused by the initial thermal stress.

The converse situation of growth of undoped Si on B-doped substrates is therefore expected to lead to a *flattening* of the composite, at least initially, and a reduction of applied stresses at deposition temperature. From this point of view the paucity of misfit dislocations is not surprising. This hypothesis, however, does not seem to explain why the misfit dislocation density appears fixed at $\approx 30/\mathrm{cm}$. Neither does it rationalize the observation that the composite is essentially flat after cooling to room temperature, for if the misfit just compensates the thermal curvature at growth temperature, it should bend appreciably after the crystal becomes isothermal, as slip will not occur at room temperature. It should be remarked that the actual physical situation must be more complicated than my ad hoc discussion has indicated. Even if one knew the actual temperature distribution through and across the growing wafer, it would be a very difficult task to calculate the stresses as a function of time and position.

Finally, I mention two possibilities that might formally account for some of the apparent difficulties. The first is that the lattice parameter of B-doped Si is not a unique function of the B concentration, so that in fact our B-doped substrates had a lattice parameter which was much closer to that of pure Si than we had supposed. The second possibility is that the elastic or plastic properties of Si:B at the saturation limit are drastically different from those just below the saturation limit. Although neither of these seems particularly likely to me, they are worthy of further investigation

in view of the absence of an entirely convincing explanation of the experimental observations.

8.5. Summary and speculations

I have briefly sketched what current models of misfit
relief predict for dislocation morphologies in epitaxial
composites with differing lattice parameters. For the
systems Si/Al$_2$O$_3$, Ge/Si, and Si/Si, there appears to be
good evidence that the predictions of such models are
not met in any simple way and that the misfit dislocation density is often smaller than anticipated.

One may not, however, conclude yet that the
models are intrinsically at fault. The reason is that
neither experimentalists nor theoreticians have paid
detailed attention to the real experimental situation
that exists in these crystal growth systems. For the
case of Si/Si, it seems clear that the effects of thermal
curvature will have to be studied experimentally in
detail and that these effects will have to be systematically incorporated into appropriate models. Whether
some such analogous effects exist for the vacuum
depositions of Ge/Si or Si/Al$_2$O$_3$ remains a moot question at the moment.

Still more speculatively, one might consider that the
models of misfit relief are in fact misleading. This is
not said out of sheer perversity but because in covalent systems such as Ge and Si, it typically requires
more energy to stretch bonds than to bend them.
Models of pseudomorphic growth such as I have
sketched implicitly assume that the elastic energy in
pseudomorphic layers is stored exclusively in bond
stretching and none in angular distortions of the
bonds. In fact, currently accepted theories of the elastic constants of diamond and zincblende crystals indicate that bond-stretching constants are about three
times as large as bond-bending constants in these
systems (Martin, 1970), so that the above assumption
appears seriously in error. No model of coherent overgrowths invoking bond bending instead of bond
stretching has yet been formulated. It appears quite
certain, however, that such a model must lead to
different atomic geometries from those of van der
Merwe–type models, and it also seems quite possible
that modes of dislocation formation and motion are
significantly different from those now expected.
Whether such a model could help to rationalize some
of the experimental results described remains a problem for the future.

Si epitaxy, on insulators, on Si, and on other semiconductors, will undoubtedly remain one of the principal ingredients of the electronic industry in the foreseeable future. Because of this, it is clear to me that
much work, both experimental and theoretical, will
have to be carefully and systematically performed

before one can justly claim that problems of the type I
have tried to describe are under control.

ACKNOWLEDGMENTS. I am grateful to Y. S. Chiang, N.
Goldsmith, P. H. Robinson, and S. H. McFarlane for their
unstinting cooperation in pursuing the experimental work
described in Section 8.4. M. S. Abrahams and G. H. Olsen
have provided much instruction and heated discussion on all
the conceptual matters discussed in this chapter. I am particularly grateful to G. H. Olsen for his critical reading of the
manuscript and to N. Goldsmith for his willingness to com-
municate his wide knowledge of fact and fancy of silicon
technology. Finally, I wish to thankfully acknowledge the
generosity of A. Howie and P. Linnington in making their
observations on Si/Al$_2$O$_3$ available prior to publication and to
G. R. Booker and A. G. Cullis for their similar kindnesses.

References

Abrahams, M. S. and C. J. Buiocchi, *Appl. Phys. Letters 27,*
325 (1975).
Abrahams, M. S., L. R. Weisberg, C. J. Buiocchi, and J.
Blanc, *J. Mat. Sci. 4:* 223 (1969).
Abrahams, M. S., C. J. Buiocchi, J. F. Corboy, Jr., and G. W.
Cullen, *Appl. Phys. Letters 28,* 275 (1976a).
Abrahams, M. S., C. J. Buiocchi, R. T. Smith, J. F. Corboy,
Jr., J. Blanc, and G. W. Cullen, *J. Appl. Phys. 47,* 5139
(1976b).
Abrahams, M. S., J. Blanc, C. J. Buiocchi, and W. E. Ham,
J. Appl. Phys., 49, 652 (1978).
Ang, C. Y. and H. M. Manasevit, *Solid State Electronics 8,*
994 (1965).
Blanc, J., unpublished data (1973).
Blanc, J. and M. S. Abrahams, *J. Appl. Phys. 47,* 5151
(1976).
Bloem, J. and A. H. Goemans, *J. Appl. Phys. 43:* 1281
(1972).
Booker, G. R., private communication (1973).
Briante, J. D., J. M. Corbett, and F. W. Boswell, *Thin Solid
Films 14:* 305 (1972).
Brooks, H., *Metal Interfaces,* Cleveland: American Society
for Metals (1952) p. 20.
Cullis, A. G. and G. R. Booker, *7th Inter. Conf. Electron
Microscopy,* p. 423, Grenoble, 1970.
Cullis, A. G., Ph.D. thesis, Oxford University (1972).
Eversteyn, F. C., P. J. W. Severin, C. H. J. v. d. Brekel, and
H. J. Peek, *J. Electrochem. Soc. 117:* 925 (1970).
Frank, F. C. and J. H. van der Merwe, *Proc, Royal Soc.
(London) A 198:* 216 (1949).
Ham, W. E., M. S. Abrahams, C. J. Buiocchi, and J. Blanc,
J. Electrochem. Soc. 124, 634 (1977).
Horn, F. H., *Phys. Rev. 97:* 1521 (1955).
Jesser, W. A., *J. Appl. Phys. 41:* 39 (1970).
Jesser, W. A. and D. Kuhlmann-Wilsdorf, *Phys. Stat. Sol.
19:* 95 (1967).
Kasper, E. and H.-J. Herzog, *Thin Solid Films 44,* 357
(1977).
Linnington, P., Ph.D. thesis, University of Cambridge (1974).
Martin, R. M., *Phys. Rev. B1:* 4005 (1970).

Matthews, J. W. (Ed.), *Epitaxial Growth,* New York; Academic Press (1975).

Matthews, J. W., S. Mader, and T. B. Light, *J. Appl. Phys. 41:* 3800 (1970).

Olsen, G. H. and W. A. Jesser, *Acta Metall. 19:* 1009, 1299 (1971).

Ruth, R. P., A. J. Hughes, J. L. Kenty, H. M. Manasevit, D. Medellin, A. C. Thorsen, Y. T. Chan, C. R. Viswanathan, and M. A. Ring, Final Report, Contract No. DAAH01-70-C-1311, Research, Development, Engineering and Missile Systems Laboratory, U.S. Army Missile Command (1973).

Sugita, Y., M. Tamura, and K. Sugarawa, *J. Appl. Phys. 40:* 3089 (1969A).

Sugita, Y., M. Tamura, and K. Sugarawa, *J. Vac. Sci. Technol. 6:* 585 (1969B).

Tamura, M. and Y. Sugita, *J. Appl. Phys. 44:* 3442 (1973).

INDEX